NITRIC OXIDE

NOVEL ACTIONS, DELETERIOUS EFFECTS, AND CLINICAL POTENTIAL

ANNALS OF THE NEW YORK ACADEMY OF SCIENCES
Volume 962

NITRIC OXIDE

NOVEL ACTIONS, DELETERIOUS EFFECTS, AND CLINICAL POTENTIAL

Edited by
Chuang C. Chiueh, Jau-Shyong Hong, and Seng Kee Leong

The New York Academy of Sciences
New York, New York
2002

Library of Congress Cataloging-in-Publication Data

Nitric oxide : novel actions, deleterious effects, and clinical potential / edited by Chuang C. Chiueh, Jau-Shyong Hong, and Seng Kee Leong.
p. ; cm. — (Annals of the New York Academy of Sciences ; v. 962)
"Proceedings of … a satellite symposium to the Annual Meeting of the Society for Neuroscience, New Orleans, LA, November 4, 2000"—Contents p.
Includes bibliographical references and index.
ISBN 1-57331-366-1 (cloth : alk. paper) — ISBN 1-57331-367-X (paper : alk. paper)
1. Nitric oxide—Physiological effect—Congresses.
[DNLM: 1. Nitric Oxide—pharmacology—Congresses. 2. Nitric Oxide—adverse effects—Congresses. 2. Nitric Oxide—therapeutic use—Congresses. QV 126 N73145 2002] I. Chiueh, Chuang C. II. Hong, Jau-Shyong. III. Leong, Seng Kee. IV. Society for Neuroscience. Meeting (2000 : New Orleans, La.) V. Series.
Q11 .N5 vol. 962
[QP535.N1]
500 s—dc21
[612/ .015]

2002002395

K-M Research/PCP
Printed in the United States of America
ISBN 1-57331-366-1 (cloth)
ISBN 1-57331-367-X (paper)
ISSN 0077-8923

ANNALS OF THE NEW YORK ACADEMY OF SCIENCES

Volume 962
May 2002

NITRIC OXIDE

NOVEL ACTIONS, DELETERIOUS EFFECTS, AND CLINICAL POTENTIAL

Editors
CHUANG C. CHIUEH, JAU-SHYONG HONG, AND SENG KEE LEONG

This volume is the result of **Nitric Oxide: The Good, the Bad, and the Beautiful**, a satellite symposium to the Annual Meeting of the Society for Neuroscience, New Orleans, Lousiana, held on November 4, 2000, and sponsored by the Taiwanese Bioscientists of America.

CONTENTS

Part I. Novel Actions

Part II. Deleterious Effects

Part III. Clinical Potential

Preconditioning-Mediated Neuroprotection

Role of Nitric Oxide, cGMP, and New Protein Expression

TSUGUNOBU ANDOH,[a] P. BOON CHOCK,[b] AND CHUANG C. CHIUEH[c]

[a]*Department of Applied Pharmacology, Faculty of Pharmaceutical Sciences, Toyama Medical and Pharmaceutical University, 2630 Sugitani, Toyama, Japan*

[b]*Laboratory of Biochemistry, National Heart, Lung and Blood Institute, NIH Building 50, Bethesda, Maryland, U.S.A*

[c]*Unit on Neurodegeneration and Neuroprotection, Laboratory of Clinical Science, National Institute of Mental Health, NIH Clinical Center, Bethesda, Maryland, U.S.A.*

ABSTRACT: Preconditioning adaptation induced by transient ischemia can increase brain tolerance to oxidative stress, but the underlying neuroprotective mechanisms are not fully understood. Recently, we developed a human brain-derived cell model to investigate preconditioning mechanism in SH-SY5Y neuroblastoma cells.[1] Our results demonstrate that a non-lethal serum deprivation-stress for 2 h (preconditioning stress) enhanced the tolerance to a subsequent lethal oxidative stress (24 h serum deprivation) and also to 1-methyl-4-phenyl–pyridinium (MPP^+).[2] Two-hour non-lethal preconditioning stress increased the expression of neuronal nitric oxide (NOS1/nNOS) mRNA, Fos, Ref-1, NOS protein, and then nitric oxide (·NO) production. As well as MnSOD expression, the ·NO-cGMP-PKG pathway mediated the preconditioning-induced upregulation of antiapoptotic protein Bcl-2 and the downregulation of adaptor protein $p66^{shc}$. We also propose that cGMP-mediated preconditioning-induced adaptation against oxidative stress may be due to the synthesis of a new protein, such as thioredoxin (Trx) since the protective effect can be blocked by Trx reductase inhibitor.[3] The antioxidative potency of Trx was approximately 100 and 1,000 times greater than GSNO and GSH, respectively. These results suggest that ·NO-cGMP-PKG signaling pathway plays an important role in the preconditioning-induced neuroprotection, and perhaps cardioprotection, against oxidative stress.

KEYWORDS: NOS1/nNOS; GSNO; cGMP; Bcl-2; Trx; MnSOD; preconditioning mechanism; neuroprotection

Address for correspondence: Tsugunobu Andoh, Department of Applied Pharmacology, Faculty of Pharmaceutical Sciences, Toyama Medical and Pharmaceutical University, 2630 Sugitani, Toyama 930-0194, Japan.
andoht@ms.toyama-mpu.ac.jp
P. Boon Chock: pbc@helix.nih.gov
Chuang C. Chiueh: chiueh@helix.nih.gov

INTRODUCTION

Preconditioning procedure is a compensatory means of protecting brain and other cells against a subsequent lethal oxidative stress.[4–7] It has been reported that the expression of heat shock protein 70 (HSP70), heme oxygenase-1 (HO-1/HSP 32) and metallothioneins may mediate preconditioning-induced adaptation in animal models of ischemia. However, the underlying adaptation mechanisms are not fully understood. We aimed to look for preconditioning-induced adaptive cytoprotective proteins in human brain cells against oxidative stress evoked by reactive oxygen species.

Recently, we had shown that serum deprivation-stress induced the production of both ·OH and ·NO in human neurotrophic SH-SY5Y cells.[1] However, serum deprivation-induced apoptosis was prevented by ·OH scavengers (i.e., sodium salycilate), but not by the inhibitors of nitric oxide synthase (NOS) (i.e., L-NAME and 7-NI). Furthermore, there is a significant correlation between the increase of ·OH, but not ·NO, and the severity of cytotoxicity reflected by apoptotic cell death. These observations indicate that ·OH rather than ·NO mediated lethal oxidative stress caused by serum deprivation. Furthermore, low concentrations (μM) of ·NO donor compounds (e.g., *S*-nitroso-L-glutathione/GSNO) inhibited oxidative stress-induced neuronal damage *in vitro*[1,2] and also *in vivo*.[8] Interestingly, serum deprivation-stress increases the expression of neuronal NOS (NOS1) mRNA within 2 h, leading to the expression of NOS1 protein and the production of ·NO within 12 to 36 h of the induction of NOS1 mRNA.[1]

To investigate the role of endogenous ·NO in neuroprotection against oxidative stress, we used a preconditioning procedure (a non-lethal serum-free stress for 2 h) to induce the expression of NOS1. Our data reaffirm that human NOS1 is inducible. In addition to the expression of NOS1, the production of ·NO and the activation of guanylyl cyclase, preconditioning enhances the tolerance to neurotoxicity induced by lethal oxidative stress, such as serum deprivation and MPP^+.[1–3] In this paper, we hypothesize a putative role for the ·NO-cGMP-PKG signaling pathway in preconditioning-induced upregulation of the expression of antiapoptotic and cytoprotective proteins. Our preliminary and unpublished data[2,3,9] suggest that the activation of ·NO-cGMP-PKG pathway by preconditioning-induced adaptation could lead to new gene expression and new protein synthesis that eventually would enhance cellular antioxidative deference and tolerance to oxidative stress-induced apoptosis.

NEW VISTA OF cGMP-MEDIATED ·NO FUNCTIONING IN THE CNS

It has been previously proposed that, via *S*-nitrosylation or transnitrosylation, ·NO in relatively high concentrations may directly bind to reactive thiyl residues in some cysteine containing proapoptotic protein caspases involved in the cascade of apoptotic cell death.[10,11] The inhibition of caspases by ·NO might lead to the observed preconditioning-induced neuroprotection. However, in our pilot study we observed that guanylyl cyclase inhibitors could prevent the neuroprotection induced by preconditioning procedure.[1–3,9] With this new information, we propose that gene induction and new protein synthesis mediated by ·NO-cGMP-PKG signaling might

(1) Preconditioning Stress

⇩

AP-1, Ref-1

⇩

(2) Induction of NOS1 or NOS3

(i) **·NO**

(ii) **cGMP**

(iii) **PKG**

⇩

(3) New Protein Synthesis

Trx

Bcl-2, MnSOD, BDNF or HSP

⇩

Antioxidation

Antiapoptosis

⇩

(4) Adaptation for Survival

(i) Neuroprotection

(ii) Cardioprotection

SCHEME 1. The role of new protein expression in cGMP-mediated hormesis during preconditioning-induced neuroprotection and cardioprotection: a novel theory of molecular biology of ·NO. **(1)** Preconditioning stress is a non-lethal stress that can induce a compensatory mechanism in neurons or other cells for protection against a subsequent lethal oxidative stress in both cell cultures and animal models. Deprivation of nutrient (serum) or energy (2-deoxy-D-glucose) can induce preconditioning cytoprotection in cell cultures. A brief non lethal ischemia in animal heart or brain can also induce a delayed preconditioning cardio- and neuroprotection, respectively. **(2)** Contrary to the current prominent theory, NOS1/nNOS and NOS3/eNOS are inducible in both mild/non-lethal and severe/lethal oxidative stress. A brief 2-hour preconditioning stress fully induced NOS1 and NOS3 that does not link to the previously predicted ·NO cytotoxicity.[1] Moreover, NOS induction increased the synthesis of ·NO and cGMP resulting in the activation of PKG kinases. **(3)** Preconditioning induced stress proteins (Ref-1, Hsp70, or Hsp32/HO-1), cytoprotective proteins (Bcl-2, MnSOD, BDNF, or GDNF), and redox protein (Trx). Andoh *et al.*[2,9] was the first to link the production of ·NO and cGMP to the preconditioning expression of Trx. It is noteworthy that Trx (less than 1 μM) is one of the most potent thiol containing antioxidants that is able to suppress ·OH generation, lipid peroxidation, and apoptosis caused by MPP^+ (see TABLE 1).[3] **(4)** Based on recently obtained results, Andoh *et al.*[1–3,9] proposed that the ·NO/cGMP/PKG-dependent Trx expression may mediate preconditioning-induced compensatory cardio- and neuroprotection using a common mechanism. This new hypothesis is consistent with an early idea of the neurobiology of ·NO and ·OH proposed by Chiueh.[26] However, it is somewhat contradictory to a prior notion that ·NO derived from NOS1/nNOS mediates neurotoxicity in the brain. In fact, neither ·NO nor peroxynitrite caused significant dopamine injury;[8] intranigral fusion of ·NO and *S*-nitrosothiols (i.e., GSNO and SNAP) enhanced protection of brain dopamine neurons against oxidative stress *in vivo*.[27]

be one of the important pathways that mediates compensatory cytoprotection induced by non-lethal preconditioning stress (see SCHEME 1). Induction of antioxidative and antiapoptotic genes and their cytoprotective proteins could have utility in future gene therapy as an alternative procedure to the viral vector-transfected gene therapy.

To investigate the molecular events of the proposed ·NO-cGMP-PKG pathway that leads to cyto- and neuroprotection, we first developed the preconditioning human cell model.[1] In our experiments, we used human brain-derived neuroblastoma SH-SY5Y cells that can synthesize, take up, and release dopamine and serotonin. A two-hour non-lethal serum deprivation (preconditioning stress) time-dependently increased the expression of NOS1 from the induction of gene to the production of ·NO and cGMP. Preconditioning inhibition of subsequent lethal oxidative stress-induced apoptosis could be acting through ·NO-mediated gene regulation, such as the upregulation of Bcl2 and the downregulation of $p66^{shc}$. Since Bcl-2 can bind to cytoplamic cytochrome c and caspase 9 to prevent the activation of caspase-3 and, subsequently, Apaf-1 in the mitochondrial-mediated apoptotic pathway,[11] preconditioning induction of Bcl-2 could protect cells from oxidant-induced apoptosis during a prolonged serum deprivation stress. Furthermore, the knock out of cell membrane adaptor protein $p66^{shc}$ significantly increased the life span of mice.[12] Thus, the decrease in the expression of this adaptor protein in preconditioned cells would lead to cell survival or tolerance during oxidative stress.

Currently, we are investigating the role of the ·NO-cGMP-PKG pathway in the expression of cytoprotective proteins, such as MnSOD, BDNF/GDNF, and Trx (SCHEME 1). It has been reported previously that oxidative stress can induce antioxidative proteins, either MnSOD[13–15] or Trx[16,17] and, stress proteins (e.g., HO-1 or Hsp70)[4–7] via unknown mechanisms. MnSOD is known to convert superoxide anion ($O_2^{\cdot-}$) to H_2O_2 in mitochondria and, thus, minimize the generation of ·OH and related oxidative stress. Unexpectedly, HO-1/Hsp32 is not detected during the preconditioning stress in neurotrophic SH-SY5Y cells (unpublished results of S.Y. Lee and C.C. Chiueh) since HO-1 is located in astroglia. Our preliminary experiments have

TABLE 1. The antioxidative potency of thiol antioxidants

	ED_{50} (μM)	
Compounds	Peroxidation of brain lipid	MPP^+-induced neurotoxicity
Trx	n.d.	less than 0.3
GSNO	less than 30	less than 30
GSH	greater than 2,000	n.d.

ABBREVIATIONS: MPP^+, 1-methyl-4-phenylpyridinium; Trx, thioredoxin; GSNO, *S*-nitroso-L-glutathione; GSH, L-glutathione; n.d., not determined.

NOTES: ED_{50} values were calculated from prior results of Rauhala *et al.*[8] and unpublished data of Andoh *et al.*[9] To measure brain lipid peroxidation, rat brain homogenate was incubated with 1 μM ferrous citrate at 37°C for 2 h. A trypan blue exclusion method was used to determine cell death in human brain-derived SH-SY5Y neuroblastoma cells after incubating with MPP^+ (1 mM) with or without the pretreatment with thiol antioxidants at 37°C for 24 h. Freshly prepared thiol antioxidants (0 to 500 or 5,000 μM) were coincubated to plot their dose–response curve.

demonstrated that serum deprivation for a non-lethal two-hour period caused a delayed increase in the simultaneous expression of MnSOD and Trx.[1–3,9]

Interestingly, preconditioning-induced new protein synthesis of Trx can be prevented by a systemic inhibition of guanylyl cyclase, cGMP, and PKG.[2,3,9] Moreover, the inhibition of reductase of Trx completely prevents the preconditioning-induced neuroprotection. Human Trx is a redox protein that contains two reactive cysteine residues (-Cys-Gly-Pro-Cys-) and that exists in both reduced and oxidized form.[18] Trx reductase converts oxidized Trx back to the active form of reduced Trx. Reduced Trx inhibits both lipid peroxidation and apoptosis caused by reactive oxygen species. Significantly, Trx is far more potent than the classic antioxidant GSH and *S*-nitroso-L-glutathione (GSNO) by about two to three orders of magnitude, respectively (see TABLE 1). Our data indicate that reconditioning induction of Ref-1, Bcl-2, MnSOD, and Trx, in fact, increases cellular antioxidative defense system in neurons and other brain cells against oxidative stress-evoked necrotic and apoptotic cell death.

With this new information in mind, we extended this project to examine the efficacy of preconditioning induction of NOS1 and related cGMP-mediated expression of cytoprotective proteins against severe oxidative stress (e.g., ·OH generation, lipid peroxidation, and apoptosis) caused by a Parkinsonism producing neurotoxin 1-methyl-4-phenyl-1,2,3,6-tetrahydropyridine (MPTP).[19] The preconditioning procedure, in fact, drastically increased tolerance of neurotrophic SH-SY5Y cells against MPP^+ toxicity approximately 30-fold.[2,3,9] Consistently, the transfection of human NOS1 to cells also enhanced the protection against severe apoptosis caused by MPP^+. These new results support our working hypothesis that ·OH could mediate oxidative stress whereas ·NO induce neuroprotection in the brain.

After examining this preconditioning mechanism in human cell models we proposed to further investigate the preconditioning mechanism, including the neuron-glia interactions, in animal models. It is highly likely that this ·NO-cGMP-PKG system could also participate in adaptative cytoprotection *in vivo* in, not only the central nervous system, but also the cardiovascular system. In fact, the perfusion of a ·NO donor, GSNO, protected oxidative stress-induced nigral damage in rat brain[8] and also ischemic-reperfusion injury in the heart of laboratory animals.[20] Since the expressions of NOS1[21,22] and NOS3[23,24] are induced by ischemia in animal models, a non-lethal ischemia may increase production of ·NO and cGMP, leading to cytoprotection in ischemic brain or heart. However, it has been very difficult to apply preconditioning stress to human brain. Therefore, we sought new ways to regulate neuroprotective genes by drugs or chemical preconditioning procedures. Recently, it has been demonstrated that selegiline or *l*-deprenyl—a potent MAO-B inhibitor—can prevent MPP^+-induced neuronal cell death without affecting MAO-B.[25] We hypothesized that selegiline could induce a chemical preconditioning via the activation of the proposed ·NO-cGMP-PKG–mediated expression of cytoprotective proteins. The development of clinical preconditioning procedures and/or drugs that induce cytoprotective genes may provide a new therapeutic approach for protection of neurons and other brain cells against oxidative stress. This preconditioning induction of cytoprotective and antiapoptotic genes could enhance the patient's quality-of-life during progressive neurodegeneration, such as ischemia/reperfusion, head trauma, hemorragic stroke, Hellervorden Spatz syndrome, Parkinson's disease, and Alzheimer's dementia.

ACKNOWLEDGMENTS

We thank Dennis L. Murphy for his encouragement and support during the early phase of this study at the NIMH Laboratory of Clinical Science. Tsugunobu Andoh is a recipient of the JSPS-NIH fellowship.

REFERENCES

1. ANDOH, T., S.Y. LEE & C.C. CHIUEH. 2000. Preconditioning regulation of bcl-2 and p66^{shc} by human NOS1 enhances tolerance to oxidative stress. FASEB J. **14:** 2144–2146.
2. ANDOH, T., P.B. CHOCK & C.C. CHIUEH. 2001. The role of neuronal nitric oxide synthase (NOS1) and related cyclic GMP-PKG signaling pathway on preconditioning-mediated neuroprotection. Soc. Neurosci. Abstr. **27:** 208.7.
3. CHIUEH, C.C., T. ANDOH & P.B. CHOCK. 2001. Protection against MPTP-induced toxicity via preconditioning induction of cytoprotective genes and proteins: Ref-1, NOS1, and Trx. Soc. Neurosci. Abstr. **27:** 688.3.
4. MURRY, C.E., R.B. JENNINGS & K.A. REIMER. 1986. Preconditioning with ischemia: a delay of lethal cell injury in ischemic myocardium. Circulation **74:** 1124–1136.
5. KATO, H., Y. LIU, K. KOGURE & K. KATO. 1994. Induction of 27-KDa heat shock protein following cerebral ischemia in a rat model of ischemic tolerance. Brain Res. **634:** 235–244.
6. EMERSON, M.R., F.E. SAMSON & T.L. PAZDERNIK. 2000. Effects of hypoxia preconditioning on expression of metallothionein-1,2 and heme oxygenase-1 before and after kainic acid-induced seizures. Cell Mol. Biol. **46:** 619–626.
7. BAKER, J.E., P. HOLMAN, B. KALYANARAMAN, *et al.* 1999. Adaptation to chronic hypoxia confers tolerance to subsequent myocardial ischemia by increased nitric oxide production. Ann. N.Y. Acad. Sci. **874:** 236–253.
8. RAUHALA, P., A.M. LIN & C.C. CHIUEH. 1998 Neuroprotection by *S*-nitrosolutathione of brain dopamine neurons from oxidative stress. FASEB J. **12:** 165–173.
9. ANDOH, T., P.B. CHOCK & C.C. CHIUEH. 2002. The role of thioredoxin in protection against oxidative stress-induced apoptosis in SH-SY5Y cells. J. Biol. Chem. In press.
10. DIMMELER, S., J. HAENDELER, M. NEHLS & A.M. ZEIHER. 1997. Suppression of apoptosis by nitric oxide via inhibition of interleukin-1β-converting enzyme (ICE)-like and cysteine protease protein (CPP)-32-like proteases. J. Exp. Med. **185:** 601–607.
11. KRUIDERING, M. & G.I. EVAN. 2000. Caspase-8 in apoptostis: the beginning of "the end"? IUBMB Life **50:** 85–90.
12. MIGLIACCIO, E., M. GIORGIO, S. MELE, *et al.* 1999. The p66^{shc} adaptor protein controls oxidative stress response and life span in mammals. Nature **402:** 243–245.
13. SHULL, S., N.H. HEINTZ, M. PERIASAMY, *et al.* 1991. Differential regulation of antioxidant enzymes in response to oxidants. J. Biol. Chem. **266:** 24398–24403.
14. WARNER, B.B., L. STUART, S. GEBB & J.R. WISPE. 1996. Redox regulation of manganese superoxide dismutase. Am. J. Physiol. **271:** L150–L158.
15. GUTTERIDGE, J.M. & B. HALLIWELL. 2000. Free radical and antioxidants in the year 2000. A historical look to the future. Ann. N.Y. Acad. Sci. **899:** 136–147.
16. HIGASHKUBO, A., N. TANAKA, N. NODA, *et al.* 1999. Increase in thioredoxin activity of intestinal epithelial cells mediated by oxidative stress. Biol. Pharm. Bull. **22:** 900–903.
17. SACHI, Y., K. HIROTA, H. MASUTANI, *et al.* 1995. Induction of ADF/TRX by oxidative stress in keratinocytes and lymphoid cells. Immunol. Lett. **44:** 189–193.
18. TANAKA, T., H. NAKAMURA, A. NISHIYAMA, *et al.* 2001. Redox regulation by thioredoxin superfamily; protection against oxidative stress and aging. Free Radical Res. **33:** 851–855.
19. CHIUEH, C.C., H. MIYAKE & M.T. PENG. 1993. Role of dopamine autoxidation, hydroxyl radical generation, and calcium overload in underlying mechanisms involved in MPTP-induced parkinsonism. Adv. Neurol. **60:** 251–258.

20. KONOREV, E.A., J. JOSEPH, M.M. TARPEY & B. KALYANARAMAN. 1996. The mechanism of cardioprotection by *S*-nitrosoglutathione monoethyl ester in rat isolated heart during cardioplegic ischaemic arrest. Br. J. Pharmacol. **119:** 511–518.
21. HIGUCHI, Y., H. HATTORI, R. HATTORI & K. FURUSHO. 1996. Increased neurons containing neuronal nitric oxide synthase in the brain of a hypoxic-ischemic neonatal rat model. Brain Dev. **18:** 369–375.
22. BLACK, S.M., M.A BEDOLLI, S. MARTINEZ, *et al.* 1995. Expression of neuronal nitric oxide synthase corresponds to regions of selective vulnerability to hypoxia-ischaemia in the developing rat brain. Neurobiol. Dis. **2:** 145–155.
23. SHI, Y., K.A. PRITCHARD, JR., P. HOLMAN, *et al.* 2000. Chronic myocardial hypoxia increases nitric oxide synthase and decreases caveolin-3. Free Radical Biol. Med. **29:** 695–703.
24. BLOCH, W., U. MEHLHORN, A. KRAHWINKEL, *et al.* 2001. Ischemia increases detectable endothelial nitric oxide synthase in rat and human myocardium. Nitric Oxide **5:** 317–333.
25. WU, R.M., C.C. CHIUEH, A. PERT & D.L. MURPHY. 1993. Apparent antioxidant effect of *l*-deprenyl on hydroxyl radical formation and nigral injury elicited by MPP^+ *in vivo*. Eur. J. Pharmacol. **243:** 241–247.
26. CHIUEH, C.C. 1994. Neurobiology of ·NO and ·OH: basic research and clinical relevance. Ann. N.Y. Acad. Sci. **738:** 279–281.
27. RAUHALA, P., T. ANDOH, K. YEH & C.C. CHIUEH. 2002. Contradictory effects of sodium nitroprusside and S-nitroso-penicillamine on oxidative stress in brain dopamine neurons *in vivo*. Ann. N.Y. Acad. Sci. **962:** this volume.

NO-Mediated Chemoresistance in C6 Glioma Cells

DING-I. YANG, JIU-HAW YIN, SNIGDHA MISHRA,
RAJ MISHRA, AND CHUNG Y. HSU

Department of Neurology and Center for the Study of Nervous System Injury, Washington University School of Medicine, St. Louis, Missouri, U.S.A.

ABSTRACT: Expression of inducible nitric oxide synthase (iNOS) in malignant glioma and other tumors has been extensively documented. Massive production of NO by iNOS has been shown to exert tumoricidal effects. However, NO may enhance vasodilation and promote neovascularization, thereby facilitating tumor growth. Compared to the effects of NO on tumor cell death and survival, correlation between NO and cytotoxicity of chemotherapeutic reagents in glioma have been less well characterized. Another gene product often linked to tumor malignancy is hypoxia-inducible factor-1 (HIF-1). HIF-1 is a transcription factor that renders malignant tumors adaptive to hypoxic stress during massive vascularization and tumor invasion. Interestingly, HIF-1 also contributes to iNOS induction under hypoxia. We have characterized the interrelationship between iNOS, HIF-1 and chemoresistance. We note that increased NO synthesis by cytokine exposure or iNOS overexpression neutralized the cytotoxicity of 1,3-bis(2-chloroethyl)-1-nitrosourea (BCNU) and 1-(2-chloroethyl)-3-cyclohexyl-1-nitrosourea (CCNU), but not cisplatin, in rat C6 glioma cells. Both BCNU and CCNU are chloroethylnitrosoureas that kill tumor cells via carbamoylating and alkylating actions. Further studies indicated that iNOS only neutralized carbamoylating action of chloroethylnitrosoureas. Expression of iNOS may inhibit HIF-1 activity under hypoxia in C6 glioma cells transfected with a VEGF promoter-driven luciferase gene. Pretreatment of C6 cells with *N*-acetyl-L-cysteine (NAC), an antioxidant, nullified the inhibitory effect of iNOS on HIF-1 binding. That NO generated by iNOS expression inhibits HIF-1 activity in hypoxic C6 cells reveals a negative feedback loop in the HIF-1→iNOS cascade. Together these results suggest a complicated role of NO in malignant tumor growth, survival and invasion.

KEYWORDS: alkylation; angiogenesis; carbamoylation; chemoresistance; chemotherapy; hypoxia-inducible factor; hypoxia; iNOS; nitric oxide

ACQUIRED CHEMORESISTANCE AGAINST CHLOROETHYLNITROSOUREAS IN GBM

Glioblastoma multiform (GBM) is the most common type of primary brain tumor accounting for more than 40% of neoplasm in the central nervous system.[1] It carries poor prognosis despite multimodality approaches consisting of surgical resection,

Address for correspondence: Chung Y. Hsu, M.D., Ph.D., Department of Neurology, Washington University School of Medicine, 660 South Euclid Avenue, Box 8111, St. Louis, MO 63110, U.S.A. Voice: 314-362-3304; fax: 314-362-9462.
hsuc@neuro.wustl.edu

radiotherapy, and chemotherapy.[2] The life expectancy of patients with GBM is usually less than one year from the time of diagnosis. To date, the five-year survival remains at 1%.[3] Surgery alone gave a median survival length of 14 weeks. Radiotherapy following surgical resection extended this figure to 36 weeks. Combination of surgery, radiotherapy, and chemotherapy prolonged life expectancy to approximately 14 months.[4]

Chloroethylnitrosourea is a family of chemotherapeutic reagents that includes 1,3-bis(2-chloroethyl)-1-nitrosourea (BCNU) and 1-(2-chloroethyl)-3-cyclohexyl-1-nitrosourea (CCNU). Chloroethylnitrosoureas kill tumors via carbamoylating and alkylating actions.[5] In aqueous solution BCNU decomposes to form two electrophilic species, namely 2-chloroethyl diazohydroxide and 2-chloroethyl isocyanate. The former carries alkylating (chloroethylating) activity, whereas the later possesses predominantly the carbamoylating potential. 2-Chloroethyl isocyanate also has secondary alkylating activity via the formation of 2-chloroethylamine.[6] Like BCNU, CCNU also generates 2-chloroethyl diazohydroxide and 2-chloroethyl diazonium ions on degradation in aqueous solution. However, the carbamoylating moiety of CCNU is cyclohexyl isocyanate,[5] which, unlike 2-chloroethyl isocyanate, does not carry the aminoethylating activity.[7]

Despite the fact that BCNU is the most commonly used adjunct chemotherapy for GBM, in part due to its lipophilic character, BCNU unfortunately did not substantially prolong median survival of GBM patients, even though the proportion of patients living more than 18 months increased from 5 to 15% with chemotherapy.[8–10] A recent analysis of results from two *brain tumor study group* protocols reaffirms a modest effect of adjunct chemotherapy in increasing long-term survival.[11] GBM develops chemoresistance against BCNU. The underlying mechanisms of chemoresistance against chloroethylnitrosoureas are not fully understood but may depend on the tumor-killing actions of these reagents. Several mechanisms have been proposed that may account for BCNU chemoresistance in GBM. These include an increase in the synthesis of the reduced form of glutathione,[12,13] O6-alkylguanine-DNA alkyltransferase expression,[14] enhancement of DNA mismatch repair,[15] and multidrug resistance genes.[16] Interventions that enhance chemosensitivity of glioma cells to BCNU may improve clinical outcomes.

EXPRESSION OF iNOS AND HIF-1 IN MALIGNANT BRAIN TUMORS

NO is a short-lived free radical serving a number of physiological functions including neurotransmission[17] vasodilatation,[18,19] and macrophage cytotoxicity.[20,21] In pathological states, NO may contribute to microbial killing[22] and neuronal degeneration.[23] NO is synthesized through the enzymatic conversion of L-arginine and molecular oxygen to L-citrulline by nitric oxide synthases (NOS).[24] Three NOS isoforms have been identified and are named after the tissues from which they were first cloned. The neuronal NOS (nNOS)[25,26] and endothelial NOS (eNOS)[27,28] are constitutively expressed, resulting in continuous basal synthesis of NO to mediate multiple physiological processes. Inducible NOS (iNOS)[21] was first cloned from mouse macrophages and, unlike the other two constitutive NOS isoforms, is expressed only under stimulation mostly by inflammatory signals, such as cytokines and/or endotoxin to massively produce NO. Another unique feature distinguishing iNOS from nNOS and eNOS is that iNOS monomer binds tightly to calmodulin and its activation is relatively insensitive to intracellular calcium level.[21] This is in contrast to nNOS and eNOS whose activation is dependent upon calcium/calmodulin.[29] NO released from macrophages has the antimicrobial activity toward certain pathogens.[30,31] NO produced by iNOS in a number of disease models has been demonstrated to exert either cytotoxic[32–34] or cytoprotective[35–37] effects, depending on the cell types and experimental paradigms selected.

NO exhibits tumoricidal activity both *in vitro*[38] and *in vivo*.[39,40] However, NO may also alter vascular reactivity or promote neovascularization,[41] contributing to invasiveness of brain tumors via enhancing blood flow to facilitate nutrient supply and removal of metabolic wastes that may be in favor of tumor growth. In this respect, NO appears to illustrate a deleterious role in promoting tumor growth. Massive production of NO may be derived from the expression of iNOS[21] in response to exogenous stimuli, such as cytokine exposure. Lipopolysaccharide, a bacterial endotoxin, in combination with interferon-γ induces iNOS expression in rat C6 glioma cells at both mRNA and protein levels.[42] Expression of iNOS has also been demonstrated in human glioblastoma cells[43] and in a variety of different brain tumors or peritumor areas, with its mRNA levels higher in malignant gliomas than meningiomas.[44] Induction of iNOS has been noted in brain tumors[43,45] and the extent of iNOS expression correlates with the degree of malignancy.[44] NO may regulate vascular reactivity,[46] permeability,[47] and angiogenesis.[41] These effects may contribute to invasion of brain tumors via enhancing blood flow, thereby facilitating nutrient supply and removal of metabolic wastes. In this respect, NO may promote tumor growth despite its tumoricidal action.

In addition to iNOS, other genes that may be involved in GBM malignancy include hypoxia-inducible factor-1 (HIF-1). HIF-1, a key regulator of mammalian oxygen homeostasis, is a heterodimeric protein complex consisting of alpha (HIF-1α) and beta (HIF-1β) subunits.[48] HIF-1β has been independently identified as an aryl hydrocarbon receptor nuclear translocator (ARNT).[49] HIF-1α expression is tightly regulated by the cellular oxygen tension[50] whereas the expression of HIF-1β subunit is oxygen-independent. Hypoxic conditions prevent ubiquitination and subsequent proteasomal degradation of HIF-1α in cells.[51] The expression of this basic helix–loop–helix transcription factor is increased in most cells in response to low oxygen tension.[52,53] The downstream genes that can be transactivated by HIF-1 include those involved in angiogenesis, such as vascular endothelial growth factor (VEGF) and erythropoietin (EPO).[54] Since malignant tumor growth and invasion require neovascularization, HIF-1 has been proposed to play a key role in the pathophysiology of cancer. Indeed, the oxygen content in a tumor is a clinically important indicator of tumor malignancy,[55] suggesting a pivotal role for HIF-1 in adaptive responses of tumors to hypoxia.[54,56]

Expression of iNOS is in part regulated by HIF-1. Thus, HIF-1 is essential for iNOS induction under hypoxic conditions.[57,58] However, the correlation between HIF-1 and NO has not been fully delineated. NO donors can inhibit[59,60] or activate[61] HIF-1 activity. The conflicting roles of NO donors on HIF-1 activity may be related to their compounding pharmacological actions other than being a NO source. Given the close association linking GBM malignancy to clinically relevant iNOS and

HIF-1 expression, we predicted that there is a complex correlation between NO, HIF-1, and GBM chemoresistance. Intracerebral implantation of C6 glioma cells into the rat brain resulted in tumor growth characteristic of GBM. Lipopolysaccharide (LPS), a bacterial endotoxin, in combination with interferon-γ (IFN-γ) or tumor necrosis factor-α induces iNOS expression in rat C6 glioma cells at the mRNA and protein levels.[62] Induction of HIF-1 in C6 glioma cells can be achieved under hypoxic condition with 1% oxygen.[63] Using C6 glioma cells as an *in vitro* model, we have explored the correlation between iNOS, HIF-1, and chemoresistance against BCNU in GBM.

iNOS EFFECTS ON GLIOMA CHEMORESISTANCE AGAINST BCNU

Despite extensive studies on iNOS expression in brain tumors, the interaction of NO derived from iNOS and the tumor-killing effects of chemotherapeutic drugs has received relatively little attention. To address this issue, we have investigated the iNOS effects on several chemotherapeutic drugs commonly used in treating brain tumors in an *in vitro* culture system.[64] We observed a strong protective effect of iNOS expression against the cytotoxicity of BCNU and CCNU, both chloroethylnitrosoureas, but not that of cisplatin, in rat C6 glioma cells. Suppression of BCNU cytotoxicity associated with iNOS overexpression could be abolished by pharmacological inhibition of NOS or coexpression of an antisense RNA against iNOS. Further identification of the specific cytotoxic action that is sensitive to iNOS expression was accomplished by applying three 1,2-bis(sulfonyl)hydrazine derivatives with carbamoylating or alkylating action,[7] temozolomide with alkylating (methylating) potential,[65] as well as 2-chloroethyl isocyanate and cyclohexyl isocyanate, the respective carbamoylating metabolite of BCNU and CCNU. Our results suggest that the protection exerted by iNOS expression is confined within the carbamoylating potential of chloroethylnitrosoureas, but cannot be extended to alkylating action. Expression of iNOS also failed to protect C6 glioma cells from cytotoxicity of temozolomide, a methylating agent. Interestingly, overexpression of iNOS indeed slightly potentiated the cytotoxicity of compound 1 and temozolomide.

Cytokines have been shown to induce iNOS expression in human glioblastoma cells.[43] Inflammation that develops following surgery and/or radiation may lead to cytokine induction of iNOS expression. BCNU is a common adjunct therapy following surgical resection and radiation in the treatment of GBM. In our studies, we have noted that L-NAME, a NOS inhibitor, can restore cell-killing effect of BCNU in C6 glioma cells. Although it remains uncertain as to whether iNOS expression in GBM may alter the efficacy of chloroethylnitrosourea *in vivo,* our finding raises the possibility that pharmacological modulation of iNOS activity to reduce cellular NO content may potentially reduce chemoresistance of glioma cells to BCNU therapy. Several iNOS inhibitors, both selective (e.g., aminoguanidine[66] and *N*-iminoethyl-L-lysine, NIL[67]) and non-selective (e.g., L-NAME) are available. Application of NOS inhibitors in the management of various neurological disorders including migraine[68] is within reach in clinical settings. Therapeutic agents including antiinflammatory agents may also enhance the tumor-killing efficacy of BCNU by preventing iNOS induction in glioma.

Animal studies are certainly needed to establish clinical relevance of our experimental results. These studies nevertheless disclose a novel *in vitro* effect of iNOS in neutralizing the immediate cytotoxic carbamoylating action of chloroethylnitrosoureas, but not their cytostatic alkylating effect in C6 glioma cells. This iNOS effect may at least in part underlie the molecular mechanism of glioma chemoresistance to BCNU, which continue to be an important group of chemotherapy drugs in the treatment of GBM.

iNOS EFFECT ON HIF-α ACTION IN C6 GLIOMA CELLS

NO donors have been shown to inhibit[59,60] or increase[61] HIF-1α activation. Although such discrepancies may be due to the use of different experimental systems including different cell types and variable hypoxic/normoxic conditions, one major concern is that NO donors are chemically distinct in their mechanisms in increasing NO formation. Following the release of NO, SNP decomposes to ferrocyanide, ferricyanide, iron ions, and cyanide, each of which may have biological effects on HIF-1 binding activity.[69] Other chemical NO donors can cause changes in cellular redox after NO release; for example, *S*-nitrosoglutathione (GSNO) decomposes to NO and the oxidized glutathione (GSSG).[70] A high concentration of GSSG alone is a prominent source of oxidative stress. Another commonly used NO donor, SIN-1, generates peroxynitrite,[71] a strong oxidant that can damage cellular enzymes, membranes, and subcellular organelles through the nitration of tyrosine residues on proteins. Finally, any chemical NO donor applied at a sufficiently high concentration may prove cytotoxic. To determine whether NO derived from iNOS expression may inhibit HIF-1 activation, we applied a molecular biological approach to directly transfect an iNOS-expressing construct into C6 glioma cells so that a more specific role of NO on HIF-1 transactivation of a downstream gene could be demonstrated. Direct iNOS expression *in situ* avoids the compounding effects of NO donors as well as the unpredictable actions of their side products on HIF-1 binding activity. We found that excessive formation of NO inhibits HIF-1-mediated transcription of a downstream gene. Because the cellular redox state dictates HIF-1α binding activity[72,73] and iNOS expression with resultant NO production is expected to enhance oxidative stress, the effect of an antioxidant on iNOS inhibition of HIF-1α activation under hypoxia was also explored. Pretreatment of C6 cells with *N*-acetyl-L-cysteine (NAC), a sglutathione precursor and an antioxidant, followed by incubation under hypoxia partially reversed the iNOS-mediated inhibition of luciferase activity that was induced by HIF-1 activation.

The molecular mechanism of NO-mediated inhibition of HIF-1 binding remains unknown, but several scenarios are plausible. NO stimulates soluble guanylyl cyclase[74] to generate cyclic GMP (cGMP), which in turn activates cGMP-dependent protein kinases. NO may also bind to ferric heme oxygen and superoxide,[75] possibly affecting the signal pathway for HIF-1 activation. NO reacting with superoxide anion with resultant peroxynitrite formation[76] could in turn generate nitrosylated tyrosine residues in proteins and interrupt signal transduction pathways mediated by tyrosine phosphorylation. NO has also been shown to directly nitrosylate the sulfhydryl groups of cysteine residues on numerous proteins thereby inhibiting their enzymatic activity.[77] Further experimental investigation is required to delineate which of these mechanisms are involved in NO inhibition of HIF-1 DNA binding. That antioxidant NAC may counteract iNOS-mediated inhibition of HIF-1 activity is also interesting. At least two explanations may be proposed to account for the NAC effect. It has been reported in aortic endothelial cells that hypoxia decreases GSH/GSSG ratio while enhancing NO synthesis, leading to nitrosothiol formation.[78] Hypoxic conditions coupled with an exogenous glutathione precursor (NAC) may increase total intracellular glutathione concentrations with a lowered GSH/GSSG ratio to favor the formation of GSNO, thus preventing the direct interaction of NO with HIF-1α. Alternatively, GSH/GSSG may stabilize the HIF-1 protein to enhance its DNA binding ability. NAC, a glutathione precursor and free radical scavenger, has been shown to stabilize the HIF-1α protein under hypoxia, thereby facilitating the HIF-1 DNA-binding activity by an increased GSH/GSSG ratio in the fetal alveolar epithelial cells.[79] Presently, the exact molecular mechanism by which NAC reverses NO inhibition of HIF-1 DNA binding remains to be defined.

CONCLUSION

We have demonstrated that expression of iNOS may in part underlie the mechanisms of chemoresistance against carbamoylating, but not alkylating action of chloroethylnitrosoureas in GBM. Expression of iNOS indeed slightly potentiates the alkylating cytotoxicity. On the other hand, expression of iNOS is capable of abolishing the HIF-1-mediated activation of downstream genes under hypoxic conditions. Provided that HIF-1 activity contributes to the adaptive response of GBM (and other types of malignant tumors) to hypoxic stress, NO production secondary to iNOS expression may conceivably interfere with HIF-dependent angiogenesis and hence tumor growth. This could be a contributing factor whereby expression of iNOS may kill tumor cells. Paradoxically, NO generated by eNOS, and perhaps that derived from iNOS under pathophysiological conditions such as in malignant tumors, may itself promote angiogenesis, blood vessel dilation, and enhance microvascular permeability. The complex roles of NO in acquired chemoresistance and its multiple actions on angiogenesis during tumor growth and invasion necessitate further experimentation in order to better understand the NO effects on tumor growth and its tumoricidal activity. This may help us to design more effective therapeutic avenues targeted at manipulating NO production in managing GBM.

ACKNOWLEDGMENTS

This work is supported in part by Pharmacia/Washington University Biomedical Research Program and Lowell B. Miller Memorial Reserach Grant from National Brain Tumor Foundation.

REFERENCES

1. KLEIHUES, P., *et al.* 1995. Histopathology, classification and grading of gliomas. Glia **15:** 211–221.
2. FINE, H.A. 1994. The basis for current treatment recommendations for malignant gliomas. J. Neurooncol. **20:** 111–120.
3. DAVIS, F.G., *et al.* 1998. Survival rates in patients with primary malignant brain tumors stratified by patient age and tumor hisitological type: an analysis based on surveillance, epidemiology, and end results (SEER) data, 1973–1991. J. Neurosurg. **88:** 1–10.
4. RAJKUMAR, S.V., *et al.* 1999. Phase I evaluation of pre-irradiation chemotherapy with carmustine and cisplatin and accelerated radiation therapy in patients with high-grade gliomas. Neurosurgery **44:** 67–73.
5. WHEELER, G.P., *et al.* 1974. Interrelationships of some chemical, physicochemical, and biological activities of several 1-(2-haloethyl)-1-nitrosoureas. Cancer Res. **34:** 194–200.
6. BECKER, K. & R.H. SCHIRMER. 1995. 1,3-Bis(2-chloroethyl)-1-nitrosourea as thiol-carbamoylating agent in biological systems. Meth. Enzymol. **251:** 173–188.
7. PENKETH, P.G., *et al.* 2000. Comparison of DNA lesions produced by tumor-inhibitory 1,2-bis(sulfonyl)hydrazines and chloroethylnitrosoureas. Biochem. Pharmacol. **59:** 283–291.
8. FINE, H.A., *et al.* 1993. Meta-analysis of radiation therapy with and without adjuvant chemotherapy for malignant gliomas in adults. Cancer **71:** 2585–2597.
9. GREEN, S.B., *et al.* 1983. Comparisons of carmustine, procarbazine, and high-dose methylprednisolone as additions to surgery and radiotherapy for the treatment of malignant glioma. Caner Treat. Rep. **67:** 121-132.
10. WALKER, M.D., *et al.* 1980. Randomized comparisons of radiotherapy and nitrosoureas for the treatment of malignant glioma after surgery. N. Engl. J. Med. **303:** 1323–1329.
11. DEANGELIS, L.M., *et al.* 1998. Malignant glioma: who benefits from adjuvant chemotherapy? Ann. Neurol. **44:** 691–695.
12. EVANS, C.G., *et al.* 1987. Glutathione and related enzymes in rat brain tumor cell resistance to 1,3-bis(2-chloroethyl)-1-nitrosourea and nitrogen mustard. Cancer Res. **47:** 2525–2530.
13. ALI-OSMAN, F., *et al.* 1989. Decreased DNA interstrand cross-linking and cytotoxicity induced in human brain tumor cells by 1,3-bis(2-chloroethyl)-1-nitrosourea after *in vitro* reaction with glutathione. Cancer Res. **49:** 5954–5958.
14. ROLHION, C., *et al.* 1999. O(6)-methylguanine-DNA methyltransferase gene (MGMT) expression in human glioblastomas in relation to patient characteristics and p53 accumulation. Int. J. Cancer **84:** 416–420.
15. SANDERSON, R.J. & D.W. MOSBAUGH. 1998. Fidelity and mutational specificity of uracil-initiated base excision DNA repair synthesis in human glioblastoma cell extracts. J. Biol. Chem. **273:** 24822–24831.
16. NUTT, C.L., *et al.* 2000. Differential expression of drug resistance genes and chemosensitivity in glial cell lineages correlate with differential response of oligodendrogliomas and astrocytomas to chemotherapy. Cancer Res. **60:** 4812–4818.
17. SCHUMAN, E.M. & D.V. MADISON. 1991. A requirement for the intercellular messenger nitric oxide in long-term potentiation. Science **254:** 1503–1506.
18. FURCHGOTT, R.F. & J.V. ZAWADZKI. 1980. The obligatory role of endothelial cells in the relaxation of arterial smooth muscle by acetylcholine. Nature **288:** 373–376.
19. HUANG, P.L., *et al.* 1995. Hypertension in mice lacking the gene for endothelial nitric oxide synthase. Nature **377:** 239–242.
20. LOWENSTEIN, C.J., *et al.* 1992. Cloned and expressed macrophage nitric oxide synthase contrasts with the brain enzyme. Proc. Natl. Acad. Sci. U.S.A. **89:** 6711–6715.
21. XIE, Q.-W., *et al.* 1992. Cloning and characterization of inducible nitric oxide synthase from mouse macrophages. Science **256:** 225–228.
22. MACMICKING, J., *et al.* 1997. Nitric oxide and macrophage function. Annu. Rev. Immunol. **15:** 323–350.

23. ZHANG, J., *et al.* 1994. Nitric oxide activation of poly(ADP-ribose) synthetase in neurotoxicity. Science **263:** 687–689.
24. PALMER, R.M., *et al.* 1987. Nitric oxide release accounts for the biological activity of endothelium-derived relaxing factor. Nature **327:** 524–526.
25. BREDT, D.S. & S.H. SNYDER. 1990. Isolation of nitric oxide synthetase, a calmodulin-requiring enzyme. Proc. Natl. Acad. Sci. U.S.A. **87:** 682–685.
26. SCHMIDT, H.H., *et al.* 1991. Purification of a soluble isoform of guanylyl cyclase-activating-factor synthase. Proc. Natl. Acad. Sci. U.S.A. **88:** 365–369.
27. FORSTERMANN, U., *et al.* 1991. Calmodulin-dependent endothelium-derived relaxing factor/nitric oxide synthase activity is present in the particulate and cytosolic fractions of bovine aortic endothelial cells. Proc. Natl. Acad. Sci. U.S.A. **88:** 1788–1792.
28. POLLOCK, J.S., *et al.* 1991. Purification and characterization of particulate endothelium-derived relaxing factor synthase from cultured and native bovine aortic endothelial cells. Proc. Natl. Acad. Sci. U.S.A. **88:** 10480–10484.
29. SCHMIDT, H.H., *et al.* 1992. Ca^{2+}/calmodulin-regulated nitric oxide synthases. Cell Calcium **13:** 427–434.
30. MACMICKING, J.D., *et al.* 1995. Altered responses to bacterial infection and endotoxic shock in mice lacking inducible nitric oxide synthase. Cell **811:** 641–650.
31. WEI, X.Q., *et al.* 1995. Altered immune responses in mice lacking inducible nitric oxide synthase. Nature **375:** 408–411.
32. KUROSE, I., *et al.* 1996. Increased nitric oxide synthase activity as a cause of mitochondrial dysfunction in rat hepatocytes: roles for tumor necrosis factor-a. Hepatology **24:** 1185–1192.
33. PERESLENI, T., *et al.* 1996. Antisense oligodeoxynucleotides to inducible NO synthase rescue epithelial cells from oxidative stress injury. Am. J. Physiol. **270:** F971–F977.
34. HSU, C.Y., *et al.* 1997. Enhancement of apoptosis in cerebral endothelial cells by selected inflammatory signals. Ann. N.Y. Acad. Sci. **843:** 148–153.
35. RAUHALA, P., *et al.* 1998. Neuroprotection by *S*-nitrosoglutathione of brain dopamine neurons from oxidative stress. FASEB J. **12:** 165–173.
36. TZENG, E., *et al.* 1997. Adenoviral transfer of the inducible nitric oxide synthase gene blocks endothelial cell apoptosis. Surgery **122:** 255–263.
37. ZHAO, H., *et al.* 1998. B-cell chronic lymphocytic leukemia cells express a functional inducible nitric oxide synthase displaying anti-apoptotic activity. Blood **92:** 1031–1043.
38. STUEHR, D.J. & C.F. NATHAN. 1989. Nitric oxide. A macrophage product responsible for cytostasis and respiratory inhibition in tumor target cells. J. Exp. Med. **169:** 1543–1555.
39. FARIAS-EISNER, R., *et al.* 1994. Nitric oxide is an important mediator for tumoricidal activity *in vivo.* Proc. Natl. Acad. Sci. U.S.A. **91:** 9407–9411.
40. DINAPOLI, M.R., *et al.* 1996. The altered tumoricidal capacity of macrophages isolated from tumor-bearing mice is related to reduce expression of the inducible nitric oxide synthase gene. J. Exp. Med. **183:** 1323–1329.
41. ANDRADE, S.P., *et al.* 1992. Inhibitors of nitric oxide synthase selectively reduce flow in tumor-associated neovasculature. Br. J. Pharmacol. **107:** 1092–1095.
42. FEINSTEIN, D.L., *et al.* 1994. Induction of nitric oxide synthase in rat C6 glioma cells. J. Neurochem. **62:** 315–321.
43. FUJISAWA, H., *et al.* 1995. Inducible nitric oxide synthase in a human glioblastoma cell line. J. Neurochem. **64:** 85–91.
44. ELLIE, E., *et al.* 1995. Differential expression of inducible nitric oxide synthase mRNA in human brain tumours. Neuroreport **7:** 294–296.
45. HARA, E., *et al.* 1996. Expression of heme oxygenase and inducible nitric oxide synthase mRNA in human brain tumors. Biochem. Biophys. Res. Commun. **224:** 153–158.
46. SWAROOP, G.R., *et al.* 1998. Effects of nitric oxide modulation on tumour blood flow and microvascular permeability in C6 glioma. Neuroreport **9:** 2577–2581.
47. NAKANO, S., *et al.* 1996. Increased brain tumor microvessel permeability after intracarotid bradykinin is mediated by nitric oxide. Cancer Res. **56:** 4027–4031.

48. SEMENZA, G.L. 2001. HIF-1 and mechanisms of hypoxia sensing. Curr. Opin. Cell Biol. **13:** 167–171.

49. MALTEPE, E., *et al.* 1997. Abnormal angiogenesis and responses to glucose and oxygen deprivation in mice lacking the protein ARNT. Nature **386:** 403–407.

50. WANG, G.L., *et al.* 1995. Hypoxia-inducible factor 1 is a basic-helix-loop-helix-PAS heterodimer regulated by cellular O_2 tension. Proc. Natl. Acad. Sci. U.S.A. **92:** 5510–5514.

51. KALLIO, P.J., *et al.* 1999. Regulation of the hypoxia-inducible transcription factor 1alpha by the ubiquitin-proteasome pathway. J. Biol. Chem. **274:** 6519–6525.

52. SEMENZA, G.L. & G.L. WANG. 1992. A nuclear factor induced by hypoxia via *de novo* protein synthesis binds to the human erythropoietin gene enhancer at a site required for transcriptional activation. Mol. Cell Biol. **12:** 5447–5454.

53. WANG, G.L. & G.L. SEMENZA. 1993. General involvement of hypoxia-inducible factor 1 in transcriptional response to hypoxia. Proc. Natl. Acad. Sci. U.S.A. **90:** 4304–4308.

54. SEMENZA, G.L. 1998. Hypoxia-inducible factor 1: master regulator of O_2 homeostasis. Curr. Opin. Genet. Dev. **8:** 588–594.

55. DACHS, G.U. & D.J. CHAPLIN. 1998. Microenvironmental control of gene expression: implications for tumor angiogenesis, progression, and metastasis. Semin. Radiat. Oncol. **8:** 208–216.

56. SEMENZA, G.L. 2000. Expression of hypoxia-inducible factor 1: mechanisms and consequences. Biochem. Pharmacol. **59:** 47–53.

57. JUNG, F., *et al.* 2000. Hypoxic regulation of inducible nitric oxide synthase via hypoxia inducible factor-1 in cardiac myocytes. Circ. Res. **86:** 319–325.

58. MELILLO, G., *et al.* 1997. Functional requirement of the hypoxia-responsive element in the activation of the inducible nitric oxide synthase promoter by the iron chelator desferrioxamine. J. Biol. Chem. **272:** 12236–12243.

59. SOGAWA, K., *et al.* 1998. Inhibition of hypoxia-inducible factor 1 activity by nitric oxide donors in hypoxia. Proc. Natl. Acad. Sci. U.S.A. **95:** 7368–7373.

60. HUANG, L.E., *et al.* 1999. Inhibition of hypoxia-inducible factor 1 activation by carbon monoxide and nitric oxide. Implications for oxygen sensing and signaling. J. Biol. Chem. **274:** 9038–9044.

61. KIMURA, H., *et al.* 2000. Hypoxia response element of the human vascular endothelial growth factor gene mediates transcriptional regulation by nitric oxide: control of hypoxia-inducible factor-1 activity by nitric oxide. Blood **95:** 189–197.

62. FEINSTEIN, D.L., *et al.* 1994. Induction of nitric oxide synthase in rat C6 glioma cells. J. Neurochem. **62:** 315–321.

63. YIN, J.H., *et al.* 2000. iNOS expression inhibits hypoxia-inducible factor-1 activity. Biochem. Biophys. Res. Commun. **279:** 30–34.

64. YIN, J.H., *et al.* 2001. Inducible nitric oxide synthase neutralizes carbamoylating potential of 1,3-bis(2-chloroethyl)-1-nitrosourea in C6 glioma cells. J. Pharmacol. Exp. Ther. **297:** 308–315.

65. DENNY, B.J., *et al.* 1994. NMR and molecular modeling investigation of the mechanism of activation of the antitumor drug temozolomide and its interaction with DNA. Biochemistry **33:** 9045–9051.

66. CORBETT, J.A. & M.L. MCDANIEL. 1996. The use of aminoguanidine, a selective iNOS inhibitor, to evaluate the role of nitric oxide in the development of autoimmune diabetes. Methods **10:** 21–30.

67. SALVEMINI, D., *et al.* 1995. Dual inhibition of nitric oxide and prostaglandin production contributes to the antiinflammatory properties of nitric oxide synthase inhibitors. J. Clin. Invest. **96:** 301–308.

68. LASSEN, L.H., *et al.* 1997. Nitric oxide synthase inhibition in migraine [letter]. Lancet **349:** 401–402.

69. BULTER, A.R. & C. GLIDEWELL. 1996. Recent chemical studies of sodium nitroprusside relevant to its hypotensive action. Chem. Soc. Rev. **221:** 670–674.

70. WILLIAMS, D.L. 1996. *S*-nitrosothiols and role of metal ions in decomposition to nitric oxide. Meth. Enzymol. **268:** 299–308.

71. HOGG, N., *et al.* 1992. Production of hydroxyl radicals from the simultaneous generation of superoxide and nitric oxide. Biochem. J. **281:** 419–424.
72. HADDAD, J.J. & S C. LAND. 2000. O(2)-evoked regulation of HIF-1alpha and NF-kappaB in perinatal lung epithelium requires glutathione biosynthesis. Am. J. Physiol. Lung Cell Mol. Physiol. **278:** L492–503.
73. HUANG, L.E., *et al.* 1996. Activation of hypoxia-inducible transcription factor depends primarily upon redox-sensitive stabilization of its alpha subunit. J. Biol. Chem. **271:** 32253–32259.
74. STONE, J.R. & M.A. MARLETTA. 1994. Soluble guanylate cyclase from bovine lung: activation with nitric oxide and carbon monoxide and spectral characterization of the ferrous and ferric states. Biochemistry **33:** 5636–5640.
75. STAMLER, J.S. 1994. Redox signaling: nitrosylation and related target interactions of nitric oxide. Cell **78:** 931–936.
76. BECKMAN, J.S., *et al.* 1990. Apparent hydroxyl radical production by peroxynitrite: implications for endothelial injury from nitric oxide and superoxide. Proc. Natl. Acad. Sci. U.S.A. **87:** 1620–1624.
77. TROY, C.M., *et al.* 2000. Caspase-2 mediates neuronal cell death induced by beta-amyloid. J. Neurosci. **20:** 1386–1392.
78. MOTTERLINI, R., *et al.* 2000. Endothelial heme oxygenase-1 induction by hypoxia. Modulation by inducible nitric-oxide synthase and *S*-nitrosothiols. J. Biol. Chem. **275:** 13613–13620.
79. HADDAD, J.J., *et al.* 2000. Antioxidant/pro-oxidant equilibrium regulates HIF-1alpha and NF-kappa B redox sensitivity. Evidence for inhibition by glutathione oxidation in alveolar epithelial cells. J. Biol. Chem. **275:** 21130–21139.

Role of Nitric Oxide in Myocardial Preconditioning

BUDDHADEB DAWN AND ROBERTO BOLLI

Experimental Research Laboratory, Division of Cardiology, University of Louisville, Louisville, Kentucky, U.S.A.

ABSTRACT: Ischemic preconditioning (PC) is a polygenic defensive cellular adaptive phenomenon whereby brief ischemic stimuli render the heart resistant to subsequent similar stress. The late phase of ischemic PC lasts for three to four days, protects against both myocardial stunning and infarction, and thus has considerable clinical relevance. Diverse signaling molecules released by a sublethal ischemic stress initiate a complex signal transduction cascade that modulates the expression of cardioprotective genes. Nitric oxide (NO), generated by the endothelial NO synthase (NOS) and acting via the formation of reactive oxygen species, activates the ε isoform of protein kinase C (PKC), which activates the Src family of protein tyrosine kinases (Src and Lck) and transcription factors (nuclear factor-κB, and possibly others), with resultant upregulation of the inducible NOS (iNOS) gene and protein expression. iNOS, and other cardioprotective proteins, including cyclooxygenase-2 and aldose reductase, confer resistance to subsequent ischemic stress. This delayed protection can also be mimicked, in the absence of ischemia, by administering NO-releasing agents, a situation that can be potentially exploited for cardioprotection in clinical situations. Identification of this novel bifunctional (trigger and mediator) role played by NO in cardiac protection, not only advances the knowledge regarding its signaling functions, but also offers a potential therapeutic strategy for patients with coronary artery disease. The purpose of this review is to summarize the current evidence in support of this critical role played by NO in ischemic PC.

KEYWORDS: nitric oxide; ischemic preconditioning; protein kinase C; protein tyrosine kinases; nuclear factor-κB

INTRODUCTION

Brief episodes of myocardial ischemia induce a powerful defensive response that renders the heart protected against subsequent ischemic episodes. This phenomenon, termed ischemic preconditioning (PC), has two distinct components: an *early* phase that lasts from a few minutes to two to three hours, and a *late* phase that develops after 12 h and lasts for three to four days.[1–3] These two phases are different not only in duration of protection, but also in signaling mechanisms. Early PC is initiated by receptor-ligand interaction, with subsequent G-protein-mediated activation of intracellular targets, whereas late PC involves a complex-signal transduction process

Address for correspondence: Roberto Bolli, M.D., Division of Cardiology, University of Louisville, Louisville, KY 40292, U.S.A. Voice: 502-852-1837; fax: 502-852-6474. rbolli@louisville.edu

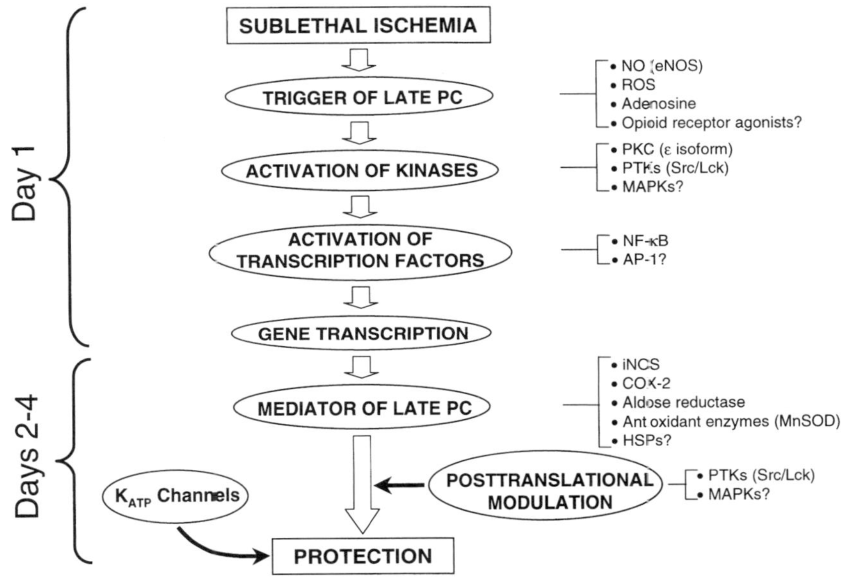

FIGURE 1. Schematic representation of the cellular mechanisms underlying late PC. A sublethal ischemic challenge causes release of chemical signals (NO, ROS, adenosine, and possibly opioid receptor agonists) that serve as triggers for the development of late PC. NO and $\cdot O_2^-$, both generated during brief episodes of ischemia-reperfusion, react to form $\cdot ONOO^-$. $\cdot ONOO^-$ then activates PKCε either directly or via its reactive byproducts, such as ·OH. A complex signal transduction cascade is then activated that includes PKC (specifically, the ε isoform), PTKs (specifically, Src and/or Lck), and probably other as-yet-unknown kinases. A similar activation of PKC and downstream kinases can be elicited pharmacologically by a wide variety of agents, including NO donors, adenosine A1 or A3 receptor agonists, or δ_1-opioid receptor agonists. The recruitment of PKC and downstream kinases leads to activation of NF-κB and almost certainly other transcription factors, resulting in increased transcription of multiple cardioprotective genes and synthesis of multiple cardioprotective proteins that serve as comediators of protection 2–4 days after the PC stimulus. The mediators of late PC identified thus far include iNOS, COX-2, aldose reductase, and Mn-SOD. Among the products of COX-2, PGE2, and/or PGI2 appear to be the most likely effectors of COX-2–dependent protection. In addition, the occurrence of cardioprotection on days 2–4 requires the activity of PTKs and possibly p38 MAPKs, potentially because iNOS and other mediators need to undergo posttranslational modulation to confer protection against ischemia. Opening of K_{ATP} channels is also essential for the protection against infarction (but not against stunning) to become manifest. The exact interrelationships among iNOS, COX-2, aldose reductase, Mn-SOD, and K_{ATP} channels are unknown, although recent evidence suggests that COX-2 may be downstream of iNOS (i.e., COX-2 is activated by NO).

leading to *de novo* synthesis and activation of multiple enzymes and effector proteins.[3,4] Both early and late PC have been investigated intensely with a view to developing therapeutic strategies for cardiac protection against ischemia in clinical situations. However, late PC, by virtue of its longer duration of protection, appears clinically more relevant. Recent research has identified several of the key signaling molecules involved in late PC, including nitric oxide (NO),[3,5–7] protein kinase C (PKC) ε isoform,[8] two members of the Src family of protein tyrosine kinases (PTKs), Src and Lck,[9] nuclear factor-κB (NF-κB),[10] inducible nitric oxide synthase (iNOS),[7,11–13] cyclooxygenase-2 (COX-2),[14] and aldose reductase.[15] Additionally, it has also been possible to mimic this protective phenomenon by administration of pharmacologic agents, most notably, NO-releasing compounds.[16,17] The importance of deciphering the signaling pathway (summarized in FIGURE 1) involved in late PC is underscored by the fact that these agents may be used therapeutically to induce a protracted preconditioned state of the myocardium in patients with coronary artery disease.

The role of NO in cardiac and vascular physiological and pathological processes has long been known.[18,19] However, its role in ischemic PC was recognized only recently.[7] The purpose of this essay is to summarize the current evidence that supports the crucial role of NO in cardiac protection afforded by ischemic PC. Although NO plays a role in early PC, its effects are more defined and proven in late PC. This review primarily focuses on the role of NO in late PC.

PATHOPHYSIOLOGY OF LATE PC

The phenomenon of ischemic PC was first described by Murry *et al.*[20] who noted a marked reduction in infarct size in dogs when a 40-min coronary occlusion was preceded by brief episodes of ischemia (four 5-min episodes of ischemia interspersed with 10-min reperfusion periods). Subsequently a delayed (or late) phase of protection that becomes manifest approximately 12 h after the initial ischemic stimulus and lasts for 3–4 days was discovered.[21,22] Late PC confers protection against both myocardial stunning and infarction. Time-course studies performed in conscious pigs demonstrated that the protection was absent at six hours after the last ischemic episode, partially manifested at 12 h, fully expressed at 24 h, disappeared within six days, but could be reinduced by similar ischemic stimuli after 10 days.[22,23] A similar time course has been demonstrated in rabbits proving that this protection is not unique to any species. Another interesting observation about late PC is that this is an *all-or-none* phenomenon.[24] When rabbits were subjected to a sequence of six 4-min coronary occlusion/4-min reperfusion cycles for four consecutive days, the magnitude of PC effect did not increase from the second to the fourth day. Furthermore, a sharp threshold exists for induction of this phenomenon. Specifically, only one or two cycles of ischemia/reperfusion failed to elicit this protection, whereas the protection afforded by three cycles of ischemia/reperfusion could not be further enhanced by increasing the number of cycles to six or twelve.[25] Although pigs have been used extensively for cardiovascular investigations, we performed several of our studies in a conscious rabbit model for the following reasons: (1) late PC against infarction could not be demonstrated in pigs;[26] (2) optimal heart size and coronary distribution in rabbits; and (3) substantially lower cost.

MECHANISM OF LATE PC

Despite rapid advances in our knowledge of this complex phenomenon, the mechanism of late PC remains incompletely understood. The signaling sequence of late PC can be broadly divided into three segments: factors that *trigger* this event, *mediators* of this protection, and the bridging *signaling pathway*[3] (summarized in FIG. 1). As mentioned already, multiple agents have been shown to play distinct roles in the initiation, signal transduction, and execution of this phenomenon and a detailed description of each is beyond the scope of this review. We limit this discussion to the specific roles played by NO and as they relate to other specific molecules.

NO AS A *TRIGGER* OF LATE PC

Several molecules have been proposed to initiate the development of the late phase of ischemic PC, including adenosine, reactive oxygen species (ROS), and NO. Although activation of adenosine A1 receptors induced delayed protection against infarction in rabbits,[27] pharmacologic blockade of adenosine receptors could not prevent the development of late PC against stunning in this species.[28] Moreover, it is unclear at this time whether the A1 or the A3 receptor (or both) is responsible for this infarct-sparing effect of adenosine.[29] An essential role of ROS generated during the PC ischemia in the development of late PC is supported by the observation that administration of antioxidant therapy (superoxide dismutase [SOD] plus catalase plus mercaptopropionyl glycine [MPG]) during the PC ischemia completely prevented the development of late PC against myocardial stunning on day 2 in conscious pigs.[30] In conscious rabbits, the development of late PC against stunning was found to be blocked by MPG alone, but not by SOD alone or catalase alone, implicating MPG-sensitive oxidants (peroxynitrite [$ONOO^-$] and/or hydroxyl radical [·OH]) in the initiation of late PC.[31] Moreover, intracoronary administration of an ROS-generating solution in the absence of ischemia was sufficient to induce a late PC effect against stunning that was equivalent to that observed after ischemic PC.[32] These data, although supportive of the notion that ROS play an essential role in triggering late PC, failed to identify the source of these ROS produced during PC ischemia.

We postulated that NO is one possible source of the ROS that trigger late PC. This proposal was based on several considerations: first, production of NO by eNOS is stimulated by calcium in presence of NADPH, both of which are produced in the endothelium of the coronary vasculature during an ischemic insult. Second, reperfusion following ischemia not only provides oxygen necessary for rapid NO synthesis, but also generates superoxide anion [$\cdot O_2^-$], which reacts with NO to produce the peroxynitrite anion [$ONOO^-$], which then protonates to generate the hydroxyl radical [·OH] or some other oxidant with similar reactivity.[33] Third, late PC is blocked by MPG and thus is triggered by MPG-sensitive ROS; that is, ·OH and/or ONOO.[33,34]

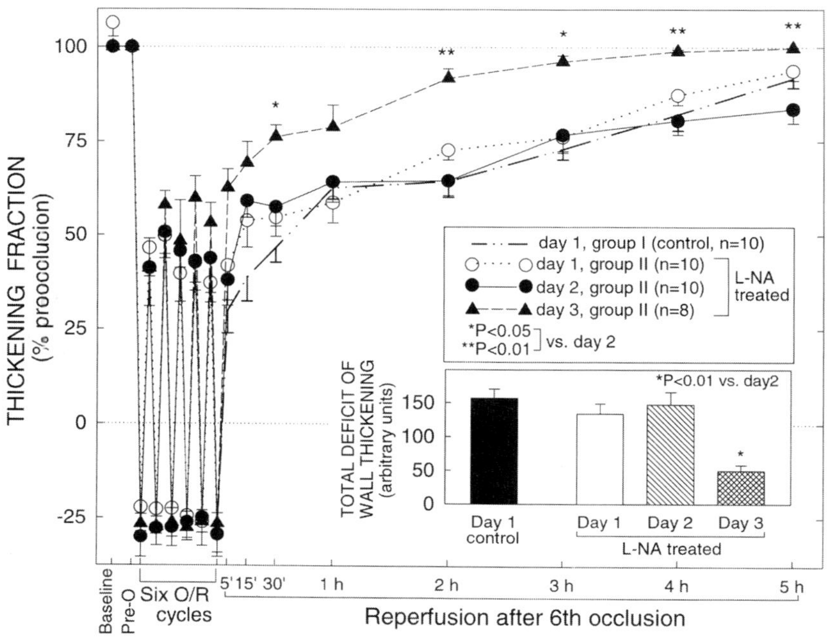

FIGURE 2. Systolic thickening fraction, expressed as a percentage of preocclusion values, in the ischemic-reperfused region in L-NA–treated rabbits before administration of L-NA (baseline), 9 min after the end of the infusion of L-NA (immediately before the first occlusion) (preocclusion [Pre-O]), 3 min into each coronary occlusion (O), 3 min into each reperfusion (R), and at selected times during the 5-h reperfusion interval following the sixth occlusion. Conscious rabbits underwent a sequence of six 4-min coronary occlusion/4-min reperfusion cycles for three consecutive days (days 1, 2, and 3). L-NA was given on day 1 prior to the first occlusion/reperfusion cycle. ○ indicates measurements taken on day 1; ●, measurements taken on day 2; and ▲, measurements taken on day 3. To facilitate comparisons, the data pertaining to day 1 of the control group are also shown (*thick interrupted line without symbols*). Thickening fraction was assessed by a 10-MHz pulsed Doppler crystal placed on the epicardium. Percent systolic thickening fraction was calculated as the ratio of net systolic thickening to end-diastolic wall thickness, multiplied by 100. Data are mean ± SEM. In contrast to control rabbits, in which the recovery of wall thickening was markedly enhanced on day 2 compared with day 1 (data not shown), in L-NA-treated rabbits the recovery of wall thickening was similar on days 1 and 2, indicating that administration of L-NA on day 1 abrogated the development of late PC against stunning on day 2. The expected late PC effect became apparent on day 3, as a result of the ischemic PC stimulus applied on day 2 (when rabbits were not treated with L-NA). **Inset.** Total deficit of wall thickening during the 5-h reperfusion period following the sixth reperfusion. The total deficit of wall thickening is an integrative measure of the overall severity of myocardial stunning, and is calculated by measuring the area between the wall-thickening versus time line and the baseline (100% line) during the recovery phase. Notice that the total deficits were similar on day 1 and day 2, indicating that L-NA prevented late PC against stunning. The deficit of wall thickening decreased on day 3, indicating that the ischemic stimulus on day 2 induced a late PC effect on day 3. (Reproduced from Ref. 5 with permission of the American Heart Association.)

Hence, we hypothesized that NO generated during the first ischemic stress triggers the development of late PC 24 h later.[5] This hypothesis was tested with the administration of N^{ω}-nitro-L-arginine (L-NA), a potent nonselective inhibitor of all three NOS isoforms, in conscious rabbits undergoing a sequence of six 4-min occlusion/4-min reperfusion cycles on three consecutive days.[5] Administration of L-NA before the PC ischemia blocked the development of the late PC protection against both

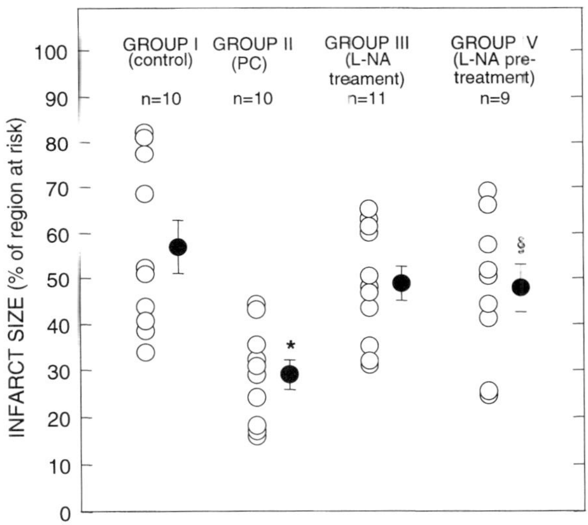

FIGURE 3. Effect of the NOS inhibitor L-NA on late PC against infarction in conscious rabbits. All rabbits underwent a 30-min coronary occlusion followed by three days of reperfusion. Rabbits in group I received neither ischemic PC nor drug treatment. Rabbits in group II were preconditioned 24 h earlier with a sequence of six 4-min coronary occlusion/4-min reperfusion cycles. Rabbits in group III underwent the same protocol as group II, except that they received L-NA prior to the six 4-min occlusion/reperfusion cycles. Rabbits in group IV received L-NA without ischemic PC 24 h before the 30-min occlusion. In group III, administration of L-NA prior to the PC ischemia abrogated the infarct-sparing effect of late PC, indicating that the development of late PC against infarction is triggered by NO. In group IV (administration of L-NA without PC), infarct size did not differ from that observed in controls, demonstrating that administration of L-NA in itself did not exert a delayed deleterious effect on myocardial infarction. Measurement of infarct size: at the conclusion of the studies, rabbits were euthanized, and the size of the occluded-reperfused coronary vascular bed (ischemic area) was determined by tying the coronary artery at the site of the previous occlusion and by perfusing the aortic root with phthalo blue (5% solution). The left ventricle was sliced into 6–7 pieces, incubated in triphenyltetrazolium chloride (1% solution, pH 7.4), fixed in formalin, and photographed. Transparencies were projected onto a paper screen, and the borders of the infarcted, ischemic-reperfused, and nonischemic myocardium were traced. Corresponding areas were measured by computerized planimetry and infarct size was expressed as a percentage of the region at risk of infarction. ○, individual rabbit; ●, mean ± SEM. *$p < 0.05$ vs. group I (controls); §$p < 0.05$ vs. group II. (Reproduced from Ref. 6 with permission of the American Physiological Society.)

myocardial stunning[5] and infarction[6] 24 h later (see FIGURES 2 and 3), indicating that NO generated during the initial PC ischemia is necessary for the development of this cardioprotective response. In subsequent studies, pretreatment with NO donors (diethylenetriamine/NO [DETA/NO] and *S*-nitroso *N*-penicillamine [SNAP]), in the absence of ischemia, induced a delayed protective effect against both myocardial stunning (see FIGURE 4) and infarction (see FIGURE 5).[16] A similar protective effect against myocardial stunning[17] and infarction[35] was subsequently observed with pretreatment with nitroglycerin (NTG) (see FIGURES 6 and 7), and was not affected by nitrate tolerance.[35] Moreover, this late PC effect of NO releasing agents (DETA/NO, SNAP, and NTG) was completely abrogated by the coadministration of MPG[16] or chelerythrine,[17] implying that MPG-sensitive oxidants ($ONOO^-$ and/or ·OH) serve as second messengers of NO-dependent signaling with subsequent activation of PKC. Taken together, these studies[5,6,16,17] indicate that the generation of NO during the PC ischemia serves as a trigger for late PC against myocardial stunning, as well as infarction via the formation of $ONOO^-$ and/or secondary ROS. Importantly, the

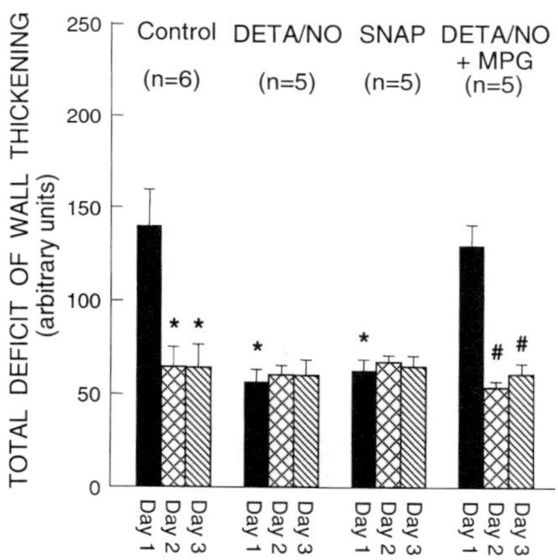

FIGURE 4. Total deficit of wall thickening (WTh) after the 6th reperfusion on days 1, 2, and 3 in the control, DETA/NO, SNAP, and DETA/NO+MPG groups (groups I, II, III, and IV, respectively). Conscious rabbits underwent a sequence of six 4-min coronary occlusion/4-min reperfusion cycles for three consecutive days (days 1, 2, and 3). Twenty-four hours before the first sequence of occlusion/reperfusion, rabbits received either no treatment (control), the NO donor DETA/NO, the NO donor SNAP, or DETA/NO in conjunction with the antioxidant MPG. Pretreatment with either DETA/NO or SNAP resulted in an attenuation of the total deficit of WTh on day 1 equivalent to that induced by ischemic PC on day 2 in control rabbits. This effect was abrogated by MPG, indicating that DETA/NO-induced late PC against stunning is mediated by MPG-sensitive oxidants (e.g., $ONOO^-$ and/or ·OH). The total deficit of WTh (mean ± SEM, arbitrary units) in each group are depicted. *$p < 0.05$ vs. control day 1; #$p < 0.05$ vs. day 1. (Reproduced from Ref. 16 with modification with permission of the American Heart Association.)

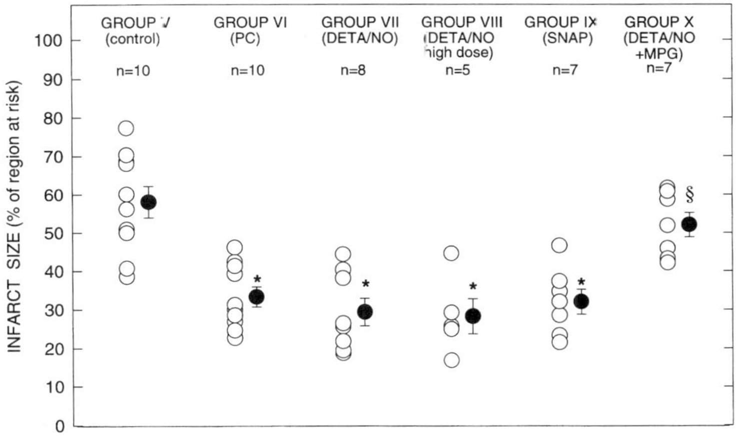

FIGURE 5. Myocardial infarct size in groups V (control group), VI (ischemic PC group), VII (DETA/NO group), VIII (DETA/NO high dose group), IX (SNAP group), and X (DETA/NO+MPG group). Conscious rabbits underwent a 30-min coronary occlusion and three days of reperfusion. Twenty-four hours before the 30-min occlusion, they received either ischemic PC (six 4-min occlusion/reperfusion cycles), DETA/NO (at two doses), SNAP, or DETA/NO in conjunction with MPG. Pretreatment with either DETA/NO or SNAP resulted in a reduction of infarct size equivalent to that induced by ischemic PC in group VI. This effect was abrogated by MPG, indicating that NO-induced late PC against infarction is mediated by MPG-sensitive oxidants. Infarct size is expressed as a percentage of the region at risk of infarction. ○, individual rabbit; ●, mean ± SEM. *$p < 0.05$ vs. group V (control group); §$p < 0.05$ vs. group VII (DETA/NO group). (Reproduced from Ref. 16 with permission of the American Heart Association.)

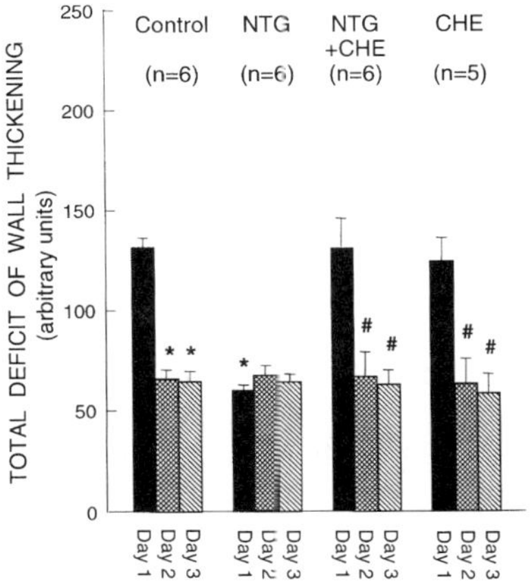

FIGURE 6. Total deficit of wall thickening (WTh, mean ± SEM, arbitrary units) after sixth reperfusion on days 1, 2, and 3 in control rabbits, and rabbits pretreated with nitroglycerin (NTG), nitroglycerin+chelerythrine (NTG + CHE), and chelerythrine alone (CHE). Pretreatment with nitroglycerin alleviated the total deficit in WTh equivalent to that observed in the control group on days 2 and 3. The late PC-mimicking effect of nitroglycerin was completely abrogated by coadministration of chelerythrine, indicating that activation of PKC is essential for NO donor-induced late PC. *$p < 0.05$ vs. control day 1; #$p < 0.05$ vs. day 1. (Reproduced from Ref. 17 with modification with permission of the American Physiological Society.)

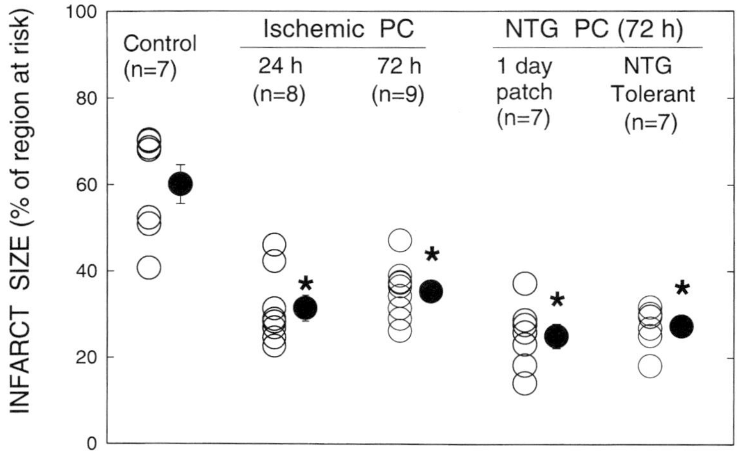

FIGURE 7. To determine the effect of continuous nitroglycerin (NTG) treatment on NTG-induced delayed infarct-sparing effects, rabbits underwent a continuous 28-day transdermal (TD)-NTG patch treatment (NTG tolerant group). The early hypotensive effects of TD-NTG did not persist during daily application of TD-NTG patches, indicating tolerance. The development of nitrate tolerance was confirmed in a separate group of five rabbits by the finding that on days 14 and 28 of continuous TD-NTG treatment, the hypotensive response to i.v. NTG was markedly attenuated compared with that observed immediately before the application of the first TD-NTG patch. Nitrate tolerant rabbits underwent a 30-min coronary occlusion 72 h after the end of this continuous 28-day TD-NTG patch treatment. Infarct size was significantly smaller compared with controls and was indistinguishable from that measured in the late ischemic PC groups, indicating that tolerance to the hemodynamic actions of NTG does not extend to the PC effects of this drug. ○, individual rabbit; ●, mean ± SEM. $*p < 0.05$ vs. control. (Reproduced from Ref. 35 with permission of the American Heart Association.)

ability of NTG to induce late PC in humans has recently been demonstrated,[36] paving the way for novel applications of nitrates for the protection of ischemic myocardium.

Although all three isoforms of NOS have been detected in the myocardium, the source of increased NO formation specifically during the PC ischemia is likely to be eNOS, which has been identified, not only in endothelial cells, but also in cardiac myocytes.[18,19] The finding that the development of late PC against stunning is blocked by pretreatment with the nonselective NOS antagonist L-NA, but not of the relatively selective iNOS antagonists aminoguanidine (AG) and *S*-methylisothiourea (SMT), is also congruent with this concept.[11] Ischemic PC could stimulate NO formation by eNOS via ischemia-induced increases in intracellular calcium and NADPH and/or via increased shear stress during the repeated reactive hyperemias. A third mechanism, release of bradykinin with activation of endothelial B_2 receptors,[37] seems implausible because in recent unpublished studies we have found that late PC against stunning is not blocked by pretreatment with HOE 140, a bradykinin B_2

receptor antagonist, at doses that completely block the hypotensive effects of 0.45 μg/kg of bradykinin.

In summary, NO, via formation of secondary oxidant species, such as $ONOO^-$ and/or ·OH, triggers the development of late PC both against myocardial stunning and myocardial infarction. This effect is not blocked by nitrate tolerance.

NO AS A MEDIATOR OF LATE PC

The fact that the protective effect of late PC requires 12–24 h to develop and lasts for 3–4 days[23,38] suggests that synthesis and degradation of cardioprotective protein(s) are involved in the manifestation of this natural phenomenon. This notion was confirmed by a study in which the development of late PC was blocked by cycloheximide, a protein synthesis inhibitor.[39] Several proteins have been proposed as possible mediators of the protection afforded by late PC, including iNOS,[11,12] cyclooxygenase-2 (COX-2),[14] antioxidant enzymes,[21] K_{ATP} channels,[40] aldose reductase,[15] and heat stress proteins.[4] It seems likely that no one of these proteins acts in isolation to effect cardioprotection, but rather it is a concerted effort. The following discussion focuses on the role played by NO in mediation of late PC protection.

Pharmacological Evidence that NO Mediates Late PC

NO, or its second messenger, cGMP, exert multiple actions on the myocardium, including antagonism of the effects of beta-adrenergic stimulation, inhibition of calcium influx into myocytes, decrease in myocardial contractility, reduction in myocardial oxygen consumption, and opening of K_{ATP} channels.[13,19] Several of these actions can protect the myocardium during ischemic or hypoxic stress. Thus, we hypothesize that the protective effects of late PC are mediated by increased NO formation secondary to ischemia-induced alterations of NOS gene expression and increased NOS activity. In two separate studies in rabbits, administration of the nonselective NOS inhibitor L-NA prior to the second ischemic challenge on day 2 completely abrogated the late PC protection against stunning (see FIGURE 8)[11] as well as infarction (see FIGURE 9),[12] indicating that the protection afforded by late PC is a NOS-dependent phenomenon. Furthermore, administration of two structurally-unrelated iNOS inhibitors, AG and SMT, on day 2 also abrogated late PC against myocardial stunning (FIG. 8),[11] suggesting that the specific isoform involved in mediating the protective effects of late PC is iNOS. The late PC-blocking effect of SMT could be reversed with concomitant administration of L-arginine, indicating that SMT acted specifically by inhibiting iNOS activity. Administration of AG prior to the 30-min occlusion on day 2 (FIG. 9) abrogated the infarct-sparing effects of late PC, again confirming that iNOS is the isoform involved in the protection against cell death.[12] These NOS inhibitors had no appreciable effect on either myocardial stunning or myocardial infarction in nonpreconditioned hearts. Taken together, these results[11,12] indicate that the cardioprotective effects of late PC are mediated by the activity of NOS, and strongly implicate iNOS as the isoform involved.

In a subsequent study, the role of cofactors involved in NO synthesis by NOS was examined. Tetrahydrobiopterin (BH_4), synthesized by a pathway catalyzed by

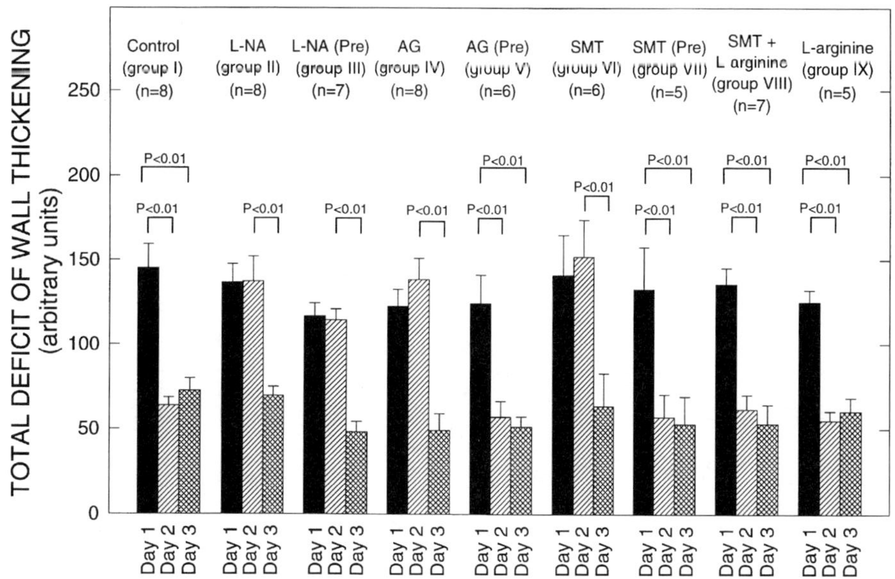

FIGURE 8. Total deficit of systolic wall thickening after the sixth reperfusion on days 1, 2, and 3 in nine groups of conscious rabbits. All groups underwent a sequence of six 4-min occlusion/4-min reperfusion cycles for three consecutive days (days 1, 2, and 3). Rabbits in group I received no treatment. Rabbits in groups II (L-NA), IV (AG), and VI (SMT) received L-NA, AG, and SMT, respectively, before the first coronary occlusion on day 2. Rabbits in groups III (L-NA-pre), V (AG-pre), and VII (SMT-pre) received the same doses of L-NA, AG, and SMT, respectively, before the first coronary occlusion on day 1. Rabbits in group VIII received the same dose of SMT in conjunction with L-arginine before the first coronary occlusion on day 2. Rabbits in group IX received L-arginine alone before the first coronary occlusion on day 2. Data are mean ± SEM. *Pre* indicates treatment on day 1. Administration of either a nonselective (L-NA) or an iNOS-selective (AG and SMT) NOS inhibitor on day 2 completely abrogated late PC against myocardial stunning (groups II, IV, and VI, respectively), indicating that NOS (specifically, iNOS) is a mediator of this cardioprotective phenomenon. The abrogation of late PC by SMT was completely reversed by L-arginine (group VIII), indicating that SMT acted specifically by inhibiting NOS activity. (Reproduced from Ref. 11 with permission of the American Heart Association.)

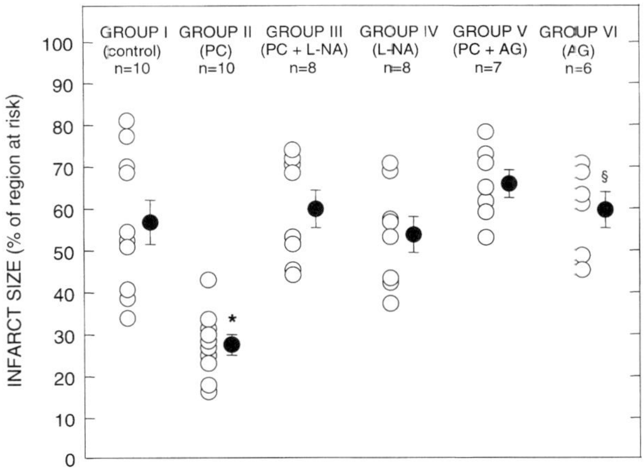

FIGURE 9. Effect of the nonselective NOS inhibitor L-NA and the relatively-selective iNOS inhibitor aminoguanidine (AG) on late PC against infarction in conscious rabbits. All groups underwent a 30-min coronary occlusion followed by three days of reperfusion. Group II was preconditioned 24 h earlier with six 4-min occlusion/reperfusion cycles. Group III underwent the same protocol, except that L-NA was given prior to the 30-min occlusion. Group IV received L-NA as group III, but was not preconditioned 24 h earlier. Group V underwent the same protocol as group II, except that AG was given before the 30-min occlusion. Group VI received AG without undergoing ischemic PC 24 h earlier. Administration of either L-NA or AG completely abrogated the infarct-sparing effect of late PC, indicating that NOS (specifically, iNOS) is a mediator of late PC against infarction. Infarct size in groups IV and VI was not different from controls, indicating that L-NA and AG did not exert a delayed deleterious effect on infarct size independent of PC. Infarct size is expressed as a percentage of the region at risk of infarction. ○, individual rabbit; ●, mean ± SEM. *$p < 0.05$ vs. group I (control); §$p < 0.05$ vs. group II (PC group). (Reproduced from Ref. 12 with permission of the American Heart Association.)

sepiapterin reductase, is an essential cofactor for the formation of NO by both iNOS and eNOS. However, in contrast to iNOS, the activity of eNOS is thought to be pulsatile rather than continuous.[18,19] Thus, if the late PC protection is mediated by continuous NO generation by iNOS, this would also require continuous synthesis of BH_4, and inhibition of sepiapterin reductase should block the mediation of late PC protection by iNOS. This concept was tested by Takano *et al.*[41] in a conscious rabbit model of late PC against myocardial stunning. Administration of *N*-acetylserotonin (NAS), an inhibitor of sepiapterin reductase, prior to the second ischemic challenge on day 2 completely abrogated the protective effects of late PC against myocardial stunning, indicating that intact BH_4 biosynthesis is necessary for the protection of late PC to become manifest.[41] However, the same dose of NAS given on day 1 prior to the first ischemic challenge had no effect on late PC.[41] Since the preexisting tissue stores of BH_4 are sufficient to support a burst of eNOS activity on day 1, thereby triggering the development of late PC, these findings strongly support the hypothesis that the isoform involved in triggering late PC is eNOS, and that iNOS is a mediator of late PC.

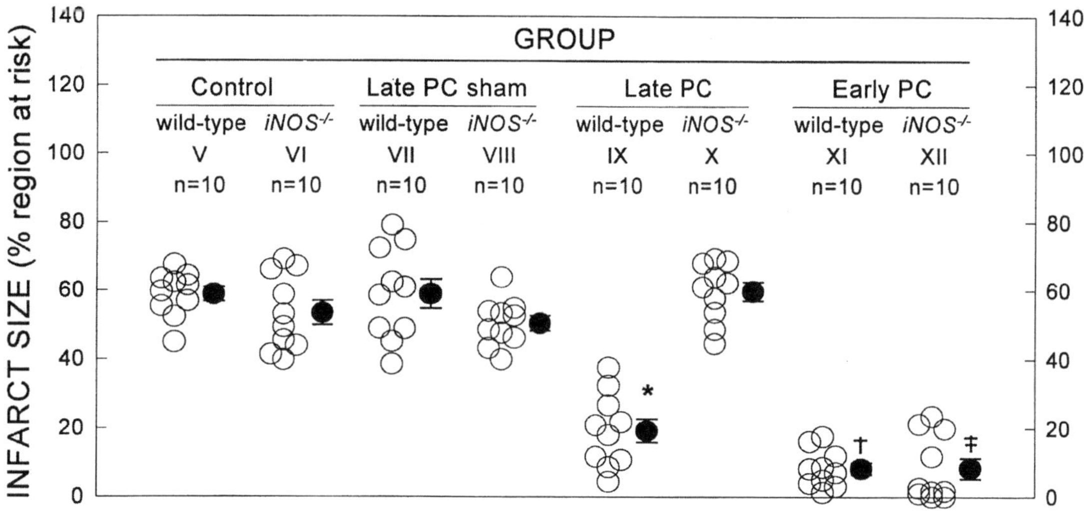

FIGURE 10. Myocardial infarct size in wild-type and *iNOS*$^{-/-}$ mice. Mice were subjected to 30 min of coronary occlusion and 24 h of reperfusion, in the absence of prior ischemic PC (groups V and VI, control groups), 24 h after a sham open chest procedure without ischemia-reperfusion (groups VII and VIII, late PC sham groups), 24 h after ischemic PC (six 4-min occlusion/4-min reperfusion cycles) (groups XI and X, late PC groups), or 10 min after the same ischemic PC protocol (groups XI and XII, early PC groups). Infarct size is expressed as a percentage of the region at risk of infarction. Late PC produced a significant reduction in infarct size in wildtype mice but not in *iNOS*$^{-/-}$ mice, whereas early PC produced marked infarct size reduction in both. ○, individual mice; ●, mean ± SEM. *$p < 0.05$ vs. group VII; †$p < 0.05$ vs. group V; ‡$p < 0.05$ vs. group VI. (Reproduced from Ref. 13 with permission of the National Academy of Sciences.)

Studies in Mice with Targeted Disruption of the iNOS Gene

All *in vivo* studies with pharmacologic manipulations are inherently limited by the selectivity of the agent for the target enzyme and its unavoidable nonspecific actions. The pharmacologic inhibitors used in the studies summarized above, AG and SMT, are only relatively specific for iNOS (IC_{50} ratio of approximately 30:1 for iNOS vs. eNOS[42]). Moreover, AG is known to exert a number of nonspecific actions in addition to inhibition of NOS, including increasing angiotensin II-mediated release of prostacyclins, inhibition of oxidative modification of low density lipoproteins, inhibition of formation of glycosylation end-products in diabetes, and inhibition of histamine metabolism, polyamine catabolism, aldose reductase, catalase, and other iron- and copper-containing enzymes.[42] Thus, conclusive evidence that late PC is mediated by iNOS cannot be provided by pharmacological studies alone.

Accordingly, we performed a series of experiments in the murine model of late PC using *iNOS*$^{-/-}$ mice.[13] In nonpreconditioned *iNOS*$^{-/-}$ mice, infarct size (63.5 ± 2.6% of the region at risk) was similar to that in wildtype mice from the same background, indicating that iNOS activity does not affect infarct size in the absence of PC (see FIGURE 10).[13] However, when *iNOS*$^{-/-}$ mice were subjected to ischemic PC 24 h earlier, infarct size (55.0 ± 4.1% of the region at risk) was not decreased compared with nonpreconditioned *iNOS*$^{-/-}$ mice, indicating that the late PC effect was completely abrogated (FIG. 10).[13] Interestingly, disruption of the iNOS gene had no effect on the early phase of ischemic PC (30-min occlusion performed 10 min after PC ischemia) (FIG. 10).[13] Furthermore, administration of the NO donor DETA/NO to wildtype mice induced an infarct-sparing effect 24 h later equivalent to that observed in mice preconditioned with ischemia demonstrating that NO can mimic the late phase of ischemic PC in mice. However, DETA/NO failed to induce a late PC effect in *iNOS*$^{-/-}$ mice (infarct size: 51.3 ± 6% of risk region) indicating that iNOS serves as an obligatory mediator even when NO is provided exogenously.[13]

Measurement of iNOS Transcript Levels

Results from the studies summarized above imply but do not prove that iNOS activity increases 24 h after initial PC ischemia. iNOS is regulated by multiple mechanisms, including transcriptional upregulation, transcript stability, translational efficiency, and posttranslational modification.[43,44] To elucidate the molecular basis for this increased iNOS activity, the steady-state levels of iNOS mRNA were measured by quantitative RNA dot blot hybridization.[45] A robust increase in steady-state iNOS mRNA levels was found in the ischemic-reperfused region three hours after the six occlusion-reperfusion cycles (+287 ± 8% vs. nonischemic region), and this was completely abolished by pretreatment with L-NA, indicating that the ischemia-induced increase in *iNOS* transcript levels is triggered by increased NO formation on day 1.[45] These results support a complex paradigm in which NO generated during the PC ischemia causes increased *iNOS* transcription or transcript stability, which results in cardioprotection 24 h later. Thus, it appears that two isoforms of NOS are activated sequentially, with initial activation of eNOS causing the subsequent activation of iNOS.

Measurement of iNOS Activity

To confirm the translation of this ischemia-induced increase in *iNOS* transcript into an increased activity at the protein level, specific NOS isoform activities in the rabbit myocardium were measured next. On day 1, a marked increase in calcium-dependent constitutive NOS (cNOS, i.e., eNOS and/or nNOS) activity was observed at 30 min after the six occlusion/reperfusion cycles (+290 ± 47% vs. nonischemic tissue) (see FIGURE 11), without any change in iNOS activity.[46] Twenty-four hours after ischemic PC, however, there was a marked increase in iNOS activity in

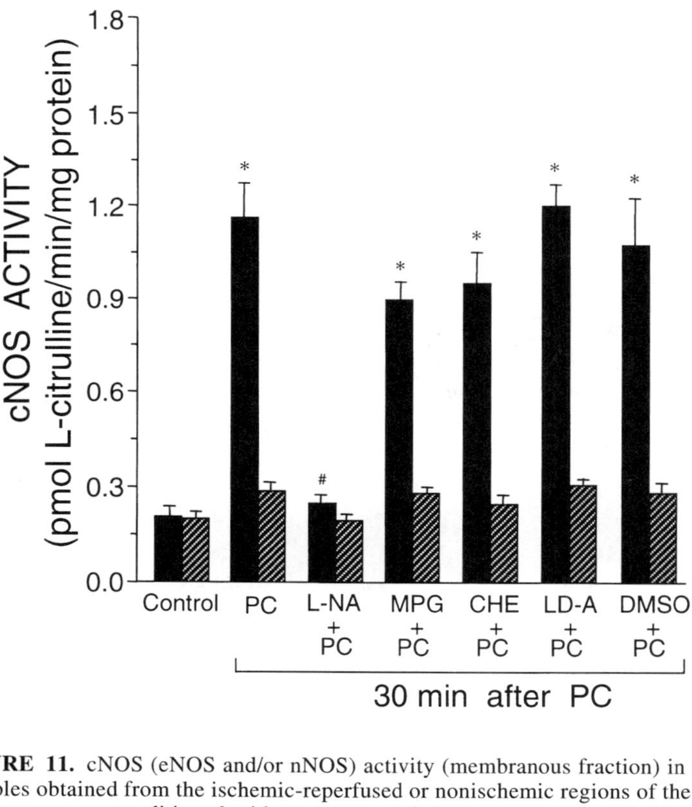

FIGURE 11. cNOS (eNOS and/or nNOS) activity (membranous fraction) in myocardial samples obtained from the ischemic-reperfused or nonischemic regions of the heart of rabbits that were preconditioned with a sequence of six 4-min coronary occlusion/4-min reperfusion cycles without any treatment (PC, group II), after pre-treatment with L-NNA (L-NNA + PC, group VII), MPG (MPG + PC, group VIII), CHE (CHE + PC, group IX), LD-A (LD-A + PC, group X), or vehicle (DMSO + PC, group XI). Control rabbits (group I) did not receive any treatment, nor did they undergo occlusion/reperfusion cycles. Data are mean ± SEM. ■, Ischemic zone; ▨, non-ischemic zone; $^*p < 0.05$ vs. control; $^\#p < 0.05$ vs. PC; $n = 5$ in each group. (Reproduced from Ref. 46 with modification with permission of the American Physiological Society.)

the ischemic/reperfused region (+131 ± 21% vs. the nonischemic region) (see FIGURE 12), with no change in cNOS activity.[46] Measurements of myocardial levels of NO_2 and NO_3 (NO_x), the stable byproducts of NO, increased in parallel by 34 ± 8% versus nonischemic tissue at 60 min after ischemic challenge on day 1 and by 36 ± 7% versus nonischemic tissue on day 2, confirming increased NO generation. Importantly, administration of the NOS inhibitor L-NA, the ROS scavenger MPG,

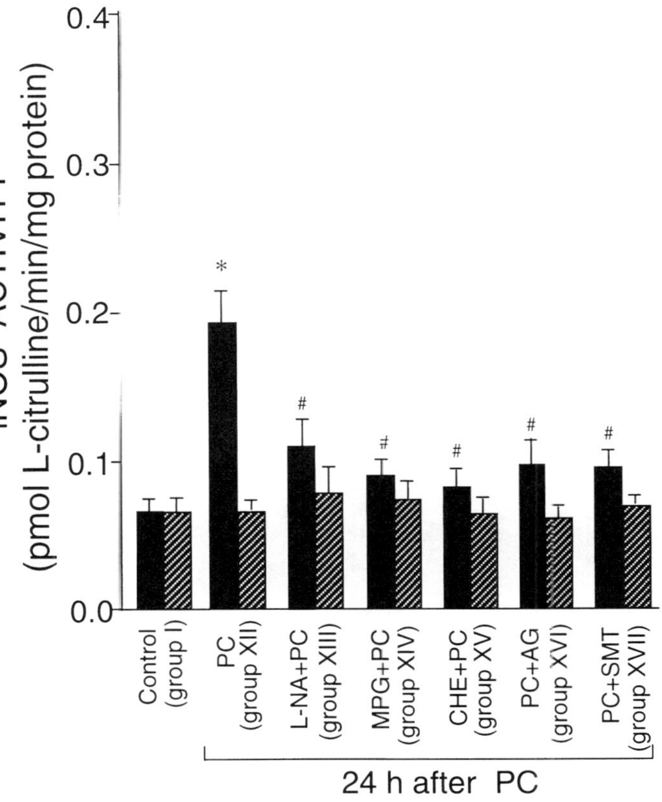

FIGURE 12. Calcium-independent NOS (iNOS) activity (cytosolic fraction) in myocardial samples obtained from the ischemic-reperfused or nonischemic regions of the heart of rabbits preconditioned 24 h earlier (day 1) with six 4-min coronary occlusion/4-min reperfusion cycles (groups XII–XVII). Before the PC ischemia (day 1), rabbits received no treatment (PC, group XII), L-NNA (L-NNA + PC, group XIII), MPG (MPG + PC, group XIV), or CHE (CHE + PC, group XV). In groups XVI and XVII, no treatment was given on day 1. On day 2 (24 h after the PC ischemia), rabbits received either AG (PC + AG, group XVI) or SMT (PC + SMT, group XVII). Control rabbits did not receive any treatment and did not undergo coronary occlusion/reperfusion cycles. Data are mean ± SEM. ■, Ischemic zone; ▨, non-ischemic zone; *$p < 0.05$ vs. control; #$p < 0.05$ vs. PC; $n = 7$ in PC group; $n = 5$ in other groups (Reproduced from Ref. 46 with modification with permission of the American Physiological Society.)

the PKC inhibitor chelerythrine, and the PTK inhibitor lavendustin A (LD-A) on day 1 completely abrogated both the increase in iNOS activity and NO_2/NO_3 levels at 24 h after ischemic PC, indicating that the upregulation of iNOS on day 2 is the result of a sequence of events occurring on day 1, which is triggered by the generation of NO and involves the formation of ROS, the activation of PKC, and the activation of PTKs.[46]

Measurement of iNOS Protein Levels

Since an increase in enzymatic activity may result from mere activation of a preexisting protein, iNOS protein levels were measured in mouse hearts for direct documentation of new enzyme synthesis. In wildtype control mice, 67% of total tissue iNOS protein was found in the cytosolic fraction and 33% in the membranous fraction.[13] The total (cytosolic + membranous fractions) tissue content of iNOS in the

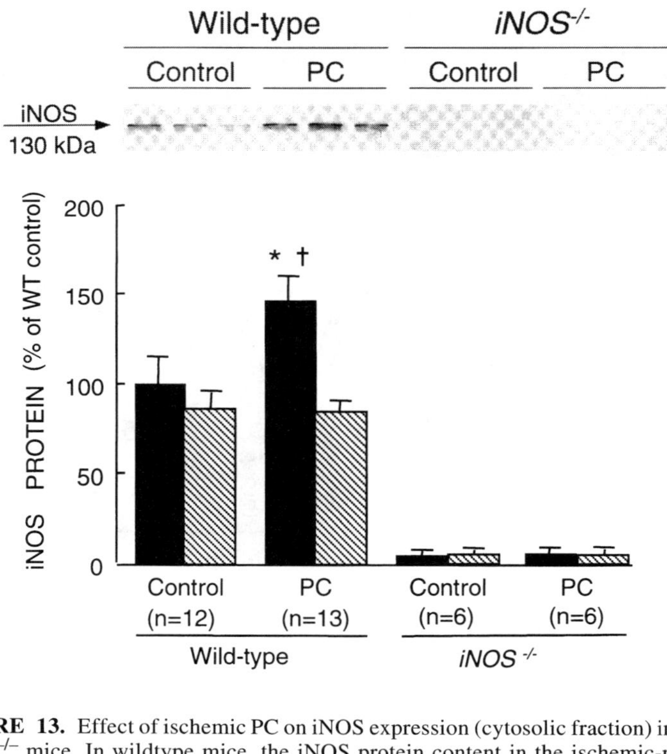

FIGURE 13. Effect of ischemic PC on iNOS expression (cytosolic fraction) in wildtype and *iNOS*$^{-/-}$ mice. In wildtype mice, the iNOS protein content in the ischemic-reperfused region was increased by 47 ± 12% 24 h after ischemic PC. In *iNOS*$^{-/-}$ mice, immunoreactive iNOS protein was undetectable, either after sham surgery or after ischemic PC. **Top:** representative Western blot showing the 130 kDa iNOS protein band; **bottom:** quantitative measurement of iNOS levels. Data are mean ± SEM. ■, Ischemic zone; ▧, non-ischemic zone; *$p < 0.05$ vs. control; †$p < 0.05$ vs. nonisochemic zone. (Reproduced from Ref. 13 with permission of the National Academy of Sciences.)

ischemic-reperfused region increased 41 ± 6% above control levels ($p < 0.05$) 24 h after ischemic PC (see FIGURE 13).[13] In *iNOS*$^{-/-}$ mice, iNOS protein signals were undetectable under control conditions as well as after ischemic PC (FIG. 13). In contrast to iNOS, the expression of eNOS did not increase 24 h after ischemic PC, either in wildtype or in *iNOS*$^{-/-}$ mice. No nNOS was detected in control or preconditioned heart samples by using commercially available nNOS monoclonal or polyclonal antibodies. For comparison, in a separate group of seven wildtype mice given lipopolysaccharide (12.5 mg/kg i.p.) 6 h earlier, the iNOS protein content in the cytosolic fraction of LV samples increased 853 ± 131% above control levels; that is, about 18 times more than that after ischemic PC. This study[13] was the first to definitively demonstrate an increase in iNOS protein level in the myocardium following ischemic PC.

iNOS-Dependent Protection

Recent studies have revealed that the delayed infarct-sparing effects elicited by exercise, by the δ_1-receptor agonist TAN-670, and by NO donors (SNAP and NTG) are absent in *iNOS*$^{-/-}$ mice, indicating that iNOS serves as an obligatory mediator for late PC protection induced by different stimuli.[47,48] Considerable evidence indicates that the manifestation of late PC protection involves additive or synergistic interaction among several proteins. The mechanism whereby iNOS-derived NO protects against ischemia, although not fully elucidated, appears to involve the activation of guanylate cyclase, since protection against both stunning and infarction is abolished by pretreatment with ODQ, a guanylate cyclase inhibitor.[49]

THE SIGNALING PATHWAY OF LATE PC

NO generated during the initial ischemic challenge, in concert with other potential triggers of this phenomenon, sets off a chain of signaling events that culminates in increased transcription of various cardioprotective genes, including *iNOS*. Among the possible signaling intermediaries, recent data identify PKC (the ε isoform), the Src family of protein tyrosine kinases (Src and Lck), and NF-κB as critical components of this pathway.

Protein Kinase C

The involvement of PKC in the genesis of late PC protection was first demonstrated by pharmacologic intervention studies in which pretreatment with the PKC inhibitors staurosporin or polymyxin B blocked the delayed protection of late PC against infarction in rabbits.[50] Direct evidence that ischemia activates PKC *in vivo* came from a series of studies in which Ping *et al.* showed ischemia-induced selective translocation and activation of only two PKC isoforms (PKCε and PKCη, both members of the subfamily of novel PKC isoforms) with no significant changes in total myocardial PKC activity or its subcellular distribution.[8] Inhibition of translocation of PKCε (but not of PKCη) by chelerythrine blocked the development of late PC against myocardial stunning, indicating that isoform-selective translocation of PKCε specifically is necessary for the development of late PC.[8] Furthermore, administration of

L-NA prior to ischemic PC on day 1 blocked the translocation of PKCε, whereas administration of NO donors, in the absence of ischemia, induced translocation and activation of PKCε quantitatively similar to that induced by ischemic PC, indicating that generation of NO during PC ischemia is responsible for this selective activation of PKCε.[51] Recruitment of PKCε was also found to be essential for the development of NO-donor-induced late PC, since this protection was abrogated by coadministration of chelerythrine with NO donors.[51] The finding that transgenic mice expressing constitutively active PKCε in the heart are protected against infarction in a manner analogous to that afforded by late PC confirmed that the role of PKCε is not species-specific.[52]

Taken together, these studies[8,51,52] indicate that PKCε serves as a critical signaling molecule in the development of NO-dependent late PC, both after an ischemic stimulus (endogenous NO) and after NO donor treatment (exogenous NO). The exact mechanism whereby NO activates PKCε appears to involve generation of NO-derived ROS. This hypothesis is supported by the fact that the antioxidant MPG, given prior to the ischemic PC protocol, blocked the activation of PKCε, indicating an essential role of ROS in this phenomenon.[16] Most likely, NO reacts with $\cdot O_2^-$ to form $ONOO^-$, which then generates secondary ROS that activate PKCε, either by direct oxidative modification[53] or via activation of phospholipase D.[54]

Protein Tyrosine Kinases

PTKs are a heterogenous family of receptor and nonreceptor kinases with more than 1,000 known members that participate in intracellular signaling. Several lines of evidence suggest that specific subfamilies of nonreceptor PTKs are involved in signaling downstream of activation of PKCε. Inhibition of nonreceptor PTKs, especially the Src family members, with LD-A abrogated late PC protection against myocardial stunning in conscious rabbits.[55] It was also found that ischemic PC selectively activates two members of the Src family PTKs (Src and Lck) and that this activation was blocked by chelerythrine, indicating that these kinases are downstream of PKC.[9] Subsequent studies in mice have demonstrated that Lck serves as a direct substrate of PKCε,[56] and that targeted disruption of the *Lck* gene abolishes the development of late PC.[57] Interestingly, administration of LD-A 24 h after the PC stimulus also blocked the increase in iNOS activity on day 2.[55] Although it is known that the activity of iNOS can be modulated at the post-translational level by tyrosine phosphorylation,[44] the precise signaling link between activation of PTKs and upregulation of iNOS activity remains unclear at this time.

Mitogen-Activated Protein Kinases

The MAPK family of kinases which includes the p44/p42 MAPKs (also called extracellular signal-regulated kinases [ERKs]), the p38 MAPKs, and the p46/p54 MAPKs (or c-jun *N*-terminal kinases [JNKs]) is another potential downstream target of PKC phosphorylation. This notion is supported by the observation that ischemic PC activates the p44/p42 mitogen-activated protein kinases (MAPKs) and MEK1 and MEK2, the direct activators of p44/p42 MAPKs[58] in rabbit hearts. Furthermore, this selective nuclear activation of p44/p42 MAPKs is blocked by chelerythrine, indicating that it is mediated by, and downstream of, PKC.[58] In recent studies, using

PKCε transgenic mouse lines, we have obtained additional evidence that the ε isoform of PKC plays a critical role in modulating the activity of p44/p42 MAPKs *in vivo*.[58] The precise role of p44/p42 MAPKs in ischemic PC, and whether they contribute to increased iNOS activity, however, remains to be elucidated.

Nuclear Factor-κB

Involvement of NF-κB, a potent oxidant-sensitive transcription factor, in late PC signaling was shown in a study by Xuan *et al.*[10] in which a marked increase in NF-κB DNA-binding activity was found after the PC ischemia in rabbit hearts. Pretreatment with the NF-κB inhibitor diethyldithiocarbamate resulted in inhibition of increased NF-κB DNA-binding activity and abrogation of the protective effects of late PC,[10] indicating that NF-κB plays a critical role in the development of delayed cardioprotection. In the absence of ischemia, administration of NO releasing agents also activated NF-κB, whereas pretreatment with agents known to block late PC, including L-NA, MPG, chelerythrine, and LD-A, abolished the ischemia-induced increase in NF-κB DNA-binding activity, indicating that the activation of NF-κB after the initial PC ischemia is the result of NO and ROS generation and is dependent on subsequent signaling via activation of PKC and PTKs.[10] More recent studies performed in the murine model further elucidated the mechanism of NF-κB activation following ischemic PC. It was found that ischemic PC induces both serine phosphorylation (via PKC-dependent activation of IKKα and IKKβ) and tyrosine phosphorylation (independent of IKK) of IκBα resulting in activation of NF-κB.[59] The finding that overexpression of PKCε in adult cardiac myocytes results in serine and tyrosine phosphorylation of IκBα, and activation of NF-κB, IKKα, and IKKβ, strongly suggests a signaling link between PKCε and NF-κB. Interestingly, ischemia-induced NF-κB activation is absent in $Lck^{-/-}$ mice, suggesting that Lck plays a crucial role in this event.[57]

SUMMARY

Our understanding of the phenomenon of late PC has evolved considerably over the last decade. It is clear now that the manifestation of this protective response is the result of a very complex signaling process leading to activation of multiple genes in a coordinated manner. Although NO plays pivotal roles both in the genesis and mediation of the late PC response, several other inducible proteins, including COX-2, K_{ATP} channels, and aldose reductase are also necessary to mediate this protection. A summary of our current understanding of the pathophysiology of the late phase of ischemic PC is illustrated in FIGURE 1. A brief ischemic stress causes a burst of NO production (probably via eNOS) as well as $\cdot O_2^-$ production. NO and $\cdot O_2^-$ could then react to form $ONOO^-$, which could then activate PKCε, either directly or via secondary byproducts such as ·OH. Activated PKCε triggers a complex signal transduction cascade which involves PTKs, NF-κB, and most likely other as-yet-unknown kinases and transcription factors, leading to increased transcription of the protective proteins, including iNOS, during the second ischemic stress. Two members of the Src family of PTKs (Src and Lck) are likely to be involved during PC ischemia. PTKs also appear to be involved in the mediation of protection, possibly by modulating iNOS activity. Thus, NO plays two completely different roles in late PC: following the PC ischemia, NO generated by eNOS initiates the development of the cardioprotective mechanism, whereas 24 h later, NO generated by iNOS protects against myocardial stunning and infarction.

CONCLUSIONS

Brief episodes of ischemia evoke a powerful cytoprotective response in the heart termed ischemic PC. This phenomenon seems to be part of a universal response of myocardium to different stressful situations, since similar protection can be demonstrated by administration of pharmacological (adenosine, NO-donors, various receptor agonists) and noxious (ROS, cytokines, heat stress) agents. Although the exact details of the intricate molecular mechanism underlying this phenomenon remain to be elucidated, it is clear that NO plays a prominent role both in initiating and in mediating this cardioprotective response. NO, and other potential initiators, appear to trigger late PC by activating the ε isoform of PKC, which mobilizes a signal transduction cascade that involves PTKs, NF-κB, and almost certainly other as-yet-unknown elements, and culminates in upregulation of *iNOS* and other cardioprotective genes. The results summarized in this review support a novel pathophysiological and molecular-genetic paradigm in which NO acts as a molecular signal that modulates cardiac gene expression in response to ischemia and possibly other stresses, resulting in long-lasting defensive adaptations mediated by *de novo* synthesis of cardioprotective proteins. This novel function of NO could have far-reaching implications not only for ischemic PC but also for many other cardiac pathophysiologic situations that are associated with enhanced NOS activity. This information can also be potentially translated into therapeutic strategies in patients with coronary artery disease to precondition the heart against ischemic injury for a prolonged period with pharmacological (e.g., NO-releasing agents such as nitrates) or genetic approaches. Indeed, the ability of NTG to induce late PC in humans has recently been demonstrated,[36] paving the way for novel applications of nitrates for the protection of ischemic myocardium.

ACKNOWLEDGMENTS

This work was supported in part by NIH RO1 grants HL-43151 and HL-55757 (R.B.), AHA Scientist Development Grant 0130146N (B.D.), the Medical Research Grant Program of the Jewish Hospital Foundation, Louisville, KY, and the Commonwealth of Kentucky Research Challenge Trust Fund.

REFERENCES

1. Bolli, R. 1996. The early and late phases of preconditioning against myocardial stunning and the essential role of oxyradicals in the late phase: an overview. Basic Res. Cardiol. **91:** 57–63.

2. DOWNEY, J.M., M.V. COHEN, K. YTREHUS, *et al.* 1994. Cellular mechanisms in ischemic preconditioning: the role of adenosine and protein kinase C. Ann. N.Y. Acad. Sci. **723:** 82–98.
3. BOLLI, R. 2000. The late phase of preconditioning. Circ. Res. **87:** 972–983.
4. MARBER, M.S. & D.M. YELLON. 1996. Myocardial adaptation, stress proteins, and the second window of protection. Ann. N.Y. Acad. Sci. **793:** 123–141.
5. BOLLI, R., Z.A. BHATTI, X.L. TANG, *et al.* 1997. Evidence that late preconditioning against myocardial stunning in conscious rabbits is triggered by the generation of nitric oxide. Circ. Res. **81:** 42–52.
6. QIU, Y., A. RIZVI, X.L. TANG, *et al.* 1997. Nitric oxide triggers late preconditioning against myocardial infarction in conscious rabbits. Am. J. Physiol. **273:** H2931–H2936.
7. BOLLI, R., B. DAWN, X.L. TANG, *et al.* 1998. The nitric oxide hypothesis of late preconditioning. Basic Res. Cardiol. **93:** 325–338.
8. PING, P., J. ZHANG, Y. QIU, *et al.* 1997. Ischemic preconditioning induces selective translocation of protein kinase C isoforms epsilon and eta in the heart of conscious rabbits without subcellular redistribution of total protein kinase C activity. Circ. Res. **81:** 404–414.
9. PING, P., J. ZHANG, Y.T. ZHENG, *et al.* 1999. Demonstration of selective protein kinase C-dependent activation of Src and Lck tyrosine kinases during ischemic preconditioning in conscious rabbits. Circ. Res. **85:** 542–550.
10. XUAN, Y.T., X.L. TANG, S. BANERJEE, *et al.* 1999. Nuclear factor-kappaB plays an essential role in the late phase of ischemic preconditioning in conscious rabbits. Circ. Res. **84:** 1095–1109.
11. BOLLI, R., S. MANCHIKALAPUDI, X.L. TANG, *et al.* 1997. The protective effect of late preconditioning against myocardial stunning in conscious rabbits is mediated by nitric oxide synthase. Evidence that nitric oxide acts both as a trigger and as a mediator of the late phase of ischemic preconditioning. Circ. Res. **81:** 1094–1107.
12. TAKANO, H., S. MANCHIKALAPUDI, X.L. TANG, *et al.* 1998. Nitric oxide synthase is the mediator of late preconditioning against myocardial infarction in conscious rabbits. Circulation **98:** 441–449.
13. GUO, Y., W.K. JONES, Y.T. XUAN, *et al.* 1999. The late phase of ischemic preconditioning is abrogated by targeted disruption of the inducible NO synthase gene. Proc. Natl. Acad. Sci. U.S.A. **96:** 11507–11512.
14. SHINMURA, K., X.L. TANG, Y. WANG, *et al.* 2000. Cyclooxygenase-2 mediates the cardioprotective effects of the late phase of ischemic preconditioning in conscious rabbits. Proc. Natl. Acad. Sci. U.S.A. **97:** 10197–10202.
15. SHINMURA, K., S.-Q. LIU, X.L. TANG, *et al.* 2000. Aldose reductase is an obligatory mediator of the late phase of ischemic preconditioning. Circulation **102**(Suppl. II): II-120. Abstract.
16. TAKANO, H., X.L. TANG, Y. QIU, *et al.* 1998. Nitric oxide donors induce late preconditioning against myocardial stunning and infarction in conscious rabbits via an antioxidant-sensitive mechanism. Circ. Res. **83:** 73–84.
17. BANERJEE, S., X.L. TANG, Y. QIU, *et al.* 1999. Nitroglycerin induces late preconditioning against myocardial stunning via a PKC-dependent pathway. Am. J. Physiol. **277:** H2488–H2494.
18. GROSS, S.S. & M.S. WOLIN. 1995. Nitric oxide: pathophysiological mechanisms. Annu. Rev. Physiol. **57:** 737–769.
19. KELLY, R.A., J.L. BALLIGAND & T.W. SMITH. 1996. Nitric oxide and cardiac function. Circ. Res. **79:** 363–380.
20. MURRY, C.E., R.B. JENNINGS & K.A. REIMER. 1986. Preconditioning with ischemia: a delay of lethal cell injury in ischemic myocardium. Circulation **74:** 1124–1136.
21. KUZUYA, T., S. HOSHIDA, N. YAMASHITA, *et al.* 1993. Delayed effects of sublethal ischemia on the acquisition of tolerance to ischemia. Circ. Res. **72:** 1293–1299.
22. SUN, J.Z., X.L. TANG, A.A. KNOWLTON, *et al.* 1995. Late preconditioning against myocardial stunning. An endogenous protective mechanism that confers resistance to postischemic dysfunction 24 h after brief ischemia in conscious pigs. J. Clin. Invest. **95:** 388–403.

23. TANG, X.L., Y. QIU, S.W. PARK, *et al.* 1996. Time course of late preconditioning against myocardial stunning in conscious pigs. Circ. Res. **79:** 424–434.
24. TESCHNER, S., Y. QIU, X.L. TANG, *et al.* 1996. Late preconditioning against myocardial stunning in conscious rabbits: a dose-related or all-or-none phenomenon? Circulation **94**(Suppl. I): I-423. Abstract.
25. QIU, Y., C. MALDONADO, X.L. TANG, *et al.* 1995. Late preconditioning against myocardial stunning in conscious rabbits. Circulation **92**(suppl. I): I-715. Abstract.
26. QIU, Y., X.L. TANG, S.W. PARK, *et al.* 1997. The early and late phases of ischemic preconditioning: a comparative analysis of their effects on infarct size, myocardial stunning, and arrhythmias in conscious pigs undergoing a 40-minute coronary occlusion. Circ. Res. **80:** 730–742.
27. BAXTER, G.F., M.S. MARBER, V.C. PATEL, *et al.* 1994. Adenosine receptor involvement in a delayed phase of myocardial protection 24 hours after ischemic preconditioning. Circulation **90:** 2993–3000.
28. MALDONADO, C., Y. QIU, X.L. TANG, *et al.* 1997. Role of adenosine receptors in late preconditioning against myocardial stunning in conscious rabbits. Am. J. Physiol. **273:** H1324–H1332.
29. TAKANO, H., R. BOLLI, R.G. BLACK, JR., *et al.* 2001. A(1) or A(3) adenosine receptors induce late preconditioning against infarction in conscious rabbits by different mechanisms. Circ. Res. **88:** 520–528.
30. SUN, J.Z., X.L. TANG, S.W. PARK, *et al.* 1996. Evidence for an essential role of reactive oxygen species in the genesis of late preconditioning against myocardial stunning in conscious pigs. J. Clin. Invest. **97:** 562–576.
31. TANG, X.L., A. RIZVI, Y. QIU, *et al.* 1997. Evidence that the hydroxyl radical triggers late preconditioning against myocardial stunning in conscious rabbits. Circulation **96**(suppl. I): I-255. Abstract.
32. TAKANO, H., X.L. TANG, Y. QIU, *et al.* 1997. Intracoronary administration of oxygen radicals induces late preconditioning against myocardial stunning in conscious rabbits. Circulation **96**(suppl. I): I-256. Abstract.
33. BECKMAN, J.S., T.W. BECKMAN, J. CHEN, *et al.* 1990. Apparent hydroxyl radical production by peroxynitrite: implications for endothelial injury from nitric oxide and superoxide. Proc. Natl. Acad. Sci. U.S.A. **87:** 1620–1624.
34. CROW, J.P. & J.S. BECKMAN. 1995. Reactions between nitric oxide, superoxide, and peroxynitrite: footprints of peroxynitrite *in vivo.* Adv. Pharmacol. **34:** 17–43.
35. HILL, M., H. TAKANO, X.-L. TANG, *et al.* 2001. Nitroglycerin induces early and late preconditioning against myocardial infarction in conscious rabbits despite development of nitrate tolerance. Circulation. **104:** 694–699.
36. LEESAR, M.A., M.F. STODDARD, B. DAWN, *et al.* 2001. Delayed preconditioning-mimetic action of nitroglycerin in patients undergoing coronary angioplasty. Circulation **103:** 2935–2941.
37. MONCADA, S. & A. HIGGS. 1993. The L-arginine-nitric oxide pathway. N. Engl. J. Med. **329:** 2002–2012.
38. BAXTER, G.F., F.M. GOMA & D.M. YELLON. 1997. Characterisation of the infarct-limiting effect of delayed preconditioning: timecourse and dose-dependency studies in rabbit myocardium. Basic Res. Cardiol. **92:** 159–167.
39. RIZVI, A., X.L. TANG, Y. QIU, *et al.* 1999. Increased protein synthesis is necessary for the development of late preconditioning against myocardial stunning. Am. J. Physiol. **277:** H874–H884.
40. TAKANO, H., X.L. TANG & R. BOLLI. 2000. Differential role of K(ATP) channels in late preconditioning against myocardial stunning and infarction in rabbits. Am. J. Physiol. Heart Circ. Physiol. **279:** H2350–H2359.
41. TAKANO, H., X.L. TANG, Y. QIU, *et al.* 1998. The tetrahydrobiopterin synthesis inhibitor *N*-acetylserotonin abrogates late preconditioning against myocardial stunning in conscious rabbits. Circulation **98**(Suppl. I): I-586. Abstract.
42. SOUTHAN, G.J. & C. SZABO. 1996. Selective pharmacological inhibition of distinct nitric oxide synthase isoforms. Biochem. Pharmacol. **51:** 383–394.
43. WANG, Y. & P.A. MARSDEN. 1995. Nitric oxide synthases: gene structure and regulation. Adv. Pharmacol. **34:** 71–90.

44. PAN, J., K.L. BURGHER, A.M. SZCZEPANIK, *et al.* 1996. Tyrosine phosphorylation of inducible nitric oxide synthase: implications for potential post-translational regulation. Biochem. J. **314:** 889–894.
45. JONES, W.K., M.P. FLAHERTY, X.L. TANG, *et al.* 1999. Ischemic preconditioning increases iNOS transcript levels in conscious rabbits via a nitric oxide-dependent mechanism. J. Mol. Cell Cardiol. **31:** 1469–1481.
46. XUAN, Y.T., X.L. TANG, Y. QIU, *et al.* 2000. Biphasic response of cardiac NO synthase isoforms to ischemic preconditioning in conscious rabbits. Am. J. Physiol. Heart Circ. Physiol. **279:** H2360–H2371.
47. GUO, Y., W. BAO, X.-L. TANG, *et al.* 2000. Pharmacological preconditioning with adenosine A1 and opioid δ1 receptor agonists is iNOS dependent. Circulation **102**(Suppl. II): II-121. Abstract.
48. GUO, Y., W.K. JONES, W. BAO, *et al.* 1999. Targeted disruption of iNOS gene abrogates NO donor-induced late preconditioning. Circulation **100**(Suppl. I): I-562. Abstract.
49. KODANI, E., X.-L. TANG, Y.-T. XUAN, *et al.* 2000. Role of cyclic guanosine monophosphate in nitric oxide-dependent late preconditioning in conscious rabbits. Circulation **102**(Suppl. II): II-270. Abstract.
50. YTREHUS, K., Y. LIU & J.M. DOWNEY. 1994. Preconditioning protects ischemic rabbit heart by protein kinase C activation. Am. J. Physiol. **266:** H1145–H1152.
51. PING, P., H. TAKANO, J. ZHANG, *et al.* 1999. Isoform-selective activation of protein kinase C by nitric oxide in the heart of conscious rabbits: a signaling mechanism for both nitric oxide-induced and ischemia-induced preconditioning. Circ. Res. **84:** 587–604.
52. PING, P., J. ZHANG, Y. ZHENG, *et al.* 2000. Cardiac-targeted transgenesis of active PKCε renders the heart resistant to infarction. Circulation **102**(Suppl. II): II-24. Abstract.
53. GOPALAKRISHNA, R. & W.B. ANDERSON. 1989. Ca^{2+}- and phospholipid-independent activation of protein kinase C by selective oxidative modification of the regulatory domain. Proc. Natl. Acad. Sci. U.S.A. **86:** 6758–6762.
54. NATARAJAN, V., M.M. TAHER, B. ROEHM, *et al.* 1993. Activation of endothelial cell phospholipase D by hydrogen peroxide and fatty acid hydroperoxide. J. Biol. Chem. **268:** 930–937.
55. DAWN, B., Y.T. XUAN, Y. QIU, *et al.* 1999. Bifunctional role of protein tyrosine kinases in late preconditioning against myocardial stunning in conscious rabbits. Circ. Res. **85:** 1154–1163.
56. SONG, C., J. ZHANG, J. GAO, *et al.* 2000. Lck tyrposine kinase is a direct substrate of PKCε in cardiac myocytes. Circulation **102**(Suppl. II): II-211. Abstract.
57. PING, P., Y. GUO, J. ZHANG, *et al.* 2000. Targeted deletion of Lck tyrosine kinase abrogates the NF-κB activation and the infarct-sparing effect induced by late preconditioning. Circulation **102**(Suppl. II): II-211. Abstract.
58. PING, P., J. ZHANG, X. CAO, *et al.* 1999. PKC-dependent activation of p44/p42 MAPKs during myocardial ischemia-reperfusion in conscious rabbits. Am. J. Physiol. **276:** H1468–H1481.
59. ZHANG, J., R. BOLLI, X.-L. TANG, *et al.* 1999. Direct demonstration of both serine and tyrosine phosphorylation of IκB-α during ischemic preconditioning in conscious rabbits. Circulation **100**(Suppl. I): I-56. Abstract.

Regulation of Caspases by Nitric Oxide

PETER K.M. KIM,[a] YOUNG-GUEN KWON,[b] HUN-TAEG CHUNG,[c] AND YOUNG-MYEONG KIM[b,d]

[a]*Department of Surgery, University of Pittsburgh School of Medicine, Pittsburgh, Pennsylvania, U.S.A.*

[b]*Vascular System Research center, Kangwon National University, Chunchon, Kangwon-do, Korea*

[c]*Department of Immunology, College of Medicine, Wonkwang University, Iksan, Chonbug, Korea*

[d]*Department of Molecular and Cellular Biochemistry, College of Medicine, Kangwon National University, Chunchon, Kangwon-do, Korea*

ABSTRACT: Nitric oxide can prevent or induce apoptosis depending on its concentration, cell type, and the oxidative milieu. Nitric oxide inhibits apoptosis and inflammation by *S*-nitrosylation of the active site cysteine of caspases, the central effector molecules of cell death as well as maturation of IL-1β and IL-18. The ability of nitric oxide to *S*-nitrosylate caspases depends on multiple factors including the presence of free iron and intracellular redox potential. There are no known direct effects of nitric oxide on promoting caspase activation or activity. However, nitric oxide has been shown to promote apoptotic pathways in numerous cell types through the indirect activation of caspases. In this article we review the relationship of nitric oxide and caspase activity, modulation of this effect by iron, and clinical implications for the use of nitric oxide in regulating inflammation and apoptosis.

KEYWORDS: nitric oxide; caspase; apoptosis; inflammation; iron

BIOLOGICAL CHEMISTRY OF NITRIC OXIDE AND *S*-NITROSYLATION

Nitric oxide (NO) is a pleiotropic, ubiquitous modulator of cellular function. NO is synthesized from L-arginine by a family of NO synthase (NOS) isoenzymes.[1] The inducible NOS (iNOS) isoform is expressed in many cells including cytokine-stimulated macrophages and hepatocytes where they can produce large amounts of NO.[2] The NO molecule is a lipophilic and highly diffusible solute whose actions depend on both its concentration and form within the cell. The biological chemistry of NO hinges around its unpaired electron, which makes NO a free, but stable radical.[3,4] In biological systems NO has a short halflife, on the order of seconds. NO is a neutral molecule and has weak chemical reactivity with thiols at neutral pH.

Address for correspondence: Young-Myeong Kim, Vascular System Research Center, Kangwon National University, Chunchon, Kangwon-do, Korea. Voice: +82-33-250-8830; fax: +82-33-250-8807.
ymkim@kangwon.ac.kr

Although NO alone does not interact readily with proteins or nucleic acids, reactive NO^+ can be generated by the loss of one electron from the NO radical.

S-nitrosylation involves the transfer of an NO^+ group to an active site thiol on cysteine amino acids, if the cysteine residues are in the reduced (free sulfhydryl) state. NO can combine directly with local electron-accepting species like oxygen, transition metal ions, or superoxide radicals to generate potent *S*-nitrosylating species (e.g., N_2O_3, NO^+, and $ONOO^-$) that engage in nitrosation of sulfur-containing molecules. NO^- in its singlet or high-energy state can also react with a thiol group.[5] Formation of such *S*-nitrosylated proteins, or *S*-nitrosothiols, is an important, posttranslational modification of protein structure that can modulate protein function. This ubiquitous regulatory mechanism is comparable to phosphorylation and has been shown to affect many processes ranging from signal transduction to DNA repair.[6] Examples include the activation or inhibition of several enzymes that ar involved with numerous cellular processes including apoptosis. *S*-nitrosylation by NO has become an increasingly identified posttranslational modification of important intracellular proteins including Ras, ornithine decarboxylase, transglutaminase, c-Jun *N*-terminal kinase/stress-activated protein kinase, and especially caspases, as we discuss below.[7–10] *S*-nitrosylation of these and other proteins may modulate their activity and apoptotic potential.

NO INHIBITS APOPTOSIS BY *S*-NITROSYLATION OF CASPASES

Programmed cell death, also known as apoptosis, is a regulated, energy-dependent process of cell shrinkage, plasma membrane blebbing, chromatin condensation, and DNA fragmentation. NO can be both proapoptotic and antiapoptotic depending on its local concentration and cell type.[11,12] Apoptosis is biologically initiated by ligation of specific death receptors of the tumor necrosis factor receptor (TNF-R) family such as CD95/Fas/Apo-1 and the receptor for TRAIL.[13] Ligand binding of the trimerized receptor recruits nitracellular adaptor molecules like FADD/MORT1 and TRADD into the death-inducing signaling complex (DISC) which then recruits and activates protease zymogens called procaspases.[14–17]

We and others have shown that inhibition of apoptosis by NO can be mediated at the level of caspases.[18–24] Caspases are a family of cysteine proteases, consisting of 14 isoforms that play an essential role in the apoptotic signaling cascade. Caspases are the mammalian counterpart of *ced-3,* a protease required for programmed cell death in the nematode *Cænorhabditis elegans,*[25] and they are the central initiator and effector molecules of apoptosis.[26] As cysteine aspartyl proteases, caspases propagate death signals by cleaving a number of cellular protein substrates. These substrates include inactive procaspase zymogens that are cleaved into active proteases leading ultimately to apoptosis.[27] Active caspases cleave specifically after aspartate residues that exist in specific amino acid sequences.[28] Caspases-8, -9, and -10 are initiator caspases that can transduce apoptotic signals by directly activating the downstream executioner caspases-3, -6, and -7.[27] Inhibition of caspase activation by NO was shown for both death ligand-dependent and ligand-independent apoptosis. These studies imply that both pathways involve NO-regulated caspases.

All caspases contain a single cysteine at the enzymatic catalytic site. Expression of caspases containing a dominant-negative mutation of the active site cysteine to alanine also reversed the inhibitory effects of NO.[29,30] This thiol is also susceptible to redox modification and can be effectively modified by *S*-nitrosylation in the presence of NO (see FIGURE 1). NO was shown to modulate caspase protease activity through *S*-nitrosylation of this active site cysteine.[19,20,30,31] This inhibition of apoptosis was reversed by the reduction of the nitrosylated caspase with a strong reducing agent dithiothreitol (DTT),[18] which indicated that the proteolytic activity is modulated by a reversible *S*-nitrosylation of the active site cysteine of caspases.[32]

Seven members of recombinant caspases tested have been shown to be susceptible to reversible inhibition by NO through this redox modification.[19] The ratio of enzyme subunit to *S*-nitrosylation is stoichiometrically found to be 1 : 1 in caspase-3 as well as caspase–8 following treatment with the NO donor *S*-nitroso-*N*-acetylpenicillamine (SNAP).[33] This suggests that caspases are *S*-nitrosylated at one thiol of the protein on the active site cysteine, although evidence exists for more than one *S*-nitrosylation site in caspase-3.[29]

By inhibiting caspase activity through *S*-nitrosylation, NO can inhibit apoptosis in hepatocytes[18,31,33] and endothelial cells[34] through the blockage of caspase proteolytic activation as well as direct suppression of caspase activity. Procaspase-8 is recruited to the death-inducing signaling complex of plasma membrane receptors by the FADD/MORT1 protein. Procaspase-8 is autoactivated when two procaspase-8

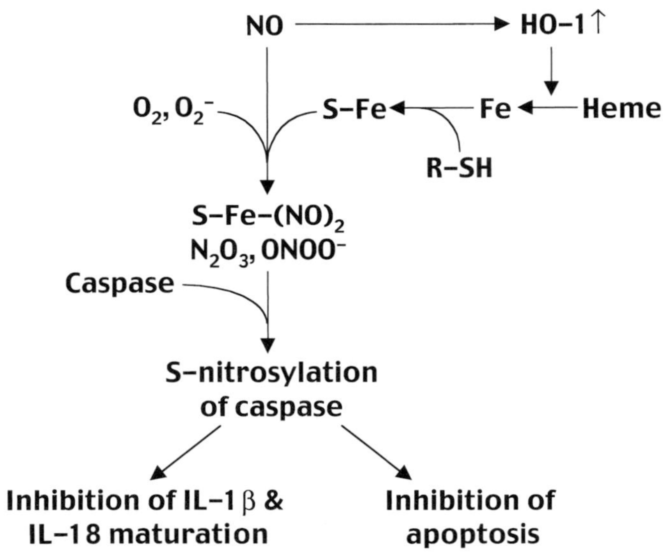

FIGURE 1. Possible pathways of the inhibition of caspase by *S*-nitrosylation. Major reactive nitrogen species for *S*-nitrosylation and the biological events of caspase *S*-nitrosylation ar shown. NO, nitric oxide; HO-1, heme oxygenase-1, S-Fe-$(NO)_2$, dinitrosyl iron complexes; S–Fe, iron-sulfur complexes; $ONOO^-$, peroxynitrite.

molecules are brought close together to cleave and activate each other.[35] Caspase-8 functions primarily upstream of Bcl-2 nd cytochrome c in the death signaling pathway.[36] The activity of caspase-8 can be suppressed by NO.[19,37] *S*-nitrosylated caspases inhibit the cleavage of Bid and Bcl-2 and block release of mitochondrial cytochrome c.[38] By preventing mitochondrial cytochrome c release, inhibition of caspase activity by *S*-nitrosylation suppresses a key step in the positive forward and feedback amplification of apoptotic signaling.

It is not known for certain whether NO inhibits autoactivation of procaspase-8, but in human leukemic T lymphocytes (Jurkat cells) treated with Fas ligand, the activation of the downstream procaspase-3 has been shown to depend on the denitrosylation of the active site cysteine.[32] This suggested that *S*-nitrosylation of the caspase active site might also function to limit caspase activation in the resting, nonapoptotic state.

NO INHIBITS THE MATURATION OF IL-1β AND IL-18 BY *S*-NITROSYLATION OF CASPASE-1

In addition to the central effect of caspases in apoptosis, inhibition of caspases like caspase-1 also has a role in the regulation of inflammatory responses, such as lipoploysaccharide (LPS)-induced interferon (IFN)-γ production.[39,40] Procytokine processing by caspase-1 (ICE : IL-1β converting enzyme) is required for the maturation and release of the critical inflammatory mediators IL-1β and IFN-γ-inducing factor (IGIF or IL-18) from activated macrophages.[41] IL-1β stimulates proinflammatory responses in neutrophils, endothelial cells, synovial cells, and osteoclasts,[42] whereas IFN-γ promotes the activation of macrophages and natural killer cells and contributes to the regulation of T helper cell immune responses.[43]

In stimulated RAW264.7 macrophages, inhibition of NO synthesis by N^G-monomethyl-L-arginine resulted in threefold increased intracellular caspase activity.[44] Addition of the NO donor SNAP inhibited caspase-1 activity in cells as well as recombinant caspase activity and cleavage of pro-IL-1β and pro-IL-18. The inhibition of caspase-1 was reversed by the addition of DTT consistent with *S*-nitrosylation of the active site cysteine as the mechanism for caspase-1 inhibition. Mice deficient in iNOS exhibited higher IL-1β and IL-18 plasma levels in response to LPS injection, suggesting an *in vivo* role for regulation of caspase-1 by NO.[44] These results also support a potential therapeutic benefit to reduction of inflammation by NO in septic shock.

NO AND IRON LEVELS

A significant mechanism for the biological role of NO is its reaction with positively charged metal ions in complexes of enzymes that contain iron.[45] In many cells, NO reacts with heme iron in guanylyl cyclase, the enzyme that makes guanosine 3′:5′ cyclic monophosphate (cGMP), to form an iron-nitrosyl complex.[2,46] *S*-nitrosylating species can be generated by the interaction of NO with an iron–sulfur complex.[47] Therefore, the capacity of NO to *S*-nitrosylate proteins like caspases depends on the abundance of proteins with iron–sulfur complexes and the availability of other thiol targets such as glutathione and free cysteine.

The reaction product of NO and iron–sulfur complexes, dinitrosyl–iron complexes, has been shown to carry out *S*-nitrosylation of caspase through the formation of NO.[37] Similarly, *S*-nitrosylation of caspase occurs effectively in iron-rich hepatocytes.[33] NO-induced apoptosis has been shown to occur in cells that are deficient in intracellular iron. In RAW 264.7 macrophages, NO induces apoptosis by caspase activation. However, when pretreated with free iron, the cellular response to NO is reversed, and the cells are protected from apoptosis.[37] The increase in non-heme iron within the cells is associated with greater iron–nitrosyl complex formation as well as increases in nitrosothiol formation upon NO exposure. Iron chelation by deferoximine in PC12 cells prevented NO-induced cell death suggesting that the effects of NO on apoptosis depend upon iron mobilization within the cell.[48] These studies indicate that cellular content of iron–sulfur complexes, such as heme-containing proteins, is a critical factor for NO-mediated *S*-nitrosylation and inhibition of apoptosis *in vivo*.

Heme oxygenase-1 (HO-1, Hsp32) is an enzyme that metabolizes iron-containing heme to biliverdin, carbon monoxide, and free iron. NO upregulated HO-1 in hepatocytes and pulmonary endothelial cells.[49,50] Conversion of heme to non-heme iron by NO-mediated induction of HO-1 increased the formation of dinitrosyl–iron complexes and protected hepatocytes from apoptosis, probably through *S*-nitrosylation of caspases.[33,49,50] Regulation of NO production and cellular iron level might be a useful tool for prevention of unwanted apoptotic death in the liver and other organs.

NO INHIBITS CASPASES AND APOPTOSIS IN HEPATOCYTES *IN VITRO* AND *IN VIVO*

Caspases mediate cellular death in numerous disease processes of the gut and liver that signal via Fas or TNFα such as inflammatory bowel disease and liver failure.[51,52] Injection of mice with anti-Fas antibody, TNFα with D-galactosamine (D-GalN), or LPS with D-GalN induced massive liver failure by apoptosis and necrosis resulting in death.[53–55] Coinjection of caspase inhibitors with anti-Fas antibody blocked caspase activity in the liver and limited hepatocellular damage.[56] *In vivo,* NO donors can inhibit apoptosis of hepatocytes induced by TNFα and D-GalN.[57] Adenoviral gene transfer of iNOS into hepatocytes produced NO and inhibited apoptosis by TNF and actinomycin D via inhibition of caspases.[58] Transgenic animals expressing the iNOS gene specifically in the live protected against low dose challenge with LPS and D-GalN.[59] Activation of the transcription factor NF-κB has been shown to stimulate iNOS expression and NO production to protect mouse hepatocytes from TNFα and Fas-mediated apoptosis by caspase inhibition.[60] Indeed, NO derivatives of aspirin (NCX-4016) and ursodeoxycholic acid (NCX-1000) have been used to prevent liver injury *in vivo.* NCX-4016 protected against concanavalin A-induced, T-lymphocyte mediated hepatotoxicity,[61] and NCX-1000 inhibited Fas-mediated liver injury.[62] In HepG2 hepatocellular carcinoma cell lines, inhibition of caspase activity by NCX-1000 was reversed by the addition of the reducing agent

DTT, suggesting *S*-nitrosylation of cysteine at the active site of caspase was the main mechanism for inhibition of apoptosis by NO.

NO INDUCES APOPTOSIS AND CASPASE ACTIVATION

Cytotoxic effects of NO and peroxynitrite have been shown to result from DNA damage and concomitant upregulation of p53 tumor suppressor.[63] NO-mediated increases in p53 activated apoptosis by increasing the ratio of proapoptotic Bax to antiapoptotic Bcl-XL, cytochrome c release, and caspase activation. Cytochrome c is released by the mitochondria in response to proapoptotic signals (see FIGURE 2). With Apaf-1 and procaspase-9, cytochrome c is a key component of apoptosome complex formation that results in cleavage and activation of death effector caspase-9.[64]

NO also has the direct effect on proteins in the mitochonria that result in release of cytochrome c, caspase activation, and apoptosis. NO can induce cytochrome c release by initiating loss of the mitochondrial membrane potential.[65] Low concentrations of NO reversibly inhibit cytochrome oxidase in competition with oxygen. Higher concentration of NO can react with peroxide and other oxygen radicals to form peroxynitrite that irreversibly inhibits mitochondrial respiration by oxidizing and damaging mitochondrial complexes I, II, IV, and V.[66] Peroxynitrite can cause mitochondrial damage including mitochondrial swelling, oxidation of mitochondrial

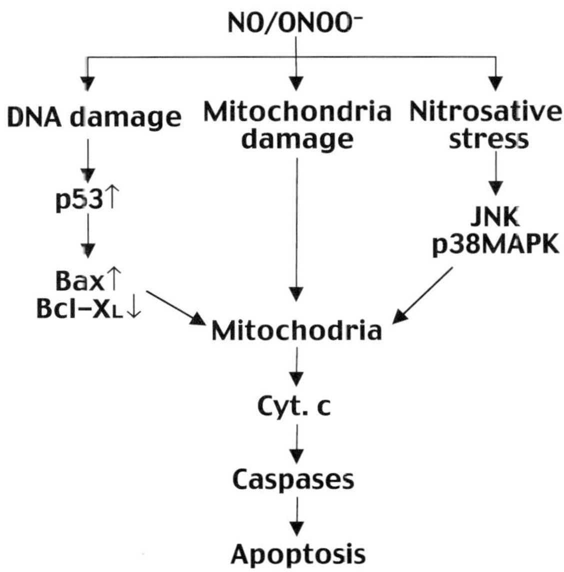

FIGURE 2. Possible signal cascades of caspase activation induced by NO. Major event of NO-mediated caspase activation is cytochrome c release from mitochondria. This pathway may be occurred under the poor redox condition such as low concentration of GSH and non-heme iron.

membrane lipid, and depolarization with alterations in the permeability transition, and thus resulting in cytochrome c release and caspase activation (FIG. 2).

NO can modulate multiple pathways that lead to cell death including those mediated by ceramide and stress kinases. By enhancement of neutral sphingomyelinase activity and inhibition of acid ceramidase activity, NO donors also increase cellular ceramide levels[67,68] that can induce several death pathways including release of cytochrome c from mitochondria, activation of caspases-9 and -3, an activation of stress-related protein kinases.[69] NO-dependent activation of procaspase-3, c-Jun *N*-terminal kinase/stress-activated protein kinases (JNK/SAPK), and p38 mitogen activated protein kinase (MAPK) has been shown to induce apoptotic death (FIG. 2).[70–73] The suppression of JNK/SAPK and p38 MAPK by protein kinase C transfection protected RAW264.7 macrophages from sodium nitroprusside-mediated apotosis by the inhibition of caspase-3 activation and activity.[74] The factors affecting NO-mediated caspase activation can be associated with the redox state and transition metal complexes within cells and expression of antiapoptotic genes.[37,49,75] Therefore, the threshold of NO level that induces caspase activation is different from one cell to another.

CLINICAL IMPLICATION OF NO-DEPENDENT REGULATION OF CASPASE ACTIVITY

Biological system requires homeostasis maintaining the balance between cell proliferation and apoptotic cell death. Abnormal alteration of this balance is linked to many human diseases such as cancer caused by the suppression of apoptosis and atherosclerosis, multi-organ failure, neurodegenerative disorders, and autoimmune diseases caused by dysregulated apoptosis. NO can alter the homeostatic balance within the cell by regulating the activation and activity of caspases, which are the central mediator of apoptosis in multiple biological and pathological processes. In some cells including hepatocytes, NO reversibly inhibits caspase activity by *S*-nitrosylation of the active site cysteine. In other cell types, NO turns on death pathways that involve the activation of caspases, probably through cytochrome c release from mitochondria. Since caspase-dependent cell death is central to numerous disease processes, these studies support the hope that modulation of caspase activity by NO may offer a promising new therapy on the biomedical horizon for the human common ills, which have only recently been understood at a molecular level. Thus, generation or delivery of appropriate amounts of NO to pertinent target sites might be useful as a therapeutic strategy for various diseases affected by alteration of apoptotic cell death.

ACKNOWLEDGMENTS

This work was supported by grants from Vascular System Research Center (Y.M.K) and Ministry of Health and Welfare Grant HMP-99-01-0001 (Y.G.K.). P.K.M. Kim is supported by grants from the American College of Surgeons as well as the Society for Surgical Oncology.

REFERENCES

1. NATHAN, C. & Q.W. XIE. 1994. Nitric oxide synthases: roles, tolls, and controls. Cell **78:** 915–918.
2. MONCADA, S., R.M. PALMER & E.A. HIGGS. 1991. Nitric oxide: physiology, pathophysiology, and pharmacology. Pharmacol. Rev. **43:** 109–142.
3. WINK, D.A. & J.B. MITCHELL. 1998. Chemical biology of nitric oxide: insights into regulatory, cytotoxic, and cytoprotective mechanisms of nitric oxide. Free Rad. Biol. Med. **25:** 434–456
4. STAMLER, J.S. 1994. Redox signaling: nitrosylation and related target interactions of nitric oxide. Cell **78:** 931–936.
5. LIPTON, S.A. & J.S. STAMLER. 1994. Actions of redox-related congeners of nitric oxide at the NMDA receptor. Neuropharmacology **33:** 1229–1233.
6. BROILLET, M.C. 1999. *S*-nitrosylation of proteins. Cell Mol. Life Sci. **55:** 1036–1042.
7. TENG, K.K., D.K. ESPOSITO, G.D. SCHWARTZ, *et al.* 1999. Activation of c-Ha-Ras by nitric oxide modulates survival responsiveness in neuronal PC12 cells. J. Biol. Chem. **274:** 37315–37320.
8. BAUER, P.M., J.M. FUKUTO, G.M. BUGA, *et al.* 1999. Nitric oxide inhibits ornithine decarboxylase by *S*-nitrosylation. Biochem. Biophys. Res. Commun. **262:** 355–358.
9. BERNASSOLA, F., A. ROSSI & G. MELINO. 1999. Regulation of transglutaminases by nitric oxide. Ann. N.Y. Acad. Sci. **887:** 83–91.
10. PARK, H.S., S.H. HUH, M.S. KIM, *et al.* 2000. Nitric oxide negatively regulates c-Jun *N*-terminal kinase/stress-activated protein kinase by means of *S*-nitrosylation. Proc. Natl. Acad. Sci. U.S.A. **97:** 14382–14387.
11. KIM, P.K.M., R. ZAMORA, P. PETROSKO & T.R. BILLIAR. 2001. The regulatory role of nitric oxide in apoptosis. Internat. Immunopharmacol. **1**(8): 1421–1441.
12. CHUNG, H.T., H.O. PAE, B.M. CHOI, *et al.* 2001. Nitric oxide as a bioregulator of apoptosis. Biochem. Biophys. Res. Commun. **282:** 1075–1079.
13. ASHKENAZI, A. & V.M. DIXIT. 1998. Death receptors: signaling and modulation. Science **281:** 1305–1308.
14. KISCHKEL, F.C., S. HELLBARDT, I. BEHRMANN, *et al.* 1995. Cytotoxicity-dependent APO-1 (Fas/CD95)-associated proteins form a death-inducing signaling complex (DISC) with the receptor. EMBO J. **14:** 5579–5588.
15. HSU, H., J. XIONG & D.V. GOEDDEL. 1995. The TNF receptor 1-associated protein TRADD signals cell death and NF-kappa B activation. Cell **81:** 495–504.
16. CHINNAIYAN, A.M., K. O'ROURKE, M. TEWARI & V.M. DIXIT. 1995. FADD, a novel death domain-containing protein, interacts with the death domain of Fas and initiates apoptosis. Cell **81:** 505–512.
17. MUZIO, M., A.M. CHINNAIYAN, F.C. KISCHKEL, *et al.* 1996. FLICE, a novel FADD-homologous ICE/CED-3-like protease, is recruited to the CD95 (Fas/APO-1) death-inducing signaling complex. Cell **85:** 817–827.
18. KIM, Y.M., R.V. TALANIAN & T.R. BILLIAR. 1997. Nitric oxide inhibits apoptosis by preventing increases in caspase-3-like activity via two distinct mechanisms. J. Biol. Chem. **272:** 31138–31148.
19. LI, J., T.R. BILLIAR, R.V. TALANIAN & Y.M. KIM. 1997. Nitric oxide reversibly inhibits seven members of the caspase family via *S*-nitrosylation. Biochem. Biophys. Res. Commun. **240:** 419–424.
20. DIMMELER, S., J. HAENDELER, M. NEHLS & A.M. ZEIHER. 1997. Suppression of apoptosis by nitric oxide via inhibition of interleukin-1beta-converting enzyme (ICE)-like and cysteine protease protein (CPP)-32-like proteases. J. Exp. Med. **185:** 601–607.
21. DIMMELER, S., J. HAENDELER, A. SAUSE & A.M. ZEIHER. 1998. Nitric oxide inhibits APO-1/Fas-mediated cell death. Cell Growth Differ. **9:** 415–422.
22. TENNETI, L., D.M. D'EMILIA & S.A. LIPTON. 1997. Suppression of neuronal apoptosis by *S*-nitrosylation of caspases. Neurosci. Lett. **236:** 139–142.
23. MANNICK, J.B., X.Q. MIAO & J.S. STAMLER. 1997. Nitric oxide inhibits Fas-induced apoptosis. J. Biol. Chem. **272:** 24125–24128.

24. MOHR, S., B. ZECH, E.G. LAPETINA & B. BRUNE. 1997. Inhibition of caspase-3 by *S*-nitrosation and oxidation caused by nitric oxide. Biochem. Biophys. Res. Commun. **238:** 387–391.
25. ELLIS, H.M. & H.R. HORVITZ. 1986. Genetic control of programmed cell death in the nematode C elegans. Cell **44:** 817–829.
26. WOLF, B.B. & D.R. GREEN. 1999. Suicidal tendencies: apoptotic cell death by caspase family proteinases. J. Biol. Chem. **274:** 20049–20052.
27. THORNBERRY, N.A. & Y. LAZEBNIK. 1998. Caspases: enemies within. Science **281:** 1312–1316.
28. ALNEMRI, E.S., D.J. LIVINGSTON, D.W. NICHOLSON, *et al.* 1996. Human ICE/CED-3 protease nomenclature. Cell **87:** 171.
29. ZECH, B., M. WILM, R. VAN ELDIK & B. BRUNE. 1999. Mass spectrometric analysis of nitric oxide-modified caspase-3. J. Biol. Chem. **274:** 20931–20936.
30. ROSSIG, L., B. FICHTLSCHERER, K. BREITSCHOPF, *et al.* 1999. Nitric oxide inhibits caspase-3 by *S*-nitrosation *in vivo.* J. Biol. Chem. **274:** 6823–6826.
31. LI, J., C.A. BOMBECK, S. YANG, *et al.* 1999. Nitric oxide suppresses apoptosis via interrupting caspase activation and mitochondrial dysfunction in cultured hepatocytes. J. Biol. Chem. **274:** 17325–17333.
32. MANNICK, J.B., A. HAUSLADEN, L. LIU, *et al.* 1999. Fas-induced caspase denitrosylation. Science **284:** 651–654.
33. KIM, Y.M., T.H. KIM, H.T. CHUNG, *et al.* 2000. Nitric oxide prevents tumor necrosis factor alpha-induced rat hepatocyte apoptosis by the interruption of mitochondrial apoptotic signaling through *S*-nitrosylation of caspase-8. Hepatology **32:** 770–778.
34. KWON, Y.G., J.K. MIN, K.M. KIM, *et al.* 2001. Sphingosine 1-phosphate protects human umbilical vein endothelial cells from serum-deprived apoptosis by nitric oxide production. J. Biol. Chem. **276:** 10627–10633.
35. MUZIO, M., B.R. STOCKWELL, H.R. STENNICKE, *et al.* 1998. An induced proximity model for caspase-8 activation. J. Biol. Chem. **273:** 2926–2930.
36. MARTIN, D.A., R.M. SIEGEL, L. ZHENG & M.J. LENARDO. 1998. Membrane oligomerization and cleavage activates the caspase-8 (FLICE/MACHalpha1) death signal. J. Biol. Chem. **273:** 4345–4349.
37. KIM, Y.M., H.T. CHUNG, R.L. SIMMONS & T.R. BILLIAR. 2000. Cellular non-heme iron content is a determinant of nitric oxide-mediated apoptosis, necrosis, and caspase inhibition. J. Biol. Chem. **275:** 10954–10961.
38. KIM, Y.M., T.H. KIM, D.W. SEOL, *et al.* 1998. Nitric oxide suppression of apoptosis occurs in association with an inhibition of Bcl-2 cleavage and cytochrome c release. J. Biol. Chem. **273:** 31437–31441.
39. GHAYUR, T., S. BANERJEE, M. HUGUNIN, *et al.* 1997. Caspase-1 processes IFN-gamma-inducing factor and regulates LPS-induced IFN-gamma production. Nature **386:** 619–623.
40. GU, Y., K. KUIDA, H. TSUTSUI, *et al.* 1997. Activation of interferon-gamma inducing factor mediated by interleukin-1beta converting enzyme. Science **275:** 206–209.
41. FANTUZZI, G. & C.A. DINARELLO. 1999. Interleukin-18 and interleukin-1 beta: two cytokine substrates for ICE (caspase-1). J. Clin. Immunol. **19:** 1–11.
42. DINARELLO, C.A. 1996. Biologic basis for interleukin-1 in disease. Blood **87:** 2095–2147.
43. BELARDELLI, F. 1995. Role of interferons and other cytokines in the regulation of the immune response. APMIS **103:** 161–179.
44. KIM, Y.M., R.V. TALANIAN, J. LI & T.R. BILLIAR. 1998. Nitric oxide prevents IL-1beta and IFN-gamma-inducing factor (IL-18) release from macrophages by inhibiting caspase-1 (IL-1beta-converting enzyme). J. Immunol. **161:** 4122–4128.
45. COOPER, C.E. 1999. Nitric oxide and iron proteins. Biochim. Biophys. Acta **1411:** 290–309.
46. MURAD, F. 1994. Regulation of cytosolic guanylyl cyclase by nitric oxide: the NO-cyclic GMP signal transduction system. Adv. Pharmocol. **26:** 19–33.
47. BOESE, M., P.I. MORDVINTCEV, A.F. VANIN, *et al.* 1995. *S*-nitrosation of serum albumin by dinitrosyl-iron complex. J. Biol. Chem. **270:** 29244–29249.

48. DESOLE, M.S., L. SCIOLA, S. SIRCANA, *et al.* 1998. Protective effect of deferoxamine on sodium nitroprusside-induced apoptosis in PC12 cells. Neurosci Lett. **247:** 1–4.
49. KIM, Y.M., H.A. BERGONIA, C. MULLER, *et al.* 1995. Loss and degradation of enzyme-bound heme induced by cellular nitric oxide synthesis. J. Biol Chem. **270:** 5710–5713.
50. YEE, E.L., B.R. PITT, T.R. BILLIAR & Y.M. KIM. 1996. Effect of nitric oxide on heme metabolism in pulmonary artery endothelial cells. Am. J. Physiol. **271:** L512–L518.
51. PINKOSKI, M.J., T. BRUNNER, D.R. GREEN & T. LIN. 2000. Fas and Fas ligand in gut and liver. Am. J. Physiol. Gastrointest. Liver Physiol. **278:** G354–G366.
52. BRADHAM, C.A., J. PLUMPE, M.P. MANNS, *et al.* 1998. Mechanisms of hepatic toxicity. I. TNF-induced liver injury. Am. J. Physiol. **275:** G387–G392.
53. OGASAWARA, J., R. WATANABE-FUKUNAGA, M. ADACHI, *et al.* 1993. Lethal effect of the anti-Fas antibody in mice. Nature **364:** 806–809.
54. PFEFFER, K., T. MATSUYAMA, T.M. KUNDIG, *et al.* 1993. Mice deficient for the 55 kd tumor necrosis factor receptor are resistant to endotoxic shock, yet succumb to L. monocytogenes infection. Cell **73:** 457–467.
55. LEIST, M., F. GANTNER, S. JILG & A. WENDEL. 1995. Activation of the 55 kDa TNF receptor is necessary and sufficient for TNF-induced liver failure, hepatocyte apoptosis, and nitrite release. J. Immunol. **154:** 1307–1316.
56. RODRIGUEZ, I., K. MATSUURA, C. ODY, *et al.* 1996. Systemic injection of a tripeptide inhibits the intracellular activation of CPP32-like proteases *in vivo* and fully protects mice against Fas-mediated fulminant liver destruction and death. J. Exp. Med. **184:** 2067–2072.
57. SAAVEDRA, J.E., T.R BILLIAR, D.L. WILLIAMS, *et al.* 1997. Targeting nitric oxide (NO) delivery *in vivo.* Design of a liver-selective NO donor prodrug that blocks tumor necrosis factor-alpha-induced apoptosis and toxicity in the liver. J. Med. Chem. **40:** 1947–1954.
58. TZENG, E., T.R. BILLIAR, D.L. WILLIAMS, *et al.* 1998. Adenovirus-mediated inducible nitric oxide synthase gene transfer inhibits hepatocyte apoptosis. Surgery **124:** 278–283.
59. MOJENA, M., S. HORTELANO, A. CASTRILLO, *et al.* 2001. Protection by nitric oxide against liver inflammatory injury in animals carrying a nitric oxide synthase-2 transgene. FASEB J. **15:** 583–585.
60. HATANO, E., B.L. BENNETT, A.M. MANNING, *et al.* 2001. NF-kappaB stimulates inducible nitric oxide synthase to protect mouse hepatocytes from TNF-α. Gastroenterology **120:** 1251–1262.
61. FIORUCCI, S., L. SANTUCCI, E. ANTONELLI, *et al.* 2000. NO-aspirin protects from T cell-mediated liver injury by inhibiting caspase-dependent processing of Th1-like cytokines. Gastroenterology **118:** 404–421.
62. FIORUCCI, S., A. MENCARELLI, B. PALAZZETTI, *et al.* 2001. An NO derivative of ursodeoxycholic acid protects against Fas-mediated liver injury by inhibiting caspase activity. Proc. Natl. Acad. Sci. U.S.A. **98:** 2652–2657.
63. MESSMER, U.K. & B. BRUNE. 1996. Nitric oxide-induced apoptosis: p53-dependent and p53-independent signalling pathways. Biochem. J. **319**(1): 299–305.
64. HENGARTNER, M.O. 2000. The biochemistry of apoptosis. Nature **407:** 770–776.
65. PACKER, M.A. & M P. MURPHY. 1995. Peroxynitrite formed by simultaneous nitric oxide and superoxide generation causes cyclosporin-A-sensitive mitochondrial calcium efflux and depolarisation. Eur. J. Biochem. **234:** 231–239.
66. BROWN, G.C. 1999. Nitric oxide and mitochondrial respiration. Biochim. Biophys. Acta **1411:** 351–369.
67. TAKEDA, Y., M. TASHIMA, A. TAKAHASHI, *et al.* 1999. Ceramide generation in nitric oxide-induced apoptosis. Activation of magnesium-dependent neutral sphingomyelinase via caspase-3. J. Biol. Chem. **274:** 10654–10660.
68. HUWILER, A., S. DORSCH, V.A. BRINER, *et al.* 1999. Nitric oxide stimulates chronic ceramide formation in glomerular endothelial cells. Biochem. Biophys. Res. Commun. **258:** 60–65.

69. HUWILER, A., J. PFEILSCHIFTER & B.H. VAN DEN. 1999. Nitric oxide donors induce stress signaling via ceramide formation in rat renal mesangial cells. J. Biol. Chem. **274:** 7190–7195.
70. JUN, C.D., B.M. CHOI, R. HOON, *et al.* 1994. Synergistic cooperation between phorbol ester and IFN-gamma for induction of nitric oxide synthesis in murine peritoneal macrophages. J. Immunol. **153:** 3684–3690.
71. SO, H.S., R.K. PARK, M.S. KIM, *et al.* 1998. Nitric oxide inhibits c-Jun *N*-terminal kinase 2 (JNK2) via *S*-nitrosylation. Biochem. Biophys. Res. Commun. **247:** 809–813.
72. JUN, C.D., C.D. OH, H.J. KWAK, *et al.* 1999. Overexpression of protein kinase C isoforms protects RAW 264.7 macrophages from nitric oxide-induced apoptosis: involvement of c-Jun *N*-terminal kinase/stress-activated protein kinase, p38 kinase, and CPP-32 protease pathways. J. Immunol. **162:** 3395–3401.
73. CALLSEN, D. & B. BRUNE. 1999. Role of mitogen-activated protein kinases in *S*-nitrosoglutathione-induced macrophage apoptosis. Biochemistry **38:** 2279–2286.
74. JUN, C.D., H.O. PAE, H.J. KWAK, *et al.* 1999. Modulation of nitric oxide-induced apoptotic death of HL-60 cells by protein kinase C and protein kinase A through mitogen-activated protein kinases and CPP32-like protease pathways. Cell Immunol. **194:** 36–46.
75. KIM, Y.M., M.E. DE VERA, S.C. WATKINS & T.R. BILLIAR. 1997. Nitric oxide protects cultured rat hepatocytes from tumor necrosis factor-alpha-induced apoptosis by inducing heat shock protein 70 expression. J. Biol. Chem. **272:** 1402–1411.

The Significance of Nitric Oxide Production in the Brain After Injury

AHMAD KHALDI,[a] C.C. CHIUEH,[b] M. ROSS BULLOCK,[a] AND J.J. WOODWARD[c]

[a]*Department of Neurosurgery, Medical College of Virginia, Virginia Commonwealth University, Richmond, Virginia, U.S.A.*

[b]*Laboratory of Clinical Sciences, National Institute of Mental Health, National Institutes of Health, Bethesda, Maryland, U.S.A.*

[c]*Department of Physiology and Neuroscience, Medical University of South Carolina, Charleston, South Carolina, U.S.A.*

ABSTRACT: Glutamate toxicity has been implicated in many aspects of brain injury including traumatic, ischemic, and hemorrhagic damage. We have used *in vitro* as well as *in vivo* methods to measure NO production and to examine the role of NO in glutamate toxicity. In building our recombinant system, we used human kidney embryonic cells, HEK 293, as host for transfection of nNOS and NMDA receptor proteins. Cells cotransfected with NMDA and nNOS were more resistant to glutamate toxicity. This resistance correlated with NO production as measured by citrulline assay. Meanwhile, the production of NO did not significantly change the response of the NMDA receptor as seen by calcium studies. Moreover, *in vivo,* NO production was directly correlated with brain tissue oxygen tension in subarachnoid hemorrhage patients. These data and others point toward the importance of NO production in the response of brain to injury.

KEYWORDS: nitric oxide production; brain injury; NMDA receptor

INTRODUCTION

It has been shown that there is an increase in nitric oxide (NO) production during brain insult, such as in brain injury, ischemic damage, and hemorrhagic events.[1] However, the role of NO production during brain insult is currently debated. The production of NO might lead to either toxicity or neuroprotection.[2,3] Some of the proposed mechanisms for toxicity include the reaction of NO with superoxide and the production of peroxynitrite, a known toxic compound.[4] Others argue that NO might react with thiol groups[5,6] on the *N*-methyl-D-aspartic acid (NMDA) receptor, and reduce toxicity by modifying the NMDA receptor response.[7,8] In the following studies, we have shown, *in vitro,* as well as *in vivo,* data that support the claim that NO production during brain insult might be part of the physiological response to injury. Furthermore, our *in vitro* data support the claim that NO production might

Address for correspondence: Ahmad Khaldi, Department of Neurosurgery, Medical College of Virginia, Virginia Commonwealth University, P.O. Box 980631, Richmond, VA 23298-0631, U.S.A. Voice: 804-828-1318; fax: 804-828-0374.

lead to reduced toxicity, since cells expressing nitric oxide synthase are more resistant to insult.

IN VITRO DATA

Coupling Between NMDA Receptor and nNOS

It has been shown that the stimulation of the NMDA receptor in brain neurons leads to activation of neuronal nitric oxide synthase (nNOS).[9] Stimulation of the NMDA receptor by agonists (glutamate and glycine) allows the movement of sodium and calcium ions into the cell.[10] The increase in intracellular calcium leads to activation of nNOS via calmodulin binding.[11] With the appropriate cofactors and substrates, the enzyme converts L-arginine to L-citrulline releasing, in turn, a molecule of nitric oxide.[12]

We have used molecular biology techniques and a recombinant expression system in which the individual components could be controlled. HEK 293 cells presented an ideal host, since they lack nNOS or NMDA receptor proteins. We transfected HEK 293 cells with nNOS, NMDA R1 subunit, and/or NMDA R2A subunit and examined cell toxicity 48 hours after transfection. Expression and function of nNOS were verified by visualization (immunoflorescence staining, NADPH diaphorase staining and Western blotting) and by functional activity studies (nitrite and citrulline assays).[13] The citrulline assay was used as a direct measurement of nNOS activity.[14] As shown in FIGURE 1, the percent of citrulline production is increased in cells coexpressing NMDA-R/nNOS over cells that only express NMDA-R (R1/R2A).

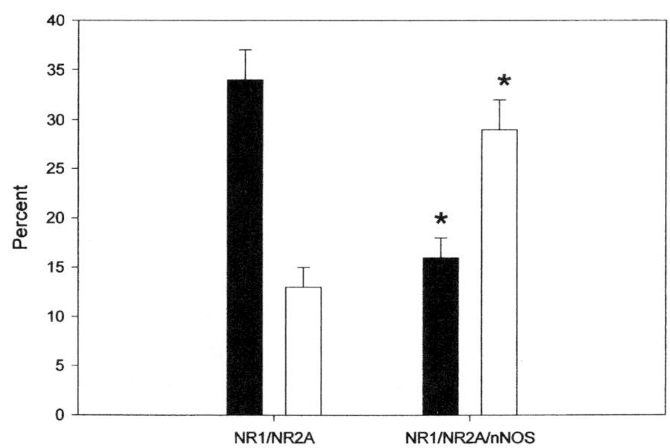

FIGURE 1. Correlation between nNOS activity and glutamate toxicity. Citrulline production was used as an indicator of nNOS activity. Cells were incubated with radio-labeled arginine for five minutes and then stimulated with NMDA (1 mM) and glycine (1 mM). Glutamate toxicity was evaluated 48 hours after transfection in NMDA R1-GFP positive cells. Percent dead was a fraction of GFP-positive cells that were stained with propidium iodide in a single field. Data represent, mean ± from three experiments; *, value significantly different from control (NR1/NR2A) tranfected cells; ANOVA followed by student's *t*-test; $p < 0.05$. ■, toxicity; □, citrulline production.

Verification of NMDA receptor expression was similar to nNOS verification studies in that immunoflorescent staining and functional activity assays were also used. The function of the NMDA receptor was verified by measuring the increase in intracellular calcium after NMDA/glycine stimulation.[13]

Measuring Toxicity of NMDA Transfected Cells

Persistent stimulation of the NMDA receptor in neurons causes sustained increases in intracellular calcium and subsequent cell death.[15] NMDA-dependent increases in the levels of reactive oxygen species (ROS) may be involved in this toxicity.[16] However, the involvement of NO in glutamate toxicity in neurons is unclear.

In HEK293 cells, we observed glutamate toxicity in cells transfected with the NMDA receptor subunits. This toxicity was blocked by the NMDA antagonist (ketamine 500 nM), suggesting that the cell death results in accumulation of glutamate in the media and subsequent NMDA receptor activation. These findings are consistent with previous studies showing that NMDA receptor transfected cells are more susceptible to glutamate toxicity.[17]

The Effect of nNOS Expression on Cell Toxicity

An NRI-GFP fusion protein was used to identify cells expressing the NMDA NR1 subunit. After 48 hours of transfection, cells were evaluated for viability by propidium iodide staining (100 ng/ml). The percentage of cell death was calculated as the fraction of GFP-positive cells that were also red (positive for propidium iodide) divided by the total number of cells that expressed GFP. As seen in FIGURE 1, the percentage of cell death of NMDA-receptor (NR1/NR2A) transfected cells is significantly lower in cells that were cotransfected with NMDA-receptor subunits (NR1/NR2A) and nNOS. Furthermore, as measured by the citrulline assay, there was an inverse correlation between NO production and toxicity.

These findings are consistent with earlier studies showing that NADPH-diaphorase-positive neurons are selectively spared in Huntington's disease.[18,19] In addition, Dawson *et al.*, showed cultured cortical neurons in which NADPH-positive neurons were resistant to NMDA-induced toxicity.[20] However, others have shown that NOS inhibitors do not always protect against NMDA toxicity in cultured neurons, arguing against the requirement for nNOS in glutamate toxicity.[21] In present study, HEK 293 cells cotransfected with NMDA-R/nNOS were less susceptible to glutamate toxicity, suggesting that NO reduces glutamate toxicity.

To verify that nNOS-positive cells were resistant to glutamate toxicity, a nNOS-GFP construct was used to identify nNOS-transfected cells. Cells cotransfected with nNOS-GFP/NMDA-R (NR1/NR2A) did not show an increase in glutamate toxicity when compared to nNOS-GFP–transfected cells. As seen in FIGURE 2, nNOS transfected cells were just as resistant to toxicity as cells that were cotransfected with nNOS and NMDA receptor. The mechanism underlying the reduced toxicity in nNOS transfected cells is unclear. The level of NMDA NR1 subunit expression was not reduced by the addition of nNOS as seen by Western blot analysis. Levels of nNOS proteins were also not significantly different in cells coexpressing NMDA receptor subunits. Finally, the NMDA receptor response in cells cotransfected with nNOS was not different from that of cells transfected with NMDA receptors alone,

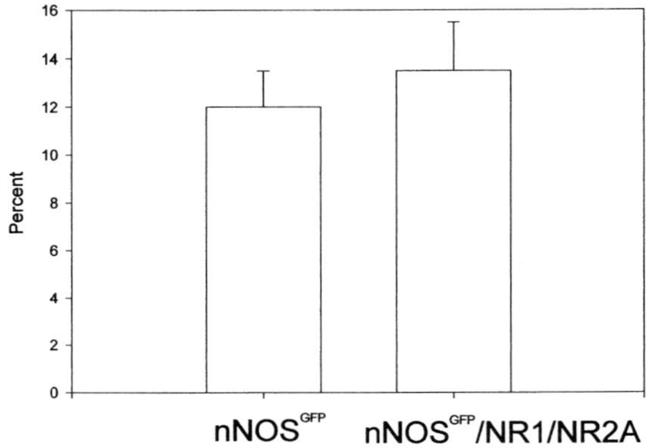

FIGURE 2. nNOS-transfected cells are more resistant to glutamate toxicity. Cells were transfected with nNOS-GFP or nNOS-GFP/NR1/NR2A. Toxicity was evaluated 48 hours after transfection using propidium iodide, PI. GFP fluorescent cells indicated the total number of transfected cells in a field. The percentage of dead nNOS-GFP-transfected cells was calculated as a fraction of the total number of transfected cells that were positive for PI. Data represent the mean ± SE from three different experiments. Each experiment consisted of at least four fields of 20 cells each. No significant difference was observed between the two groups. (One-way ANOVA).

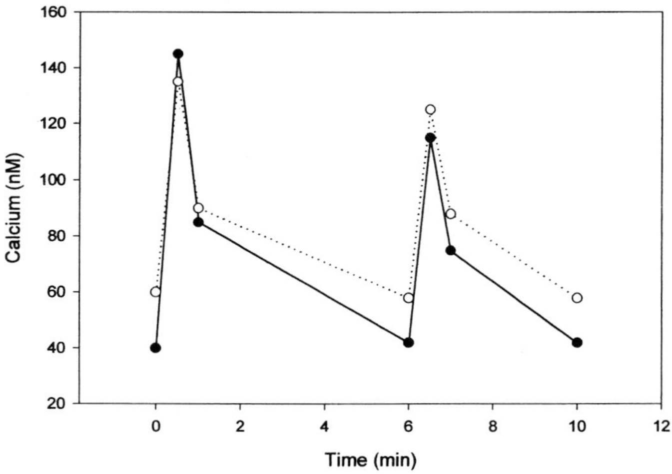

FIGURE 3. Average responses to NMDA (100 μM) and glycine (10 μM) stimulation in NMDA receptor–transfected cells and NMDA/nNOS–transfected cells. Cells were loaded with fura-PE3 and intracellular calcium levels were measured. Cells were stimulated with NMDA/glycine for 30 seconds followed by a six-minute recovery period. A second NMDA/glycine stimulation was applied for an additional 30 seconds. Points on the graph indicate an average from at least 12 cells. No significant differences were observed between the two groups at any time. ●, NR1/NR2A; ○, NR1/NR2A/nNOS. (One-way ANOVA).

as seen in FIGURE 3. This suggests that nNOS expression may alter events subsequent to NMDA receptor activation that are normally involved in toxicity.

NO has been shown to cause *S*-nitrosylation of various proteins through interaction with reactive thiols.[5,22,23] Thus, the mechanism for nNOS protection may be due to an inactivation of proteins, such as caspases that are normally involved in the excitotoxic cascade that follows NMDA receptor activation.[22] Inhibition of these caspases by NO donors, such as NO-NSAID (a nitric oxide–releasing nonsteriodal antiinflammatory) may reduce glutamate toxicity.[24]

IN VIVO DATA

To investigate the potential role of NO in human brain injury, we collected microdialysis samples from subarachnoid hemorrhage (SAH) patients and measured their nitrite and nitrate levels as indicators of NO production. In addition, we used an oxygen tension probe and measured brain tissue oxygen tension. As seen in FIGURE 4, there was a positive correlation ($r^2 = 0.326$) between local brain tissue oxygen tension and NO production, which is consistent with the vasodilatory effect of NO.[1]

ROLE OF NO IN INJURY

The role of NO in brain injury is controversial and complex. NO may directly inactivate reactive oxygen species generated during brain injury or stroke and thus decrease the extent of toxicity following reperfusion.[8,25,26] It is also clear that, as shown in the present study, nNOS-positive cells are less sensitive to NMDA-mediated toxicity. Thus, the use of NOS inhibitors during treatment of brain injury should be

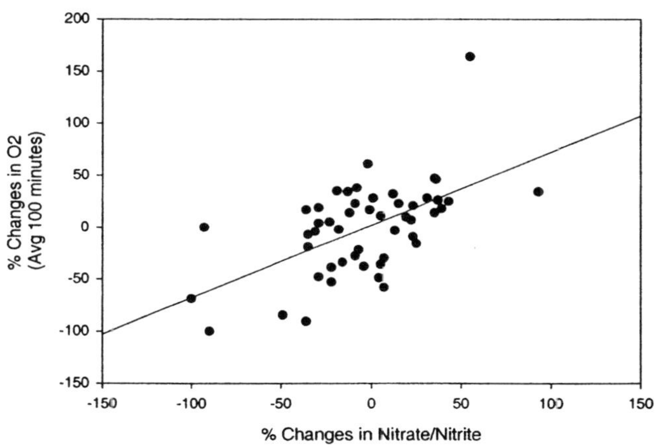

FIGURE 4. Percent changes in dialysate nitrate and nitrite and brain tissue oxygen tension for subarachnoid hemorrhage patients ($r^2 = 0.326$; $p < 0.001$).

considered with caution. A controlled NO production might present a normal pathophysiologic response of the tissue to damage and inhibition of this response may interfere with the normal repair process. NO-dependent vasodilators such as nitroglycerin may provide protection by enhancing blood flow, as well as by releasing NO, which reacts with thiol groups, leading to a decrease in glutamate toxicity.[22]

REFERENCES

1. KHALDI, A., A. ZAUNER, M. REINERT, *et al.* 2001. Measurement of nitric oxide and brain tissue oxygen tension in patients after severe subarachnoid hemorrhage. Neurosurgery **49:** 33–40.
2. LIPTON, S.A., Y. CHOI, Z. PAN, *et al.* 1993. A redox-based mechanism for a neuroprotective and neurodestructive effects of nitric oxide and related nitroso-compounds. Nature **364:** 626–632.
3. IADECOLA, C. 1997. Bright and dark sides of nitric oxide in ischemic brain injury. Trend. Neurosci. **20:** 132–139.
4. BECKMAN, J., J. CHEN, J. CROW & Y. YE. 1994. Reactions of nitric oxide, superoxide and peroxynitrite with SOD in neurodegeneration. Prog. Brain Res. **103:** 371–380.
5. STAMLER, J.S. 1994. Redox signaling: nitrosylation and related target interactions of nitric oxide. Cell **78:** 931–936.
6. LIPTON, S.A., D. SINGEL & J.S. STAMLER. 1994. NO in CNS. Prog. Brain Res. **103:** 359–364.
7. LIPTON, S.A. & J.S. STAMLER. 1994. Actions of redox-related congeners of nitric oxide at the NMDA receptor. Neuropharmacology **33:** 1229–1233.
8. CHIUEH, C.C. 1999. Neuroprotective properties of nitric oxide. Ann. N.Y. Acad. Sci. **890:** 301–311.
9. GARTHWAITE, J., S. CHARLES & R. CHESS-WILLIAMS. 1988. EDRF releases on activation of NMDA receptor suggests role as intercellular messenger in the brain. Nature **336:** 385–387.
10. MACDERMOTT, A.B., M.L. MAYER, G.L. WESTBROOK, *et al.* 1986. NMDA receptor activation increases cytoplasmic calcium concentration in cultured spinal cord neurons. Nature **321:** 519–522.
11. GRIFFITH, O.W. & D.J. STUEHR. 1995. Nitric oxide synthase: properties and catalytic mechanism. Annu. Rev. Physiol. **57:** 707–736.
12. MONCADA, S. & A. HIGGS. 1993. The L-arginine-nitric oxide pathway. N. Engl. J. Med. **329:** 2002–2012.
13. KHALDI, A. 1995. Examination of the NMDA Receptor-nNOS Coupling Using a Recombinant Expression System. Master Thesis, Virginia Commonwealth University, Richmond.
14. CHANDLER, L.J., G. SUTTON, D. NORWOOD, *et al.* 1997. Chronic ethanol increases NMDA stimulated nitric oxide formation but not receptor density in cultured cortical neurons. Mol. Pharmacol. **51:** 733–740.
15. GARTHWAITE, G. & J. GARTHWAITE. 1994. NO does not mediate glutamate neurotoxicity, nor is it neuroprotective, in rat brain slices. Neuropharmacology **33:** 1431–1438.
16. GUNASEKA, P.G., A.G. KANTHASAMY, J.L. BOROWITZ & G.E. ISOM. 1995. NMDA receptor activation produces concurrent generation of nitric oxide and reactive oxygen species: implication of cell death. J. Neurochem. **65:** 2016–2021.
17. ANEGAWA, N.J., D.R. LYNCH, T.A. VERDOORN & D.B. PRITCHETT. 1995. Transfection of NMDA receptors in non-neuronal cell line leads to cell death. J. Neurochem. **64:** 2004–2012.
18. FERRANTE, R.J., N.W. KOWALI, M.F. BEAL, *et al.* 1985. Selective sparing of a class of striatal neurons in Huntington's disease. Science **230:** 561–563.
19. KOH, J., S. PETERS & D.W. CHOI. 1986. Neurons containing NADPH-diaphorase are selectively resistant to quinolinate toxicity. Science **234:** 73–76.
20. DAWSON, V.L., T.M. DAWSON, D.A. BARTLEY, *et al.* 1993. Mechanisms of NO-medicated neurotoxicity in primary brain cultures. J. Neurosci. **13:** 2651–2661.

21. HEWETT, S.J., J.A. CORBETT, M.L. MCDANIEL & D.W. CHOI. 1993. Inhibition of NO formation does not protect murine cortical cell cultures from NMDA neurotoxicity. Brain Res. **625:** 337–341.
22. LIPTON, S.A. 1999. Neuronal protection and destruction by NO. Cell Death Differ. **6:** 943–951.
23. STAMLER, J.S., D. SIMON, J. OSBORNE, *et al.* 1992. S-nitrosylation of proteins with nitric oxide: synthesis and characterization of biological active compounds. Proc. Natl. Acad. Sci. U.S.A. **89:** 444–448.
24. FIORUCCI, S. 2001. NO-releasing NSAIDs are caspase inhibitors. Trends Immunol. **22:** 232–235.
25. CHIUEH, C.C. 1994. Neurobiology of ·NO and ·OH: basic research and clinical relevance. Ann. N.Y. Acad. Sci. **738:** 279–281.
26. PLUTA, R.M., R. RAK, D. WINK, *et al.* 2001. Effects of nitric oxide on reactive oxygen species production and infarction size after brain reperfusion injury. Neurosurgery **48:** 884–893.

Contradictory Effects of Sodium Nitroprusside and *S*-Nitroso-*N*-Acetylpenicillamine on Oxidative Stress in Brain Dopamine Neurons *in Vivo*

PEKKA RAUHALA,[a] TSUGUNOBU ANDOH,[b]
KEVIN YEH,[c] AND CHUANG C. CHIUEH[c]

[a]*Institute of Biomedicine, Pharmacology, University of Helsinki, Helsinki, Finland*

[b]*Department of Applied Pharmacology, Toyama Medical and Pharmaceutical University, Toyama, Japan*

[c]*Laboratory of Clinical Science, National Institute of Mental Health, NIH Clinical Center, Bethesda, Maryland, U.S.A.*

ABSTRACT: To investigate whether nitric oxide (·NO) is neurotoxic or neuroprotective in the brain, we compared the *in vivo* role of *S*-nitroso-*N*-acetylpenicillamine (SNAP) with that of sodium nitroprusside (SNP) on ferrous citrate-induced oxidative stress and neuronal loss in the rat nigrostriatal dopaminergic system. It is known that light irradiation releases ·NO from its donor compounds; these irradiated ·NO donors were used as sham controls in this study. Intranigral infusion of ferrous citrate (4.2 nmol) into the rat midbrain substantia nigra compacta area caused acute lipid peroxidation in the substantia nigra and chronic dopamine depletion in the caudate nucleus. Coinfusion of freshly prepared SNAP (0–8.4 nmol) or ·NO (about 2 nmol), but not SNP, rescued iron-induced dopamine depletion in the rat brain *in vivo*. In fact, SNP produced prooxidative effects similar to ferrous citrate both *in vivo* and *in vitro*, since SNP is a redox iron complex. Consistently, ·NO and SNAP inhibited, whereas SNP potentiated, ·OH generation and lipid peroxidation evoked by ferrous citrate *in vitro*. We previously reported that freshly prepared, but not irradiated, *S*-nitroso-L-glutathione (GSNO) protected brain dopamine neurons against oxidative stress *in vivo*. As well as these antioxidative properties, our recent reports (see Ref. 1) indicate that ·NO/GSNO activated guanylyl cyclase, increased cGMP and that could lead to PKG-mediated expression of MnSOD, Bcl-2, and thioredoxin for preconditioning neuroprotection against 1-methyl-4-phenylpyridinium (MPP^+).[1] In conclusion, ·NO and *S*-nitrosothiols (e.g., GSNO and SNAP) can scavenge reactive oxygen species and activate the heme moiety of guanylyl cyclase, resulting in protection of brain dopamine neurons through both antioxidative and antiapoptotic mechanisms.

KEYWORDS: sodium nitroprusside; *S*-nitroso-*N*-acetylpenicillamine; oxidative stress; substantia nigra; lipid peroxidation; hydroxyl radical; caudate nucleus; iron

Address for correspondence: Pekka Rauhala, Institute of Biomedicine, Pharmacology, P.O. Box 63, University of Helsinki 00014, Finland. Fax: 358-9-191-25364.
pekka.rauhala@helsinki.fi
Tsugunobu Andoh: andoht@ms.toyama-mpu.ac.jp
C.C. Chiueh: chiueh@helix.nih.gov

INTRODUCTION

Nitric oxide (·NO), an endogenous compound with a reactive nitrogen radical, is generated by nitric oxide synthase (NOS). ·NO plays critical roles in the cardiovascular system, such as the endothelium-derived relaxing factor (EDRF), the inhibitor of platelet aggregation, and cardioprotection.[2–6] These biological effects of ·NO are not derived from activation of specific receptors, but rather act by the interaction of ·NO with heme- and cysteine-containing enzymes.[7,8] The interaction of ·NO with heme iron activates guanylyl cyclase, leading to vasorelaxation[9] and upregulation of the synthesis of new proteins.[10–12] Through the process of redox modification, ·NO modulates cysteine residues and thus regulates the activity of various cysteine-containing enzymes such as c-fos/c-Jun (AP-1), caspases, cysteine proteases, iron-regulating proteins (IRP1 and 2), aconitase, and HIV-1 protease.[10,13–15] Increasing evidence suggests that the induction of NOS1 may lead to the cGMP-mediated expression of cytoprotective genes and proteins in the central nervous system for preconditioning-induced neuroprotection.[1,10]

·NO Controversy: Antioxidant or Prooxidant

Chemically, the relatively less active ·NO radicals can scavenge the more reactive oxygen species, lipid peroxyl radicals, and thiyl radicals.[15,16] Furthermore, ·NO scavenges GS· and $O_2^{-\cdot}$, yielding *S*-nitroso-L-glutathione (GSNO) and peroxynitrite ($ONOO^-$), respectively. Interestingly, $ONOO^-$ only causes weak oxidative stress in brain preparations that can be blocked by GSNO and ·NO.[17] Carbon dioxide prevents $ONOO^-$-induced protein oxidation but not the formation of 3-nitrotyrosine.[18] Finally, the interactions of ·NO with bioactive iron or small molecular weight iron complexes inhibits the Fenton reaction, resulting in a decrease in the generation of reactive oxygen species (·OH) and peroxyl lipid radicals (LOO·). Furthermore, the antioxidative potency of GSNO is approximately 100 times that of the classic antioxidant GSH.[17] The physiological functions of ·NO and *S*-nitrosothiols at low micromolar concentrations are, thus, different from the results obtained from experiments using millimolar concentrations of ·NO donor compounds. At physiological concentrations, ·NO radicals act like antioxidants rather than prooxidants.

The Controversial Role of ·NO in CNS: Neuroprotection or Neurotoxicity

Site-specific iron overload in the brain has been suggested to cause neurodegenerative disorders such as Parkinson's disease and Hallervorden-Spatz disease through the generation of ·OH and the initiation of brain lipid peroxidation.[19–21] The increase in non-heme iron concentration in the substantia nigra of Parkinson's patients suggests that accumulation of iron complexes may promote degeneration of nigrostriatal neurons.[21–23] Based on the reactivity of ·NO with iron and reactive oxygen species, we proposed that ·NO and related donor compounds may protect against oxidative stress induced by small molecular weight iron complexes in the dopaminergic nigrostriatal system.[15] We recently demonstrated that GSNO protects nigrostriatal dopamine neurons against ferrous citrate induced oxidative stress and injury.[17] Moreover, Andoh *et al.*, revealed that GSNO can also protect against MPP^+-induced apoptosis in SH-SY5Y cells via a cGMP-mediated expression of new antioxidative and antiapoptotic proteins.[10–12] These new findings are at odds with earlier reports that ·NO mediates neurotoxicity in the brain—reports based on *in vitro* findings obtained from high concentrations of SNP and GSNO. In fact, light-irradiated SNP and GSNO at high micromolar concentrations (greater than 500 μM) that can no longer release ·NO are still toxic to cells and neurons in cultures, indicating clearly that ·NO does not mediate the observed toxicity (A. Khaldi, P. Rauhala, and C.C. Chiueh, unpublished observations). To resolve these controversial roles of ·NO in the CNS, we reinvestigated the mechanisms of action of two different compounds of ·NO donors, SNP and SNAP, using both *in vivo* and *in vitro* brain preparations.

METHODS

Intranigral Drug Administration

Sprague-Dawley rats (male 250–350 g, Taconic Farms, Germantown, NY) were anesthetized with chloral hydrate (400 mg/kg, i.p.) and then prepared for intranigral infusion of sterilized drug solutions.[24] Ferrous citrate (4.2 nmol, in 1 μl Ringer's solution) with or without ·NO (approximately 2 nmol) or ·NO donor compounds (e.g., 4.2 to 16.8 nmol SNAP and SNP) were slowly coinfused for five minutes into the substantia nigra. Rats were kept in plastic cages with filter cage tops and with free access to water and food during the experimental period.

Measurement of Nigral Lipid Peroxidation and Striatal Dopamine

Animals were sacrificed by decapitation 4 h, 24 h, or 7 days after unilateral intranigral administration of drugs. Both the control and infused sides of midbrain substantia nigra compacta areas and caudate nuclei were dissected for the measurement of lipid peroxidation and dopamine levels, respectively. A fluorescence assay procedure based on the formation of fluorescent crosslinked amino acids with malondialdehyde species was used to measure lipid peroxidation in freshly dissected substantia nigra samples.[25,26] Dopamine concentrations in the right and left side caudate nuclei were measured by using an HPLC-EC system as described elsewhere.[27]

Salicylate Trapping of ·OH

·OH generation was monitored with a modified salicylate hydroxylation procedure.[23,28] The reaction mixture contained 210 μM ferrous citrate, 1 mM ·OH trapping agent sodium salicylate, and 1 ml of filtered Ringer's solution in a plastic tube that was irradicated under a table lamp for 2 h. This ·OH generating system does not require the addition of H_2O_2. ·OH adducts of salicylic acid, 2,3- and 2,5-dihydroxybenzoic acid in 20 μl samples were measured using an HPLC-EC system.

Materials

SNAP and SNP were purchased from RBI (Natick, MA, USA); ferrous ammonium sulfate from Sigma (St. Louis, MO, USA) and citric acid from Reagents, Inc. (Charlotte, NC, USA). Saturated ·NO solution (approximately 2 mM) was prepared by bubbling ice-cold Ringer's solution first with nitrogen gas for 30 min to remove dissolved oxygen in the air-tight bottle of Ringer's solution. Then this oxygen free Ringer's solution was saturated with purified ·NO gas (filtering through 1 N NaOH solution before use to remove HNO_2) for 60 min (Matheson GAS Products, East Rutherford, NJ, USA). ·NO solutions were freshly prepared and protected from light-irradiation before use. ·NO exhausted SNP and SNAP solutions were irradiated under a table lamp for at least 48 h to remove the molecular ·NO and used as sham controls. Sodium ferricyanide and *N*-acetylpenicillamine are not the proper sham controls, since there are not the true brakedown products of the corresponding ·NO donor compounds.

RESULTS

·OH Mediates Neurotoxic effects of SNP and Ferrous Citrate in Vivo: *A Mechanism in Common*

We have previously demonstrated that intranigral infusion of ferrous citrate (0 to 8.4 nmol) acutely causes lipid peroxidation in the substantia nigra, since ferrous citrate increases the generation of ·OH, and also initiates long lasting brain lipid peroxidation.[19] Subsequently, intranigral infusion of ferrous citrate damages nigral neurons, reflected by decreases in dopamine levels in the nerve terminal area in the same side caudate nucleus five to seven days later. In fact, SNP is as toxic as ferrous citrate; SNP enhanced nigral oxidative stress including the increased generation of ·OH (see FIGURE 1A), and brain lipid peroxidation (FIG. 1B). SNP also caused dopamine depletion (FIG. 1C). We previously suggested that SNP is a redox iron complex that can generate reactive oxygen species including ·OH and lipid radicals when it is mixed with ascorbate.[29] Interestingly, light-exposed and ·NO-exhausted SNP still causes oxidative stress that is completely prevented by hemoglobin, ·NO and desferoxamine in *in vitro* preparations.

Opposite Effects of SNP and SNAP on Iron-Induced Oxidative Stress in Brain Dopamine Neurons in Vivo

Redox cycling of ferrous citrate iron complexes in the presence of oxygen can generate ·OH without the need of adding H_2O_2.[19,25] Intranigral infusion of ferrous citrate (4.2 nmol) acutely induced lipid peroxidation in the substantia nigra that lasts for more than a week, leading to neuronal loss—reflected in the lowering of dopamine levels in the nerve terminal region caudate nucleus. This may be due to the fact that ferrous citrate increased the generation of reactive oxygen species, such as ·OH that were augmented by SNP and blocked by SNAP concentration dependently *in vitro* (see FIGURE 2A). Moreover, coinfusion of freshly prepared SNAP (0–8.4 nmol), but not SNP protected substantia nigra against acute lipid peroxidation (FIG. 2B) and chronic dopamine depletion produced by ferrous citrate *in vivo* (FIG. 2C). In fact, SNP potentiated the oxidative stress and dopamine depletion caused by ferrous citrate. Sodium ferricyanide does not enhance iron-induced lipid peroxidation since it is not a breakdown product of irradiated SNP. As discuss above irradiated SNP still generates ·OH and causes significant oxidative stress. *N*-acetylpenicillamine, a breakdown product of SNAP did not inhibit lipid peroxidation in

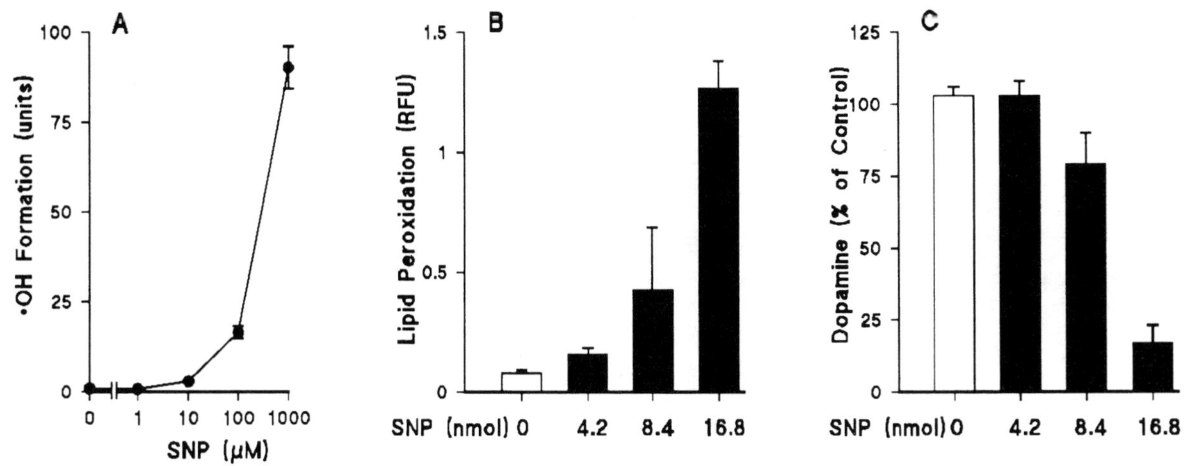

FIGURE 1. prooxidant and neurotoxic effects of SNP *in vitro* and *in vivo.* **(A)** Effects of SNP on ·OH generation *in vitro.* A salizylate hydroxylation trapping method was used to monitor ·OH generation during a two-hour incubation period of SNP (0 to 1,000 μM) in Ringer's solution with the ascorbate (100 uM) to promote the redox cycling of the iron moiety of SNP (dinitroferricyanide/sodium nitroprusside). **(B)** Acute effects of SNP on lipid peroxidation in rat midbrain substantia nigra compacta area *in vivo.* SNP (0–16.8 nmol) was infused unilaterally into substantia nigra and acute lipid peroxidation was measured 24 h after drug treatment. Each *bar* shows a mean ± SEM value ($n = 3$–4) of relative fluorescent units (RFU). Light irradiated SNP can no longer release ·NO but still cause neurotoxicity. **(C)** Chronic effects of intranigral infusion of SNP (0–16.8 nmol) on striatal dopamine levels ipsilateral to the side of SNP infusion. Caudate nucleus of control and infused sides were dissected and processed for the HPLC measurement of dopamine levels seven days after intranigral infusion of SNP. Each *bar* shows a mean ± SEM value ($n = 3$–4) of dopamine levels (percentage of control) in the caudate nucleus.

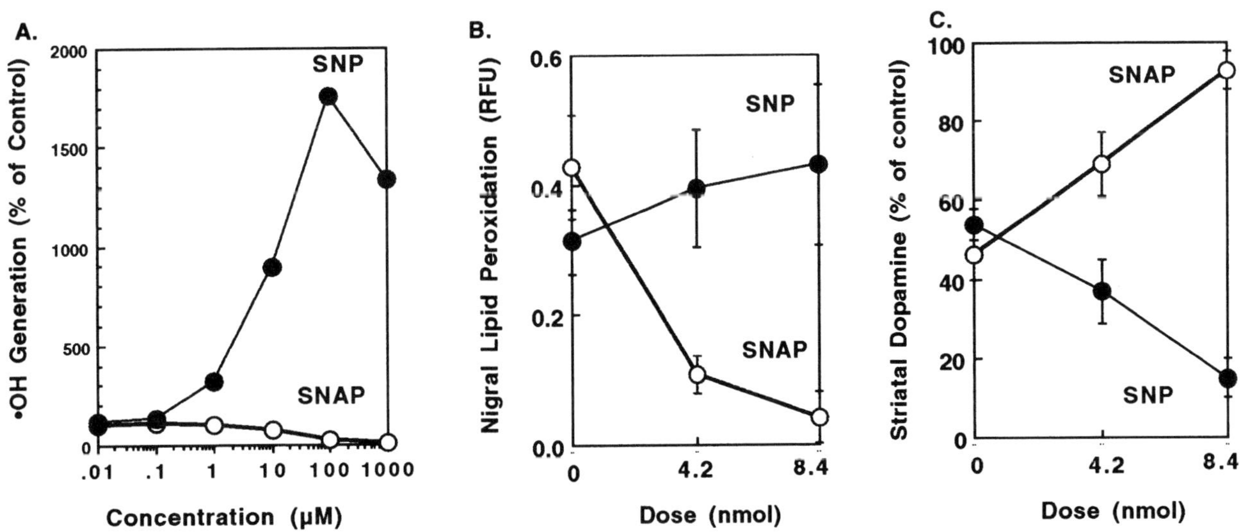

FIGURE 2. A tale of two ·NO donor compounds: contradictory effects of SNP and SNAP on the ferrous citrate induced ·OH generation *in vitro* and dopamine depletion in rat nigrostriatal dopaminergic system *in vivo*. **(A)** Effect of SNAP and SNP (0 to 1,000 µM) on ferrous citrate (210 µM) induced ·OH generation *in vitro*. ·OH generation during a two-hour incubation period was trapped by salicylate hydroxylation procedure. **(B)** Effects of SNP and SNAP (0–8.4 nmol) on ferrous citrate (4.2 nmol) induced acute lipid peroxidation in rat midbrain substantia nigra compacta area. Brain lipid peroxidation reflected by the production of fluorescent adduct of malondiladehyde with amino acids was measured in midbrain tissue four hours after intranigral infusion of drug solutions. Each *bar* shows mean ± SEM (n = 4–8) of relative fluorescent units (RFU). **(C)** Effects of intranigral infusion of SNP and SNAP (0–8.4 nmol) on ferrous citrate-induced dopamine depletion in the nerve terminal area of caudate nucleus. The caudate nuclei of both hemispheres (control and infusion sides) were dissected and processed for the measurement of dopamine levels seven days after intranigral infusion of drugs. Each *bar* shows a mean ± SEM value (n = 5–7) of dopamine levels (percent of control) in the caudate nucleus.

brain homogenates (data not shown). By inhibiting iron-induced lipid peroxidation in the midbrain substantia nigra, freshly prepared SNAP dose-dependently and completely reversed the *in vivo* dopaminergic toxicity caused by intranigral infusion of 4.2 nmol ferrous citrate. Similar to the effects of GSNO, this unique neuroprotective effect of SNAP subsided when the SNAP solution was previously irradiated under a table lamp for two days to remove the ·NO.

·NO Mediates Neuroprotection but not Neurotoxicity in the Brain

We previously used the same *in vivo* procedure to investigate the possible neurotoxic effects of $ONOO^-$ and the results were disappointedly negative.[17] The infusion of up to about 30 nmol of $ONOO^-$ in 1 ul failed to cause any lipid peroxidation in the substantia nigra and dopamine depletion in the striatum. The above experiments with freshly prepared and irradiated ·NO donor compounds clearly suggest that ·NO does not mediate neurotoxicity in midbrain dopaminergic nigral neurons. In fact, freshly prepared ·NO releasing compounds (e.g., GSNO and SNAP) protect brain neurons against severe oxidative stress *in vivo*. We thus used an ·NO solution to demonstrate that ·NO can concentration-dependently and completely inhibit the generation of reactive oxygen species (·OH trapped by salicylate hydroxylation) evoked by ferrous citrate (see FIGURE 3A). To directly evaluate the antioxidative, neuroprotective action of ·NO, we intranigrally infused 1 μl of saturated ·NO in Ringer's solution (about 2 nmol) to see whether ·NO is friend or foe to brain dopamine neurons. Contrary to a prominent hypothesis, that ·NO mediates neurotoxicity, we found that ·NO prevented iron-induced brain lipid peroxidation (FIG. 3B) and, thus, protected nigrostriatal dopaminergic neurons against ferrous citrate-induced oxidative nigral injury *in vivo* (FIG. 3C). The neuroprotective efficacy of ·NO is less than SNAP since the half-life of ·NO is relative short after infusion into the midbrain substantia nigra. The present *in vivo* beneficial results of SNAP and ·NO are consistent with our earlier reports that freshly prepared GSNO, but not ·NO-exhausted GSNO, protects nigral neurons against severe oxidative stress *in vivo*.[17]

It is noteworthy that GSNO (30 μM) can inhibit the catalytic activity of proapoptotic caspase-3 to cleave DEVD-pNa substrate.[1,11] For example, serum withdrawal in SH-SY5Y neuroblastoma cell cultures resulted in a fourfold increase in cytoplasmic caspase-3 activity over controls (serum free group 0.51 ± 0.04 unit over control group 0.13 ± 0.04 unit). Freshly prepared GSNO, but not light-irradiated GSNO (30 μM), significantly decreased the serum-free enhanced caspase 3 activity (fresh GSNO group 0.18 ± 0.02 unit verse irradiated GSNO group 0.50 ± 0.03). Therefore, in addition to preventing oxidative stress (e.g., ·OH generation and lipid peroxidation), ·NO can also activate guanylyl cyclase, increasing cGMP that could lead to the inhibition of caspases and, thus, prevent the apoptosis caused by oxidative stress.

DISCUSSION

Antioxidative Properties of ·NO

These *in vivo* results support previous *in vitro* findings that SNAP, GSNO, and ·NO are potent antioxidants, and that ·NO released from freshly prepared *S*-nitrosothiols (i.e., GSNO and SNAP) mediates antioxidation and neuroprotection

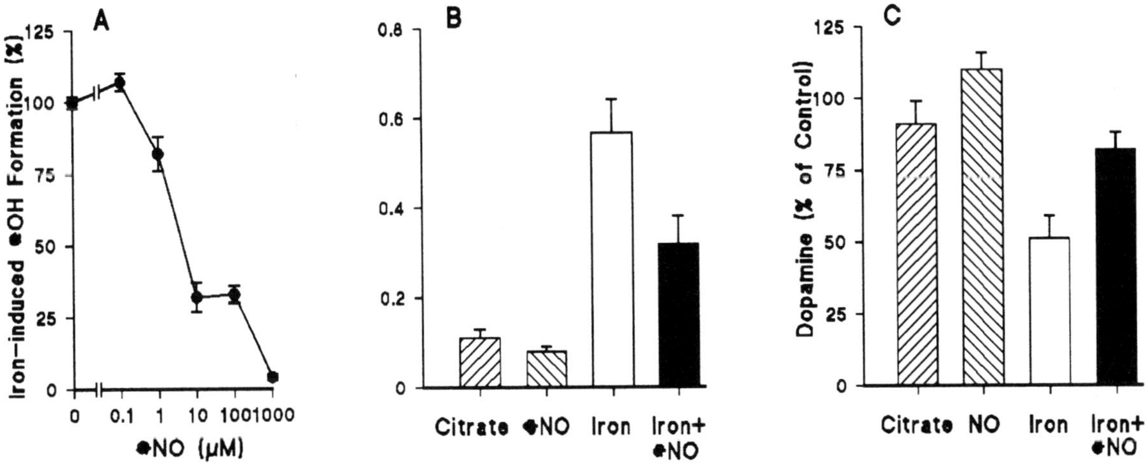

FIGURE 3. Antioxidant and neuroprotective effects of ·NO on the ferrous citrate induced ·OH generation *in vitro* and in rat nigrostriatal dopaminergic system *in vivo.* (**A**) Effect of ·NO on ferrous citrate (210 uM) induced ·OH generation *in vitro.* A salizylate hydroxylation trapping method was used to monitor ·OH generation during a two-hour incubation. This redox cycling system of ferrous citrate and isocitrate complexes do not require the addition of H_2O_2 and ascorbate. (B) Effects of ·NO on iron-induced brain lipid peroxidation *in vivo.* ·NO (about 2 nmol) was coin fused with ferrous citrate (4.2 nmol) into rat substantia nigra compacta area and acute lipid peroxidation was measured four hours after drug treatment. Acute peroxidation of brain polyunsaturated fatty acids in midbrain samples were assayed by measuring florescent adducts of malondialdehyde crosslinked with amino acids. Each *bar* shows a mean ± SEM value ($n = 4–8$) of relative fluorescent units (RFU). (**C**) The role of ·NO on iron-induced striatal dopamine depletion. ·NO (about 2 nmol) was coinfused with ferrous citrate (4.2 nmol) into rat substantia nigra compacta area of one side. Seven days later, caudate nucleus of both hemisphere were dissected and processed for the measurement of dopamine levels. Each *bar* shows a mean ± SEM value ($n = 4–8$) of dopamine levels (percent of control) in the caudate nucleus.

in the brain. This novel neuroprotective action of ·NO in brain dopamine neurons contradicts an early hypothesis that ·NO mediates neurotoxicity in the brain.[15] In fact, ·NO can bind to iron complexes and prevent the generation of ·OH radicals caused by the redox cycling of iron complexes, such as hemoglobin, ferrous citrate, and SNP. We have also proposed that ·NO may scavenge thiyl radicals such as GS· and Cys·, and convert these reactive thiyl species into potent antioxidants, such as GSNO and *S*-nitrosocysteine. In the presence of physiological concentrations of CO_2, ·NO scavenges $O_2^{-\cdot}$ and immediately converts $ONOO^-$ to less reactive and nontoxic metabolites. Furthermore, GSNO can also completely neutralize $ONOO^-$.

Earlier *in vitro* studies indicate that ·NO may interrupt the chain reaction of lipid peroxidation.[25,26,30,31] In our *in vivo* studies, we have also shown that authentic ·NO and ·NO released from either GSNO or SNAP block iron-induced lipid peroxidation of the poly-unsaturated fatty acids in the midbrain.[17,32] In this study, we directly demonstrated that intracerebral infusion of ·NO solution leads to inhibition of lipid peroxidation and also cyto- and neuroprotection. ·NO scavenges peroxyl lipid radicals, such as LOO· to LOO-NO and, thus, stops the propagation of brain lipid peroxidation caused by ·OH generated by iron complexes in the central nervous system.

A Novel Neuroprotective Mechanism: Cyclic GMP Mediates Antiapoptotic Activity of ·NO and S*-nitrosothiols*

In addition to scavenging lipid peroxyl and thiyl radicals by ·NO, *S*-nitrosothiols also release nitrosonium ions (NO^+) and sulfhydryl groups for modification to cysteine containing proteins including hemoglobin, caspases and HIV-1 protease.[8,9,13–16] MPP^+ causes apoptosis in human neurotrophic SH-SY5Y cells, since MPP^+ releases mitochondrial cytochrome c and activates proapoptotic proteins, such as caspase-9 and -3. Caspase-9 is a protease containing reactive cysteine residues for activating caspase–3 and then activates Apaf-1 to induce apoptosis. Our recent observations indicate that MPP^+-induced apoptosis can be prevented by GSNO,[10–12] since GSNO, but not light-irradiated GSNO, clearly inactivated caspase-3 at low micromolar concentrations. We have also hypothesized that endogenously synthesized ·NO from NOS1 can upregulate antiapoptotic protein Bcl-2 and antioxidative protein (e.g., MnSOD and Trx) via a cGMP-mediated mechanism in a neuroblastoma cell line derived from human brain.[33]

Is 3-Nitrotyrosine a Reliable Toxic Marker for Peroxynitrite?

As well as scavenge superoxide anions ($O_2^{-\cdot}$) to generate peroxynitrite, ·NO also scavenges glutathionyl radicals (GS·), leading to the generation of a more potent antioxidant GSNO that is at least three orders of magnitude more potent than the classic antioxidant GSH. Furthermore, GSNO concentration-dependently blocks the weak oxidative stress caused by peroxynitrite in brain homogenates.[17] Moreover, possibly due to the presence of CO_2, the intracerebral infusion of peroxynitrite fails to induce any significant damage to brain dopaminergic neurons in the substantia nigra compacta area *in vivo*. Despite these negative *in vivo* results, ·NO derived reactive nitrogen species have been suggested to be neurotoxic, and to be responsible for every known neurodegenerative brain disorder. Peroxynitrite has received a lot of attention and accumulation of 3-nitrotyrosine has been used as an indicator of peroxynitrite-mediated nitrosative stress. However, 3-nitrotyrosine can be formed by multiple pathways, such as nitrites, leucocytes, and myeloperoxidase enzyme.[34,35] Therefore, tyrosine nitration should not be considered a specific marker of $ONOO^-$ formation. It is obvious that it is necessary to use the appropriate sham controls to rule out possible artifacts when one uses ·NO donor compounds in both *in vitro* and *in vivo* experiments. For example, SNP is known to be a cytotoxic agent, however, the underlying toxic mechanism is predominantly mediated by ·OH, but not ·NO, since the light-irradiated SNP can no longer release ·NO but still evokes cytotoxicity. This toxic effect of SNP can be blocked by antioxidants, including ·OH scavengers such as salicylate, GSNO, ·NO, and methylene blue—a scavenger of both ·OH and ·NO.

·NO Mediates Neuroprotection Against Oxidative Stress in Vivo

Both the current results and our previous reports show that ·NO, SNAP, and GSNO have similar antioxidative and neuroprotective properties against small molecular weight iron-induced oxidative stress and brain injury *in vivo*. These ·NO compounds also protect oxidative damages caused by hemoglobin[32] and HIV-1 protease.[14] However, after releasing its molecular ·NO, the irradiated or oxidized *S*-nitrosothiols can no longer prevent oxidative stress and brain injury, suggesting that ·NO released from *S*-nitrosothiols mediates the antioxidative protection against oxidative stress. New information suggests that *S*-nitrothiols and ·NO can also modify the reactive cysteine residues of enzymes and proteases. Acting through this unique mode of action, ·NO binds to a heme moiety and also to reactive thiyl residues to regulate enzyme activity, via either downregulation (e.g., caspases and viral protease) or upregulation (guanylyl cyclase). GSNO may inactivate cysteine-containing proteases via (1) *S*-nitrosylation, (2) transnitrosylation, or (3) formation of disulfite bridges, such as *S*-glutathionylation. Thus, GSNO is far more potent than other ·NO donors in its inhibition of caspases in *in vitro* preparations. Moreover, our new data indicate that ·NO binds to the heme moiety of guanylyl cyclase and generates cGMP that increases new protein synthesis resulting in the expression of antiapoptotic protein Bcl-2 and also antioxidative proteins of MnSOD and Trx (Andoh *et al.*, this volume).

In conclusion, the neuroprotective mechanisms of ·NO and *S*-nitrosothiols consist of both antioxidative and antiapoptotic events that include cGMP-mediated new protein synthesis. (1) ·NO prevents the redox cycling of iron complexes (i.e., ferrous citrate, SNP, and hemoglobin). (2) ·NO scavenges reactive oxygen species (·OH and $O_2^{-\cdot}$) and reactive thiyl species (GS· and Cys·). (3) ·NO terminates the brain lipid peroxidation chain reaction by scavenging LOO·. (4) ·NO inhibits the expression of proapoptotic caspases and $p66^{shc}$. (5) ·NO activates guanylyl cyclase and cGMP-mediated induction of antiapoptotic protein Bcl-2 and other cytoprotective proteins (MnSOD, BDNF/GDNF, and Trx) during preconditioning-induced neuroprotection. This new view of ·NO/cGMP-mediated preconditioning-induced neuro- and cardioprotection is discussed by other authors in this volume.[33,36] Our *in vivo* data have provided relevant evidence to support a new hypothesis that ·NO and *S*-nitrosothiols can protect against the oxidative stress/injury evoked by free radicals generated by iron complexes, through the mechanisms of both antioxidation and antiapoptosis. We further propose that interactions of ·NO between iron complexes, with peptidyl cysteine residues, and the activation of cGMP-mediated new protein synthesis of

MnSOD, Bcl-2 and Trx may be beneficial in slowing such progressive neurodegenerative conditions as the iron-induced oxidative stress and/or caspases-induced apoptosis in Parkinson's disease, head trauma, hemorrhagic stroke, and perhaps the Hallervorden-Spatz syndrome as well.

ACKNOWLEDGMENTS

We appreciate the encouragement and support of Dennis L. Murphy (NIMH). Pekka Rauhala was supported in part by an Orion Fellowship and Paulo Foundation and Tsugunobu Andoh by the Japanese Society for the Promotion of Science, respectively.

REFERENCES

1. CHIUEH, C.C., T. ANDOH & P.B. CHOCK. 2001. Protection against MPTP-induced toxicity via preconditioning induction of cytoprotective genes and proteins: Ref-1, NOS1, and Trx. Soc. Neurosci. Abstr. **27:** 208.7.
2. MARTIN, W., G.M. VILLANI, D. JOTHIANANDAN & R.F. FURCHGOTT. 1985. Selective blockade of endothelium-dependent and glycerol trinitrate-induced relaxation by hemoglobin and by methylene blue in the rabbit aorta. J. Pharmacol. Exp. Ther. **232:** 708–716.
3. IGNARRO, L.J., G.M. BUGA, K.S. WOOD, *et al.* 1987. Endothelium-derived relaxing factor produced and released from artery and veins is nitric oxide. Proc. Natl. Acad. Sci. U.S.A. **84:** 9265–9269.
4. MONCADA, S., R.M.J. PALMER & E.A. HIGGS. 1991. Nitric oxide: physiology, pathophysiology, and pharmacology. Pharmacol. Rev. **43:** 109–142.
5. GUO, Y., W.K. JONES, Y.T. XUAN, *et al.* 1999. The late phase of ischemia preconditioning is abrogated by targeted disruption of the inducible NO synthase gene. Proc. Natl. Acad. Sci. U.S.A. **96:** 11507–11512.
6. BAKER, J.E., P. HOLMAN, B. KALYANARAMAN, *et al.* 1999. Adaptation to chronic hypoxia confers tolerance to subsequent myocardial ischemia by increased nitric oxide production. Ann. N.Y. Acad. Sci. **874:** 236–253.
7. HENRY, Y., M. LEPOIVRE, J.-C. DRAPIER, *et al.* 1993. EPR characterization of molecular targets for NO in mammalian cells and organelles. FASEB J. **7:** 1124–1134.
8. STAMLER, J.S. 1994. Redox signaling: nitrosylation and related target interactions of nitric oxide. Cell **78:** 931–936.
9. ARNOLD, W.P., C.K. MITTAL, S. KATSUKI & F. MURAD. 1977. Nitric oxide activates guanylate cyclase and increases guanosine 3′:5′-cyclic monophosphate levels in various tissue preparations. Proc. Natl. Acad. Sci. U.S.A. **74:** 3203–3207.
10. ANDOH, T., S.Y. LEE & C.C. CHIUEH. 2000. Preconditioning regulation of bcl-2 and $p66^{shc}$ by human NOS1 enhances tolerance to oxidative stress. FASEB J. **14:** 2144–2146.
11. ANDOH, T., P.B. CHOCK & C.C. CHIUEH. 2001. The role of neuronal nitric oxide synthase (NOS1) and related cyclic GMP-PKG signaling pathway on preconditioning-mediated neuroprotection. Soc. Neurosci. Abstr. **27:** 688.3.
12. ANDOH, T., P.B. CHOCK & C.C. CHIUEH. 2002. The role of thioredoxin in protection against oxidative stress-induced apoptosis in SH-SY5Y cells. J. Biol. Cem. In press.
13. MARSHALL, H.E., K. MERCHANT & J.S. STAMLER. 2000. Nitrosation and oxidation in the regulation of gene expression. FASEB J. **14:** 1889–1900.
14. LEE, S.Y., D.A. DAVIS, K.M. PEARSON, *et al.* 2002. *S*-nitroso-L-glutathione modifies dimeric interface of HIV-1 protease and inhibits drug-resistant mutant. Submitted.
15. CHIUEH, C.C. 1994. Neurobiology of ·NO and ·OH: basic research and clinical relevance. Ann. N.Y. Acad. Sci. **738:** 279–281.

16. CHIUEH, C.C. & P. RAUHALA. 1999. The redox pathway of *S*-nitrosoglutathione, glutathione and nitric oxide in cell to neuron communications. Free Rad. Res. **31:** 641–650.
17. RAUHALA, P., A.M.-Y. LIN & C.C. CHIUEH. 1998. Neuroprotection by *S*-nitroglutathione of brain dopamine neurons form oxidative stress. FASEB J. **12:** 165–173.
18. BERLETT, B.S., R.L. LEVINE & E.R. STADTMAN. 1998. Carbon dioxide stimulates peroxynitrite-mediated nitration of tyrosine residues and inhibits oxidation of methionine residues of glutamine syntetase: both modifications mimic effects of adenylation. Proc. Natl. Acad. Sci. U.S.A. **95:** 2784–2789.
19. SZIRAKI, I., K.P. MOHANAKUMAR, P. RAUHALA, *et al.* 1998. Manganese: a transition metal protects nigrostriatal neurons from oxidative stress in the iron-induced animal model of Parkinsonism. Neuroscience **85:** 1101–1111.
20. CHIUEH, C.C. 2001. Iron-overload, oxidative stress and axonal dystrophy in brain disorders. Ped. Neurol. **25:** 138–147.
21. SOFIC, E., P. RIEDERER, H. HEINSEN, *et al.* 1988. Increased iron (III) and total iron content in post mortem substantia nigra of parkisonian brain. J. Neural Transm. **74:** 199–205.
22. DEXTER, D.T., F.R. WELLS, F. AGID, *et al.* 1987. Increased nigral iron content in postmortem parkinsonian brain. Lancet **2:** 1219–1220.
23. CHIUEH, C.C., H. MIYAKE & M.T. PENG. 1993. Role of dopamine autoxidation, hydroxyl radical generation, and calcium overload in underlying mechanisms involved in MPTP-induced parkinsonism. Adv. Neurol. **60:** 251–258.
24. SUN, C.J., J.N. JOHANNESSEN, W. GESSNER, *et al.* 1988. Neurotoxic damage to the nigrostriatal system in rats following intranigral administration of $MPDP^+$ and MPP^+. J. Neural Transm. **74:** 75–86.
25. MOHANAKUMAR, K.P., A. DE BARTOLOMEIS, R.-M. WU, *et al.* 1994. Ferrous-citrate complex and nigral degeneration: evidence for free-radical formation and lipid peroxidation. Ann. N.Y. Acad. Sci. **738:** 392–399.
26. RAUHALA, P. & C.C. CHIUEH. 2000. Effects of atypical antioxidative agents, *S*-nitrosoglutathione and manganese, on brain lipid peroxidation induced by iron leaking form tissue disruption. Ann. N.Y. Acad. Sci. **899:** 238–254.
27. CHIUEH, C.C., Z. ZUKOWSKA-GROJEC, K.L. KIRK & I.J. KOPIN. 1983. 6-Fluorocatecholamines as false adrenergic neurotransmitters. J. Pharmacol. Exp. Ther. **225:** 529–533.
28. CHIUEH, C.C., G. KRISHNA, P. TULSI, *et al.* 1992. Intracranial microdialysis of salicylic acid to detect hydroxyl radical generation through dopamine autooxidation in the caudate nucleus: effects of MPP^+. Free Rad. Biol. Med. **13:** 581–583.
29. RAUHALA, P., A. KHALDI, K.P. MOHANAKUMAR & C.C. CHIUEH. 1998. Apparent role of hydroxyl radicals in oxidative brain injury induced by sodium nitroprusside. Free Rad. Biol. Med. **24:** 1065–1073.
30. KANNER, J., S. HAREL & R. GRANIT. 1991. Nitric oxide as an antioxidant. Arch. Biochem. Biophys. **289:** 130–136.
31. GORBUNOV, N.V., A.N. OSIPOV, B.W. DAY, *et al.* 1995. Reduction of ferrylmyoglobin and ferrylhemoglobin by nitric oxide: a protective mechanism against ferryl hemoprotein-induced oxidations. Biochemistry **34:** 6689–6699.
32. BERGEN, P.A., P. RAUHALA, C.M. SPOONER & C.C. CHIUEH. 1999. Hemoglobin and iron-evoked oxidative stress in the brain: protection by bile pigments, manganese and *S*-nitrosoglutahione. Free Rad. Res. **31:** 631–640.
33. ANDOH, T., P.B. CHOCK & C.C. CHIUEH. 2002. Preconditioning-mediated neuroprotection: role of nitric oxide, cGMP and protein expression. Ann. N.Y. Acad. Sci. **962:** this volume.
34. GUNTHER, M.R., L.C. HIS, J.F. CURTIS, *et al.* 1997. Nitric oxide trapping of the tyrosyl radical of prostaglandin H synthase-2 leads to tyrosine iminoxyl radical and nitrotyrosine formation. J. Biol. Chem. **272:** 17086–17090.
35. EISERICH, J.P., M. HRISTOVA, C.E. CROSS, *et al.* 1998. Formation of nitric oxide-derived inflammatory oxidants by myeloperoxidase in neutrophils. Nature **391:** 393–397.

36. DAWN, B. & R. BOLLI. 2002. Role of nitric oxide in myocardial preconditioning. Ann. N.Y. Acad. Sci. **962:** this volume.

L-Citrulline Recycle for Synthesis of NO in Cerebral Perivascular Nerves and Endothelial Cells

TONY J.F. LEE AND JUN-GE YU

Department of Pharmacology, Southern Illinois University, School of Medicine, Springfield, Illinois, U.S.A.

ABSTRACT: Recycle of L-citrulline to form L-arginine in cerebral perivascular nerves has been well described, providing direct evidence that nitric oxide (NO) is synthesized and released from these nerves to act as the transmitter for vasodilation. NO is also synthesized and released from cerebral endothelial cells, involving L-citrulline conversion to L-arginine. Evidence for the presence of enzymes involved in the conversion, however, has not been shown. The presence of nitric oxide synthase (NOS), argininosuccinate synthetase (ASS), and argininosuccinate lyase (ASL), and their coexistence with NADPH-diaphorase (NADPHd), a marker for NOS, in endothelial cells of middle cerebral arteries and the circle of Willis of the pig, therefore, were examined using combined immunohistochemical and histochemical techniques. NOS-, ASS-, and ASL-immunoreactivities were found in almost all endothelial cells of all cerebral arteries examined. All ASS-, ASL-, and NOS-immunoreactive (I) endothelial cells also stained positively for NADPHd, suggesting that ASS, ASL, and NOS were colocalized in endothelial cells of middle cerebral arteries and the circle of Willis. These results provide morphological evidence that cerebral vascular endothelial cells like cerebral perivascular nerves contain enzymes necessary for recycling L-citrulline to L-arginine to synthesize NO via an argininosuccinate (AS) pathway.

KEYWORDS: argininosuccinate lyase; argininosuccinate synthetase; NADPH diaphorase; nitric oxide synthase (NOS); colocalization; cerebral arteries; endothelial cells; pig

INTRODUCTION

NO is synthesized in the presence of nitric oxide synthase (NOS) from L-arginine with L-citrulline as a byproduct in various types of cells.[1,2] We have demonstrated biochemically and pharmacologically that L-citrulline in cerebral perivascular nerves is actively converted to form L-arginine to synthesize NO[3,4] (see FIGURE 1). The known pathway of L-citrulline conversion to L-arginine in the urea cycle is via the production of argininosuccinate (AS).[5] In this pathway, AS synthetase (ASS) catalyzes the formation of AS from L-citrulline and aspartate. AS is then cleaved to produce L-arginine and fumarate by AS lyase (ASL) (FIG. 1). Our studies using

Address for correspondence: Tony J.-F. Lee, Ph.D., Southern Illinois University, School of Medicine, Department of Pharmacology, P.O. Box 19629, Springfield, IL 62794-9629, U.S.A. Voice: 217-785-2197; fax: 217-524-0145.

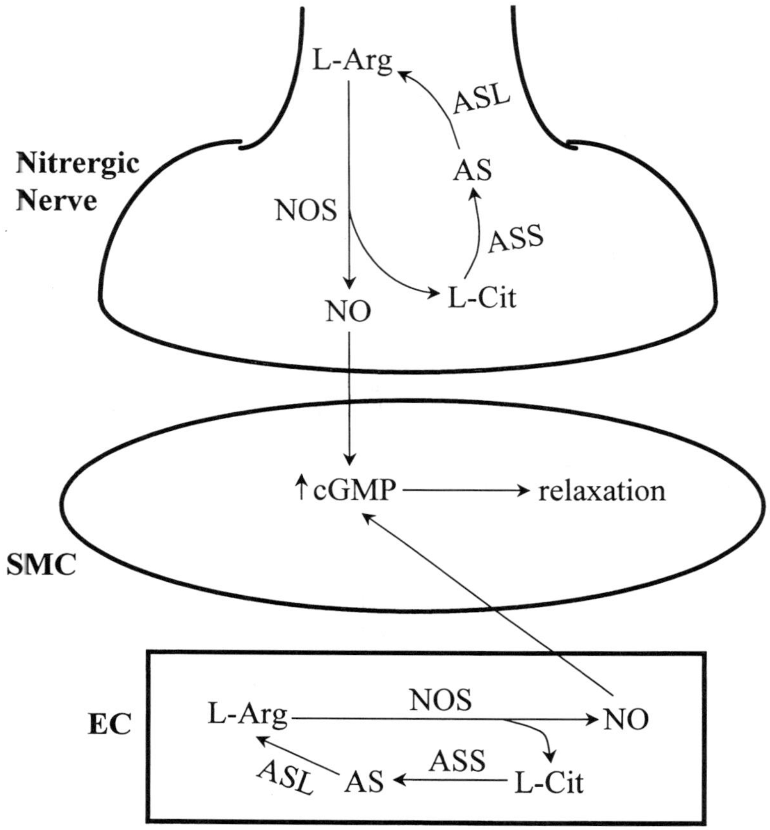

FIGURE 1. Conversion of L-citrulline (L-Cit) to L-arginine (L-Arg) to synthesize NO in the presence of nitric oxide synthase (NOS) in cerebral endothelial cells (EC) and cerebral perivascular nitrergic nerves. The conversion in both cells is catalyzed by argininosuccinate synthetase (ASS) and argininosuccinate lyse (ASL). NO induces relaxation of smooth muscle cells (SMC) by increasing in cGMP synthesis. AS (argininosuccinate).

sequential double labeling techniques; that is, immunoperoxidase or immunofluorescence combined with NADPH diaphorase (NADPHd) histochemical staining, have demonstrated that dense NOS-immunoreactive (I), ASS-I, ASL-I, and NADPHd-reactive fibers in porcine cerebral arteries and veins are completely coincident fibers. ASS, ASL, and NOS are axoplasmic enzymes in these perivascular nerves. Immunoreactivities of these three enzymes are also found to coincide in the sphenopalatine ganglion (SPG), a major source of cerebral nitrergic neurons.[6,7] These results suggest that ASS, ASL, and NOS are colocalized in the same neurons,[8,9] providing morphological evidence for the presence of necessary enzymes catalyzing the conversion of L-citrulline to L-arginine for the synthesis of more NO in cerebral perivascular nitrergic nerves. Thus, an active citrulline-arginine cycle exists in cerebral perivascular nerves. Recycling L-citrulline to form L-arginine, therefore, provides a means to maintain an adequate supply of NO during neurovascular transmission. This finding provides strong evidence that NO is synthesized and released in perivascular nitrergic nerves.

The physiological significance of endothelial cells in regulating vascular tone and resistance is well recognized.[10–12] NO has been shown to be one of the vasodilating substances released from the endothelial cells.[13,14] Results from pharmacological studies have indicated that L-citrulline is converted to L-arginine in cerebral vascular endothelial cells.[4] The pathway of L-citrulline conversion to L-arginine remains unclear. The present study was, therefore, designed to examine if ASS, ASL, and NOS are present and are colocalized in the same endothelial cells in porcine cerebral arteries. The results indicate that ASS-, ASL-, and NOS-immunoreactivities are colocalized in the same endothelial cells, suggesting that L-citrulline conversion to L-arginine in cerebral vascular endothelial cells like that in cerebral perivascular nerves is via AS pathway.

MATERIALS AND METHODS

Chemicals

NADPH (reduced form), Triton X-100, and nitro blue tetrasodium were obtained for Sigma Chemical Co. (St Louis, MO). Biotinylated goat antirabbit IgG and fluorescein avidin D were purchased from Vector Labs (Burlingame, CA).

Porcine Cerebral Arteries

Fresh heads of adult pigs ($n = 5$) of either sex were obtained at a local slaughter house. The entire brain was removed and middle cerebral arteries and circle of Willis were dissected under a dissecting microscope in oxygenated (95% O_2/5% CO_2) Krebs' bicarbonate solution (pH 7.4) containing 196.87 mM NaCl, 5.16 mM KCl, 1.33 mM $CaCl_2$, 25.56 mM $NaHCO_3$, 1.22 mM $MgSO_4.7H_2O$, 1.01 mM dextrose, 0.34 mM EDTA, and 0.28 mM ascorbic acid[15] at room temperature.

Fixative

The periodate–paraformaldehyde–picric acid–lysine (PPFL) solution was prepared as follows: (1) Solution A: To 0.2 M lysine-HCl in distilled water add 0.1 M dibasic sodium phosphate until pH is 7.4. The solution was then diluted to 0.1 M lysine with 0.1 M sodium phosphate buffer, pH 7.4 (resulting phosphate concentration was 0.05 M). Just before use, three parts of the lysine–phosphate buffer were mixed with 1 part 8% paraformaldehyde. (2) Solution B: three parts of saturated aqueous solution of picric acid were combined with two parts of 20% paraformaldehyde, then the mixture was diluted by three volumes of 0.1 M phosphate buffer pH 7.4. Finally three parts of Solution A were mixed with one part of Solution B. Sodium *m*-periodate (16.05 mg/10 ml) was then added to the final solution.

Immunohistochemistry

The cerebral arteries were fixed in PPPL fixative for four hours. Arteries were then washed in PBS, and kept in PB containing 20% sucrose for 24 h at 4°C. Arteries were cross-sectioned at 8 μm thickness with a cryostat microtome and placed on Vectabond (Vector Labs)-coated slides and air-dried for two hours. Immunohistochemistry for the sections was carried out using indirect immunofluorescence method.[8,9,16] Specific antisera to purified ASS, ASL, and eNOS were raised in the rabbits. The procedures of enzyme purification, characterization, and specificity of antisera against ASS and ASL are described elsewhere.[8,9,17–20] The sections were washed in 0.01 M PBS, and blocked with 2.5% normal goat serum diluted in 0.05% Triton X-100/PBS for 30 min at room temperature. After washing in PBS, the specimens were incubated in rabbit antiserum against ASS in a dilution of 1:2,000, ASL 1:2,000, or eNOS 1: 1,000 (Affinity BioReagents) for 24–48 h at 4°C. After another wash in PBS, specimens were incubated in affinity purified biotinylated goat antirabbit IgG antibody (1:200 diluted in PBS, Vector Labs) for four hours at 4°C, then washed with PBS and incubated for another four hours at 4°C with fluorescein avidin D (Vector Labs). The specimens were then rinsed and coverslipped with Vectashield mounting medium for photography under a fluorescence microscope fitted with proper filters. After taking photographs of immunofluorescence, the specimens were washed with PBS and processed for NADPHd histochemistry as described below.

NADPH-Diaphorase Histochemical Staining

Following fixation, incubation with specific antibodies in demonstrating immunofluorescence labeling and photographing as described above, specimens were incubated in 0.1 M phosphate buffer (pH 8.0) containing 1 mg/ml of NADPH (reduced form), 0.1 mg/ml of nitro blue tetrasodium and 0.2% Triton X-100 at 37°C for one hour.[3] The specimens were rinsed with PBS and mounted with Gel/Mount (Biomeda, Foster City, CA) and examined under a Zeiss light microscope.

Controls

The specificity of the immunolabeling was determined by omitting ASS, ASL, or eNOS antisera, as well as by using non-immunized normal rabbit serum or non-immunized normal goat serum. No labeling was found in these control specimens. For control of NADPHd activity, NADPH was omitted from the incubation medium. This resulted in elimination of NADPHd reaction product.[3,8,9]

RESULTS

Almost all endothelial cells in porcine middle cerebral arteries and the circle of Willis (10 sections from each artery from five animals) were found to be immunoreactive (I) for eNOS, ASS, and ASL (see FIGURE 2). All eNOS-I, ASS-I, and ASL-I endothelial cells were stained positively for NADPHd. eNOS-I, ASS-I, and ASL-I fibers were also found in the adventitia. All three immunoreactive fibers were coincident with NADPHd fibers; a result consistent to those found in the whole-mount preparations of porcine cerebral arteries.[8,9] NOS-, ASS-, and ASL-immunoreactivities or NADPHd reactivities were not found in smooth muscle cells. For controls, eNOS-, ASS-, and ASL-immunoreactivities were not observed in the vessel walls after omitting primary antibody or incubating tissues with normal serum from non-immunized rabbits or goats (data not shown). NADPHd reactivities were not observed when NADPH was omitted from the incubation medium.

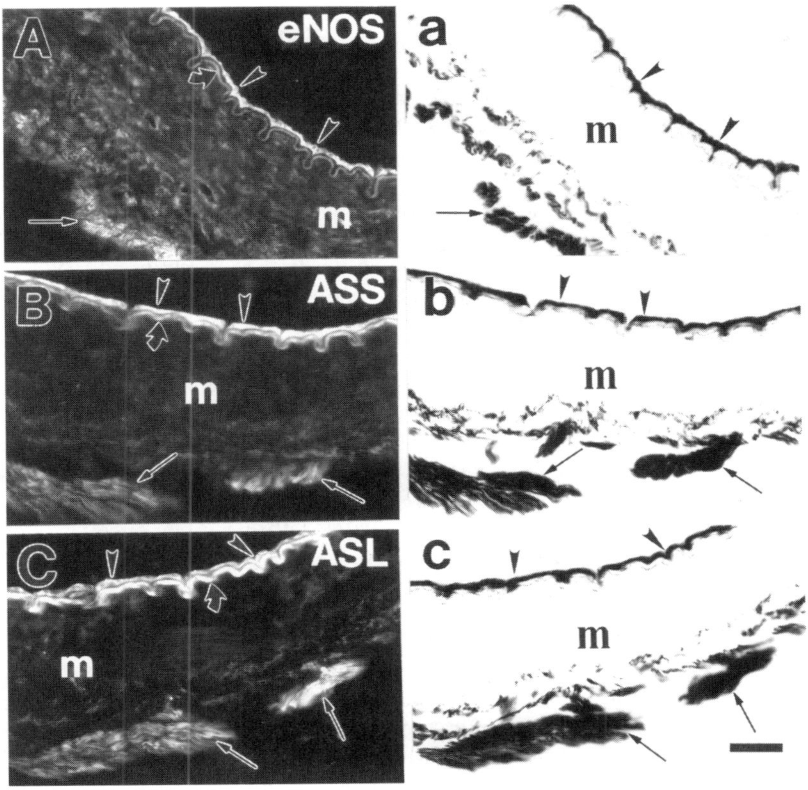

FIGURE 2. Photomicrographs showing immunofluorescence labeling of eNOS (**A**), ASS (**B**), and ASL (**C**) in cross sections of whole-mount preparations of the circle of Willis, followed by histochemical staining of NADPHd in each corresponding preparation (**a**, **b**, and **c**, respectively). Almost all endothelial cells were immunostained for eNOS, ASS, and ASL (*arrowheads* in **A**, **B**, and **C**); and all eNOS-I, ASS-I, and ASL-I endothelial cells were also stained positively for NADPHd (*arrowheads* in **a**, **b**, and **c**). eNOS-I, ASS-I, and ASL-I fibers were also found in the adventitia of the arteries (*arrows* in **A**, **B**, and **C**), and they were almost completely coincident with NADPHd-positive fibers (*arrows* in **a**, **b**, and **c**). NOS-, ASS-, and ASL-immunoreactivities and NADPHd-reactivities were not found in smooth muscle cells (*m*). Note the presence of spontaneous fluorescence in internal elastic lamina (*curved arrows*). *Scale bar* = 50 μm.

DISCUSSION

The present immunohistochemical and histochemical studies demonstrate that eNOS-, ASS-, and ASL-immunoreactivities are present in all endothelial cells of porcine middle cerebral arteries and the circle of Willis examined; and that they are completely colocalized with NADPHd-reactivities. These results suggest that ASS and ASL are colocalized with NOS in the same endothelial cells of cerebral arteries.

NO is synthesized from L-arginine with L-citrulline as a byproduct catalyzed by NOS in various types of cells including endothelial cells.[1,2,21] Several studies have further shown that cultured endothelial cells can recycle L-citrulline to form L-arginine.[22,23] A similar conclusion has been reached in endothelial cells in cerebral arteries.[4] This was based on *in vitro* tissue bath studies showing that, in the presence of active muscle tone, cerebral arteries without endothelial cells did not response upon application of nitro-L-arginine (NLA), L-arginine or L-citrulline.[4] However, in arteries with intact endothelial cells and in the presence of active muscle tone, NLA further raised or constricted the active muscle tone, presumably by blocking the spontaneous release of NO from the endothelial cells.[4,23,24] The NLA-induced constriction was reversed by L-arginine and L-citrulline to the original basal tone. The former is due to a direct increase in L-arginine substrate concentration for NOS, and the latter is believed due to an increase in L-arginine concentration indirectly following its conversion from L-citrulline.[4]

According to the urea cycle, L-citrulline is converted to L-arginine via AS pathway involving ASS and ASL.[5] It is, therefore, hypothesized that L-citrulline conversion to form L-arginine is mediated by AS pathway involving ASS and ASL in cerebral endothelial cells, like that in cerebral perivascular nerves.[8,9] In the present study, almost all endothelial cells are NOS-I, ASS-I, and ASL-I, and are positively stained for NADPHd reactivities. NADPHd has been shown to be a marker for NOS,[25–27] although some reports indicate that NADPHd staining in certain cells is not because of NOS.[28,29] Our studies, however, indicate that NOS-I and NADPHd fibers are completely coincident fibers in porcine cerebral arteries and veins,[8,9] indicating that in the pig cerebral arteries, NADPHd is a reliable marker for NOS. This is further supported by results of the present study in that almost all adventitial NOS-I, ASS-I, and ASL-I fibers are coincident with NADPHd fibers. It should be noted that antibodies against eNOS was used in the present study. The presence of eNOS-immunoreactivities in central nervous system has been reported.[30] Therefore, the complete coincidence of NOS-I, ASS-I, and ASL-I endothelial cells with NADPHd positive endothelial cells, suggesting that ASS, ASL, and NOS are colocalized in the same endothelial cells. This colocalization provides a mechanism for converting L-citrulline to L-arginine for synthesis of NO in the same endothelial cells.

Simultaneous localization of NOS-, ASS-, and ASL-immunoreactivities in the same preparations was not possible because the available primary antibodies to these three proteins were raised in the same species (rabbit). Therefore, the combined immunofluorescence and NADPHd histochemistry method was used as an alternative to investigate possible colocalization of NOS, ASS, and ASL.[8,9]

In summary, the present studies demonstrate that endothelial cells in cerebral arteries are equipped with enzymes necessary for recycling L-citrulline to L-arginine in order to synthesize NO. This conversion appears to be via AS pathway. Thus, cerebrovascular endothelial cells like cerebral perivascular nerves[3,4,8,9] can maintain adequate concentration of L-arginine, by recycling L-citrulline, in providing an *unlimited* supply of NO to regulate cerebral vascular tone (FIG. 1). The exact physiological and pathophysiologic significance of this L-citrulline recycle to form L-arginine remains to be determined.

ACKNOWLEDGMENTS

This work was supported by NIH grants HL47574 and HL27763, AHA/IHA, and SIU CRC-EAM.

REFERENCES

1. PALMER, R.M.J. & S. MONCADA. 1989. A novel citrulline-forming enzyme implicated in the formation of nitric oxide by vascular endothelial cells. Biochem. Biophy. Res. Commun. **158:** 348-352.
2. BREDT, D.S. & S.H. SNYDER. 1990 Isolation of nitric oxide synthetase, a calmodulin-requiring enzyme. Proc. Natl. Acad. Sci. U.S.A. **87:** 682–685.
3. CHEN, F.Y. & T.J.F. LEE. 1995 Arginine synthesis from citrulline in perivascular nerves of cerebral artery. J. Pharmacol. Exp. Ther. **273:** 895–901.
4. LEE, T.J.F., S. SARWINSKI, T. ISHINE, *et al.* 1996. Inhibition of cerebral neurogenic vasodilation by L-glutamine and nitric synthase inhibitors and its reversal by L-citrulline. J. Pharmacol. Exp. Ther. **276:** 353–358.
5. RATNER, S. 1973. Enzymes of arginine and urea synthesis. Adv. Enzym. **39:** 1–90.
6. KIMURA, T., J.G. YU, L. EDVINSSON & T.J.F. LEE. 1997. Cholinergic, nitric oxidergic innervation in cerebral arteries of the cat. Brain Res. **773**(1–2): 117–124.
7. EDVINSSON, L., T. ELSAS, N. SUZUKI, *et al.* 2001. Origin and co-localization of nitric oxide synthase, CGRP, PACAP, and VIP in the cerebral circulation of the rat. Microsc. Res. Tech. **53**(3): 221–228.
8. YU, J.G., W.E. O'BRIEN & T.J.F. LEE. 1997. Morphological evidence for L-citrulline conversion to L-arginine via the argininosuccinate pathway in porcine cerebral perivascular nerves. J. Cereb. Blood Flow Metab. **17:** 884–893.
9. YU, J.G., T. ISHINE, T. KIMURA, *et al.* 1997. L-Citrulline conversion to L-arginine in sphenopalatine ganglia and cerebral perivascular nerves in the pig. Am. J. Physiol. **273:** H2192–H2199.
10. FARACI, F.M. 1992. Regulation of the cerebral circulation by endothelium. Pharmacol. Ther. **56:** 1–22.
11. IADECOLA, C., D.A. PELLIGRINO, M.A. MOSKOWITZ & N.A. LASSEN. 1994. Nitric oxide synthase inhibition and cerebrovascular regulation. J. Cereb. Blood Flow Metab. **14:** 175–192.
12. AKOPOV, S., R. SERCOMBE & J. SEYLAZ. 1996. Cerebral vascular reactivity: role of endothelium/platelet/leukocyte interactions. Cerebrovasc. Brain Metab. Rev. **8:** 11–94.
13. PALMER, R.M.J., A.G. FERRIGE & S. MONCADA. 1987. Nitric oxide release accounts for the biological activity of endothelium-derived relaxing factor. Nature **327:** 524–526.
14. MONCADA, S., R.M.J. PALMER & E.A. HIGGS. 1991. Nitric oxide: physiology, pathology and pharmacology. Pharmacol. Rev. **43:** 109–142.
15. LEE, T.J.F., W.R. HUME, C. SU & J.A. BEVEN. 1978. Neurogenic vasodilation of cat cerebral arteries. Circ. Res. **42:** 535–542.
16. YU, J.G. & T.J.F. LEE. 1992. Choline acetyltransferase immunoreactivities in cerebral blood vessels. Chin. J. Pharmacol. Toxicol. **6:** 28–34.

17. O'BRIEN, W.E. 1979. Isolation and characterization of argininosuccinate synthetase from human liver. Biochemistry **18:** 5353–5356.
18. O'BRIEN, W.E. & R.H. BARR. 1981. Argininosuccinate lyase: purification and characterization from human liver. Biochemistry **20:** 2056–2060.
19. ANT-RAMOS, L.R., W.E. O'BRIEN & S.R. VINCENT. 1992. Immunochemical localization of argininosuccinate synthetase in the rat brain in relation to nitric oxide synthase-containing neurons. Neuroscience **51:** 773–789.
20. LEE, T.J.F., J.G. YU, W.E. O'BRIEN, *et al.* 1997. L-Citrulline conversion to L-arginine in porcine cerebral arteries and veins. J. Cereb. Blood Flow Metab. **17**(Suppl. 1): S347.
21. BUNIATIAN, H.CH. & M.A. DAVTIAN. 1966. Urea synthesis in brain. J. Neurochem. **13:** 743–753.
22. HECKER, M., W.C. SESSA, H.J. HARRIS, *et al.* 1990. The metabolism of L-arginine and its significance for biosynthesis of endothelium-derived relaxing factor: cultured endothelial cells recycle L-citrulline to L-arginine. Proc. Natl. Acad. Sci. U.S.A. **87:** 8612–8616.
23. WU, W. & C.J. MEININGER. 1993. Regulation of L-arginine synthesis from L-citrulline by L-glutamine in endothelial cells. Am. J. Physiol. **265:** H1965–H1971.
24. LEE, T.J.F. & S. SARWINSKI. 1991. Nitric oxidergic neurogenic vasodilation in cerebral arteries of the pig. Blood Vessels **28:** 407–412.
25. HOPE, B.T., G.J. MICHAEL, K.M. KNIGGE & S.R. VINCENT. 1991. Neuronal NADPH diaphorase is nitric oxide synthase. Proc. Natl. Acad. Sci. U.S.A. **88:** 2811–2814.
26. DAWSON, T.M., D.S. BREDT, M. FOTUHI, *et al.* 1991. Nitric oxide synthase and neuronal NADPH diaphorase are identical in brain and peripheral tissues. Proc. Natl. Acad. Sci. U.S.A. **88:** 7797–7801.
27. SCHMIDT, H.H.H.W., G.D. GAGNE, M. NAKANE, *et al.* 1992. Mapping of neural nitric oxide synthase in the rat suggests frequent co-localization with NADPH diaphorase but not with soluble guanylyl cyclases, and noval paraneural functions for nitrinergic signal transduction. J. Histochem. Cytochem. **40:** 1439–1456.
28. KISHIMOTO, J., E.B. KEVERNE, J. HARDWICK & P.C. EMSON. 1993. Localization of nitric oxide synthase in mouse olfactory and vermeronasal system: a histochemical, immunological and *in situ* hybridization. Eur. J. Neurosci. **5:** 1684–1694.
29. SANATER, R.M. & D. SYMONS. 1993. Distribution of NADPH-diaphorase activity in rat paravertebral, and pelvic sympathetic ganglia. Cell Tissue Res. **317:** 115–121.
30. DINERMAN, J.L., T.M. DAWSON, M.J. SCHELL, *et al.* 1994. Endothelial nitric oxide synthase localized to hippocampal pyramidal cells: implications for synaptic plasticity. Proc. Natl. Acad. Sci. U.S.A. **91:** 4214–4218.

Glial-Neuronal Transfer of Arginine and *S*-Nitrosothiols in Nitric Oxide Transmission

K.Q. DO,[a] G. GRIMA,[c] B. BENZ,[a] AND T.E. SALT[b]

[a]*Centre de Recherche en Neurosciences Psychiatriques, University of Lausanne, Prilly-Lausanne, Switzerland*

[b]*Department of Visual Science, Institute of Ophthalmology, University College London, London, United Kingdom*

ABSTRACT: The arginine-nitric oxide (Arg-NO) and the *S*-nitrosothiols systems, two less well-studied aspects of NO transmission in the central nervous system, are reviewed. A growing body of evidence suggested that they play a crucial role in NO synthesis and activity. L-Arginine, the NO precursor, is predominantly localized in glia. Together with *in vitro* and *in vivo* results of arginine release, this suggests a transfer of arginine from glia to neurons in order to supply NO synthase with its substrate. NO biosynthesis may thus involve the co-occurrence of the glial-neuronal transfer of arginine and of NOS activation. The arginine availability may shed light on the dual, beneficial and toxic effects of NO. At low arginine concentrations, neuronal NO synthase generates NO and superoxide, favouring the production of the toxin peroxynitrite. NMDA-induced excitotoxicity in neuronal cells is dependent on arginine availability and glia may play a neuroprotective role by supplying arginine. The reversible *S*-nitros(yl)ation of thiol containing molecules may represent an important cellular signal transduction mechanism, probably comparable to phosphorylation. *S*-nitrosothiols, in particular through the presence and release of *S*-nitroso-cysteinylglycine in sensory thalamus, may act as a local buffering system in NO transmission. This may represent a novel specific facilitating mechanism in order to enhance transmission of persistent stimuli.

KEYWORDS: glial-neuronal transfer; arginine; *S*-nitrosothiols; nitric oxide transmission

THE IMPORTANCE OF THE GLIAL-NEURONAL TRANSFER OF ARGININE IN PHYSIOLOGY AND PATHOLOGY

The nitric oxide (NO) precursor L-arginine (Arg) is a semi-essential amino acid. Although it can be synthesized from glutamate, its availability is limited during development.[1] Arg has attracted much interest in recent years in the context of NO biosynhesis. This has helped to clarify its role as a precursor of NO, especially in the central nervous system (CNS), where complex cellular interactions are involved in the biochemistry of Arg.

Address for correspondence: Kim Q. Do, Ph.D., Centre Universitaire de Recherche en Neurosciences Psychiatriques, Site de Cery, CH-1008 Prilly-Lausanne, Switzerland. Voice: +41-21-643 65 60; fax: +41-21-643 65 62.
Kim.Do@inst.hospvd.ch

Metabolism and Transport of Arg in the Central Nervous System

The main metabolic pathway of Arg is the urea cycle (ornithine cycle), in which arginine is converted to urea and ornithine (see Ref. 2 for a review). This cycle is important for the disposal of the toxic metabolite ammonia. It involves five enzymes: arginase cleaves urea from Arg, whereby ornithine is formed. Ornithine is then further metabolized to citrulline by ornithine transcarbamylase. In the second part of the cycle, Arg is resynthesized by argininosuccinate synthetase (ASS) and argininosuccinate lyase (ASL). ASS catalyzes the formation of argininosuccinate from citrulline and aspartate, whereas ASL cleaves arginosuccinate to yield fumarate and Arg. In addition, Arg is used for the synthesis of proteins and peptides, creatine, agmatine and the synthesis of polyamines. Although the CNS contains arginase, ASL, and ASS, it lacks ornithine transcarbamylase.[2–5] In consequence, citrulline, whose presence in the CNS has been known for a long time, derives mainly from NO synthesis, via NO synthase (NOS) activity. The expression of ASS and ASL indicates that a partial urea cycle exists in the brain, with the most likely function as an Arg-recycling pathway. After NO production, citrulline is thus synthesized back to Arg via argininosuccinate.

Several uptake systems for Arg have been identified. As a member of the cationic amino acids, Arg is mainly transported by the y^+ system (for a review, see Ref. 6). Other amino acids taken up by this system include lysine and ornithine. y^+ is Na^+- and pH independent, but is inhibited by high extracellular K^+. Moreover, y^+ can mediate both the influx and efflux of cationic amino acids, and these fluxes are trans-stimulated. In the CNS, neurons as well as glial cells express the y^+ uptake system.[7,8] Lysine, ornithine, and several Arg analogues, such as *N*-monomethyl-L-arginine, are competitive inhibitors of this uptake. Following the molecular cloning and characterization of the Arg transporter, y^+ has been renamed to CAT (for cationic amino acid transporter). Four such transporters have been cloned to date: CAT-1, CAT-2, CAT-2a, and CAT-3.[9–11] CAT-1 is the previously described REC-a locus for murine ecotropic leukemia virus susceptibility.[12] CAT-1, CAT-2, and CAT-3 exhibit characteristics originally assigned to the y^+ system. The three transporter isoforms have a high affinity for Arg (micromolar K_m values). CAT-1 and CAT-2 show a relatively ubiquitous distribution in many organs, including the brain, whereas CAT-3 is preferentially expressed in the CNS.[11] CAT-2a originates from the same ene as CAT-2 by alternative splicing. In contrast to CAT-2, CAT-2a has a low affinity for Arg (millimolar K_m values) and is specifically expressed in the liver.[9,13]

Glial-Neuronal Transfer of Arg

The CNS is able to synthesize Arg from citrulline and is, therefore, potentially independent from the periphery in this respect. At the cellular level, present data favour an intercellular shuttling of Arg and of its metabolites. Immunohistochemical studies using antibodies against ASS, ASL, and nNOS have shown a differential cellular localization of these enzymes in many regions of the brain.[14–16] Neuronal NOS (nNOS) and ASS were rarely colocalized. ASS was often expressed in neurons that are in close proximity to the nNOS-containing cells. In the cerebellum, for example, nNOS was present in granule cells and basket cells, whereas ASS immunoreactivity was restricted to another type of neuron, possibly the Lugaro cells.[16] Moreover, ASL staining was most prominent in different cells, the Golgi epithelial or Bergman glial cells.[14] L-Arg itself is found predominantly in glial cells.[17] These observations, combined with the existence of specific transport system for Arg in neurons[7] and glial cells[8] suggest a considerable shuttling of intermediates between different cell types, which carries with it the possibility of a signaling role for these intermediates. This possibility is supported by the following review of the data.

The release of endogenous Arg has been demonstrated in several *in vitro* and *in vivo* studies.[18–20] In particular, the selective electrical stimulation of the parallel fiber or mossy fiber pathways caused a release of endogenous Arg from cerebellar slices.[18] Using push–pull perfusion and recording methods in the rat ventrobasal thalamus *in vivo,* we observed the release of endogenous Arg following physiological stimulation of somatosensory afferents.[20] Moreover, the application of L-Arg onto thalamic relay neurons facilitated sensory synaptic transmission, via synthesis of NO.[20] Interestingly, studies using antibodies against Arg indicate glial cells as the source for Arg.[17,21,22] This has been described in various brain regions, such as the cerebellum, cortex, thalamus, and pituitary. In a double labeling study in the ventroposterior nucleus of the thalamus, Kharazia and coworkers found that Arg is especially concentrated in glial cells that are in close proximity to NOS expressing neural elements.[23] In particular, they observed that NADPHd-positive fibers are often apposed to Arg-positive astrocytes and oligodendrocytes. These results on the *in vitro* and *in vivo* Arg release, together with its glial localization, suggest a transfer of Arg from glial cells to neurons in order to supply NO synthase with its substrate. Similarly, the release of Arg from astrocytes in culture has been demonstrated in various studies.[24–26]

Under physiological conditions, the release of Arg from glia is mediated by the activation of glial glutamate receptors,[25,27] whereas the toxic derivative of NO and superoxide, peroxynitrite, has also been found to induce this release.[26] By taking advantage of the finding that glutamate evokes the release of Arg via stimulation of glial ionotropic non-NMDA receptors, but not NMDA receptors,[25] we investigated the role of Arg availability on the NMDA-induced NO synthesis as measured by cyclic GMP accumulation in immature rat brain slices. The effect of NMDA was blocked by the NOS inhibitor, N^G-nitro-L-Arg, and by the antagonist of ionotropic non-NMDA receptors, 6-cyano-7-nitroquinoxaline-2,3-dione (CNQX). This inhibition was overcome by the presence of exogenously applied Arg. CNQX also blocked the NMDA-evoked release of [^{3}H]-Arg from cerebellar slices. Moreover, the Arg uptake inhibitor L-lysine reduced the cyclic GMP response to NMDA significantly. Therefore, the extracellular Arg availability, which depends on the activation of ionotropic non-NMDA receptors, determines the rate of NO biosynthesis by the neuronal NOS.[27] Together with the reported release of Arg from glial cells upon activation of glial ionotropic non-NMDA receptors,[25] and the predominant glial localization of arginine, these data demonstrate that glial cells supply the NO precursor, Arg, to NOS containing neurons. They provide evidence for an essential role of the arginine transfer from glial cells to neurons for the biosynthesis of NO. Moreover, NO formation may depend on the co-occurrence of this Arg transfer and of the NOS activation (see FIGURE 1). As described below, the glial-neuronal transfer of Arg is likely to play an important role in physiological and pathological processes involving NO synthesis.

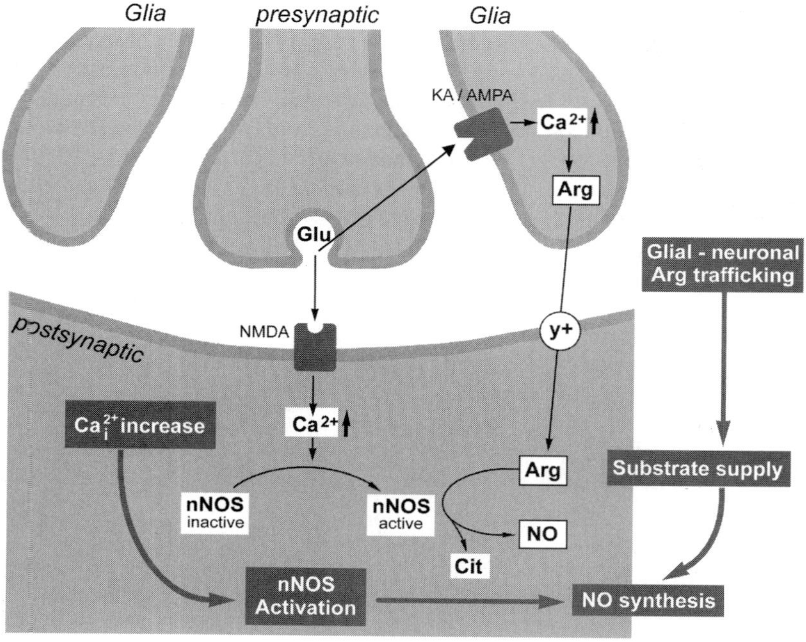

FIGURE 1. Schematic representation of the proposed model of co-occurrence of the glial-neuronal arginine (Arg) transfer and nNOS (nitric oxide synthase) activation. Glutamate (Glu) evokes the release of Arg from glia via stimulation of glial inotropic non-NMDA receptors. Arg is transported to NOS containing neurons via the y^+ uptake system. Our data[27] show that Arg availability, which depends on the activation of ionotropic non-NMDA receptors, determines the rate of NO biosynthesis by neuronal NOS. Moreover, Glu, through stimulation of postsynaptic NMDA receptors and Ca^{2+} increase, can activate neuronal NOS. NO formation may, thus, depend on the coincident occurrence of the glial-neuronal transfer of Arg and NOS activation.

Involvement of the Arg Glial-Neuronal Transfer in Physiological Processes

Because of the lack of specific modulators of the glial-neuronal Arg transfer *in vivo,* data on the role of this process are scarce. However, studies on the role of Arg availability in NO-dependent processes point to an involvement of glial cells as a source for Arg. For example, we showed that iontophoretic application of Arg in the ventrobasal thalamus *in vivo* exerts a facilitatory action on the synaptic response via the synthesis of NO.[20] Kara and Friedlander[28] demonstrated that Arg application into the visual cortex modulates visually evoked responses through the generation of NO. These findings showed that synaptic transmission is modulated by the presence of Arg. In view of the evidence presented above indicating that Arg is released by glial cells and taken up by nNOS containing neurons, where it serves for the synthesis of NO, it is very likely that this process plays an important role in many physiological aspects of neurotransmission involving NO. In particular, in the thalamus, the release and glial-neuronal transfer of Arg[20] may represent the activity of a local modulatory system, which enables rapid enhancement of synaptic transmission upon stimulation of sensory afferents: in effect a local positive feedback system.

An intercellular Arg transfer might not only be important for the NO synthesis by the nNOS. Expression of iNOS is linked to the simultaneous induction of the Arg recycling enzymes argininosuccinate synthetase and argininosuccinate lyase.[29–31] Although iNOS is predominantly expressed in microglial cells in mixed glial cell cultures, argininosuccinate synthetase seems to be induced mainly in astrocytes.[32] A transfer of Arg and/or of the recycling intermediates from astrocytes to microglial cells is also likely to occur under conditions of NO synthesis by the iNOS.

Involvement of the Arg Glial-Neuronal Transfer in Pathological Processes

Since the discovery that NMDA receptor activation and the subsequent increase in intracellular Ca^{2+} leads to the stimulation of NO synthesis, much attention has focused on the possible involvement of NO in neurotoxicity. Indeed, *in vitro* and *in vivo* data indicate that NO or a related molecule mediates excitotoxicity.[33–35] However, potent antioxidative and neuroprotective properties have been also reported for NO (for a review, see Ref. 36). In this section, we discuss the involvement of Arg in these dual effects of NO, both beneficial and toxic.

Excitotoxic neuronal cell death is characterized by an overactivation of glutamate receptors, in particular the NMDA subtype, and the stimulation of nNOS. At low L-Arg concentrations, nNOS generates NO and superoxide ($O_2^{-\cdot}$), favoring the production of the toxin peroxynitrite ($ONOO^-$). We have shown that the toxicity induced by NMDA application in neuronal cultures depends on the availability of Arg.[37] Indeed, we observed that NMDA application for five minutes in the absence of added L-Arg induces neuronal cell death, and that the presence of L-Arg during NMDA application prevents cell loss by blocking $O_2^{-\cdot}$ and $ONOO^-$ formation and by inhibiting mitochondrial depolarization. Because Arg is transferred from glial cells to neurons upon activation of glial glutamate receptors, we hypothesized that glial cells play an important modulator role in excitotoxicity by releasing L-Arg. Furthermore, L-Arg derived from astrocyte supernatant inhibits NMDA-induced toxic radical formation, mitochondrial dysfunction, and cell death.[37] These results suggest that glial cells may protect neurons from excitotoxicity by supplying L-Arg. This potential neuroprotective mechanism may lead to an alternative approach for the treatment of neurodegenerative diseases involving excitotoxic processes, such as ischemia. Indeed, under ischemic conditions, the supply of Arg from the blood would be reduced, and NO synthesis in the endothelial cells, as well as in neurons, would depend entirely on the glial Arg pool. The accumulation of glutamate is likely to mobilize this Arg and thus, to increase NO synthesis, which in turn might improve vasodilation. In addition, NO^+ could prevent further excitotoxic damage by inhibiting the NMDA receptor activation.[38,39] However, a longer lasting and more widespread insult would cause a maximal mobilization of Arg, followed by a high rate of NO synthesis. This might explain the observation that the level of NO is maximal 5–15 min after the onset of focal cerebral ischemia in rats.[40–42] The subsequent decline of NO to baseline within the following 60 min[40] might reflect the limited substrate availability. Since the glutamate level remains elevated for a longer period,[43] activation of the nNOS would be extended. Under these conditions, NO and superoxide might be generated, eventually leading to the formation of peroxynitrite. At that time, further processes, among which some are independent of NO, could be initiated. Peroxynitrite has been shown to block glutamate uptake,[44] to nitrate or hydroxylate proteins (reviewed in Ref. 45), to induce lipid peroxydation,[46] to block the mitochondrial respiratory chain irreversibly (reviewed in Ref. 47), to deplete ATP stores through the activation of the poly(ADP-ribose) synthetase,[48] and to promote apoptotic cell death.[49] Therefore, under pronounced ischemic conditions, the brain may not be able to overcome the detrimental processes, and a fatal chain reaction would be set in motion. In several studies, the application of Arg or an NO donor for 10–60 min immediately after middle cerebral artery occlusion reduced the infarct size,[50–53] The effect of Arg is known to be sustained for at least 1.5 h after cessation of its infusion.[51] Interestingly, Arg is released from astrocytes not only by glutamate receptor activation[25] or by high K^+,[24] but also by peroxynitrite.[26] This suggests that even after this toxic radical is formed, glial derived Arg may be protective, at least for a certain period. Altogether, these findings provide evidence for a protective role of glial-derived Arg in ischemia, and possibly also in other neuropathologies involving NOS activation. In addition, this model could pave the way to new potential treatment of these pathologies.

S-NITROSOTHIOLS IN CENTRAL NERVOUS SYSTEM: NOVEL MODE OF NO TRANSMISSION MEDIATED VIA *S*-NITROSO-CYSTEINYL GLYCINE

Involvement of S-*Nitrosothiols in Biological Processes*

Although the biosynthesis of NO is well established, one of the key issues that remains to be solved is if NO participates in the biological responses immediately after biosynthetic generation or if there are intermediate stages during which NO itself is trapped, transported, and released at the presumed site of NO action to exert its functions. Given the extreme lability and facile inactivation of NO by heme iron, nonheme iron, superoxide anion, oxygen and other biochemical components, it has been suggested that some intermediate molecular species, such as *S*-nitrosothiols (RSNOs), serve to stabilize NO and/or transfer it from site of origin to target effector sites.[54] The predominant bound forms of NO in human plasma are *S*-nitrosothiols, mostly *S*-nitrosoproteins.[55] RSNOs have been proposed to be involved in a variety of biological functions (reviewed in Ref. 54). These include providing a reserve pool for NO,[55] key intermediates in the action of certain nitro vasodilatating compounds, such as sodium nitroprusside and organic nitrites and nitrates,[56,57] inhibition of platelet aggregation,[58] and bronchodilatation.[59,60] Stamler and coworkers have proposed that a *S*-nitroso derivative of hemoglobin may regulate vascular tone and tissue oxygenation via the transfer of NO between different pools of thiols.[61] The reversible *S*-nitros(yl)ation of thiol-containing molecules may represent an important cellular signal transduction mechanism,[54] implicated in control of both cellular and system functions, including respiration, neuronal plasticity, vasorelaxation, skeletal muscle contractility, immunity, oxidative stress, and apoptosis.[54,61–67] In the CNS, *S*-nitrosocysteine (CysNO), when injected into the nucleus tractus solitarii, produced hypotension and bradycardia through the activation of stereoselective recognition sites.[68] CysNO, through incorporation into rat hippocampus via the L-type-like aminoacid transporter, stimulated noradrenaline release in a cyclic GMP-independent manner.[69] Chiueh and collaborators reported that *S*-nitrosoglutathione (GSNO) protects brain dopamine neurons against oxidative damage *in vitro* and *in vivo*.[70–72] In very recent reports, Gozal and collaborators reported that RSNOs play a central role in the regulation of breathing.[73]

Endogenous Presence of S-*Nitrosoglutathione*

In the rat brain we investigated if the *S*-nitrosothiol of glutathione (GSH), which is the predominant low molecular weight thiol in CNS, is present endogenously.[74] The HPLC analysis of cerebellar extract from [^{35}S]-cysteine prelabelled slices suggested that *S*-nitrosogltathione (GSNO) was indeed present in rat brain. In order to detect endogenous GSNO, a methodology based on liquid chromatography-mass spectrometry was developed. As well as, an unequivocal identification of the endogenous GSNO, this method also permitted its precise quantification using ^{15}N-labelled GSNO ([^{15}N]-GSNO) as internal standard. GSNO level in adult cerebellum amounts to 15.4 ± 1.4 pmol/mg protein.[74] In order to further study the metabolism and biological roles of GSNO, we investigated the possible existence of other endogenous RSNOs in CNS. Indeed, the diverse biological functions of various RSNOs are based on their unique chemistry. Reports on the physiological chemistry of RSNOs indicate that they can: (1) undergo a transnitrosation reaction;[75–77] (2) release NO in the presence of cuprous ion,[78,79] ascorbate, or thiols;[80] and (3) serve as a possible source of nitrosonium (NO^+) or nitroxyl (NO^-) ions.[81,82] These reactions depend heavily on the structures of RSNOs and their environment. It is, therefore, important to know precisely the nature of endogenous RSNOs and their compartmentalization *in vivo.*

Release of S-*Nitrosocyteinylglycine: A Novel Mode of NO Transmission?*

We established a specific and sensitive method for RSNO analysis, based on HPLC analysis followed by derivatization with 2,3-diaminonaphthalene,[83] in biological samples.[84] It allows a specific detection of low molecular weight *S*-nitrosothiols at the nanomolar level in brain tissue extracts. We identified the presence in rat brain of a novel endogenous RSNO, *S*-nitrosocyteinylglycine (CysNOGly).[84] NMDA induced in brain slices an increase in CsNOGly levels, which was inhibited by the NO synthase inhibitor *N*-nitro-arginine. This increase was also blocked by acivicin, an inhibitor of the γ-glutamyl transpeptidase, the glutathione degradation enzyme, which cleaves cysteinylglycine (Cys-Gly) from glutathione. These pharmacological studies suggested that the formation of CysNOGly *in vivo* was closely related both to the synthesis of NO and to glutathione metabolism. CysNOGly can stimulate guanylate cyclase *in vitro* more potently than GSNO.[84] In addition, in *in vivo* electrophysiologic studies, CysNOGly can enhance thalamic neuronal responses to sensory stimuli and ionotropic glutamate receptor activation in a manner reminiscent of NO donors and 8Br-cyclic-GMP.[84]

In terms of thalamic function, it is known that NO synthase is located in the terminals of cholinergic afferents ascending to the thalamus from the brainstem.[21,23,85] We have previously suggested that this NO synthase can be activated by stimulation of sensory afferents.[20] Taken together, these results suggest that CysNOGly plays a role in synaptic transmission in ventrobasal thalamus. The NO formed and released into the extracellular *milieu* following sensory afferent stimulation may be buffered prior to acting upon postsynaptic guanylate cyclase, and CysNOGly may act as an endogenous buffering system regulating the bioavailability of NO before releasing NO to stimulate guanylate cyclase. The NO delivery may be regulated by site-specific mechanisms, via reactions with transition metal ions,[86] or other reducing systems, such as thioredoxin or superoxide anion.[79] This mechanism is consistent with recent *in vitro* results, using spin-trapping electron paramagnetic resonance, showing that NO is readily trapped by thiols in less than 10 min and approximately 70–90% is subsequently released.[87] This mechanism could serve to restrict the locus of action of NO and so increase the local availability of NO for delivery to its target. Such a mechanism might also result in a temporal smoothing or prolongation of NO activation of guanylate cyclase. It is known that NO donor application and guanylate cyclase activation in thalamic neurones leads to a change in neuronal responses that favors transmission of sensory information as seen throughout the sensory thalamic nuclei under conditions of wakefulness or arousal.[88,89] Because CysNOGly has similar actions, it is conceivable that it too plays a role in regulating transmission through the thalamus. It is, thus, possible to speculate that under continued production of NO following prolonged sensory stimulation, CysNOGly formation occurs, serving as a local buffering system. This might act as a novel specific local facilitating mechanism in order to enhance transmission of persistent sensory stimuli.

CONCLUSION

We have focused and reviewed two less well-studied aspects of NO transmission in CNS, the Arg-NO and the *S*-nitrosothiols systems. A growing body of evidence suggested that they play a crucial role in NO synthesis and activity. NO biosynthesis may involve the coincident occurrence of the glial-neuronal transfer of Arg and of NOS activation. The Arg availability may shed light on the dual, both beneficial and toxic effects of NO. The reversible *S*-nitros(yl)ation of thiol containing molecules may represent an important cellular signal transduction mechanism, probably comparable to phosphorylation. *S*-nitrosothiols, in particular through the presence and release of CysNOGly, may act as a local buffering system in NO transmission, and represent a novel specific facilitating mechanism in order to enhance transmission of persistent stimuli.

REFERENCES

1. Wu, G. & S.M. Morris, Jr. 1998. Arginine metabolism: nitric oxide and beyond. Biochem. J. **336**(1): 1–17.
2. Ratner, S. 1973. Enzymes of arginine and urea synthesis. Adv. Enzymol. Relat. Areas Mol. Biol. **39:** 1–90.
3. Ratner, S. & K. Murakami-Murofushi. 1980. A new radiochemical assay for argininosuccinase with purified [^{14}C]argininosuccinate. Anal. Biochem. **106:** 134–147.

4. KATO, H., I. OYAMADA, M. MIZUTANI-FUNAHASHI & H. NAKAGAWA. 1976. New radioisotopic assays of argininosuccinate synthetase and argininosuccinase. J. Biochem. (Tokyo) **79:** 945–953.
5. YU, Y., K. TERADA, A. NAGASAKI, *et al.* 1995. Preparation of recombinant argininosuccinate synthetase and argininosuccinate lyase: expression of the enzymes in rat tissues. J. Biochem. (Tokyo) **117:** 952–957.
6. WHITE, M.F. 1985. The transport of cationic amino acids across the plasma membrane of mammalian cells. Biochim. Biophys. Acta **822:** 355–374.
7. WESTERGAARD, N., P.M. BEART & A. SCHOUSBOE. 1993. Transport of L-[^{3}H]arginine in cultured neurons: characteristics and inhibition by nitric oxide synthase inhbitors. J. Neurochem. **61:** 364–367.
8. SCHMIDLIN, A. & H. WIESINGER. 1994. Transport of L-arginine in cultured glial cells. Glia **11:** 262–268.
9. KIM, J.W., E.I. CLOSS, L.M. ALBRITTON & J.M. CUNNINGHAM. 1991. Transport of cationic amino acids by the mouse ecotropic retrovirus receptor. Nature **352:** 725–728.
10. WANG, H., M.P. KAVANAUGH, R.A. NORTH & D. KABAT. 1991. Cell-surface receptor for ecotropic murine retroviruses is a basic amino-acid transporter. Nature **352:** 729–731.
11. ITO, K. & M. GROUDINE. 1997. A new member of the cationic amino acid transporter family is preferentially expressed in adult mouse brain. J. Biol. Chem. **272:** 26780–26786.
12. ALBRITTON, L.M., L. TSENG, D. SCADDEN & J.M. CUNNINGHAM. 1989. A putative murine ecotropic retrovirus receptor gene encodes a multiple membrane-spanning protein and confers susceptibility to virus infection. Cell **57:** 659–666.
13. CLOSS, E.I., L.M. ALBRITTON, J.W. KIM & J.M. CUNNINGHAM. 1993. Identification of a low affinity, high capacity transporter of cationic amino acids in mouse liver. J. Biol. Chem. **268:** 7538–7544.
14. NAKAMURA, H., T. SAHEKI & S. NAKAGAWA. 1990. Differential cellular localization of enzymes of L-arginine metabolism in the rat brain. Brain Res. **530:** 108–112.
15. NAKAMURA, H., T. SAHEKI, H. ICHIKI, *et al.* 1991. Immunocytochemical localization of argininosuccinate synthetase in the rat brain. J. Comp. Neurol. **312:** 652–679.
16. ARNT-RAMOS, L.R., W.E. O'BRIEN & S.R. VINCENT. 1992. Immunohistochemical localization of argininosuccinate synthetase in the rat brain in relation to nitric oxide synthase-containing neurons. Neuroscience **51:** 773–789.
17. AOKI, E., R. SEMBA, K. MIKOSHIBA & S. KASHIWAMATA. 1991. Predominant localization in glial cells of free L-arginine. Immunocytochemical evidence. Brain Res. **547:** 190–192.
18. HANSEL, C., A. BATCHELOR, M. CUÉNOD, *et al.* 1992. Delayed increase of extrcellular arginine, the nitric oxide precursor, following electrical white matter stimulation in rat cerebellar slices. Neurosci. Lett. **142:** 211–214.
19. KENDRICK, K.M., R.C. DE LA GUEVARA-GUZMAN, J. CHRISTENSEN, *et al.* 1996. NMDA and kainate-evoked release of nitric oxide and classical transmitters in the rat striatum: *in vivo* evidence that nitric oxide may play a neuroprotective role. Eur. J. Neurosci. **8:** 2619–2634.
20. DO, K.Q., K.E. BINNS & T.E. SALT. 1994. Release of the nitric oxide precursor, arginine, from the thalamus upon sensory afferent stimulation, and its effect on thalamic neurons *in vivo*. Neuroscience **60:** 581–586.
21. KHARAZIA, V.N., P. PETRUSZ, K. USUNOFF, *et al.* 1997. Arginine and NADPH diaphorase in the rat ventroposterior thalamic nucleus. Brain Res. **744:** 151–155.
22. POW, D.V. 1994. Immunocytochemical evidence for a glial localisation of arginine, and a neuronal localisation of citrulline in the rat neurohypophysis: implications for nitrergic transmission. Neurosci. Lett. **181:** 141–144.
23. USUNOFF, K.G., V.N. KHARAZIA, J.G. VALTSCHANOFF, *et al.* 1999. Nitric oxide synthase-containing projections to the ventrobasal thalamus in the rat. Anat. Embryol. (Berl.) **200:** 265–281.
24. COSSENZA, M. & D.C. PAES. 2000. L-Arginine uptake and release by cultured avian retinal cells: differential cellular localization in relation to nitric oxide synthase. J. Neurochem. **74:** 1885–1894.

25. GRIMA, G., B. BENZ & K.Q. DO. 1997. Glutamate-induced release of the nitric oxide precursor, arginine, from glial cells. Eur. J. Neurosci. **9:** 2248–2258.
26. VEGA-AGAPITO, V., A. ALMEIDA, S.J. HEALES, *et al.* 1999. Peroxynitrite anion stimulates arginine release from cultured rat astrocytes. J. Neurochem. **73:** 1446–1452.
27. GRIMA, G., M. CUENOD, S. PFEIFFER, *et al.* 1998. Arginine availability controls the *N*-methyl-D-aspartate-induced nitric oxide synthesis: involvement of a glial-neuronal arginine transfer. J. Neurochem. **71:** 2139–2144.
28. KARA, P. & M.J. FRIEDLANDER. 1999. Arginine analogs modify signal detection by neurons in the visual cortex. J. Neurosci. **19:** 5528–5548.
29. HATTORI, Y., E.B. CAMPBELL & S.S. GROSS. 1994. Argininosuccinate synthetase mRNA and activity are induced by immunostimulants in vascular smooth muscle. Role in the regeneration or arginine for nitric oxide synthesis. J. Biol. Chem. **269:** 9405–9408.
30. NUSSLER, A.K., T.R. BILLIAR, Z.Z. LIU & S.M. MORRIS, JR. 1994. Coinduction of nitric oxide synthase and argininosuccinate synthetase in a murine macrophage cell line. Implications for regulation of nitric oxide production. J. Biol. Chem. **269:** 1257–1261.
31. NAGASAKI, A., T. GOTOH, M. TAKEYA, *et al.* 1996. Coinduction of nitric oxide synthase, argininosuccinate synthetase, and argininosuccinate lyase in lipopolysaccharide-treated rats. RNA blot, immunoblot, and immunohistochemical analyses. J. Biol. Chem. **271:** 2658–2662.
32. SCHMIDLIN, A. & H. WIESINGER. 1998. Argininosuccinate synthetase: localization in astrocytes and role in the production of glial nitric oxide. Glia **24:** 428–436.
33. DAWSON, D.A. 1994. Nitric oxide and focal cerebral ischemia: multiplicity of actions and diverse outcome. [Review] Cerebrovasc. Brain Metab. Rev. **6:** 299–324.
34. DAWSON, T.M., V.L. DAWSON & S.H. SNYDER. 1994. Molecular mechanisms of nitric oxide actions in the brain. Ann. N.Y. Acad. Sci. **738:** 76–85.
35. IADECOLA, C. 1997. Bright and dark sides of nitric oxide in ischemic brain injury. Trend. Neurosci. **20:** 132–139.
36. CHIUEH, C.C. 1999. Neuroprotective properties of nitric oxide. Ann. N.Y. Acad. Sci. **890:** 301–311.
37. GRIMA, G., B. BENZ & K.Q. DO. 2001. Glial-derived arginine, the nitric oxide precursor, protects neurons from NMDA-induced excitotoxicity, Europ. J. Neurosci. **25:** 1762–1770.
38. LEI, S.Z., Z.H. PAN, S.K. AGGARWAL, *et al.* 1992. Effect of nitric oxide production on the redox modulatory site of the NMDA receptor-channel complex. Neuron **8:** 1087–1099.
39. MANZONI, O., L. PREZEAU, P. MARIN, *et al.* 1992. Nitric oxide-induced blockade of NMDA receptors. Neuron **8:** 653–662.
40. KADER, A., V.I. FRAZZINI, R.A. SOLOMON & R.R. TRIFILETTI. 1993. Nitric oxide production during focal cerebral ischemia in rats. Stroke **24:** 1709–1716.
41. MALINSKI, T., F. BAILEY, Z.G. ZHANG & M. CHOPP. 1993. Nitric oxide measured by a porphyrinic microsensor in rat brain after transient middle cerebral artery occlusion. J. Cereb. Blood Flow Metab. **13:** 355–358.
42. TOMINAGA, T., S. SATO, T. OHNISHI & S.T. OHNISHI. 1993. Potentiation of nitric oxide formation following bilateral carotid occlusion and focal cerebral ischemia in the rat: *in vivo* detection of the nitric oxide radical by electron paramagnetic resonance spin trapping. Brain Res. **614:** 342–346.
43. BENVENISTE, H. 1991. The excitotoxin hypothesis in relation to cerebral ischemia. Cerebrovasc. Brain Metab. Rev. **3:** 213–245.
44. TROTTI, D., D. ROSSI, O. GJESDAL, *et al.* 1996. Peroxynitrite inhibits glutamate transporter subtypes. J. Biol. Chem. **271:** 5976–5979.
45. BECKMAN, J.S. & W.H. KOPPENOL. 1996. Nitric oxide, superoxide, and peroxynitrite: the good, the bad, and ugly. Am. J. Physiol. **271:** C1424–C1437.
46. RADI, R., J.S. BECKMAN, K.M. BUSH & B.A. FREEMAN. 1991. Peroxynitrite-induced membrane lipid peroxidation: the cytotoxic potential of superoxide and nitric oxide. Arch. Biochem. Biophys. **288:** 481–487.

47. BOLANOS, J.P., A. ALMEIDA, V. STEWART, *et al.* 1997. Nitric oxide-mediated mitochondrial damage in the brain: mechanisms and implications for neurodegenerative diseases. J. Neurochem. **68:** 2227–2240.
48. ZHANG, J., V.L. DAWSON, T.M. DAWSON & S.H. SNYDER. 1994. Nitric oxide activation of poly(ADP-ribose) synthetase in neurotoxicity. Science **263:** 687–689.
49. BONFOCO, E., D. KRAINC, M. ANKARCRONA, *et al.* 1995. Apoptosis and necrosis: two distinct events induced, respectively, by mild and intense insults with *N*-methyl-D-aspartate or nitric oxide/superoxide in cortical cell cultures. Proc. Natl. Acad. Sci. U.S.A. **92:** 7162–7166.
50. MORIKAWA, E., M.A. MOSKOWITZ, Z. HUANG, *et al.* 1994. L-Arginine infusion promotes nitric oxide-dependent vasodilation, increases regional cerebral blood flow, and reduces infarction volume in the rat. Stroke **25:** 429–435.
51. MORIKAWA, E., Z. HUANG & M.A. MOSKOWITZ. 1992. L-Arginine decreases infarct size caused by middle cerebral arterial occlusion in SHR. Am. J. Physiol. **263:** H1632–H1635.
52. MORIKAWA, E., S. ROSENBLATT & M.A. MOSKOWITZ. 1992. L-Arginine dilates rat pial arterioles by nitric oxide-dependent mechanisms and increases blood flow during focal cerebral ischaemia. Br. J. Pharmacol. **107:** 905–907.
53. ZHANG, F. & C. IADECOLA. 1993. Nitroprusside improves blood flow and reduces brain damage after focal ischemia. Neuroreport **4:** 559–562.
54. STAMLER, J.S. 1994. Redox signaling: nitrosylation and related target interactions of nitric oxide. [Review] Cell **78:** 931–936.
55. STAMLER, J.S., O. JARAKI, J. OSBORNE, *et al.* 1992. Nitric oxide circulates in mammalian plasma primarily as an *S*-nitroso adduct of serum albumin. Proc. Natl. Acad. Sci. U.S.A. **89:** 7674–7677.
56. IGNARRO, L.J., H. LIPPTON, J.C. EDWARDS, *et al.* 1981. Mechanism of vascular smooth muscle relaxation by organic nitrates, nitrites, nitroprusside and nitric oxide: evidence for the involvement of *S*-nitrosothiols as active intermediates. J. Pharmacol. Exp. Ther. **218:** 739–749.
57. FEELISCH, M. 1991. The biochemical pathways of nitric oxide formation from nitrovasodilators: appropriate choice of exodenous NO donors and aspects of preparation and handling of aqueous NO sulutions. J. Cardiovasc. Pharmacol. **17:** S25–S33.
58. RADOMSKI, M.W., D.D. REES, A. DUTRA & S. MONCADA. 1992. *S*-nitroso-glutathione inhibits platelet activation *in vitro* and *in vivo*. Br. J. Pharmacol. **107:** 745–749.
59. GASTON, B., J. REILLY, J.M. DRAZEN, *et al.* 1993. Endogenous nitrogen oxides and bronchodilator *S*-nitrosothiols in human airways. Proc. Natl. Acad. Sci. U.S.A. **90:** 10957–10961.
60. GASTON, B., J.M. DRAZEN, A. JANSEN, *et al.* 1994. Relaxation of human bronchial smooth muscle by *S*-nitrosothiols *in vitro*. J. Pharmacol. Exp. Ther. **268:** 978–984.
61. JIA, L., C. BONAVENTURA, J. BONAVENTURA & J.S. STAMLER. 1996. *S*-nitrosohæmoglobin: a dynamic activity of blood involved in vascular control. Nature **380:** 221–226.
62. MESSMER, U.K., E.G. LAPETINA & B. BRUNE. 1995. Nitric oxide-induced apoptosis in RAW 264.7 macrophages is antagonized by protein kinase C- and protein kinase A-activating compounds. Mol. Pharmacol. **47:** 757–765.
63. STAMLER, J.S., E.J. TOONE, S.A. LIPTON & N.J. SUCHER. 1997. (S)NO signals: translocation, regulation, and a consensus motif. Neuron **18:** 691–696.
64. MELINO, G., F. BERNASSOLA, R.A. KNIGHT, *et al.* 1997. *S*-nitrosylation regulates apoptosis. Nature **388:** 432–433.
65. KOBZIK, L., M.B. REID, D.S. BREDT & J.S. STAMLER. 1994. Nitric oxide in skeletal muscle [see comments]. Nature **372:** 546–548.
66. EU, J.P., L. LIU, M. ZENG & J.S. STAMLER. 2000. An apoptotic model for nitrosative stress. Biochemistry **39:** 1040–1047.
67. BELTRAN, B., A. ORSI, E. CLEMENTI & S. MONCADA. 2000. Oxidative stress and *S*-nitrosylation of proteins in cells. Br. J. Pharmacol. **129:** 953–960.
68. DAVISSON, R.L., M.D. TRAVIS, J.N. BATES, *et al.* 1997. Stereoselective actions of *S*-nitrosocysteine in central nervous system of conscious rats. Am. J. Physiol. **272**(5-2): H2361–H2368.

69. SATOH, S., T. KIMURA, M. TODA, *et al.* 997. Involvement of L-type-like amino acid transporters in *S*-nitrosocysteine-stimulated noradrenaline release in the rat hippocampus. J. Neurochem. **69:** 2197–2205.
70. RAUHALA, P., K.P. MOHANAKUMAR, I. SZIRAKI, *et al.* 1996. Nitric oxide, lipid peroxidation, ferrous citrate, free radical and *S*-nitrosothiols. *S*-Nitrosothiols and nitric oxide, but not sodium nitroprusside, protect nigrostriatal dopamine neurons against iron-induced oxidative stress *in vivo.* Synapse **23:** 58–60.
71. RAUHALA, P., A.M.Y. LIN & C.C. CHIUEH. 1998. Neuroprotection by *S*-nitrosoglutathione of brain dopamine neurons from oxidative stress. FASEB J. **12**(2): 165–173.
72. RAUHALA, P. & C.C. CHIUEH. 2000. Effects of atypical antioxidative agents, *S*-nitrosoglutathione and manganese, on brain lipid peroxidation induced by iron leaking from tissue disruption. Ann. N.Y. Acad. Sci. **899:** 238–254.
73. LIPTON, A.J., M.A. JOHNSON, T. MACDONALD, *et al.* 2001. *S*-nitrosothiols signal the ventilatory response to hypoxia. Nature **413:** 171–174.
74. KLUGE, I., U. GUTTECK-AMSLER, M. ZOLLINGER & K.Q. DO. 1997. *S*-nitrosoglutathione in rat cerebellum: identification and quantification by liquid chromatography-mass spectrometry. J. Neurochem. **69:** 2599–2607.
75. SCHARFSTEIN, J.S., J.F. KEANEY, JR., A. SLIVKA, *et al.* 1994. *In vivo* transfer of nitric oxide between a plasma protein-bound reservoir and low molecular weight thiols. J. Clin. Invest. **94:** 1432–1439.
76. MEYER, D.J., H. KRAMER, N. OZER, *et al.* 1994. Kinetics and equilibria of *S*-nitrosothiol-thiol exchange between glutathione, cysteine, penicillamines and serum albumin. FEBS Lett. **345:** 177–180.
77. LIU, Z., M.A. RUDD, J.E. FREEDMAN & J. LOSCALZO. 1998. *S*-Transnitrosation reactions are involved in the metabolic fate and biological actions of nitric oxide. J. Pharmacol. Exp. Ther. **284:** 526–534.
78. GORREN, A.C.F., A. SCHRAMMEL, K. SCHMIDT & B. MAYER. 1996. Decomposition of *S*-nitrosoglutathione in the presence of copper ions and glutathione. Arch. Biochem. Biophys. **330:** 219–228.
79. SINGH, R.J., N. HOGG, J. JOSEPH & B. KALYANARAMAN. 1996. Mechanism of nitric oxide release from *S*-nitrosothiols. J. Biol. Chem. **271:** 18596–18603.
80. SCORZA, G., D. PIETRAFORTE & M. MINETTI. 1997. Role of ascorbate and protein thiols in the release of nitric oxide from *S*-nitroso-albumin and *S*-nitroso-glutathione in human plasma. Free Rad. Biol. Med. **22:** 633–642.
81. ARNELLE, D.R. & J.S. STAMLER. 1995. NO^+, NO·, and NO^- donation by *S*-nitrosothiols: implications for regulation of physiological functions by *S*-nitrosylation and acceleration of disulfide formation. Arch. Biochem. Biophys. **318:** 279–285.
82. WONG, P.S., J. HYUN, J.M. FUKUTO, *et al.* 1998. Reaction between *S*-nitrosothiols and thiols: generation of nitroxyl (HNO) and subsequent chemistry. Biochemistry **37:** 5362–5371.
83. PARK, J.K.J. & P. KOSTKA. 1997. Fluorometric detection of biological *S*-nitrosothiols. Analyt. Biochem. **249**(1): 61–66.
84. SALT, T.E., H. ZHANG, B. MAYER, *et al.* 2000. Novel mode of nitric oxide neurotransmission mediated via *S*-nitroso-cysteinyl-glycine. Eur. J. Neurosci. **12:** 3919–3925.
85. VINCENT, S.R. & H. KIMURA. 1992. Histochemical mapping of nitric oxide synthase in the rat brain. Neuroscience **46:** 755–784.
86. MAYER, B., S. PFEIFFER, A. SCHRAMMEL, *et al.* 1998. A new pathway of nitric oxide cyclic GMP signaling involving *S*-nitroglutathione. J. Biol. Chem. **273:** 3264–3270.
87. SHEU, F.S., W. ZHU & P.C. FUNG. 2000. Direct observation of trapping and release of nitric oxide by glutathione and cysteine with electron paramagnetic resonance spectroscopy. Biophys. J. **78:** 1216–1226.
88. PAPE, H.C. & R. MAGER. 1992. Nitric oxide controls oscillatory activity in thalamocortical neurons. Neuron **9:** 441–448.
89. SHAW, P.J., S.L. CHARLES & T.E. SALT. 1999. Actions of 8-bromo-cyclic-GMP on neurones in the rat thalamus *in vivo* and *in vitro.* Brain Res. **833:** 272–277.

Nitric Oxide Protects Astrocytes from Oxidative Stress

SARA J. ROBB[a] AND JAMES R. CONNOR[b]

[a]*Department of Physiology, University College London, Gower Street, London, United Kingdom*

[b]*Leader Laboratory for Alzheimer's Research, The Pennsylvania State University, Department of Neuroscience and Anatomy, M.S. Hershey Medical School, Hershey, Pennsylvania, U.S.A.*

ABSTRACT: Oxidative stress in astrocytes has been observed in a number of pathological conditions, however the effect of nitric oxide on oxidatively stressed astrocytes remains unclear. The goal of this research was to determine how nitric oxide influences the mechanism of peroxide-induced oxidative stress in astrocytes. We used enriched mouse-astrocyte cultures and *tertiary*-butyl hydroperoxide to study astrocyte oxidative stress. Cellular integrity was measured at the ultrastructural level by electron microscopy and plasma membrane integrity measured by lactate dehydrogenase release. Mitochondrial swelling and disruption of cristae were observed by electron microscopy in the absence of damage to other organelles. Astrocytes exposed to peroxide eventually succumbed to necrotic loss of plasma membrane integrity. However, when astrocytes were subjected to oxidative stress in the presence of nitric oxide, mitochondrial integrity was preserved and cell death delayed. This protection of nitric oxide resembled that of iron chelation and nitric oxide reduced susceptibility usually associated with iron-loaded astrocytes exposed to oxidative stress. Taken together, these data suggest that nitric oxide can act as a powerful antioxidant for astrocytes preserving mitochondrial and cellular integrity during oxidative stress.

KEYWORDS: nitric oxide; cellular integrity; mitochondrial integrity; astrocytes; free radicals

In the past astrocytes were considered passive in the central nervous system. It is now recognized that astrocytes provide vital functions that are necessary for proper neural performance. Astrocytes maintain extracellular ion concentrations, remove neurotransmitters, participate in signalling cascades, and protect other neural cell types from oxidative stress. Oxidative stress has been implicated in a number of neurological diseases and in the normal aging process. Although the focus of research dealing with neurological disease has mainly been on the susceptibility of neurons and oligodendrocytes to oxidative stress, it has become clear that astrocytes are also vulnerable. It is reasonable to expect that oxidatively stressed astrocytes are less able to fulfil their protective functions, and may actually contribute to neural pathology.

Address for correspondence: James R. Connor, Ph.D., Leader Laboratory for Alzheimer's Research, Dept. of Neuroscience and Anatomy, Hershey Medical School, Hershey, PA 17033, U.S.A. Voice: 717-531-5408; fax: 717-531-5104.
jrc3@psu.edu.

Oxidative stress results when the level of reactive oxygen species (ROS) increases above the antioxidant level that mechanisms can handle. This may be the result of increased levels of ROS, increased metal availability, decreased levels of antioxidants, or decreased reducing equivalents and activities of detoxifying enzymes. The classic reaction associated with oxidative stress is the Fenton reaction depicted below.

$$H_2O_2 + Fe^{2+} \rightarrow \cdot OH$$

The essential components of the Fenton reaction are hydrogen peroxide (H_2O_2) and ferrous iron (Fe^{2+}). In this reaction, hydrogen peroxide reacts with Fe^{2+} to yield the highly reactive hydroxyl radical (·OH). Hydroxyl radical can induce massive cellular damage by reacting with DNA, proteins and lipids. Cells generate low levels of H_2O_2 during normal metabolic processes.[1,2] In addition, iron is available for chemical reaction, as low molecular iron species, iron bound to small chelates such as ATP and citrate.[3,4] Under normal circumstances the coexistence of H_2O_2 and iron results in little oxidative stress because cells have antioxidant mechanisms to limit oxidative stress. These include the antioxidants vitamin E and vitamin C, enzymes such as superoxide dismutase (SOD) to decrease superoxide ($O_2^{\cdot -}$) levels, catalase and glutathione peroxidase (GPx) to decrease levels of H_2O_2 and ferritin to store intracellular iron. A perturbation in the delicate balance between these components can result in oxidative stress. For example, situations that result in increased H_2O_2 or low molecular weight iron, may also result in oxidative stress and cellular damage through Fenton-mediated generation of ·OH. One mechanism that leads to increased levels of H_2O_2 is the inhibition of the mitochondrial respiratory chain.[5]

The mitochondrial electron transport chain (ETC) produces $O_2^{\cdot -}$ as a byproduct of normal metabolism.[6] FIGURE 1 is a schematic of the ETC. Electrons are passed along the ETC resulting in the production of ATP. Electrons can "leak" from Complex I onto

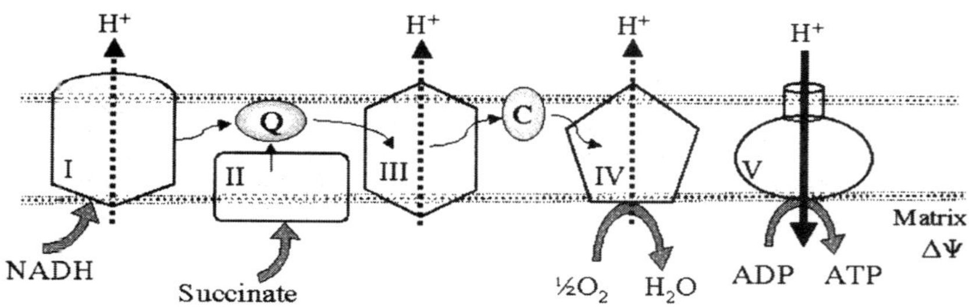

FIGURE 1. Mitochondrial electron transport chain. Electrons enter the electron transport chain at Complex I, NADH ubiquinone reductase, or Complex II, succinate ubiquinone reductase. Electrons are passed from Complex I and II to ubiquinone (Q), which in turn transfers the electrons to Complex III, ubiquinol cytochrome c reductase. Complex III reduces cytochrome c (C), which transfers the electrons to Complex IV, cytochrome c oxidase. Complex IV is the site of oxygen consumption by the ETC, which reduces oxygen to water and reoxidizes cytochrome c. Complexes I, III, and IV pump protons across the inner mitochondrial membrane to generate a potential difference ($\Delta\Psi_m$). The energy from this proton concentration gradient is used by the ATP synthase, Complex V to convert ADP to ATP, providing energy to the cell.

molecular oxygen resulting in $O_2^{-\cdot}$ production during normal respiration.[7] Clearance of $O_2^{-\cdot}$ occurs by the dismutation reaction of SOD, yielding H_2O_2. As mentioned above, increased levels of H_2O_2 can result in the Fenton reaction producing ·OH. Therefore, it is vital that H_2O_2 be converted to a less reactive species. Detoxification of H_2O_2, occurs by the actions of GPx using reduced glutathione (GSH) as a source for reducing electrons and by catalase, which uses NADPH as an electron source. The antioxidants vitamins E and C are able to react with free radicals that escape the detoxification enzymes, eliminating the cascade of damage. If levels of ROS exceed the capacity of this system oxidative stress will ensue.

Oxidative stress can result from mitochondrial damage. Generation of $O_2^{-\cdot}$ by the mitochondrial ECT is increased in situations where the ETC is inhibited.[5,8] For example, cyanide inhibition of Complex IV will prevent electron transfer to oxygen to produce water. When Complex IV is inhibited, the other complexes cannot transfer their electrons and he chain becomes reduced. Under these conditions, the electrons are thought to be more likely to reduce molecular oxygen increasing levels of $O_2^{-\cdot}$. The increase $O_2^{-\cdot}$ can result in increased levels of H_2O_2 and ultimately elevated levels of ·OH. As mentioned above, ·OH has been associated with damage to DNA, proteins and lipids. Once ·OH has been generated, it can initiate a cascade of electron transfers leaving behind oxidized species as electrons are passed from molecule to molecule. Perhaps the most damaging effect of this cascade is the oxidation of lipids in cell membranes. Lipid peroxidation can lead to loss of organelle integrity and plasma membrane integrity resulting in the demise of the cell. Mitochondria are particularly vulnerable since they are the source of many ROS and are a proximal target for oxidizing species. Damage to the mitochondrial inner membrane and the proteins in the ETC can encourage further generation of ROS further increasing oxidative stress.

Astrocytes are exposed to ROS through the generation of $O_2^{-\cdot}$ by the ETC and subsequent production of H_2O_2 as well as through the actions of monoamine oxidase B (MAO B). MAO B is located on the outer mitochondrial membrane and generates H_2O_2 during the metabolism of amine neurotransmitters.[9] Levels of ROS and oxidative stress increase with age. It has been shown that there is an age-related increase in MOA B activity and electron leakage from the mitochondrial ETC.[10,11] Each of these mechanisms will result in increased intracellular ROS generation and oxidative stress. Because the increased H_2O_2 in astrocytes occurs at the site of the mitochondrial membrane, astrocyte mitochondria were hypothesized to be the organelles most susceptible to peroxide-induced oxidative damage. Astrocytes may also be vulnerable to oxidative stress when iron levels increase because they do not usually contain ferritin.[12] In the absence of ferritin, astrocytes must rely on clearance of H_2O_2 to prevent Fenton-generated oxidative stress. An increase in intracellular iron has been observed in astrocytes in brains of patients with diseases such as Alzheimer's and Parkinson's.[13] The increased iron presence in senile plaques is accompanied by an increase in expression of heme oxygenase, which breaks down heme to liberate iron.[14] Upregulation of heme oxygenase has been associated with iron deposition into astrocyte mitochondria, which may contribute to mitochondrial oxidative damage in disease states.[15] Interestingly, astrocytes begin to express ferritin in Alzheimer's disease plaques.[16] This may be an attempt to sequester the excess iron and limit oxidative stress. In the absence of ferritin, the primary mechanism for detoxification of H_2O_2 in astrocytes is reduction of H_2O_2 by GPx.[17,18] Reduced glutathione, the electron source for GPx, is most concentrated in astrocyte cytosol with much lower concentrations in astrocyte mitochondria.[19] These low levels of GSH may leave astrocyte mitochondria very susceptible to peroxide-induced damage. Additional protection is afforded by the elimination of free radicals by vitamin E and C. Perturbations in this protective system occur in situations of neurodegenerative disease and trauma and should be understood in order to define strategies to decrease oxidative stress to astrocytes.

In order to understand the mechanism of oxidative stress in astrocytes, culture models have been employed. *In vitro* studies of oxidative stress in astrocytes frequently induce oxidative stress through the application of exogenous peroxide in the form of *tertiary*-butyl hydroperoxide (*t*-bOOH). This organic peroxide is similar to H_2O_2, in that it can participate in Fenton Chemistry, yet has the advantages of being more lipophilic, stable, and is a substrates for GPx. Application of *t*-bOOH results in oxidative stress in astrocytes appears to be uninfluenced by extracellular iron levels, suggesting the oxidative stress is intracellular rather than extracellular.[20] When astrocytes are exposed to *t*-bOOH, there is an appropriate loss of cellular GSH indicating the oxidative stress is overwhelming the actions of GPx.[19] The mechanism of peroxide–induced oxidative stress to astrocytes involves a number of cellular changes that can be divided into primary and secondary events (see FIGURE 2). Primary events associated with oxidative stress to neonatal cortical mouse astrocytes are increased intracellular calcium and a subsequent increase in the $O_2^{-\cdot}$ levels.[21] Upon addition of peroxide, there is a small transient increase in calcium. Further exposure to peroxide results in a continued elevated levels of $O_2^{-\cdot}$, increased intracellular calcium, and secondary events that eventually result in energy failure, as evidenced by the loss of cellular ATP and collapse of the mitochondrial membrane potential.[21] Astrocytes exposed to peroxide eventually succumb to necrosis accompanied by swelling of organelles and loss of plasma membrane integrity. The first organelle to exhibit necrotic swelling is the mitochondria, with mitochondrial swelling occurring prior to loss of viability.[21] This is consistent with the observation that mitochondria are a major source of free radical generation. Peroxide-induced oxidative stress in astrocytes has been shown to be highly iron dependent, suggesting that Fenton chemistry is responsible for the initiation of damage. Chelation of astrocytic iron with desferrioxamine delays the cascade of events associated with the demise of astrocytes exposed to peroxide. Iron chelation postpones the increase in intracellular calcium, increase in $O_2^{-\cdot}$, loss of mitochondrial membrane potential and ATP, as well as preventing the necrotic death observed when astrocytes with normal iron status are exposed to peroxide.[20,21]

As mentioned above, one of the primary events associated with peroxide-induced oxidative stress to astrocytes is a transient increase in intracellular calcium. Astrocytes express the constitutive form of nitric oxide synthase (NOS I), which is calcium-dependent for activation and has been shown to produce ·NO when intracellular calcium rises.[22] Additionally, astrocytes can express the inducible form of the enzyme, NOS II.[23,24] This form of NOS is seen in astrocytes under pathological conditions *in vivo,* such as following ischemia or in multiple sclerosis.[25,26] It was conceivable that ·NO may have a profound influences on oxidative stress in astrocytes as NOS II and oxidative stress are both associated with astrocytes in neural disease

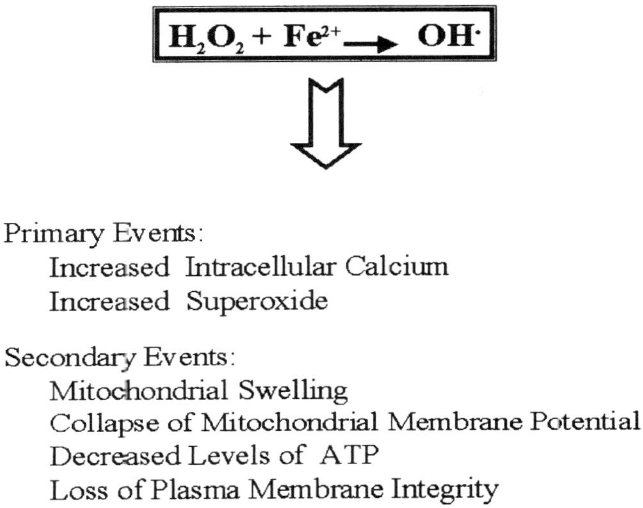

FIGURE 2. Mechanism of oxidative stress in astrocytes. Exposure of astrocytes to peroxide results in a cascade of events that can be classified as either primary or secondary events. The primary events occur upon addition of peroxide and the initiation of Fenton chemistry. They include an increase in intracellular calcium ($[Ca^{2+}]_i$) and increased superoxide ($O_2^{\cdot -}$). The secondary events occur after the cell has been challenged with oxidative stress for some time. Continued exposure to oxidative stress results in swelling of mitochondria, decreased mitochondrial membrane potential ($\Delta\Psi_m$), decreased ATP, and finally LDH release. This mechanism is iron mediated with iron chelation preventing both primary and secondary events.

and injury. In order to test this, the *in vitro* model of oxidative stress mentioned above has been used. In culture, astrocytes can be induced to express NOS II when treated with cytokines and lipopolysaccharide (LPS).[23,24] The induction of NOS II in astrocytes results in inhibition of the mitochondrial respiratory chain in astrocytes and decreased oxygen consumption.[27] We hypothesized the mitochondrial effects associated with increased levels of NO would influence the susceptibility of astrocytes to oxidative stress. In order to test this, astrocytes were treated with LPS and interferon γ (INF-γ) for 18 hours to maximize the endogenous ·NO levels. Astrocyte cultures were then subjected to oxidative stress in the form of peroxide and loss of viability monitored by LDH release for four hours. FIGURE 3A shows the LDH release from oxidatively stressed astrocytes. Astrocytes show significant loss of viability after three hours exposure to peroxide (1.5 mM *t*-bOOH) with LDH release reaching 75% at 3 hours and 92.5% at four hours. However, when astrocytes are treated with LPS and INF-γ to increase endogenous ·NO we observed a remarkable decrease in loss of viability, with LDH after three hours of peroxide only being 22.5% and 68.7% after four hours. The protective effect of LPS and INF-γ could be attributed to endogenous ·NO as the NOS inhibitor L-NAME opposed the protection and returned the level of toxicity to that seen when naïve astrocytes were exposed to

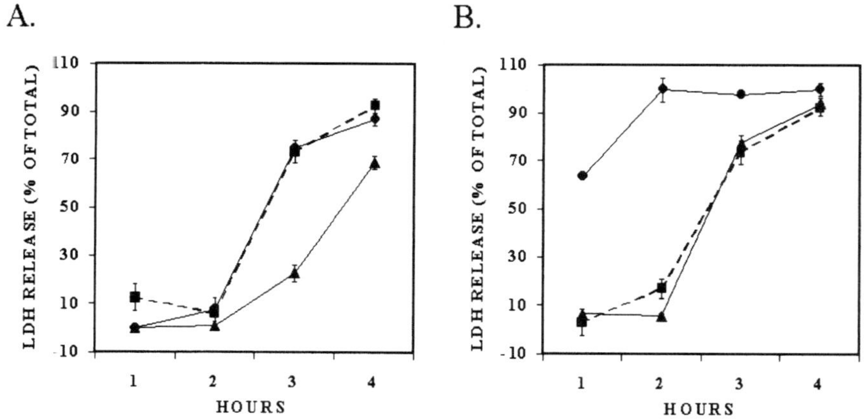

FIGURE 3. A. Nitric oxide inhibits loss of plasma membrane integrity in oxidatively stressed astrocytes. Lactate dehydrogenase (LDH) release was used to measure loss of plasma membrane integrity and loss of viability in oxidatively stressed astrocytes. **A.** Shows the typical LDH release in response to peroxide (■). Astrocyte producing nitric oxide following iNOS induction show significantly less LDH release (▲). Inclusion of the non-specific NOS inhibitor L-NAME in iNOS expressing cultures (●) eliminated the protection afforded by NO. **B.** Nitric oxide opposes iron-induced exacerbation of loss of plasma membrane integrity in oxidatively stressed astrocytes. The typical LDH release in response to peroxide is again shown as (■). When astrocytes are iron loaded, they are much more susceptible to oxidative stress, as indicated by the accelerated LDH release (●). When the nitric oxide donor SNAP is applied to iron loaded cultures, the LDH release returns to the level observed in astrocyte cultures with normal iron status (▲).

oxidative stress. As mentioned above, astrocyte susceptibility to peroxide-induced damage is determined by iron status with iron chelation ablating damage and iron loading severely enhancing the damaging effects of peroxide. FIGURE 3B shows the magnitude of enhancement of LDH release when iron loaded astrocytes are exposed to peroxide. After one hour 63.6% of cells lost viability and nearly 100% died after two hours of oxidative stress. It is evident that the combination of peroxide and high intracellular iron facilitates Fenton reaction mediated damage and oxidative stress in astrocytes. However, when iron loaded astrocytes are exposed to a ·NO donor and peroxide the toxicity mimics that of naïve astrocytes exposed to peroxide. These data strongly suggest there is an interaction between ·NO and iron that is essential in determining the toxicity associated with oxidative stress in astrocytes. Upon further examination it became evident that ·NO decreased both primary and secondary events associated with oxidative stress in astrocytes. Although endogenous ·NO did not affect the increase in intracellular calcium after addition of peroxide the subsequent events were all decreased. Astrocytes activated with LPS and INF-γ to produce ·NO responded to the peroxide addition with a transient increase in calcium, however the peroxide-mediated increase in $O_2^{-\cdot}$ was not observed, nor was the loss of $\Delta\Psi_m$. The mechanism of protection appeared to decrease damage at the site of

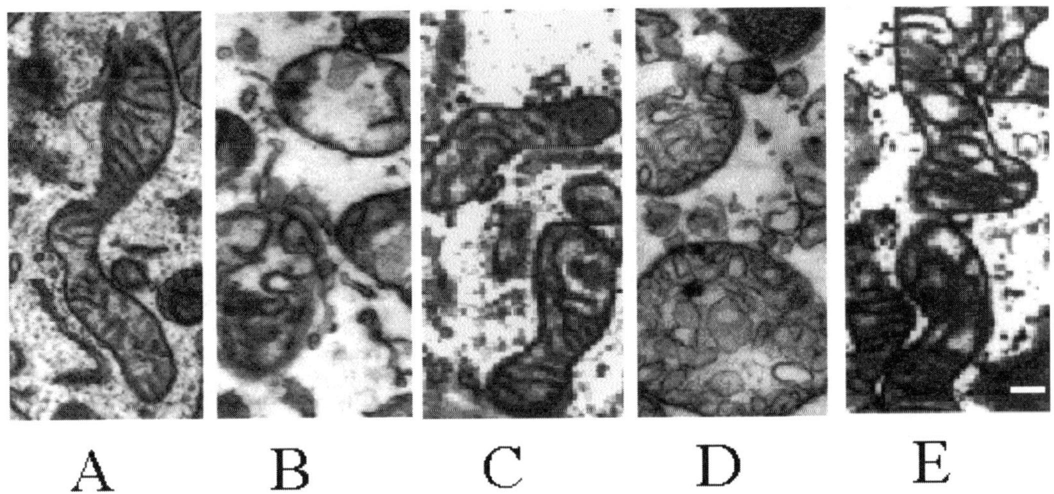

FIGURE 4. Nitric oxide opposes loss of mitochondrial integrity in oxidatively stressed astrocytes. Normal astrocyte mitochondria are elongated and can have elaborate shapes (**Panel A**). When astrocytes are exposed to peroxide (**Panel B**), they become round and begin to show damage to their ultrastructure. In astrocytes generating ·NO after iNOS induction (**Panel C**) mitochondrial damage is lessened in response to peroxide. Iron loaded astrocytes show accelerated damage to mitochondrial structure when treated with peroxide (**Panel D**), however this iron-accelerated damage is neutralized by the ·NO donor SNAP (**Panel E**). *Scale bar* = 255 nm.

mitochondria. When electron microscopy was used to examine the ultrastructure of the mitochondria it became apparent the mitochondria are the first organelle effected by peroxide-induced oxidative stress in astrocytes. Mitochondrial swelling is observed after two hours of oxidative stress. Mitochondrial damage occurs after the generation of ROS and prior to loss of viability. FIGURE 4 shows the swelling associated with two hours of oxidative stress to astrocytes. Panel A shows a mitochondrion from control astrocytes, whereas Panel B illustrates the extent of swelling and disruption following two hours of peroxide. The swelling is significantly decreased when astrocytes are treated with LPS and INF-γ prior to peroxide exposure (Panel C), an effect opposed by decreasing NO (data not shown). The relationship between ·NO and iron is also observed in these data. Iron loading of astrocytes results in severely swollen mitochondria after two hours of oxidative stress as seen in Panel D. It should also be mentioned that, under conditions of iron loading prior to oxidative stress, loss of plasma membrane integrity was also observed (FIG. 3B). The severity of swelling observed when iron loaded astrocytes were oxidatively stressed could be diminished by the inclusion of the ·NO donor SNAP (Panel E). Both endogenous and exogenous ·NO were capable of reducing the amount of mitochondrial swelling associated with oxidative stress in astrocytes.

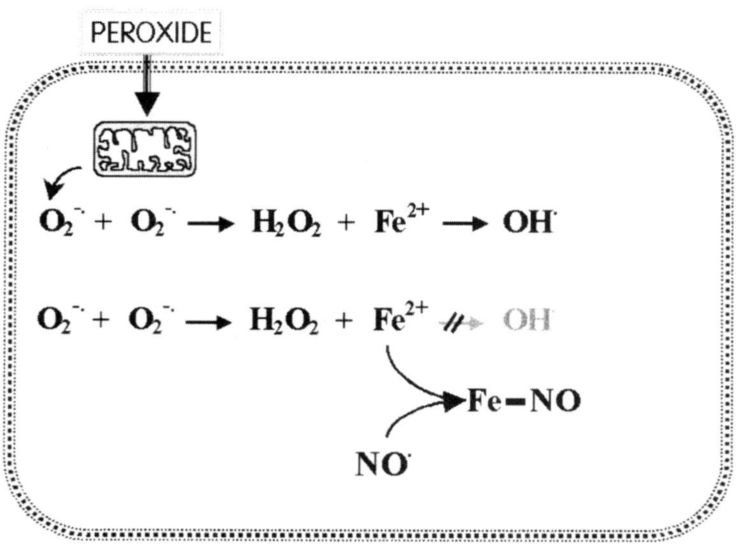

FIGURE 5. Proposed mechanism for nitric oxide decreasing peroxide-induced oxidative stress. When astrocytes are exposed to oxidative stress in the form of peroxide, a cascade of iron-dependent events ensues. Peroxide causes an immediate increase in superoxide ($O_2^{-\cdot}$) from mitochondria. The increased levels of $O_2^{-\cdot}$ contribute to intracellular peroxide levels. Peroxide reacts with iron (Fe^{2+}) to produce the highly damaging hydroxyl radical (·OH). We propose that ·NO decreases oxidative stress in astrocytes by forming nitrosyl iron (Fe^{2+}-NO) conjugates that inhibits the Fenton reaction and decreases production of ·OH.

Nitric oxide was capable of decreasing the toxicity of oxidatively stressed astrocytes by interrupting the cascade of events associated with oxidative stress. Treatment of astrocytes with peroxide results in increased calcium and increased levels of $O_2^{-\bullet}$. Endogenous nitric oxide did not alter the increase in intracellular calcium, nor the gradual increase in calcium observed while astrocytes are treated with peroxide, however, ·NO did decrease the other events associated with the progression of oxidative stress. Nitric oxide decreased the levels of $O_2^{-\bullet}$, preserved the mitochondrial membrane potential, and decreased mitochondrial swelling. Interestingly calcium chelation exacerbated the very events that ·NO decreased. When astrocytes were oxidatively stressed in calcium free media, $O_2^{-\bullet}$ levels increased greatly, the $\Delta\Psi_m$ loss was accelerated, and the cell death was enhanced. Perhaps the reason calcium depletion accelerates these events is that calcium is needed to activate NOS I and produce ·NO. The protection afforded by ·NO was similar to that seen in iron-depleted astrocytes. ·NO has a high affinity for iron and the similarities in the mode of protection of astrocytes from oxidative stress suggest an interaction between iron and ·NO. FIGURE 5 is a schematic of the mechanism by which we propose that NO decreases oxidative stress in astrocytes. When astrocytes are oxidatively stressed with peroxide the mitochondria are affected first. Superoxide levels increase at the site of mitochondria, which can lead to increased levels of peroxide. the exogenous and endogenous peroxide can react with the intracellular iron by Fenton chemistry to create a hydroxyl radical. The hydroxyl radical is then free to attack DNA, protein, and lipids. The cascade results in damage to mitochondria and ultimately cell death. If iron is chelated, the cascade is prevented, even in the presence of high levels of exogenous peroxide. Detoxification of the peroxide can occur without the danger of peroxide and iron reacting to make hydroxyl radical. We propose a second way to decrease Fenton mediated oxidative damage is by the interaction of iron and nitric oxide to make nitrosyl iron. The chelation of the astrocytic intracellular iron by ·NO results in a protection from oxidative stress, which is similar to iron chelation with mitochondria being protected and cell death delayed.

ACKNOWLEDGMENT

This work was Supported by HD30704 (NIH).

REFERENCES

1. BOVERIS, A., N. OSHINO & B. CHANCE. 1972. The cellular production of hydrogen peroxide. Biochem. J. **128:** 617–630.
2. BOVERIS, A. & B. CHANCE. 1973. The mitochondrial generation of hydrogen peroxide. General properties and effect of hyperbaric oxygen. Biochem. J. **134:** 707–716.
3. POLLACK, S., T. CAMPANA & J. WEAVER. 1985. Low molecular weight iron in guinea pig reticulocytes. Am. J. Hematol. **19:** 75–84.
4. POLLACK, S. 1992. Receptor-mediated iron uptake and intracellular iron transport. Am. J. Hematol. **39:** 113–118.
5. BOVERIS, A., E. CADENAS & A.O. STOPPANI. 1976. Role of ubiquinone in the mitochondrial generation of hydrogen peroxide. Biochem. J. **156:** 435–444.
6. CHANCE, B., H. SIES & A. BOVERIS. 1979. Hydroperoxide metabolism in mammalian organs. Physiol. Rev. **59:** 527–605.

7. BOVERIS, A. & E. CADENAS. 1975. Mitochondrial production of superoxide anions and its relationship to the antimycin insensitive respiration. FEBS Lett. **54:** 311–314.
8. TURRENS, J.F. & A. BOVERIS. 1980. Generation of superoxide anion by the NADH dehydrogenase of bovine heart mitochondria. Biochem. J. **191:** 421–427.
9. TIPTON, K.F. 1967. The sub-mitochondrial localization of monoamine oxidase in rat liver and brain. Biochim. Biophys. Acta **135:** 910–920.
10. BHASKARAN, D. & E. RADHA. 1983. Monoamine levels and monoamine oxidase activity in different regions of rat brain as a function of age. Mech. Ageing Devel. **23:** 151–160.
11. AMES, B.N., M.K. SHIGENAGA & T.M. HAGEN. 1995. Mitochondrial decay in aging. Biochim. Biophys. Acta **1271:** 165–170.
12. BENKOVIC, S.A. & J.R. CONNOR. 1993. Ferritin, transferrin, and iron in selected regions of the adult and aged rat brain. J. Compar. Neurol. **338:** 97–113.
13. JELLINGER, K., W. PAULUS, I. GRUNDKE IQBAL, *et al.* 1990. Brain iron and ferritin in Parkinson's and Alzheimer's diseases. J. Neural Transmiss. Parkinsons Dis. Demen. Section **2:** 327–340.
14. SCHIPPER, H.M., S. CISSE & E.G. STOPA. 1995. Expression of heme oxygenase-1 in the senescent and Alzheimer-diseased brain. Ann. Neurol. **37:** 758–768.
15. SCHIPPER, H.M., L. BERNIER, K. MEHINDATE & D. FRANKEL. 1999. Mitochondrial iron sequestration in dopamine-challenged astroglia: role of heme oxygenase-1 and the permeability transition pore. J. Neurochem. **72:** 1802–1811.
16. CONNOR, J.R., B.S. SNYDER, P. AROSIO, *et al.* 1995. A quantitative analysis of isoferritins in select regions of aged, parkinsonian, and Alzheimer's diseased brains. J. Neurochem. **65:** 717–724.
17. MCKENNA, O., G. ARNOLD & E. HOLTZMAN. 1976. Microperoxisome distribution in the central nervous system of the rat. Brain Res. **117:** 181–194.
18. RAPS, S.P., J.C. LAI, L. HERTZ & A.J. COOPER. 1989. Glutathione is present in high concentrations in cultured astrocytes but not in cultured neurons. Brain Res. **493:** 398–401.
19. O'CONNOR, E., A. DEVESA, C. GARCIA, *et al.* 1995. Biosynthesis and maintenance of GSH in primary astrocyte cultures: role of L-cystine and ascorbate. Brain Res. **680:** 157–163.
20. ROBB, S.J. & J.R. CONNOR. 1998. An *in vitro* model for analysis of oxidative death in primary mouse astrocytes. Brain Res. **788:** 125–132.
21. ROBB, S.J., L.D. ROBB-GASPERS, R.C. SCADUTO, *et al.* 1999. Influence of calcium and iron on cell death and mitochondrial function in oxidatively stressed astrocytes. J. Neurosci. Res. **55:** 674–686.
22. KNOWLES, R.G., M. PALACIOS, R.M. PALMER & S. MONCADA. 1989. Formation of nitric oxide from L-arginine in the central nervous system: a transduction mechanism for stimulation of the soluble guanylate cyclase. Proc. Natl. Acad. Sci. U.S.A. **86:** 5159–5162.
23. GALEA, E., D.L. FEINSTEIN & D.J. REIS. 1992. Induction of calcium-independent nitric oxide synthase activity in primary rat glial cultures. Proc. Natl. Acad. Sci. U.S.A. **89:** 10945–10949.
24. SIMMONS, M.L. & S. MURPHY. 1993. Cytokines regulate L-arginine-dependent cyclic GMP production in rat glial cells. Eur. J. Neurosci. **5:** 825–831.
25. ENDOH, M., K. MAIESE & J. WAGNER. 1994. Expression of the inducible form of nitric oxide synthase by reactive astrocytes after transient global ischemia. Brain Res. **651:** 92–100.
26. OLESZAK, E.L., E. ZACZYNSKA, M. BHATTACHARJEE, *et al.* 1998. Inducible nitric oxide synthase and nitrotyrosine are found in monocytes/macrophages and/or astrocytes in acute, but not in chronic, multiple sclerosis. Clin. Diagnos. Lab. Immunol. **5:** 438–445.
27. BROWN, G.C., J.P. BOLANOS, S.J. HEALES & J.B. CLARK. 1995. Nitric oxide produced by activated astrocytes rapidly and reversibly inhibits cellular respiration. Neurosci. Lett. **193:** 201–204.

The Roles of Nitric Oxide in Dorsal Root Ganglion Neurons

THIMMASETTAPPA THIPPESWAMY AND RICHARD MORRIS

Department of Veterinary Preclinical Sciences, University of Liverpool, Brownlow Hill and Crown Street, Liverpool, United Kingdom

ABSTRACT: The functions of nitric oxide (NO) in primary somatosensory neurons are reviewed. During the early development of these neurons the neuronal isoform of nitric oxide synthase (nNOS) is expressed during neurite extension. As their axons extend peripherally and start to take up nerve growth factor (NGF), nNOS starts to disappear from the majority of these neurons. A small number (less than 5%) continue to express nNOS, and for this small population NO may have a role in synaptic transmission. Following peripheral nerve section in adult rats, nNOS is reexpressed in many small peptidergic DRG neurons that have been axotomized. At the same time cGMP synthesis is increased in satellite glia cells. From culture studies, it was established that NGF negatively regulates nNOS synthesis in DRG neurons and that block of NO production leads to neuronal death. Further recent data is reviewed that supports the view that NO has a neuroprotective action preventing loss of DRG neurons and facilitates regeneration.

KEYWORDS: nitric oxide; nitric oxide synthase; dorsal root ganglion; primary afferent; neuroprotection; apoptosis; nerve growth factor; caspase; Bax; tumor necrosis factor α; c-jun; calcitonin gene-related peptide

INTRODUCTION

The findings that led to the present ideas were obtained from studies on primary sensory neurons and, consequently, this review is confined to a discussion of the roles of nitric oxide (NO) in these neurons.

HETEROGENEITY OF DORSAL ROOT GANGLION (DRG) NEURONS

Large DRG neurons give rise to myelinated axons and are concerned with mechanoreception, whereas smaller DRG neurons give rise to either lightly myelinated axons or unmyelinated axons.[1] These small neurons are very diverse functionally, some are concerned with mechanoreception and thermoreception, large numbers are nociceptors,[2] whereas others must mediate the sensation of itch.[3] Some are even silent in the absence of peripheral inflammation.[4]

Address for correspondence: Dr. Richard Morris, Department of Veterinary Preclinical Sciences, Brownlow Hill/Crown Street, University of Liverpool, Liverpool, L69 7ZJ, UK. Voice: 0151 794 4227; fax: 0151 794 4243.
richardm@liverpool.ac.uk

NITRIC OXIDE SYNTHASE IN ADULT DRG NEURONS

If DRG or the trigeminal ganglia from the adult rat are stained with antibodies to NOS (or using the NADPH-diaphorase histochemical reaction); then positive staining is found in the somata of less than 5% of the neurons.[5] Capillary endothelial cells are also stained, but none of the satellite glia cells or other non-neuronal cell types are stained. The use of antibodies that are specific for the neuronal isoform of NOS (nNOS), reveals that the neuronal staining is due to the expression of this isoform, whereas that in the capillary endothelial cells is the endothelial isoform (eNOS). All the nNOS ganglion neurons are small to medium sized and double staining reveals that many express the neuropeptide, calcitonin gene-related peptide (CGRP). The nNOS they express is also found in their axons, which are either unmylinated or small myelinated ones. These axons can be seen entering the superficial laminae of the spinal cord. The small size of these neurons and their central projections suggest that they may be involved in nociception.[6] Peripheral nerve ligation experiments also reveal that the enzyme is being transported peripherally.[7] Hence, for this small subpopulation of DRG neurons in the adult, NO would appear to have a specialized function and may act as a signalling molecule at both central and peripheral terminals. In the nodose ganglion more neurons stain for NOS and it is possible that NO has a role in visceral sensation.[8]

NO IN DEVELOPING DRG

The heterogeneity of DRG neurons is reflected in the complex nature of their changing growth factor dependence during development. An extreme example of this is seen in the selective loss of only down-hair receptor afferents in mice in which the neurotrophin 4 gene has been knocked out.[9] During development many more primary afferent neurons are produced than those that persist in the adult, and in transgenic mice overexpressing NGF in the epidermis, increases exceeding 100% in DRG neuron numbers are seen.[10] This reveals that neurogenisis is followed by a competitive phase in which the neurites of the primary afferents compete to take up peripherally and centrally produced trophic factors.[11] Nerve section in young neonatal rats (0–3 days old) leads to extensive neuronal death in the denervated DRG,[12] whereas a section of the same nerve in the adult leads to very limited neuronal cell death.[13] Hence, it appears as though DRG neurons go through three principle phases in which growth factors have different roles. In the early phase of neurogenisis and neurite extension, DRG neurons are not dependent on factors such as NGF for survival. As they establish peripheral contacts and take up growth factors they become dependent on them for survival and competition leads to reduction in DRG neuron numbers. In the adult, these neurons again become independent of these neurotrophic factors for survival. The expression of nNOS is also developmentally regulated. In the early stage embryo, prior to neurite extension, all DRG neurons express nNOS.[14] We found the expression of nNOS during embryonic development to be the inverse of both the epidermal expression of NGF and the ganglion neuron expression of the NGF high affinity receptor, *trkA*.[15] In DRG cultures we found that NGF negatively regulates nNOS.[16] Hence, it seems probable that as DRG neurons take up NGF from the skin they switch off nNOS production.

NO AS A NEURON-GLIAL SIGNALLING SYSTEM IN DRG

When DRG are acutely dissected and exposed to NO-donors, it is found that cGMP synthesis occurs, which is confined exclusively to the satellite glia cells surrounding the DRG neurons.[5] Virtually all these glia cells synthesize this cyclic nucleotide suggesting that they may all have the capacity to respond to NO. However, it can be questioned as to why this should be the case if so few neurons contain nNOS in the adult. Indeed it may even be argued that this second messenger system could permit responses to NO released from capillary endothelial cells. This is an interesting possibility since DRG neurons are outside the blood–brain barrier and are exposed to circulating molecules. Hence, the satellite glia cells may have some critical function in regulating the neuronal environment. Although these speculations are interesting, we have evidence that the source of NO that activates synthesis of cGMP in the satellite glia cells is indeed the neurons. If the acutely dissected ganglia are treated with capsacin, cGMP synthesis is observed in groups of satellite glia cells in patches throughout the ganglion.[17] The target ion-channel for capsaicin (VR1) is expressed on the neurons and not the capillary endothelial cells. Hence, this suggests that NO synthesized in the neurons can lead to cGMP synthesis in the satellite glia cells and this is a neuron-glial signalling system. Other data to be discussed later also supports this conclusion.

NO SYNTHESIS IN INJURED DRG NEURONS

Another feature of DRG neurons that makes them unusual is their regenerative capacities. This is mainly a consequence of their relationships with glia cells. DRG neuron soma are spherical with no dendritic processes. They give rise to a single axon that bifurcates to give rise to a branch running into a peripheral nerve that terminates in target tissues, and a branch that runs into a dorsal root (or equivalent cranial nerve root) to terminate in the spinal cord or brainstem. Along the dorsal root is a transition zone where the axons change from forming associations with Schwann cells to associations with oligodendrocytes and astrocytes. Following section of a peripheral nerve, depending on the precise nature of the damage, different extents of regeneration can occur, often leading to some degree of functional recovery. In contrast, when the dorsal root is cut no regeneration beyond the dorsal root transition zone occurs. This almost certainly reflects the relationships between the axons and glia cells and very recently it has been shown that substantial regeneration can occur through this transition zone if NGF, neurotrophin 3 or glial-cell-line-derived neurotrophic factor (GDNF) are infused into this region.[18]

Hökfelt's group has shown that section of the peripheral axons of primary afferents can lead to the synthesis of mRNA for nNOS in DRG neuron soma, *in vivo*[19] and, as a result, nNOS protein production[20] occurs principally in the small DRG neurons that express neuropeptides, such as substance P and CGRP.[20,21] NO production in axotomized DRG neurons also leads to the synthesis of cGMP in satellite glia cells.[17,22] Interestingly, not all the satellite glia cells surrounding nNOS-positive DRG neurons express cGMP, whereas most of the large nNOS-negative neurons are surrounded by cGMP-expressing glia cells.[17,22]

NO IS NEUROPROTECTIVE TO DRG NEURONS

In culture, many *adult* (over P15) DRG neurons deprived of NGF express nNOS. If this nNOS is blocked with a range of blockers, the vast majority of neurons expressing nNOS die.[23] Within six hours they show retracted neurites, cytoplasmic blebbing, and eccentric nuclei; and by 24 hours they have fragmented and are no longer apparent. In contrast, perfectly healthy looking large DRG neurons remain. The effects of blocking the synthesis of cGMP were, if anything, even more pronounced, producing even greater cell loss.[23] Initially it was difficult to establish that this cell death was apoptosis using the TUNEL staining method, but recently we have shown clear DNA laddering, and hence, the nNOS block is precipitating apoptosis. These NO-blocker actions can be partially reversed with L-arginine or 8-bromo-cGMP. Collectively these observations suggest that NO has a neuroprotective role. Several important questions now emerge. (1) Does this occur *in vivo*? (2) How is the NO-cGMP, neuron–glial signalling system involved? (3) Are other mechanisms implicated in apoptosis precipitated by nNOS block?

Does This Occur in Vivo*?*

Intraperitoneal injection of the relatively selective nNOS blocker, 1-(2-trifluoromethylphenyl) imidazole (TRIM), in animals subjected 10 days previously to unilateral sciatic nerve section, causes axotomized DRG neurons to develop an irregular shape, vacuolate, and show highly eccentric nuclei. These changes are not restricted to the small nNOS-positive neurons, but are also present in the large nNOS-negative neurons in the ganglia giving rise to the sectioned nerve. In sham-operated animals and on the contralateral side in operated animals the neurons are normal.[24] This suggests that NO is exerting a neuroprotective role for DRG neurons *in vivo*. This observation coupled with the observations on cGMP suggest that NO released from the small neurons is also providing protection for the large nNOS-negative neurons.

How is the NO-cGMP, Neuron–Glial Signalling System Involved?

The actions of cGMP are quite diverse, it is known to regulate ion channels, activate enzymes, and also cause changes in gene expression. Among these possibilities we have started to examine changes in gene expression, these could result in two different actions. One possibility is that glia genes that encode for molecules that initiate apoptosis in the neurons are suppressed by the activation of cGMP. The other possibility is that cGMP causes the activation of genes promoting neuronal survival; for example, those for growth factors, such as NGF. Although this requires much more investigation, we now have preliminary data suggesting that both processes are involved. Immunocytochemical studies in adult cultures deprived of NGF and treated with NOS-blockers, show some elevation of TNFα, but no change in several receptors bearing death domains (Fas/APO-1, TNF-R, and the low affinity NGF receptor p75).[25] Following nerve section, it has been found that satellite glia cells in DRG start to synthesize NGF.[26] We have started to examine if NO will induce growth factor synthesis in glia cells *in vitro.* Our initial PCR studies on dissociated DRG cultures did not reveal any differences between control and NO-donor treated cultures. However, immunocytochemical staining revealed that numerous non-neuronal cells in these cultures are already producing NGF. Using much more homogeneous Schwann cell cultures, a marked elevation of mRNA for NGF was found following treatment with the NO-donor compound, SIN-1, whereas the controls were virtually negative. Preliminary micro-array studies are also indicating changes in the expression of several other genes; for example, insulin-like growth-factor-binding protein-5, Jun-D and heat-shock protein 27 were all upregulated following SIN-1 or 8-Br-cGMP treatment of Schwann cells. Although this micro-array data is encouraging, more experiments under a greater range of conditions (e.g., different NO-donors, cGMP-block etc.) are required. Quantitative studies with competitive PCR or Northern blotting are also needed.

Are Other Mechanisms Implicated in Apoptosis Precipitated by the nNOS Block?

NO may alter the activity of a range of enzymes, via nitrosylation of thiol groups, including the caspases, and hence, inhibit apoptosis.[27–29] In view of this, we carried out studies of the effects of caspase inhibitors on the cell death of nNOS expressing neurons in DRG cultures following treatment with NOS blocking drugs. These rescue experiments established that a blocker with specificity towards caspase 9 was most effective.[25] Caspase 9 is activated by the breakdown of mitochondria membrane.[30] It also plays a pivotal role in the downstream activation of other caspases[31] and, hence, this may explain why block of this enzyme was most effective in these experiments. A blocker of Caspase 3 also reduced neuron loss following L-NAME treatment in these cultures, although this was less effective. Following peripheral nerve section c-jun is also found in the nuclei of axotomized DRG neurons. Since it has also been shown that NO can inhibit the activity of *c-jun N*-terminal kinase 2 by *S*-nitrosylation and as a result inhibit apoptosis,[32] this is another possible mechanism for the protective action of NO in DRG neurons.

The Bcl-2 family of proteins are also clearly of interest in the context of apoptosis. We have obtained immunocytochemical data to reveal that nNOS block in DRG neurons induces the synthesis Bax without causing changes in Bcl_2.[25] Recently Bcl-xL has been implicated in promoting DRG neuron survival by blocking the activation of caspase 3,[33] and it will be interesting to establish if NO interacts in a any way with Bcl-xL. Bax protein is considered to form an integral part of anionic pores in the mitochondrial membrane leading to the breakdown of the mitochondrial membrane potential. In turn, this results in cytochrome-c release and activation of procaspase 9, which, in turn, activates caspase 9. Hence, this data suggests that NO may be neuroprotective by inhibiting mitochondrial breakdown due to Bax formation and by inhibiting the activation of caspases 9 and 3.

SUMMARY AND CONCLUSIONS

FIGURE 1 summarizes some of these findings. In the small DRG neurons that express nNOS following nerve section, NO is neuroprotective acting intrinsically, probably by inhibiting Bax synthesis and caspase activation. Additionally, NO is neuroprotective to the large DRG neurons by activating cGMP in their satellite glia cells, which is causing the production of neuroprotective molecules or the inhibition of proapoptotic molecules.

In conclusion, it is now possible to propose that most of these findings point to a feedback relationship between NO production and NGF synthesis. In early development nNOS is present in all these ganglion neurons. As DRG neurons start to take

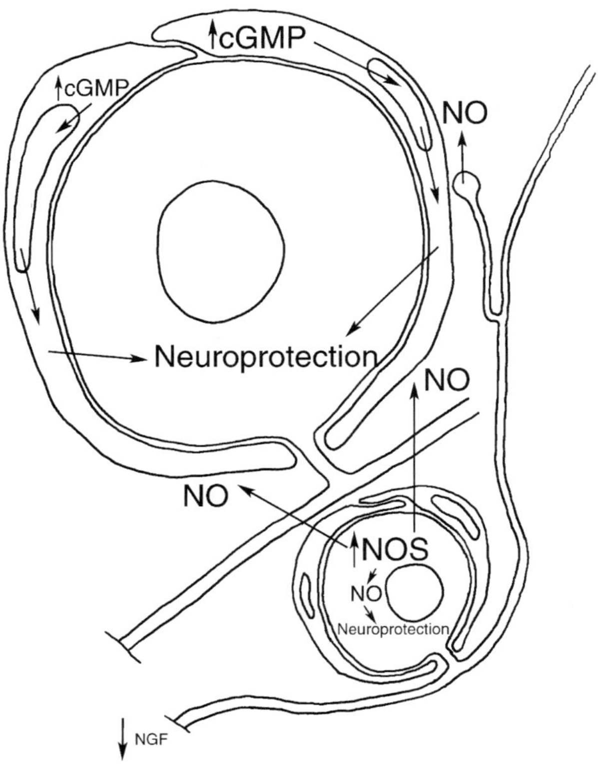

FIGURE 1. This schematic illustration summarizes the principle findings concerning the role of NO in neuroprotection for adult DRG neurons following peripheral axotomy. The axotomy results in a removal of retrograde transport of trophic factors such as NGF from target tissues. This results in an increase in nNOS in small DRG neurons. The NO produced inhibits apoptosis in these neurons intrinsically. Diffusion of NO either from these neurons or from axonal sprouts induces cGMP in satellite glia cells particularly around large DRG neurons. This in turn might result in protection of these large neurons.

up NGF from target tissues, nNOS protein production is downregulated. If in the adult DRG peripheral axons are sectioned, the neurons are deprived of retrogradely transported NGF and start once again to synthesize nNOS. This response is probably confined to neurons expressing *trkA* receptors. As a result of increased NO production, cGMP is synthesized in satellite glia cells, which in turn may start to produce NGF to promote neuronal survival and axon regeneration. This feedback relationship may extend to other neurotrophic factors and this requires further investigation.

REFERENCES

1. Lawson, S.N. 1992. Morphological and biochemical cell types of sensory neurons. *In* Sensory Neurons: Diversity, Development, and Plasticity. S.A. Scott, Ed.: 27–59. Oxford University Press, Oxford.
2. Belmonte, C. & F. Cervero. 1996. Neurobiology of Nociceptors. Oxford University Press, Oxford.
3. McMahon, S.B. & M. Koltzenburg. 1992. Itching for an explanation. Trend. Neurosci. **15:** 497–501.
4. Lynn, B. 1991. 'Silent' nociceptors in the skin. Trend. Neurosci. **14:** 95.
5. Morris, R., E. Southam, D.J. Braid, *et al.* 1992. Nitric oxide may act as a messenger between dorsal root ganglion neurones and their satellite cells. Neurosci. Lett. **137:** 29–32.
6. Meller, S.T. & G.F. Gebhart. 1993. Nitric oxide (NO) and nociceptive processing in the spinal cord. Pain **52:** 127–136.
7. Fiallos-Estrada, C.E., W. Kummer, B. Mayer, *et al.* 1993. Long-lasting increase of nitric oxide synthase immunoreactivity, NADPH-diaphorase reaction and c-JUN co-expression in rat dorsal root ganglion neurons following sciatic nerve transection. Neurosci. Lett. **150:** 169–173.
8. Morris, R., E. Southam, S.R. Gittins, *et al.* 1993. NADPH-diaphorase staining in autonomic and somatic cranial ganglia of the rat. Neuroreport **4:** 62–64.
9. Stucky, C.L., T. DeChiara, R.M. Lindsay, *et al.* 1998. Neurotrophin 4 is required for the survival of a subclass of hair follicle receptors. J. Neurosci. **18:** 7040–7046.
10. Albers, K.M., D.E. Wright & B.M. Davis. 1994. Overexpression of nerve growth factor in epidermis of transgenic mice causes hypertrophy of the peripheral nervous system. J. Neurosci. **14:** 1422–1432.
11. Alvares, D. & M. Fitzgerald. 1999. Building blocks of pain: the regulation of key molecules in spinal sensory neurones during development and following peripheral axotomy. Pain S71–S85.
12. Himes, B.T. & A. Tessler. 1989. Death of some dorsal root ganglion neurons and plasticity of others following sciatic nerve section in adult and neonatal rats. J. Compar. Neurol. **284:** 215–230.
13. Coggeshall, R.E., H.A. Lekan, T.P. Doubell, *et al.* 1997. Central changes in primary afferent fibers following peripheral nerve lesions. Neuroscience **77:** 1115–1122.
14. Bredt, D.S. & S.H. Snyder. 1994. Transient nitric oxide synthase neurons in embryonic cerebral cortical plate, sensory ganglia, and olfactory epithelium. Neuron **13:** 301–313.
15. Thippeswamy, T. & R. Morris. 1998. Evidence for developmental changes in the role of nitric oxide (NO) in dorsal root ganglion (DRG) neurones in the rat. J. Physiol. **509P:** 182P–183P.
16. Thippeswamy, T. & R. Morris. 1997. Nerve growth factor inhibits the expression of nitric oxide synthase in neurones in dissociated cultures of rat dorsal root ganglia. Neurosci. Lett. **230:** 9–12.

17. THIPPESWAMY, T. & R. MORRIS. 2001. Evidence that nitric oxide induced synthesis of cGMP occurs in a paracrine but not autocrine fashion and that the site of its release can be regulated: studies in dorsal root ganglia *in vivo* and *in vitro*. Nitric Oxide **5**(2): doi. 10. 1006/ niox. 2001. 0136 online.
18. RAMER, M.S., J.V. PRIESTLEY & S.B. MCMAHON. 2000. Functional regeneration of sensory axons into the adult spinal cord. Nature **403:** 312–316.
19. VERGE, V.M., Z. XU, X.J. XU, *et al.* 1992. Marked increase in nitric oxide synthase mRNA in rat dorsal root ganglia after peripheral axotomy: *in situ* hybridization and functional studies. Proc. Natl. Acad. Sci. U.S.A. **89:** 11617–11621.
20. ZHANG, X., V. VERGE, H.Z. WIESENFELD, *et al.* 1993. Nitric oxide synthase-like immunoreactivity in lumbar dorsal root ganglia and spinal cord of rat and monkey and effect of peripheral axotomy. J. Compar. Neurol. **335:** 563–575.
21. JAIN, R.K., T. THIPPESWAMY & R. MORRIS. 2000. Nitric oxide synthase induction in dorsal root ganglion neurones following peripheral nerve section occurs selectively in calcitonin gene-related peptide containing neurones. Eur. J. Neurosci. **12:** 433.
22. SHI, T.J., K. HOLMBERG, Z.Q. XU, *et al.* 1998. Effect of peripheral nerve injury on cGMP and nitric oxide synthase levels in rat dorsal root ganglia: time course and coexistence. Pain **78:** 171–180.
23. THIPPESWAMY, T. & R. MORRIS. 1997. Cyclic guanosine 3′,5′-monophosphate-mediated neuroprotection by nitric oxide in dissociated cultures of rat dorsal root ganglion neurones. Brain Res. **774:** 116–122.
24. THIPPESWAMY, T., R.K. JAIN, M. NAZLI, *et al.* 2001. Inhibition of neuronal nitric oxide synthase results in neurodegenerative changes in axotomised dorsal root ganglion neurones: evidence for a neuroprotective role of nitric oxide *in vivo*. Neurosci. Res. **44:** 37–44.
25. THIPPESWAMY, T. & R. MORRIS. 2001. Bax, TNFα, caspase-3 and -9 are inhibited by endogenous nitric oxide: Immunocytochemical studies on dissociated rat DRG neurones. Nitric Oxide. Eur. J. Neurosci. **14:** 1229–1236.
26. LEE, S.E., H. SHEN, G. TAGLIALATELA, *et al.* 1998. Expression of nerve growth factor in the dorsal root ganglion after peripheral nerve injury. Brain Res. **796:** 99–106.
27. MELINO, G., F. BERNASSOLA, R.A. KNIGHT, *et al.* 1997. *S*-nitrosylation regulates apoptosis. Nature **388:** 432–433.
28. MOHR, S., B. ZECH, E.G. LAPETINA, *et al.* 1997. Inhibition of caspase-3 by *S*-nitrosation and oxidation caused by nitric oxide. Biochem. Biophys. Res. Commun. **238:** 387–391.
29. TENNETI, L., D.M. D'EMILIA & S.A. LIPTON. 1997. Suppression of neuronal apoptosis by *S*-nitrosylation of caspases. Neurosci. Lett. **236:** 139–142.
30. BUDIHARDJO, I., H. OLIVER, M. LUTTER, *et al.* 1999. Biochemical pathways of caspase activation during apoptosis. Annu. Rev. Cell Dev. Biol. **15:** 269–290.
31. SLEE, E.A., M.T. HARTE, R.M. KLUCK, *et al.* 1999. Ordering the cytochrome c-initiated caspase cascade: hierarchical activation of caspases-2, -3, -6, -7, -8, and -10 in a caspase-9-dependent manner. J. Cell Biol. **144:** 281–292.
32. SO, H.S., R.K. PARK, M.S. KIM, *et al.* 1998. Nitric oxide inhibits c-Jun *N*-terminal kinase 2 (JNK2) via *S*-nitrosylation. Biochem. Biophys. Res. Commun. **247:** 809–813.
33. URASE, K., T. MOMOI, E. FUJITA, *et al.* 1999. Bcl-xL is a negative regulator of caspase-3 activation in immature neurons during development. Brain Res. Dev. Brain Res. **116:** 69–78.

Spatial and Temporal Correlation of Nitric Oxide Synthase Expression with CuZn-Superoxide Dismutase Reduction in Motor Neurons following Axotomy

WAN-HUA AMY YU

Department of Cell Biology and Anatomical Sciences, City University of New York Medical School, New York, New York, U.S.A.

ABSTRACT: Axotomized neurons expressing neuronal nitric oxide synthase (nNOS) may use nitric oxide (NO), known for its antioxidant activities and ability to scavenge free radicals, to protect against oxidative stress. This hypothesis was tested by immunohistochemical examination of superoxide dismutase (SOD) in neurons of the hypoglossal nucleus (HGN) and dorsal motor nucleus of the vagus nerve (DMV) one day to ten weeks after unilateral hypoglossal nerve crush or avulsion combined with vagus nerve crush in adult rats, and also in neurons of the anterior horn (AH) one week after unilateral sciatic nerve crush or avulsion. In the HGN, emergence of nNOS coincided temporally with reduction of CuZn-SOD immunoreactivity (ir), and the level of reduction correlated with that of nNOS induction, differing only in magnitude between nerve crush and nerve avulsion. The two nerve lesion models further revealed the concurrence of nNOS abatement with recovery of CuZn-SOD ir, and absence of nNOS abatement with persistent low CuZn-SOD ir. In the AH, reduced CuZn-SOD ir was localized in the segments containing nNOS positive neurons as a result of sciatic nerve avulsion. CuZn-SOD ir was unchanged in the absence of nNOS induction following sciatic nerve crush. DMV neurons were devoid of CuZn-SOD ir. However, increased Mn-SOD ir one and two weeks post crush was similar to that in HGN neurons. DMV neurons lacked both nNOS abatement and CuZn-SOD ir, which may explain their particular vulnerability to cell death from axotomy in comparison with other peripheral neurons. These data suggest that axotomy-induced nNOS expression is causally linked to oxidative stress, and that NO is neuroprotective, but can become neurodestructive when produced in excess.

KEYWORDS: hypoglossal neurons; vagal motor neurons; spinal motor neurons; nerve crush; nerve avulsion; oxidative stress; cell death; immunohistochemistry

INTRODUCTION

Peripheral nerve lesion in adult rats induces the expression of neuronal nitric oxide synthase (nNOS), and axotomized neurons normally devoid of nNOS (or

Address for correspondence: Dr. Wan-hua Amy Yu, Department of Cell Biology and Anatomical Sciences, City University of New York Medical School, Convent Avenue and 138th Street, New York, NY 10031. U.S.A. Voice: 212-650-6859; fax: 212-650-6812.
ayu@med.cuny.edu

expressing the enzyme at very low levels) become strongly nNOS immunoreactive (ir). Despite the similarity of the phenomenon in somatic and visceral motor as well as sensory neurons,[1–6] the issue of nitric oxide (NO) mediating neurodestructive or neuroprotective processes remains controversial. It has been argued that NO may play a pivotal role in death of motor neurons subsequent to ventral root or nerve avulsion, since the severity of neuronal loss can be alleviated by blocking nNOS induction,[7–10] or inhibiting NOS activity.[11–12] However, treatment with an nNOS inhibitor after nerve transection results in degenerative changes in dorsal root ganglion (DRG) neurons.[13] Since avulsion lesion leads to persistent nNOS expression in virtually all axotomized neurons,[14–15] whereas nerve crush or transection results in only transient nNOS expression in some axotomized neurons.[15] Low levels of NO produced for a limited time are likely to exert effects that are different from production in excessive amounts over a protracted period of time.

When generated by human neuroblastoma cells expressing nNOS as a result of serum deprivation, NO at micromolar ranges increases cell viability via cGMP-mediated up- and downregulation of the antiapoptotic bcl-2 protein and $p66^{shc}$.[16] NO also protects dopaminergic neurons against reactive oxygen species via its antioxidative activities *in vitro*[17] and *in vivo.*[18] Thus, nNOS expression may accord axotomized neurons the opportunity to use NO in resisting the oxidative stress arising from injury-induced increases in the rate of protein synthesis.[19] Consistent with this view is the finding that following nerve transection, daily treatment with melatonin (an antioxidant and free radical scavenger) attenuates nNOS expression in axotomized neurons in a dose-dependent manner.[20]

Aerobic metabolism generates superoxide anions that are converted to hydrogen peroxide and water by superoxidase dismutase (SOD) in both mitochondria (Mn-SOD) and the cytoplasm (CuZn-SOD). Although nerve lesion transiently elevates Mn-SOD mRNA and protein levels, similar changes for CuZn-SOD were absent from axotomized neurons.[21–22] To ascertain the causal linkage of nNOS expression with oxidative stress, this study was undertaken to examine the status of SOD (with special attention to CuZn-SOD) in the neurons of the hypoglossal nucleus (HGN), dorsal motor nucleus of the vagus nerve (DMV), and anterior horn (AH), taking advantage of the dissimilarity in nNOS expression in this diverse group of motor neurons following nerve lesion.[14–15]

MATERIALS AND METHODS

Adult, female Sprague-Dawley Rats (180–200 g) were anesthetized with ketamine (50 mg/kg, i.p.) and xylazine (5 mg/kg, i.m.). The right hypoglossal nerve was either crushed or avulsed, and the right vagus nerve was crushed as described elsewhere.[15] At 1, 2, and 3 days, and 1, 2, 3, 6, and 10 weeks postoperation (PO), rats ($n = 2$ for each lesion type) were perfused transcardially with 4% paraformaldehyde (PF) in phosphate buffered saline (PBS). In a separate group of anesthetized rats, the right sciatic nerve was crushed or avulsed at midthigh, and the rats ($n = 2$ for each lesion type) were perfused with PF one week PO. Two sham operated animals served as non-lesion controls. The brainstem and the lower lumbar spinal cord were dissected, postfixed in PF for an additional three hours, and cryoprotected with

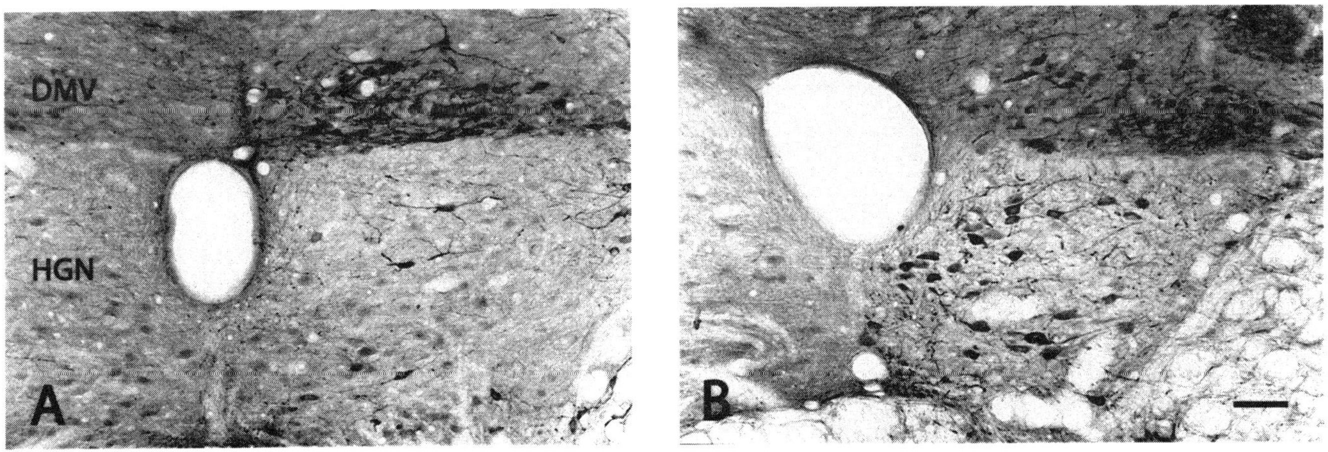

FIGURE 1. Photomicrographs showing nNOS-immunoreactive neurons in the hypoglossal nucleus (HGN) and the dorsal motor nucleus of the vagus nerve (DMV) six weeks after right hypoglossal and vagus nerve lesions. Note that axotomy induced nNOS expression has largely abated in the HGN of nerve crushed (**A**), but not in the HGN of nerve avulsed (**B**) animals, and that the size of the HGN of the latter has shrunk considerably due to cell loss. Upregulated nNOS expression remains unabated in DMV neurons. *Bar* = 100 μm.

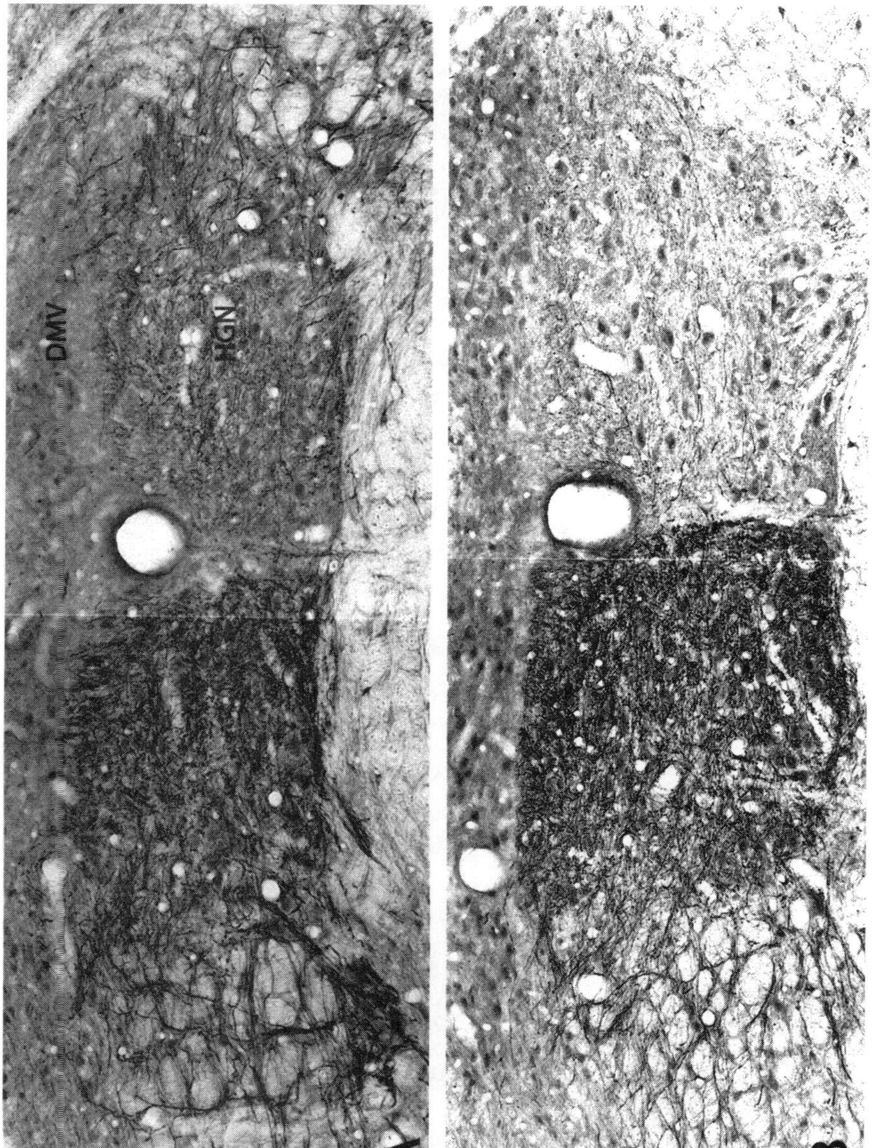
DMV
HGN
A
B

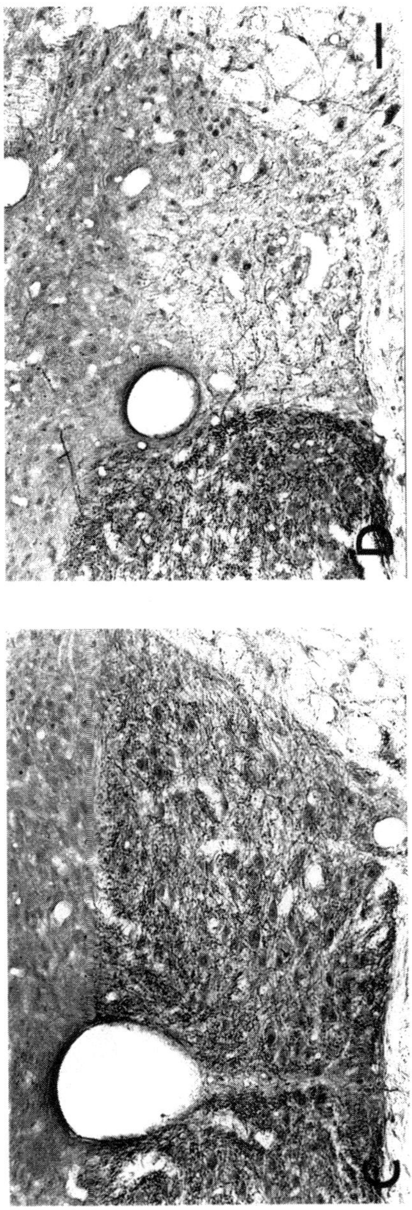

FIGURE 2. Photomicrographs showing CuZu-SOD immunoreactivity in the HGN. The staining intensity was reduced on the axotomy side relative to the contralateral, intact side three days (**A**) post lesion, further reduced to near extinction one week (**B**) after nerve avulsion, recovered in nerve crushed (**C**), but not in nerve avulsed (**D**) animals six weeks post lesion. Note that the DMV was devoid of immunoreactivity. *Bar* = 100 µm.

30% sucrose in PBS. Thirty µm-thick sections were cut with a freezing microtome in the transverse plane for the brainstem and in the transverse and horizontal planes for the spinal cord, and collected sequentially into eight wells, each containing sections spaced 240 µm apart.

Rabbit antiporcine nNOS antibody (Alexis Biochemicals, San Diego, CA) at 1:10,000 dilution, and sheep antihuman CuZn-SOD and Mn-SOD antibodies (Binding Site Ltd., Birmingham, England) at 1:2,000 and 1:3,000 dilution, respectively, were used on free-floating sections, followed by biotinylated secondary antibodies, ABC elite kit, and DAB kit (Vector Laboratories, Burlingame, CA), as described elsewhere.[15] Immunostaining controls using normal rabbit or sheep serum as a substitute for the primary antibody were completely without immunoreactivity.

RESULTS

Hypoglossal Nucleus

HGN neurons were devoid of NOS ir until three days after nerve crush or avulsion. Initially, only a few neurons were nNOS positive, but the number increased markedly one week PO, especially in nerve avulsed animals in which virtually every ipsilateral HGN neuron showed nNOS immunopositivity. Six weeks after nerve crush, the nNOS was ablated and HGN neurons were largely free from nNOS (see FIGURE 1 A). In contrast, nNOS persisted in neurons for as long as six and ten weeks after nerve avulsion amidst extensive neuronal loss, greatly shrinking the size of the HGN (FIG. 1 B).

Strong CuZn-SOD ir was localized in the HGN of non-lesion controls. Neuronal processes, including those traversing to the reticular formation from the lateral border of the HGN, were more intensely stained than neuronal perikayria. Three days after nerve crush or avulsion, CuZn-SOD ir was reduced relative to the contralateral HGN (see FIGURE 2 A). The reduction was more pronounced one week PO, especially in nerve avulsed animals in which only a few neuronal processes retained CuZn-SOD ir (FIG. 2 B). CuZn-SOD ir had almost fully recovered six weeks after nerve crush (FIG. 2 C), and was indistinguishable from the contralateral HGN 10 weeks PO. In contrast, CuZn-SOD ir was still depressed 6 and 10 weeks after nerve avulsion (FIG. 2 D). On the other hand, Mn-SOD ir, localized primarily in neuronal perikarya, was transiently elevated in the ipsilateral HGN neurons one week post nerve crush or avulsion (see FIGURE 3).

Dorsal Motor Nucleus of the Vagus Nerve

A few nNOS positive neurons were present in the non-lesion controls. The number and staining intensity of neurons with nNOS ir increased on the lesioned side two days PO, and continued to rise over time. Abatement of nNOS was not detected at six or ten weeks PO (FIG. 1). CuZn-SOD ir was absent in either controls or nerve lesioned animals (FIG. 2). However, DMV neurons were Mn-SOD ir. Its staining intensity was increased in the ipsilateral DMV neurons one and two weeks PO (FIG. 3), but had become indistinguishable from the contralateral side by three weeks PO.

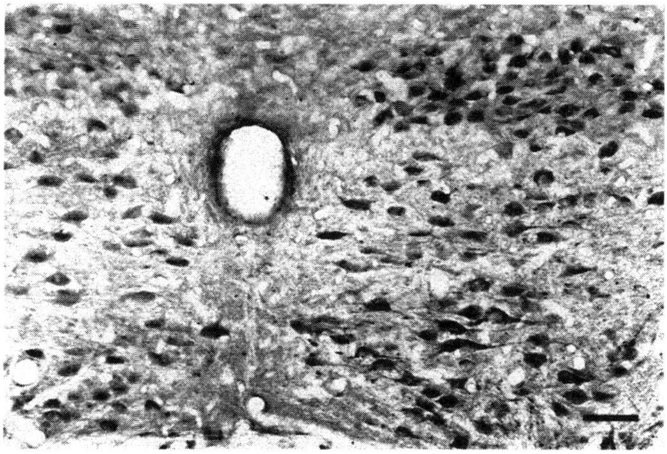

FIGURE 3. Photomicrograph showing Mn-SOD immunoreactivity in HGN and DMV neurons one week post lesion. Note the increased staining intensity on the lesion side. *Bar* = 100 μm.

Lumbar Spinal Cord

Neurons in the AH were devoid of nNOS ir in both non-lesion controls and sciatic nerve crushed animals. In contrast, many nNOS positive neurons were localized in the ipsilateral motor pool of the sciatic nerve avulsed animals. CuZn-SOD ir was reduced on the side of nerve avulsion (see FIGURE 4A), and the reduction was limited to the segments containing axotomized motor neurons (FIG. 4B, arrows). This reduction was not as pronounced as in the HGN after nerve avulsion. CuZn-SOD ir was unaltered in sciatic nerve crushed animals that had a staining intensity similar to that of the non-lesion controls.

DISCUSSION

The present study reveals a striking pattern of temporal and spatial correlation of nNOS expression with CuZn-SOD ir reduction in neurons of the HGN and AH following hypoglossal and sciatic nerve lesions, respectively. That the two cellular events have an antecedent relationship is further corroborated by the following findings. First, the time course and extent of nNOS expression run in parallel with, and conform closely to that of CuZn-SOD ir reduction, as shown in the HGN, although the magnitude of nNOS induction varies considerably following nerve crush and nerve avulsion. Reversible nNOS expression occurs in the former, but not in the latter. Second, CuZn-SOD ir reduction in the AH is limited to the segments in which nNOS was induced, since nNOS is inducible in AH neurons only by nerve avulsion, but not by nerve crush or transection.[1,6] The absence of CuZn-SOD mRNA upregulation,[22] and reduced CuZn-SOD protein levels seen in the present study could render injured neurons vulnerable to oxidative stress. Mn-SOD ir is transiently elevated in HGN and DMV neurons, indicative of elevated protein levels akin to that reported in facial motor and DRG neurons,[21,22] as are mRNA levels.[22] However, it may not be sufficient to compensate for the loss of CuZn-SOD, or to protect injured neurons from oxidative stress. In the HGN, increased Mn-SOD ir after either type of nerve lesion is evident only at one week PO, whereas the rate of protein synthesis increases significantly as early as day 2 following hypoglossal nerve transection. This increase is sustained at 25% above the rate of the contralateral HGN through postaxotomy day 24.[19] Even on postaxotomy day 35, 14 days after establishing a functional connection with the target, protein synthesis rates are still higher on the lesioned side. In order to counteract sustained high levels of free radical production due to elevated protein synthesis, neurons could either activate endogenous

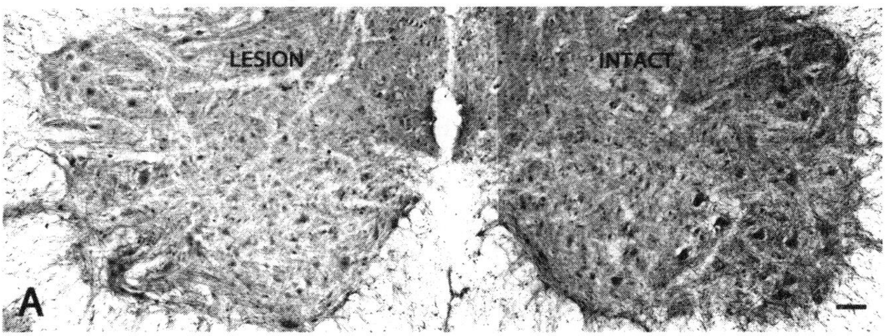

FIGURE 4. Photomicrograph showing CuZn-SOD immunoreactivity in the anterior horn one week after sciatic nerve avulsion. Note that the staining intensity was reduced on the lesion side relative to the contralateral intact side shown in a transverse section **(A)**, and limited to segments (*arrows*) containing axotomized neurons shown in a horizontal section **(B)**. *Bar* = 100 μm.

antioxidants, induce antioxidant substances, or both. Since NO scavenges superoxide anions and exerts other antioxidative effects, it is well suited to augment the function of endogenous antioxidants. The remarkably close proximity of the time course of nNOS expression in the HGN neurons in the nerve transection experiment[15] with that of elevated protein synthesis[19] (extending even to the maximum number of nNOS positive neurons with peak increases in the rate of protein synthesis), further attests to the notion that axotomized neurons express nNOS in order to generate NO to contain the oxidative stress.

It is well known that cranial motor neurons are more prone to nNOS induction than their spinal counterparts. The former readily express nNOS following nerve crush or transection, whereas the latter, with the exception of pudental motor neurons in the sacral spinal cord,[3] express nNOS only after nerve or nerve root avulsion.[1,6,14] The disparity may be explained by regional differences in the amount of SOD. As shown in the present study, nerve injury-induced reduction in CuZn-SOD ir is much more severe in the HGN than in the AH. Furthermore, Mn-SOD upregulation in the HGN is rather short lived compared with that in the AH,[21,22] different methodologies being used notwithstanding. Thus, when challenged by a sudden rise of free radical production, cranial motor neurons may not possess the same capability as spinal motor neurons to handle the oxidative stress, and hence, are more apt to recruit nNOS to produce NO for antioxidant defense.

It is noteworthy that despite abundant CuZn-SOD in the motor neurons of cranial and spinal nerves,[23,24] CuZn-SOD ir is conspicuously absent from DMV neurons. DMV neurons have been known for decades to be more susceptible than most other peripherally projecting motor neurons to delayed neuronal death from axonal injury.[25,26] This unique and specific vulnerability may be rooted in their lack of CuZn-SOD on the one hand, and persistent nNOS expression on the other; nNOS is not downregulated 10 weeks post nerve crush, consistent with the lack of nNOS abatement regardless of the type of nerve injury.[15]

Although SOD is the first line of defense against the damaging effects of superoxide anions, CuZn-SOD is a key target for protein oxidative injury,[27] and Mn-SOD is susceptible to tyrosine nitration, with the degree of enzyme inactivation paralleling the nitration.[28,29] The selective reduction of CuZn-SOD ir in axotomized neurons shown in the present study further substantiates the frailty of the cellular antioxidant defense system under super-physiological conditions.

In conclusion, this study has clearly established the association of axotomy-induced nNOS expression with oxidative stress, and confirmed the Janus-faced nature of NO, which is neuroprotective under regulated nNOS expression but neurodestructive when the production is uncontrolled and prolonged.

ACKNOWLEDGMENTS

This study was supported by a research grant (RF 62302 00 31) from the PSC-CUNY Research Award Program of the City University of New York.

REFERENCES

1. Fiallos-Estrada, C.E., W. Kummer, B. Mayer, *et al.* 1993. Long-lasting increase of nitric oxide synthase immunoreactivity, NADPH-diaphorase reaction and c-JUN co-expression in rat dorsal root ganglion neurons following sciatic nerve transection. Neurosci. Lett. **150:** 169–173.
2. Klimaschewski, L., N. Obermuller, M. Majewski, *et al.* 1996. Increased expression of nitric oxide synthase in a subpopulation of rat sympathetic neurons after axotomy—correlation with vasoactive intestinal peptide. Cell Tissue Res. **285:** 419–425.
3. Pullen, A.H. & P. Humphreys. 1999. Protracted elevation of neuronal nitric oxide synthase immunoreactivity in axotomised adult pudendal motor neurons. J. Anat. **194:** 547–565.
4. Vizzard, M.A., S.L. Erdman & W.C. de Groat. 1995. Increased expression of neuronal nitric oxide synthase (NOS) in visceral neurons after nerve injury. J. Neurosci. **15:** 4033–4045.
5. Wu, W., F.J. Liuzzi, F.P. Schinco, *et al.* 1994. Neuronal nitric oxide synthase is induced in spinal neurons by traumatic injury. Neuroscience **61:** 719–726.
6. Yu, W.H.A. 1994. Nitric oxide synthase in motor neurons after axotomy. J. Histochem. Cytochem. **42:** 451–457.
7. Chai, H., W. Wu, K.F. So, *et al.* 1999. Long-term effects of a single dose of brain-derived neurotrophic factor on motoneuron survival following spinal root avulsion in the adult rat. Neurosci. Lett. **274:** 147–150.
8. Novikov, L., L. Novikova & J.O. Kellerth. 1995. Brain-derived neurotrophic factor promotes survival and blocks nitric oxide synthase expression in adult rat spinal motoneurons after ventral root avulsion. Neurosci. Lett. **200:** 45–48.
9. Sakamoto, T., K. Watabe, T. Ohashi, *et al.* 2000. Adenoviral vector-mediated GDNF gene transfer prevents death of adult facial motoneurons. Neuroreport **11:** 1857–1860.
10. Watabe, K., T. Ohashi, T. Sakamoto, *et al.* 2000. Rescue of lesioned adult rat spinal motoneurons by adenoviral gene transfer of glial cell line-derived neurotrophic factor. J. Neurosci. Res. **60:** 511–519.
11. Ruan, R.S., S.K. Leong & K.H. Yeoh. 1995. The role of nitric oxide in facial motoneuronal death. Brain Res. **698:** 163–168.
12. Wu, W. & L. Li. 1993. Inhibition of nitric oxide synthase reduces motoneuron death due to spinal root avulsion. Neurosci. Lett. **153:** 121–124.
13. Thippeswamy, T., R.K. Jain, N. Mumtaz & R. Morris. 2001. Inhibition of neuronal nitric oxide synthase results in neurodegenerative changes in the axotomised dorsal root ganglion neurons: evidence for a neuroprotective role of nitric oxide *in vivo.* Neurosci. Res. **40:** 37–44.
14. Wu, W. 1993. Expression of nitric-oxide synthase (NOS) in injured CNS neurons as shown by NADPH diaphorase histochemistry. Exp. Neurol. **120:** 153–159.
15. Yu, W.H.A. 1997. Regulation of nitric oxide synthase expression in motoneurons following nerve injury. Dev. Neurosci. **19:** 247–254.
16. Andoh, T., S.Y. Lee & C.C. Chiueh. 2000. Preconditioning regulation of bcl-2 and p66shc by human NOS1 enhances tolerance to oxidative stress. FASEB J. **14:** 2144–2146.
17. Wink, D.A., I. Hanbauer, M.C. Krishna, *et al.* 1993. Nitric oxide protects against cellular damage and cytotoxicity from reactive oxygen species. Proc. Natl. Acad. Sci. U.S.A. **90:** 9813–9817.
18. Rauhala, P., A.M. Lin & C.C. Chiueh. 1998. Neuroprotection by *S*-nitrosoglutathione of brain dopamine neurons from oxidative stress. FASEB J. **12:** 165–173.
19. Sun, Y., G.E. Deibler & C.B. Smith. 1993. Effects of axotomy on protein synthesis in the rat hypoglossal nucleus: examination of the influence of local recycling of leucine derived from protein degradation into the precursor pool. J. Cereb. Blood Flow Metab. **13:** 1006–1012.

20. Chang, H.M., E.A. Ling, J.H. Lue, *et al.* 2000. Melatonin attenuates neuronal NADPH-d/NOS expression in the hypoglossal nucleus of adult rats following peripheral nerve injury. Brain Res. **873:** 243–251.
21. Rosenfeld, J., S. Cook & R. James. 1997. Expression of superoxide dismutase following axotomy. Exp. Neurol. **147:** 37–47.
22. Yoneda, T., S. Inagaki, Y. Hayashi, *et al.* 1992. Differential regulation of manganese and copper/zinc superoxide dismutases by the facial nerve transection. Brain Res. **582:** 342–345.
23. Moreno, S., R. Nardacci & M.P. Ceru. 1997. Regional and ultrastructural immunolocalization of copper-zinc superoxide dismutase in rat central nervous system. J. Histochem. Cytochem. **45:** 1611–1622.
24. Pardo, C.A., Z. Xu, D.R. Borchelt, *et al.* 1995. Superoxide dismutase is an abundant component in cell bodies, dendrites, and axons of motor neurons and in a subset of other neurons. Proc. Natl. Acad. Sci. U.S.A. **92:** 954–958.
25. Aldskogius, H., K.D. Barron & R. Regal. 1980. Axon reaction in dorsal motor vagal and hypoglossal neurons of the adult rat. Light microscopy and RNA-cytochemistry. J. Comp. Neurol. **193:** 165–177.
26. Lewis, P.R., P.B. Jones, S.M. Breathnach & V. Navaratnam. 1972. Regenerative capacity of visceral preganglionic neurones. Nat. New Biol. **236:** 181–182.
27. Andrus, P.K., T.J. Fleck, M.E. Gurney & E.D. Hall. 1998. Protein oxidative damage in a transgenic mouse model of familial amyotrophic lateral sclerosis. J. Neurochem. **71:** 2041–2048.
28. MacMillan-Crow, L.A., J.P. Crow, J.D. Kerby, *et al.* 1995. Nitration and inactivation of manganese superoxide dismutase in chronic rejection of human renal allografts. Proc. Natl. Acad. Sci. U.S.A. **93:** 11853–11858.
29. MacMillan-Crow, L.A., J.P. Crow & J.A. Thompson. 1998. Peroxynitrite-mediated inactivation of manganese superoxide dismutase involves nitration and oxidation of critical tyrosine residues. Biochemistry **37:** 1613–1622.

Regulation of Endothelial Nitric Oxide Synthase Activity and Gene Expression

KENNETH K. WU

Vascular Biology Research Center Institute of Molecular Medicine and Division of Hematology, University of Texas–Houston Health Science Center, Houston, Texas, U.S.A.

ABSTRACT: Endothelial nitric oxide synthase (eNOS) is constitutively expressed in endothelial cells lining the blood vessel and the heart. It plays a major role in vascular and tissue protection. Its activity is tightly controlled by an intramolecular autoinhibitory element that hinders calmodulin binding. This molecular hindrance is removed by elevated intracellular calcium levels. The catalytic activity of eNOS is augmented by phosphorylation of a C-terminal serine residue (Ser-1177 of human eNOS) through the phosphatidyl-3 kinase (PI-3K)/Akt pathway. Its activity is also enhanced by binding to heat shock protein-90. These two processes are calcium independent. The two biochemical events appear to facilitate calmodulin access to its binding site. eNOS is upregulated at the transcriptional level. Its upregulation is mediated by an increased Sp1 binding to its cognate site on eNOS promoter/enhancer region via the action of protein phosphatase 2A (PP2A). PP2A is activated by a signaling pathway including PI-3γ → Janus activated kinase 2 (Jak2) → MEK-1 → ERK1 and 2. The transcriptional and posttranslational enhancement of eNOS activity is two- to threefold above the basal level. A higher magnitude of augmentation of eNOS gene expression can be achieved by gene transfer, which confers protection against vascular diseases and ischemia-induced tissue injury in experimental animals. These findings provide new insight into the protective role of eNOS and the therapeutic potential of eNOS gene therapy.

KEYWORDS: nitric oxide; eNOS; vascular biology; gene therapy

INTRODUCTION

Endothelial nitric oxide synthase (eNOS) was originally discovered as the enzymatic activity that catalyzed the synthesis of endothelium-dependent relaxing factor, nitric oxide (NO), from L-arginine.[1–3] Its cDNA was the last of the three isoforms that were cloned and is designated NOS-3, after neuronal NOS as NOS-1 and inducible NOS as NOS-2. eNOS sequence is highly homologous to the other two isoforms. Its structure and catalytic properties are also similar to those of other isozymes (for reviews see Refs. 4–7). Human eNOS is encoded by a single copy of genome located on chromosome 7q35-36.[8] The genomic DNA encodes a 1294-aa protein with a molecular weight of about 135 kDa.[9,10] eNOS is constitutively

Address for correspondence: Kenneth K. Wu, Vascular Biology Research Center and Division of Hematology, University of Texas–Houston Medical School, Houston, Texas, U.S.A. Voice: 713-500-6801; fax: 713-500-6812.
kenneth.k.w@uth.tmc.edu

expressed in endothelial cells that line the entire vasculature, including the vessels in the brain and the heart. ENOS plays an important role in maintaining vascular tone and vascular integrity. It is considered to be a major vasoprotective molecule. Mice with homozygous eNOS deletion are prone to develop hypertension and ischemia-induced tissue damage.[11,12] The protective action of eNOS is attributed to its catalysis of NO production, a radical that diffuses extracellularly and targets subendothelial smooth muscle cells and blood platelets. NO activates soluble guanylyl cyclase and the resultant cyclic GMP synthesis induces smooth muscle cell relaxation and inhibits platelet aggregation and adhesion. NO is a highly reactive molecule and an excessive NO production causes cell damage. The catalytic action of eNOS must, therefore, be tightly regulated. Recent studies have provided convincing evidence that eNOS regulation occurs at the transcriptional and posttranslational levels. Furthermore, eNOS has certain intrinsic molecular properties that allow for its control at the basal cellular condition.

SILENCING eNOS ACTIVITY AT BASAL CELLULAR STATE

Studies from cultured endothelial cells have shown that eNOS at the basal cellular state is inactive. Evidence is now emerging that the eNOS activity is controlled by two inherent eNOS properties. One biochemical characteristic of eNOS is myristoylation at glycine-2 and palmitoylation at cystein-15/cysteine-26 which allows its localization to plasmalemmal caveolæ, where it binds to caveolin-1 (Cav-1).[13–16] Like other signaling molecules, eNOS binds to the scaffold domain of Cav-1 and its activity is blocked by this interaction.[17–19] Another molecular feature of eNOS is the presence of an approximately 50-aa sequence at the C-terminal region of the reductase domain that hinders calmodulin binding to CaM binding site.[20–23] Deletion of this element from eNOS molecule almost completely abolishes the requirement for Ca^{2+} and the deletion mutant behaves similarly to iNOS.[23]

Thus, it may be speculated from results derived mostly from *in vitro* studies, that eNOS activity in basal state is silenced because of constraints from several intrinsic structural features: (1) membrane anchoring via myristoylation of the N-terminal glycine-2, and palmitoylation of two cysteine residues; (2) Cav-1 binding motif that brings it to caveolæ with resultant sequestration of the enzyme; and (3) an "auto-inhibitory" element at the C-terminal region that hinders calmodulin binding. These structural features are possibly closely related as all these molecular constraints are concurrently released by a surge of intracellular calcium.

To understand how these structural elements pose a severe constraint to basal eNOS activity requires a brief overview of the structure–activity relationship of this enzyme, which has been extensively evaluated by biochemical and molecular genetic approaches. Purified eNOS has two biodomains: a C-terminal reductase domain that contains NADPH, FAD, FMN, and CaM binding sites and an N-terminal oxygenase domain that contains binding sites for tetrahydrobiopterin (BH_4), heme, and L-arginine.[24,25] These two domains are independently active, as evidenced by their separate expression in baculovirus systems that show intact activity. Furthermore, these two domains reconstitute eNOS activity when they are mixed in aqueous solution.[25] Site-directed mutagenesis identified cysteine-184 as heme ligand[26] and glutamate-361 as the L-arginine binding site.[27] These two binding sites have been confirmed by X-ray crystallographic structure of the oxygenase domain.[28] The X-ray structure reveals that the substrate channel and heme binding site are supported by a scaffold and that heme, as well as BH_4, play an important role in maintaining the integrity of this scaffold.[28–32] Mutation of key residues in this scaffold region collapses the catalytic activity.[33] BH_4 is absolutely required for eNOS activity.[34] It is involved in dimer formation and electron transfer, as well as active site structural integrity. Catalysis of NO synthesis at this active site requires stepwise transfer of electrons initiated by NADPH binding to its site in the reductase domain.[35] Electrons are transferred from NADPH to FAD and then to FMN from where electron is transferred to the active site heme of the oxygenase domain and converts ferric heme to ferrous that binds O_2 actively. Experimental evidence suggests that CaM bound to eNOS play a pivotal role in electron transfer from the reductase domain to the oxygenase domain.[35] However, it is unclear as to how CaM facilitates electron transfer. Since CaM binding to the reductase domain is almost completely blocked by the C-terminal autoinhibitory element, it is unclear whether the eNOS activity is blocked through an additional inactivation mechanism when it binds to Cav-1 or its binding to Cav-1 is primarily to sequester it in caveolæ. This is an important issue for further investigation since eNOS could be distributed to different compartments of endothelial cells and the caveolæ component may be involved in specialized signaling activities. In this context, a cytosolic compartment of eNOS is highly possible, and this compartment may be the factory where NO is synthesized and diffused out of the cells to act on its blood and subendothelial targets.

ACTIVATION OF eNOS UNDER PHYSIOLOGICAL STIMULATION

Endothelial cells are stimulated by various physical and chemical stimuli, including shear stress, thrombin, histamine, bradykinin, and acetylcholine. These exogenous stimuli evoke signaling transmissions through Ca^{2+} and other signaling molecules. The effect of Ca^{2+} on eNOS activation has been extensively investigated. It releases eNOS from Cav-1 and displaces the autoinhibitory element to allow for CaM access to its binding site, thereby activating eNOS catalytic activity.[23,36] eNOS has been shown to bind heat shock protein 90 (hsp90) in a calcium-independent manner.[37] Hsp90 allosterically modifies eNOS, thereby increasing eNOS activity. Evidence has been provided to show that eNOS forms a complex with hsp90 in endothelial cells stimulated with histamine, bradykinin, vascular endothelial growth factor, and fluid shear stress resulting in an approximately threefold increase in eNOS activity.[37] Furthermore, hsp90 facilitates the dissociation of eNOS from caveolin-1.[38]

It has been reported that shear stress induces phosphorylation of serine-1177 of human eNOS (or serine-1179 of bovine eNOS) resulting in a threefold increase in eNOS activity.[39,40] Phosphorylation is mediated by the phosphatidylinositol-3 kinase (PI-3K) pathway.[41] PI-3K phosphorylates and activates Akt (PkB) kinase, which in turn phosphorylates eNOS.[40,41] It is unclear as to how phosphorylation of this C-terminal serine residue increases the NO synthesis activity. It is speculated that phosphorylation reduces the hindrance of CaM binding by the autoinhibitory element.[42]

eNOS catalytic activity is influenced by BH_4 and substrate availability. Under certain pathophysiological conditions, there may be a relative inadequacy of BH_4 which causes a compromise of NO synthesis, whereas excessive BH_4 increases NO production.[43] Under some pathological conditions, L-arginine could also be a limiting factor in NO synthesis. For example, in some studies exogenous L-arginine infusion has been shown to increase NO synthesis and improve endothelial function.[44] It seems odd that L-arginine could be a limiting factor since L-arginine freely diffuses into cells and the intracellular L-arginine concentration is greater than that required for eNOS catalytic action. It has recently been suggested that there may be a separate pool of L-arginine in the cell to serve as eNOS substrate, and that this pool of substrate is supplied by an active transport of L-arginine by a specific L-arginine binding protein.[45] Further studies are needed to confirm this.

UPREGULATION OF eNOS EXPRESSION

ENOS protein and mRNA levels have been shown to be upregulated by chronic exercise, hypoxia, fluid shear stress, and lysophosphatidylcholine (lysoPC).[46–48] The mechanisms by which these physical and chemical factors increase eNOS expression in endothelial cells were unclear. We were intrigued by the ability of lysoPC to increase eNOS, because lysoPC was considered to be an important mediator of endothelial injury, inflammation and vascular diseases.[49–52] We proposed that upregulated eNOS is an adaptive vasoprotective mechanism. To determine whether lysoPC increased eNOS expression at the transcriptional level, we performed nuclear run-off analysis and observed a two- to threefold increase in nascent eNOS mRNA synthase by lysoPC.[48] These results are consistent with stimulation of eNOS transcription by lysoPC.

To elucidate the mechanism of transcriptional upregulation, we evaluated the effects of lysoPC on eNOS promoter activity, determined the *cis*-regulatory element at the 5′-flanking promoter/enhancer region of eNOS that is involved in lysoPC-induced activation and the transcriptional activators that bound to the involved regulatory element. The eNOS promoter region does not contain TATA box and is G-C rich. A canonical Sp1 site at −104/−90 is essential for basal eNOS promoter activity.[53,54] 5′-Deletion and site-directed mutation analysis identified this Sp1 element as being essential, not only for basal, but also lysoPC-induced promoter activity.[55] Electrophoretic mobility shift and supershift assays revealed selective Sp1 binding to this cognate site and a twofold augmentation of Sp1 binding to this site by lysoPC.[55] LysoPC did not increase the Sp1 protein level in endothelial cell nuclear extracts. A selective inhibitor of protein phosphatase 2A (PP2A), okadaic acid, abrogated the enhanced Sp1 binding by lysoPC, suggesting that Sp1 dephosphorylation by PP2A is the underlying mechanism for increased Sp1 binding by lysoPC.

By using dominant negative mutants of PI-3K and Ras combined with pharmacologic inhibitors of PI-3K, Janus activated kinase-2 (Jak2) and MEK-1, we discovered a distinct signaling pathway by which lysoPC augmented eNOS promoter activity. The results indicate that lysoPC activated PI-3Kγ in an pertussis toxin-sensitive pathway.[56] PI-3Kγ activated Jak2, which in turn activated MEK-1 bypassing Ras and Raf-1. MEK-1 activated ERK1 and ERK2. PI-3K inhibitors, such as wortmannin and

Ly294002, MEK-1 inhibitor PD-98059, and Jak2 inhibitor AG-490 all suppressed eNOS promoter activity and Sp1 binding. These results suggest that ERK1 and 2 increased PP2A activity, thereby increasing Sp1 binding and eNOS promoter activity. We then demonstrated by immunoprecipitation, that PP2A bound to casein kinase 2 (CK2) in nucleus.[57] Activated ERK1 and 2 bound to CK2/PP2A complex where they increased PP2A activity.[57] *In vitro* experiments with purified CK2, PP2A and ERK2 provided clear evidence that PP2A activity was controlled by CK2 and activated ERK2 increased PP2A activity by reducing the CK2 control.[57] These findings have led us to propose a model for eNOS promoter activation (see FIGURE 1). In this model, we propose that lysoPC binds to a pertussis toxin sensitive Gi-coupled receptor, from where a cascade of kinases are activated with the final activation of PP2A. It is unclear if PP2A directly dephosphorylates Sp1. We suspect that this model of eNOS

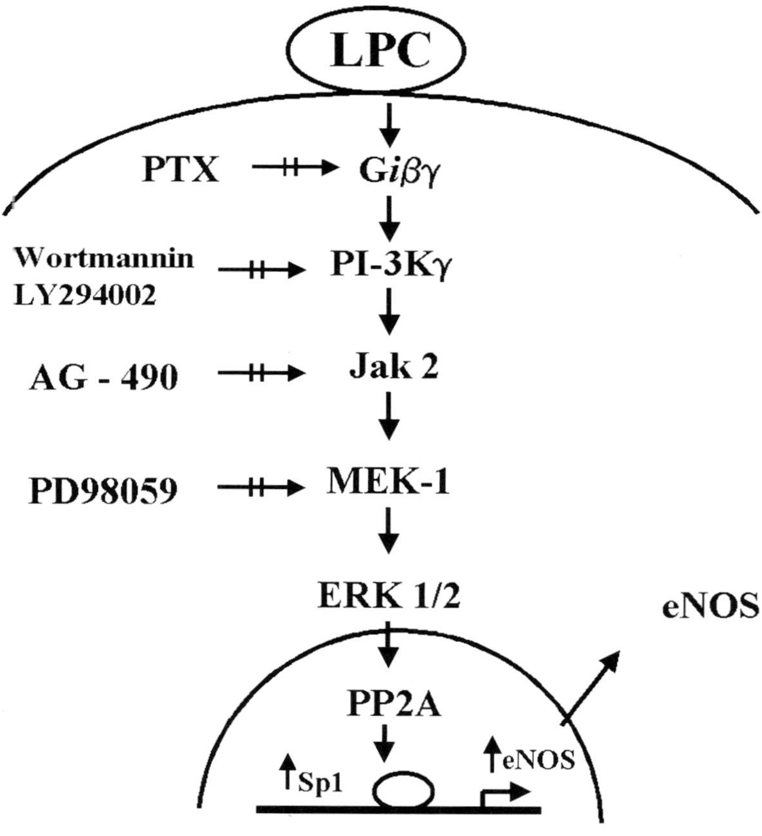

FIGURE 1. A schematic model depicting the signaling pathway by which eNOS promoter activity is upregulated by lysoPC. *Arrows* denote inhibition by selective inhibitors shown on the left hand side of the arrows. The selective involvement of PI-3Kγ was demonstrated by using a specific PI-3Kγ dominant negative mutant.[56]

transcriptional upregulation may explain the eNOS upregulation by shear stress and other stimuli.

Thus, transcriptional and posttranslational mechanisms are operative to increase eNOS activity in response to exogenous insults. The level of increase is limited to two- to threefold over the basal level of NO synthesis. This adaptive mechanism intends to counter the acute insult without producing excessive NO that may cause tissue damage. The tightly regulated NO synthesis is probably inadequate to protect vascular wall and tissues from persistent, severe insults. The endothelium becomes dysfunctional and eventually destroyed, and without the healthy endothelial protection, the vascular tissues are inflicted with perpetual inflammatory and proliferative lesions leading to complex vascular diseases, such as atherosclerosis, restenosis, and tissue infarction.

AUGMENTATION OF eNOS ACTIVITY BY GENE TRANSFER

Vascular damage reduces the capacity of endothelial cells to generate vasoprotective NO to fight against exogenous insults. One approach to compensate for the loss of this defense molecule is eNOS gene transfer. It has been shown in several animal models that eNOS transfer by viral vectors ameliorate a wide variety of disorders including injury-induced carotid artery intimal hyperplasia, atherosclerosis, reprofusion injury, pulmonary vasoconstriction, and shock (for a review, see Ref. 58). The beneficial effects of eNOS transgene on such diverse complex vascular diseases and ischemia-induced tissue damage suggest that augmented NO production at the vascular sites possess inhibitory actions against inflammation, vascular remodeling, smooth muscle proliferation, and myocardial damage. The mechanism by which NO exerts the multitude of actions is not completely understood. NO may confer beneficial effects by increasing blood flow and inducing apoptosis of proliferative smooth muscle cells. It could also have an effect on monocyte/macrophage infiltration and vascular remodeling contributed by fibroblasts and smooth muscle cells. Suppression of these cellular activities is likely to be contributed by a combination of effects of NO on protein nitration, heme function and gene expression.

In conclusion, eNOS plays a pivotal role in protecting vascular endothelium and its underlying vascular tissues. Multiple mechanisms for its upregulation are operative to increase its capacity for defending against acute insults. eNOS gene transfer protects against restenosis, atherosclerosis, pulmonary hypertension, and ischemia/reperfusion tissue damage and has important therapeutic implications.

REFERENCES

1. FURCHGOTT, R.F. & J.V. ZAWADZKI. 1980. The obligatory role of endothelial cells in the relaxation of arterial smooth muscle by acetylcholine. Nature **288:** 373–376.
2. PALMER, R.M.J., D.S. ASHTON & S. MONCADA. 1988. Vascular endothelial cells synthesize nitric oxide from L-arginine. Nature **333:** 664–666.
3. IGNARRO, L.J., G.M. BUGA, K.S. WOOD, *et al.* 1987. Endothelium-derived relaxing factor produced and released from artery and vein in nitric oxide. Proc. Natl. Acad. Sci. U.S.A. **84:** 9265–9269.

4. MARLETTA, M.A. 1994. Nitric oxide synthase: aspects concerning structure and catalysis. Cell **78:** 927–930.
5. NATHAN, C. & Q.W. XIE. 1994. Nitric oxide synthases: roles, tolls, and controls. Cell **78:** 915–918.
6. WU, K.K. 1995. Inducible cyclooxygenase and nitric oxide synthase. Adv. Pharmacol. **33:** 179–207.
7. ADERTON, W.K., C.E. COOPER & R.G. KNOWLES. 2001. Nitric oxide synthases: structure function and inhibition. Biochem. J. **357:** 593–615.
8. MARSDEN, P.A., H.H.Q. HENG, S.W. SCHERER, *et al.* 1993. Structure and chromosomal localization of the human constitutive endothelial NO synthase. J. Biol. Chem. **268:** 17478–17488.
9. LAMAS, S., P.A. MARSDEN, G.K. LI, *et al.* 1992. Endothelial nitric oxide synthase: molecular cloning and characterization of a distinct constitutive enzyme isoform. Proc. Natl. Acad. Sci. U.S.A. **89:** 6348–6352.
10. SESSA, W.C., J.K. HARRISON, C.M. BARBER, *et al.* 1992. Molecular cloning and expression of a cDNA encoding endothelial cell nitric oxide synthase. J. Biol. Chem. **267:** 15274–15276.
11. SHESELY, E.G., N. MAEDA, H.-S. KIM, *et al.* 1996. Elevated blood pressure in mice lacking endothelial nitric oxide synthase. Proc. Natl. Acad. Sci. U.S.A. **93:** 13176-13181.
12. KUHLENCORDT, P.J., R. GYURKO, F. HAN, *et al.* 2001. Accelerated atherosclerosis, aortic aneurysm formation and ischemic heart disease in apolipoprotein E/endothelial nitric oxide synthase double knockout mice. Circulation **104:** 448–454.
13. LIU, J., G. GARCIA-CARDENA & W.C. SESSA. 1995. Biosynthesis and palmitoylation of endothelial nitric oxide synthase: mutagenesis of palmitoylation sites, cysteines-15 and/or -26, argues against depalmitoylation-induced translocation of the enzyme. Biochemistry **34:** 12333–12340.
14. LIU, J., T.E. HUGHES & W.C. SESSA. 1997. The first 35 amino acids and fatty acylation sites determine the molecular targeting of endothelial nitric oxide synthase into the Golgi region of cells: a green fluorescent protein study. J. Cell Biol. **137:** 1525–1535.
15. SHAUL, P.W., E.J. SMART, L.J. ROBINSON, *et al.* 1996. Acylation targets endothelial nitric-oxide synthase to plasmalemmal caveolæ. J. Biol. Chem. **271:** 6518–6522.
16. GARCIA-CARDENA, G., P. OH, J. LIU, J.E. SCHNITZER, *et al.* 1996. Targeting of nitric oxide synthase to endothelial cell caveolæ via palmitoylation: implications for nitric oxide signaling. Proc. Natl. Acad. Sci. U.S.A. **93:** 6448–6453.
17. FERON, O., L. BELHASSEN, L. KOBZIK, *et al.* 1996. Endothelial nitric oxide synthase targeting to caveolae. Specific interactions with caveolin isoforms in cardiac myocytes and endothelial cells. J. Biol. Chem. **271:** 22810–22814.
18. JU H., R. ZOU, V.J. VENEMA, *et al.* 1997. Direct interaction of endothelial nitric-oxide synthase and caveolin-1 inhibits synthase activity. J. Biol. Chem. **272:** 18522–18525.
19. GARCIA-CARDENA, G., P. MARTASEK, B.S. MASTERS, *et al.* 1997. Dissecting the interaction between nitric oxide synthase (NOS) and caveolin. Functional significance of the NOS caveolin binding domain *in vivo*. J. Biol. Chem. **272:** 25437–25440.
20. SALERNO, J.C., D.E. HARRIS, K. IRIZARRY, *et al.* 1997. An autoinhibitory control element defines calcium-regulated isoforms of nitric oxide synthase. J. Biol. Chem. **272:** 29769–29777.
21. NISHIDA, C.R. & P.R. ORTIZ DE MONTELLANO. 1999. Autoinhibition of endothelial nitric-oxide synthase. Identification of an electron transfer control element. J. Biol. Chem. **274:** 14692–14698.
22. DAFF, S., I. SAGAMI & T. SHIMIZU. 1999. The 42-amino acid insert in the FMN domain of neuronal nitric-oxide synthase exerts control over Ca^{2+}/calmodulin-dependent electron transfer. J. Biol. Chem. **274:** 30589–30595.
23. CHEN, P.F. & K.K. WU. 2000. Characterization of the roles of the 594–645 region in human endothelial nitric oxide synthase in regulating calmodulin binding and electron transfer. J. Biol. Chem. **275:** 13155–13163.

24. SHETA, E.A., K. MCMILLAN & B.S. MASTERS. 1994. Evidence for a bidomain structure of constitutive cerebellar nitric oxide synthase. J. Biol. Chem. **269:** 15147–15153.
25. CHEN, P.F., A.L. TSAI, V. BERKA, *et al.* 1996. Endothelial nitric-oxide synthase. Evidence for bidomain structure and successful reconstitution of catalytic activity from two separate domains generated by a baculovirus expression system. J. Biol. Chem. **271:** 14631–14635.
26. CHEN, P.F., A.-L. TSAI & K.K. WU. 1994. Cysteine 184 of endothelial nitric oxide synthase is involved in heme coordination and catalytic activity. J. Biol. Chem. **269:** 25062–25066.
27. CHEN, P.F., A.L. TSAI, V. BERKA, *et al.* 1997. Mutation of Glu-361 in human endothelial nitric-oxide synthase selectively abolishes L-arginine binding without perturbing the behavior of heme and other redox centers. J. Biol. Chem. **272:** 6114–6118.
28. RAMAN, C.S., H. LI, P. MARTASEK, *et al.* 1998. Crystal structure of constitutive endothelial nitric oxide synthase: a paradigm for pterin function involving a novel metal center. Cell **95:** 939–950.
29. CRANE, B.R., A.S. ARVAI, R. GACHHUI, *et al.* 1997. The structure of nitric oxide synthase oxygenase domain and inhibitor complexes. Science **278:** 425–431.
30. CRANE, B.R., A.S. ARVAI, D.K. GHOSH, *et al.* 1998. Structure of nitric oxide synthase oxygenase dimer with pterin and substrate. Science **279:** 2121–2126.
31. FISCHMANN, T.O., A. HRUZA, X.D. NIU, *et al.* 1999. Structural characterization of nitric oxide synthase isoforms reveals striking active-site conservation. Nat. Struct. Biol. **6:** 233–242.
32. LI, H., C.S. RAMAN, C.B. GLASER, *et al.* 1999. Crystal structures of zinc-free and -bound heme domain of human inducible nitric-oxide synthase. Implications for dimer stability and comparison with endothelial nitric-oxide synthase. J. Biol. Chem. **274:** 21276–21284.
33. CHEN, P.F., V. BERKA, A.L. TSAI, *et al.* 1998. Effects of Asp-369 and Arg-372 mutations on heme environment and function in human endothelial nitric oxide synthase. J. Biol. Chem. **273:** 34164–34170.
34. CHEN, P.F., A.L. TSAI & K.K. WU. 1995. Cysteine 99 of endothelial nitric oxide synthase (NOS-III) is critical for tetrahydrobiopterin-dependent NOS-III stability and activity. Biochem. Biophys. Res. Commun. **215:** 1119–1129.
35. ABU-SOUD, H.M., L.L. YOHO & D.J. STUEHR. 1994. Calmodulin controls neuronal nitric oxide synthase by a dual mechanism: activation of intra- and inter-domain electron transfer. J. Biol. Chem. **269:** 32047–32050.
36. MICHEL, J.B., O. FERON, D. SACKS, *et al.* 1997. Reciprocal regulation of endothelial nitric oxide synthase by Ca^{2+}-calmodulin and caveolin. J. Biol. Chem. **272:** 15583–15586.
37. GARCIA-CARDENA, G., R. FAN, V. SHAH, *et al.* 1998. Dynamic activation of endothelial nitric oxide synthase by Hsp90. Nature **392:** 821–824.
38. GRATTON, J.-P., J. FONTANA, D.S. O'CONNOR, *et al.* 2000. Reconstitution of an endothelial nitric oxide synthase, hsp90 and caveolin-1 complex *in vitro.* J. Biol. Chem. **275:** 22268–22272.
39. CORSON, M.A., N.L. JAMES, S.E. LATTA, *et al.* 1996. Phosphorylation of endothelial nitric oxide synthase in response to fluid shear stress. Circ. Res. **79:** 984–991.
40. DIMMELER, S., I. FLEMING, B. FISSLTHALER, *et al.* 1999. Activation of nitric oxide synthase in endothelial cells by Akt-dependent phosphorylation. Nature **399:** 601–605.
41. FULTON, D., J.P. GRATTON, T.J. MCCABE, *et al.* 1999. Regulation of endothelium-derived nitric oxide production by the protein kinase Akt. Nature **399:** 597–601.
42. MCCABE, T.J., D. FULTON, L.J. ROMAN, *et al.* 2000. Enhanced electron flux and reduced calmodulin dissociation may explain "calcium-independent" eNOS activation by phosphorylation. J. Biol. Chem. **275:** 6123–6128.
43. ROSENKRANZ-WEISS, P., W.C. SESSA, S. MILSTEIN, *et al.* 1994. Regulation of nitric oxide synthesis by proinflammatory cytokines in human umbilical vein endothelial cells: elevations in tetrahydrobiopterin levels enhance endothelial nitric oxide synthase specific activity. J. Clin. Invest. **93:** 2236–2243.
44. LOSCALZO, J. 2000. What we know and we don't know about L-arginine and NO. Circulation **101:** 2126–2129.

45. CLOSS, E.I., J.-S. SCHELD, M. SHARAFI, *et al.* 2000. Substrate supply for nitric oxide synthase in macrophages and endothelial cells: role of cationic amino acid transporters. Mol. Pharmacol. **57:** 68–74.
46. ARNET, U.A., A. MCMILLAN, J.L. DINERMAN, *et al.* 1996. Regulation of endothelial nitric oxide synthase during hypoxia. J. Biol. Chem. **271:** 15069–15073.
47. TOPPER, J.N., J. CAI, D. FALB, *et al.* 1996. Identification of vascular endothelial gene differentially responsive to fluid mechanic stimuli: COX-2, eNOS, and Mn^{2+}SOD are selectively upregulated by steady laminar shear stress. Proc. Natl. Acad. Sci. U.S.A. **93:** 10417–10422.
48. ZEMBOWICZ, A., J.-L. TANG & K.K. WU. 1995. Transcriptional induction of endothelial nitric oxide synthase type III by lysophosphatidylcholine. J. Biol. Chem. **270:** 17006–17010.
49. NISHIZUKA, Y. 1992. Intracellular signaling by hydrolysis of phospholipids and activation of protein kinase C. Science **258:** 607–614.
50. PARTHASARATHY, S., U.P. STEINBRECHER, J. BARNETT, *et al.* 1985. Essential role of phospholipase A2 in endothelial cell-induced modification of low density lipoprotein. Proc. Natl. Acad. Sci. U.S.A. **82:** 3000–3004.
51. ASAOKA, Y., K. YOSHIDA, Y. SASAKI, *et al.* 1993. Possible role of mammalian secretary type II phospholipase A2 in T-lymphocyte activation: Implication in propagation of inflammatory reaction. Proc. Natl. Acad. Sci. U.S.A. **90:** 716–719.
52. QUINN, M.T., S. PARTHASARATHY & D. STEINBERG. 1988. Lysophosphatidylcholine: a chemotactic factor for human monocytes and its potential role in atherosclerosis. Proc. Natl. Acad. Sci. U.S.A. **85:** 2805–2809.
53. ZHANG, R., W. MIN & W.C. SESSA. 1995. Functional analysis of the human endothelial nitric oxide synthase promoter. J. Biol. Chem. **270:** 15320–15326.
54. TANG, J.-L., A. ZEMBOWICZ, X.-M. XU, *et al.* 1995. Role of Sp1 in transcriptional activation of human nitric oxide synthase-3 gene. Biochem. Biophys. Res. Commun. **213:** 673–680.
55. CIESLIK, K., A. ZEMBOWICZ, J.-L. TANG, *et al.* 1998. Transcriptional regulation of endothelial nitric oxide synthase by lysophosphatidylcholine. J. Biol. Chem. **273:** 14885–14890.
56. CIESLIK, K., C.S. ABRAMS & K.K. WU. 2001. Upregulation of endothelial nitric oxide synthase promoter by the phosphatidylcholine 3-kinase γ/Janus kinase 2/MEK-1–dependent pathway. J. Biol. Chem. **276:** 1211–1219.
57. CIESLIK, K., C-M. LEE, J.-L. TANG, *et al.* 1999. Transcriptional regulation of endothelial nitric oxide synthase by an interaction between casein kinase 2 and protein phosphatase 2A. J. Biol. Chem. **274:** 34669–34675.
58. VON DER LEYEN, H.E. & V.J. DZAU. 2001. Therapeutic potential of nitric oxide synthase gene manipulation. Circulation **103:** 2760–2765.

Nitric Oxide Transport and Storage in the Cardiovascular System

BERNARD MULLER, ANDREI L. KLESCHYOV, JACICARLOS L. ALENCAR, ANATOLY VANIN, AND JEAN-CLAUDE STOCLET

Université Louis Pasteur, CNRS UMR 7034, Faculté de Pharmacie BP 24, 67401 ILLKIRCH Cedex, France

ABSTRACT: Despite short halflife in biological fluids, nitric oxide (NO) can produce remote or long lasting effect in the cardiovascular system. Long distance transport or local storage of NO might explain these effects. In blood, recent findings suggest that in addition to being a major consumption pathway, interaction of NO with hemoglobin may permit O_2-governed transport of NO (as *S*-nitrosohemoglobin) to tissues in which NO may be released together with O_2, via transnitrosation of a transport protein. In blood vessels, two different putative NO stores have been characterized. The first is the photosensitive store, formed from endothelium-derived NO. The mechanism of NO release from this store in the body (in absence of light) and its physiological relevance are unknown. The second store is generated in conditions of high tissue NO levels, as a consequence of the inducible NO synthase activity or in various stress conditions. This NO store involves formation of protein-bound dinitrosyl iron complexes or *S*-nitrosated proteins, or both. Low molecular weight thiols can displace NO from these stores and probably transfer it to target membrane protein(s) such as K^+ channels, via transnitrosation reactions. These stores may be involved in defence mechanisms against inflammation or stress. Thus, NO transport and storage mechanisms may be implicated in a variety of NO effects. The mechanisms of their formation and of NO release and their physiologic and pathophysiologic relevance deserve further investigations.

KEYWORDS: dinitrosyl iron complexes; *S*-nitrosothiols; transnitrosation; nitric oxide

INTRODUCTION

In this paper, NO refers to any redox form of NO (·NO, NO^+, or NO^-). Early studies on the biological effects of NO have shown that it can interact with heme proteins such as hemoglobin (Hb) and soluble guanylyl cyclase. The reversible activation of guanylyl cyclase accounts for many effects of NO, whereas interaction with oxy-Hb results in the inactivation of NO and represents a major pathway for NO elimination. Despite its short halflife in biological fluids (a few seconds), recent data indicate that NO can also produce remote or prolonged effects. This is the case for instance of the

Address for correspondence: Bernard Muller, Université Louis Pasteur, CNRS UMR 7034, Faculté de Pharmacie, BP 24, 67401 ILLKIRCH Cedex, France. Voice: +33 (0)3 90 24 41 29; fax: +33 (0)3 90 24 43 13.
bmuller@pharma.u-strasbg.fr

diuretic action of inhaled NO,[1] or of long lasting hemodynamic changes caused by NO donor infusion.[2] Such effects may involve long distance transport or local storage, and subsequent release of NO.

Even though guanylyl cyclase appears to be the most sensitive molecular target of NO, many of the effects of NO are cyclic GMP-independent and result from interactions with other targets. The free radical ·NO has high affinity for iron (II) and rapidly reacts with superoxide anion. In addition, it can be oxidized to NO^+ or reduced to NO^-, both of which display differential reactivity compared to NO radical.[3] The principal reactions, underlying cyclic GMP-independent effects of NO, appear to be *S*-nitrosation of cysteine residues in proteins, and formation of protein-bound dinitrosyl–iron complexes (DNIC).[3–5] Some of the reactions triggered by NO are rapidly reversible (such as activation of guanylyl cyclase). Others lead to generation of more stable species (especially *S*-nitrosated proteins and DNIC with protein thiols) that might be involved in long lasting effects of NO or in its transport or storage.

Recently, a role for red blood cells in transport and delivery of NO to tissues has been proposed. Evidence for release of NO from preformed stores in blood vessels has also been provided by several laboratories, including ours. The current knowledge on transport and consumption of NO in red cells is briefly summarized below, before discussing formation of releasable NO stores in vessels in a more detailed fashion.

TRANSPORT AND CONSUMPTION OF NO IN RED BLOOD CELLS

It has been known for several years that interaction of NO with the oxy-Hb, with subsequent formation of met-Hb and nitrate, is a major pathway for NO inactivation and elimination.[6] In a series of recent articles,[7–9] Stamler and colleagues further documented the interactions between NO and Hb, suggesting that Hb may also act as a NO transporter. Interaction of heme-Fe(II) with NO results in formation of iron-nitrosyl-Hb ($HbFe_{II}NO$). High O_2 tension causes the allosteric transition of Hb that, in alveolar capillaries, may favour the transfer of NO moiety from heme group to the cysteine 93 residue of the β-chain, forming *S*-nitroso-Hb (SNO-Hb). Conversely, SNO-Hb releases NO when pO_2 drops. Thus, the processes of $HbFe_{II}NO$ and SNO-Hb formation and decomposition, and of NO group exchange between Fe_{II} and thiol, appear to be regulated by O_2 gradient, through allosteric transition of Hb. There is experimental evidence to support the existence of such NO transport and delivery in animal models. However, this concept has been challenged recently in human volunteers.[10,11]

The rate of NO consumption by red blood cells is much slower than predicted from the rate of NO reaction with free Hb. Under physiological conditions, multiple mechanisms may reduce the NO consumption rate in blood vessels, including flow and diffusion layer surrounding red blood cells. Transmembrane and intracellular diffusion are also rate limiting.[12] Recent investigations on the mechanisms of exportation of NO bioactivity from red blood cells reveal the role of a plasma-membrane protein the anion-exchange protein AE-1. It has been demonstrated that upon deoxygenation, the NO group is transferred through transnitrosation from the cysteine residue of Hb to vicinal cysteine residue in the cytoplasmic domain of AE-1.[13] This novel mechanism might be crucial for NO delivery in tissues.

In addition to SNO-Hb, *S*-nitrosoalbumin is also present in blood. It might also serve as a NO-transporter,[14] but this putative role still needs to be documented. Early studies have also shown that blood proteins can transfer NO in form of paramagnetic DNIC.[15] This has been demonstrated in animals maintained on drinking water with elevated content of nitrate, or after administration of model DNIC. The pathophysiologic relevance of these findings is still not completely understood.

RELEASABLE NO STORES IN BLOOD VESSELS

Evidence for the Existence of NO Stores in Blood Vessels

It has been known for a long time that precontracted blood vessels mounted in an organ bath relax when exposed to light.[16] Light-induced release of NO has been directly demonstrated in endothelium-denuded aortic strips from rats, using a porphyrinic sensor.[17] The existence of photosensitive NO stores is further supported by the finding that photosensitivity is progressively lost in endothelium-denuded blood vessels repeatedly exposed to light, but spontaneously recovers in the dark when endothelium-derived NO production is preserved, or when endothelium-denuded vessels are exposed to a NO donor.[18] In addition, photorelaxation is accompanied by an increase in cyclic GMP and is attenuated by inhibitors of guanylyl cyclase activation and NO scavengers.[19,20]

Another approach that provides indirect evidence for the presence of NO stores in blood vessels consists in studying relaxation caused by low molecular weight thiols like *N*-acetylcysteine (NAC). Low molecular weight thiols can displace NO from *S*-nitrosoproteins and $Fe(NO)_2$ from protein-bound DNIC.[5] As illustrated in FIGURE 1, these reactions may result in formation of NO-donating agents or in transnitrosation of proteins, or in transfer of $Fe(NO)_2$ groups between thiols in proteins. In experiments performed in precontracted rat aortic rings, NAC relaxed lipopolysaccharide (LPS)-treated rings, whereas it had no effect in control rings.[21] This relaxation was endothelium-independent and was observed in the presence of the NO-synthase (NOS) inhibitor L-NAME added immediately before the low molecular weight thiol. However, when L-NAME was present or L-arginine absent during preincubation with LPS, NAC caused no relaxation. Low-molecular weight thiol-induced relaxation could be demonstrated only when preincubation with LPS was run in the presence of L-arginine allowing the LPS-induced NOS-2 to produce high output NO in the vessel. NAC produced relaxation, not only in NOS 2 expressing rat aortic rings, but also in rings preexposed to some NO donors.[22] Altogether, these findings provide strong evidence to support the hypothesis that high level of tissue NO produced by NOS-2 activity or exposure of the vessels to NO donors results in formation of low-molecular weight thiol-displaceable NO stores. However, in LPS-exposed vessels, NAC-induced relaxation was not associated with an increase in cyclic GMP tissue content,[23] indicating that it was probably not caused by the free radical NO itself. Rather, it was inhibited by tetrabutylammonium, a non selective K^+ channel blocker,[23] suggesting that relaxation resulted from cyclic-GMP-independent activation of K^+ channels. Since the latter may result from *S*-nitrosation of these channels,[24] low molecular weight thiols might transfer NO from storage proteins to a cysteine of the ionic channel, via a transnitrosation reaction.

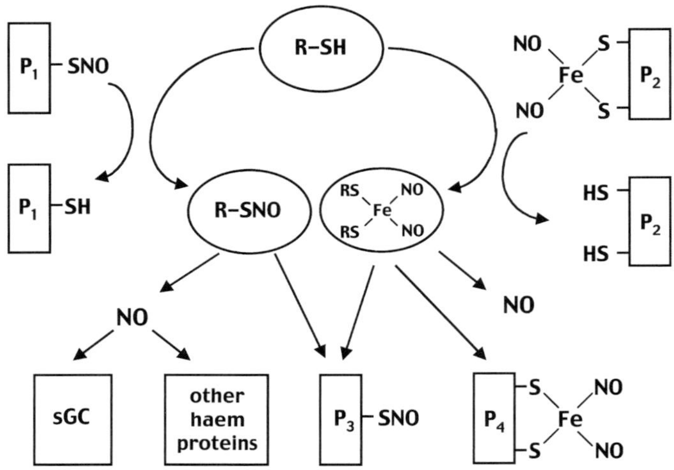

FIGURE 1. Schematic representation of the effects of low molecular weight thiols (R-SH) on *S*-nitrosated protein (Px) and on dinitrosyl–iron complexes with protein thiols. sGC, soluble guanylyl cyclase. (Adapted from Ref. 5, with permission.)

Since diethyldithiocarbamate (DETC) is capable of releasing ·NO or NO^+ from both S-nitrosothiols and DNIC, it has also been used to monitor NO stores in blood vessels, using the same approach as that used with low molecular weight thiols. In order to assess the presence of NO-stores, the aorta was removed from rats exposed to various stress conditions or injected with DNIC-thiosulfate as an NO donor; alternatively, the aorta was exposed to DNIC-thiolsulfate *in vitro*.[4,25,26] DETC caused relaxation in DNIC-exposed, but not in control precontracted aortic rings, and relaxation was blunted by a soluble guanylyl cyclase inhibitor.[4,25,26] It is likely that the DETC-displaceable NO store is identical to the NAC-displaceable store. In agreement with this view, DETC produced relaxation in LPS-exposed rat aortic rings, under the same conditions as NAC, and it abolished relaxation upon further addition of NAC.[27]

In conclusion, there are major differences between photorelaxation and relaxation produced by low molecular weight thiols or DETC. There is direct evidence that the former is caused by release of the free radical NO from stores generated from endothelium-derived NO. The latter has been found in conditions of much larger NO formation, such as those resulting from induction of NOS-2 activity, and relaxation associated with this store is cyclic GMP-independent. These differences suggest that the stores or the mechanism of NO or a NO-related releasing compound, or both, may be different.

Mechanisms of Formation, Nature, and Localization of the NO Stores

In experiments performed in the rat tail artery, exposure of vessels to both authentic NO or a low molecular weight *S*-nitrosothiol for a time as short as 300 ms induced formation of photosensitive NO stores.[18] Pretreatment of the vessels with thiol alkylating or thiol depleting compounds reduced photorelaxation,[18,20] suggesting the involvement of thiols. Since authentic NO was also able to induce the NO store formation, the molecular mechanism of *S*-nitrosation reaction remains to be elucidated. Recently, it has been reported that regeneration in the dark of the photosensitive store is either enhanced or blunted by agents that either increase or decrease the tissue glutathione level in the rat tail artery.[28] This suggests that *S*-nitrosation of intracellular glutathione by nitrosating species formed from endothelium-derived NO is involved in formation of photosensitive store.

In the rat aorta, the LPS-induced formation of a NAC-displaceable NO store is associated with generation of protein-bound DNIC characterized by its electron paramagnetic resonance (EPR) spectrum.[21,29] EPR experiments have shown that NAC can partially transform DNIC from high to low molecular weight.[21] Thus, it has been hypothesized that protein-bound DNIC represent a NO store and that the low molecular weight DNIC formed in the presence of NAC either releases NO, or transfers $Fe(NO)_2$ group, or *S*-nitrosates another target protein. The latter mechanism seems the most likely, because, as discussed above, the NAC-induced relaxation is not associated with elevation in cyclic GMP tissue content, but probably involves activation of K^+ channels,[23] via *S*-nitrosation of the channel protein. The reactions leading from NO formation by NOS2 to generation of protein-bound DNIC are unknown, as well as the extra or intracellular localization of DNIC. Furthermore, the involvement of *S*-nitrosated proteins in LPS-induced NO storage cannot be excluded. In rat aorta, when NO store formation was induced by preexposure to *S*-nitrosoglutathione, addition of $HgCl_2$ (which can displace NO from *S*-nitrosothiols[30]) produced vasorelaxation and abolished the relaxant effect of further addition of NAC,[22] supporting the view that the store could be a *S*-nitrosated protein. Since glutathione, which does not penetrate into cells, also produced vasorelaxation in aortic rings preexposed to *S*-nitrosoglutathione (unpublished observations), the store may be extracellular or at the external side of cell membranes.

ROLE OF NO STORES IN BLOOD VESSELS

The hypothesis that NO stores could be implicated in physiological responses has been proposed to explain endothelium-dependent relaxing effects that persist after blockade of NO synthase by inhibitors. This is the case, for instance, for sensory non adrenergic non cholinergic vasodilatation in rabbit small mesenteric artery,[31] and of NO-mediated (i.e., oxyHb-inhibited) relaxation in rat mesenteric and hepatic small arteries.[32] Incomplete blockade of the enzyme activity could also explain these effects. However, the part of acetylcholine- and bradykinin-induced vasodilatation[33] and of sympathetic neurogenic vasodilatation[34] that persists after administration of a NOS inhibitor in conscious rats is use dependent, consistent with the view that it is mediated by release of NO from a store that could not be replenished in the absence of NO production. Such a store might be the photosensitive one, because its formation depends on endothelial NO production. Of course, photosensitivity is not likely of direct physiological relevance, but NO might be released in the dark from photosensitive stores, by unidentified pathways.

Formation of low molecular weight thiol-displaceable vascular NO stores subsequent to the induction of NOS2 is probably involved in long lasting hyporesponsiveness to vasoconstrictors elicited by endotoxin[21,35] and by NO donors.[2,36] Release of NO from these stores might explain the enhanced vasodilatation or the increase in cyclic GMP level independent of current NOS activity that have been reported in the aorta of rats treated with LPS,[37] and in human vascular smooth muscle cells exposed to interleukin-1.[38] However, it is difficult to discriminate between the effects of NO release from preformed stores and the long lasting effects of NO possibly caused by prolonged activation of a target (such as one might expect from *S*-nitrosation of a protein like a K^+ channel). Both cyclic GMP accumulation in vessels and activation of K^+ channels seem involved in long lasting vascular hyporeactivity caused by LPS or NO donors.[2,23,36]

As well as being implicated in long lasting vascular hyporesponsiveness and hypotension caused by endotoxemia or sepsis,[35] formation of NO stores is induced in various stress conditions, including heat shock and hypoxia.[25,26] It has been suggested that formation of these stores could be a part of adaptative defence of the organism to stress; it might buffer large quantities of NO produced by NOS2, thereby preventing cytotoxic and hypotensive effects of excessive NO, on the one hand, then provide a reserve of NO to be used in case of deficiency, on the other hand. In support of this view, it has been reported that formation of DETC-releasable vascular NO stores induced by hypoxia was associated with prevention of both excessive and impaired endothelium-dependent relaxations caused by myocardial infarct or hypertension in rats, respectively.[26]

CONCLUSIONS

This paper has reviewed findings that raise a number of questions concerning the fate of NO produced by NOS, or released by NO donors. A tentative synthesis of these findings is illustrated in FIGURE 2.

There is no doubt that Hb contained in red cells plays a major role in inactivation and elimination of NO released from tissues, especially the endothelium. However, limitation of the NO consumption rate by the presence of the red cell membrane raises the question of the underlying mechanisms, and of their occurrence in other cell types. Progress in understanding the complex interactions between NO and Hb led to suggest a role for SNO-Hb in transport of NO in blood and its delivery in tissues. Furthermore, the recent discovery that AE-1 is involved in the exportation of NO from red cells opens a new field of investigations on the role of *S*-nitrosation and transnitrosation reactions in storage of NO and release of NO from theses stores in red cells (and perhaps also in other cells). The hypothesis that red cells might transport NO and release it in tissues might explain some remote or long lasting effects of NO. However, the possible physiologic and pathophysiologic relevance of NO transport and release by red cells requires further investigations, especially in human beings.

S-nitrosation reactions are probably also involved in generation of NO stores in blood vessels. These stores are different depending whether they are formed from constitutive endothelium-derived NO or during inflammatory responses. Putative

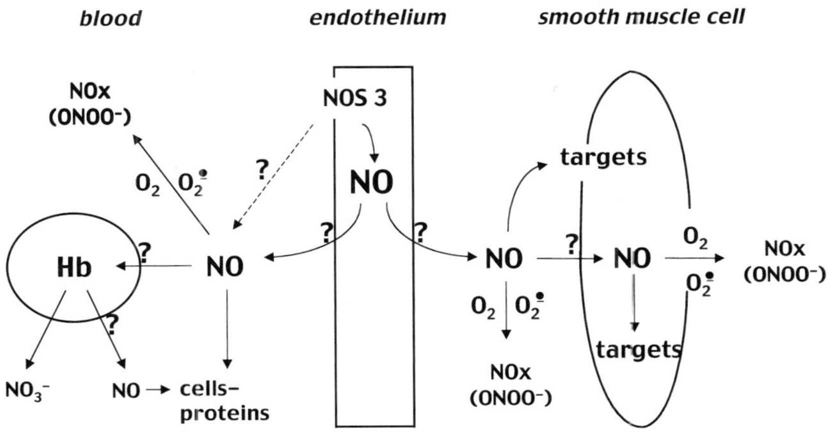

FIGURE 2. Tentative representation of the fate of NO produced by NO-synthase 3 (NOS 3) in the endothelial cells. *Question marks* indicate putative pathways that remain to be clarified (see text). Interaction of NO with hemoglobin (Hb) contained in red cells may lead to oxidation into NO_3^- or transport to remote targets. NO_x stands for an oxidized breakdown product of NO (NO_2^- or NO_3^-).

stores generated by endothelium-derived NO are photosensitive, but the mechanism(s) possibly triggering the release of NO from these stores in the body and their role are still unknown. Vascular NO stores generated in conditions of high NO output (i.e., after induction of NOS-2 activity, in animals submitted to various stress, or after exposure to NO donors) release NO upon exposure to low molecular weight thiols. *S*-nitrosation of protein thiol groups or formation of DNIC complexes with protein thiols, or both, may explain their formation in blood vessels. The mechanism(s) of NO release from these stores in the body and their role deserve further investigations. Both formation of these stores and subsequent NO release might be involved in defense mechanisms in conditions of stress or inflammation, by protecting against deleterious effects of both excessive or impaired NO production. Formation of these stores and subsequent release of NO might also be involved in long term effects of NO donors.

ACKNOWLEDGMENTS

This work was partially supported by an INTAS grant 97-524.

REFERENCES

1. TRONCY, E., M. FRANCOEUR, I. SALAZKIN, *et al.* 1997. Extra-pulmonary effects of inhaled nitric oxide in swine with and without phenylephrine. Br. J. Anæsth. **79:** 631–640.

2. EDUARDO DA SILVA-SANTOS, J. & J. ASSREUY. 1999. Long-lasting changes of rat blood pressure to vasoconstrictors and vasodilators induced by nitric oxide donor infusion: involvement of potassium channels. J. Pharmacol. Exp. Ther. **290:** 380–387.
3. STAMLER, J.S. 1994. Redox signaling: nitrosylation and related target interactions of nitric oxide. Cell **78:** 931–936.
4. VANIN, A.F. 1998. Dinitrosyl iron complexes and *S*-nitrosothiols are two possible forms for stabilization and transport of nitric oxide in biological systems. Biochemistry (Moscow) **63:** 782–793.
5. STOCLET, J.C., E. TRONCY, B. MULLER, *et al.* 1998. Molecular mechanisms underlying the role of nitric oxide in the cardiovascular system. Exp. Opin. Invest. Drugs **7:** 1769–1779.
6. WENNMALM, A., G. BENTHIN, A. EDLUND, *et al.* 1993. Metabolism and excretion of nitric oxide in humans. An experimental and clinical study. Circ. Res. **73:** 1121–1127.
7. JIA, L., C. BONAVENTURA, J. BONAVENTURA & J.S. STAMLER. 1996. *S*-nitrosohaemoglobin: a dynamic activity of blood involved in vascular control. Nature **380:** 221–226.
8. STAMLER, J.S., L. JIA, J.P. EU, *et al.* 1997. Blood flow regulation by *S*-nitrosohemoglobin in the physiological oxygen gradient. Science **276:** 2034–2037.
9. GOW, A.J. & J.S. STAMLER. 1998. Reactions between nitric oxide and haemoglobin under physiological conditions. Nature **391:** 169–173.
10. GLADWIN, M.T., J.H. SHELHAMER, A.N. SCHECHTER, *et al.* 2000. Role of circulating nitrite and *S*-nitrosohemoglobin in the regulation of regional blood flow in humans. Proc. Natl. Acad. Sci. U.S.A. **97:** 11482–11487.
11. GLADWIN, M.T., F.P. OGNIBENE, L.K. PANNELL, *et al.* 2000. Relative role of heme nitrosylation and beta-cysteine 93 nitrosation in the transport and metabolism of nitric oxide by hemoglobin in the human circulation. Proc. Natl. Acad. Sci. U.S.A. **97:** 9943–9948.
12. VAUGHN, M.W., K.T. HUANG, L. KUO & J.C. LIAO. 2000. Erythrocytes possess an intrinsic barrier to nitric oxide consumption. J. Biol. Chem. **275:** 2342–2348.
13. PAWLOSKI, J.R., D.T. HESS & J.S. STAMLER. 2001. Export by red blood cells of nitric oxide bioactivity. Nature **409:** 622–626.
14. STAMLER, J.S., O. JARAKI, J. OSBORNE, *et al.* 1992. Nitric oxide circulates in mammalian plasma primarily as an S-nitroso adduct of serum albumin. Proc. Natl. Acad. Sci. U.S.A. **89:** 7674–7677.
15. VANIN, A.F. & V.I. VARICH. 1979. Formation of nitrosyl complexes of nonheme iron (2.03 complexes) in animal tissues *in vivo.* Biofizika **24:** 666–670.
16. FURCHGOTT, R.F., S.J. EHRREICH & E. GREENBLATT. 1961. The photoactivated relaxation of rabbit aorta. J. Gen. Physiol. **44:** 499–519.
17. KUBASZEWSKI, E., A. PETERS, S. MCCLAIN, *et al.* 1994. Light-activated release of nitric oxide from vascular smooth muscle of normotensive and hypertensive rats. Biochem. Biophys. Res. Commun. **200:** 213–218.
18. MEGSON, I.L., F.W. FLITNEY, J. BATES & R.N. WEBSTER. 1995. "Repriming" of vascular smooth muscle photorelaxation is dependent on endothelium-derived nitric oxide. Endothelium **3:** 39–46.
19. FURCHGOTT, R.F., W. MARTIN, P.D. CHERRY, *et al.* 1985. Endothelium-dependent relaxation, photorelaxation and cGMP. *In* Vascular-Neuroeffector Mechanisms. J.A. Bevan, *et al.*, Eds.: 105–114. Elsevier.
20. LOVREN, F. & C.R. TRIGGLE. 1998. Involvement of nitrosothiols, nitric oxide and voltage-gated K^+ channels in photorelaxation of vascular smooth muscle. Eur. J. Pharmacol. **347:** 215–221.
21. MULLER, B., A.L. KLESCHYOV & J.C. STOCLET. 1996. Evidence for *N*-acetylcysteine sensitive nitric oxide storage as non haeme dinitrosyl-iron complexes in lipopolysaccharide-treated rat aorta. Br. J. Pharmacol. **119:** 1281–1285.
22. ALENCAR, J.L., J.C. STOCLET & B. MULLER. 2001. Formation of releasable NO stores on tissue thiols in blood vessels: differential effect of NO donors. Fundam. Clin. Pharmacol. **15**(Suppl. 1): 9P085.

23. MULLER, B., A.L. KLESCHYOV, S. MALBLANC & J.C. STOCLET. 1998. Nitric oxide-related cyclic GMP-independent relaxing effect of *N*-acetylcysteine in lipopolysaccharide-treated rat aorta. Br. J. Pharmacol. **123:** 1221–1229.
24. BOLOTINA, V.M., S. NAJIBI, J.J. PALACINO, *et al.* 1994. Nitric oxide directly activated calcium-dependent potassium channels in vascular smooth muscle. Nature **368:** 850–853.
25. MALYSHEV, I.Y., T.A. ZENINA, L.Y. GOLUBEVA, *et al.* 1999. NO-dependent mechanisms of adaptation to hypoxia. Nitric Oxide **3:** 105–113.
26. MANUKHINA, E.B.. S. YU. MASHINA, B.V. SMIRIN, *et al.* 2000. Role of nitric oxide in adaptation to hypoxia and adaptive defense. Physiol. Res. **49:** 89–97.
27. MULLER, B., A.L. KLESCHYOV, S. MALBLANC & J.C. STOCLET. 1999. Inhibition by tetrabutylammonium of cyclic GMP-independent vasorelaxation induced by interaction of thiols with nitric oxide stores. *In* Endothelium-Dependent Hyperpolarizations. P.M. Vanhoutte, Ed.: 133–141. Harwood Academic Publishers.
28. Megson, I.L., S.A. Holmes & K.S. Magid. 2000. Selective modifiers of glutathione biosynthesis and 'repriming' of vascular smooth muscle photorelaxation. Br. J. Pharmacol. **130:** 1575–1580.
29. KLESCHYOV, A.L., B. MULLER, T. KERAVIS, *et al.* 2000. Adventitia-derived nitric oxide in rat aortas exposed to endotoxin: cell origin and functional consequences. Am. J. Physiol. **279:** H2743–2751.
30. SAVILLE, B. 1958. A scheme for the colorimetric determination of microgram amounts of thiols. Analyst **83:** 670–672.
31. KAKUYAMA, M., P. VALLANCE & A. AHLUWALIA. 1998. Endothelium-dependent sensory NANC vasodilatation: involvement of ATP, CGRP and a possible NO store. Br. J. Pharmacol. **123:** 310–316.
32. CHAUHAN, S.D, R.J. MACALLISTER, L.H. CLAPP & A. AHLUWALIA. 2000. Evidence that NO may contribute to "EDHF-like" responses in rat mesenteric and hepatic small arteries. (Abstr.) Br. J. Pharmacol. **129:** 1P.
33. DAVISSON, R.L, J.N. BATES, A.K. JOHNSON & S.J. LEWIS. 1996. Use-dependent loss of acetylcholine- and bradykinin-mediated vasodilation after nitric oxide synthase inhibition. Evidence for preformed stores of nitric oxide-containing factors in vascular endothelial cells. Hypertension **28:** 354–360.
34. DAVISSON, R.L., R.A. SHAFFER, A.K. JOHNSON & S.J LEWIS. 1996. Use-dependent loss of active sympathetic neurogenic vasodilation after nitric oxide synthase inhibition inconscious rats. Evidence for the presence of preformed stores of nitric oxide-containing factors. Hypertension **28:** 347–353.
35. STOCLET, J.C, B. MULLER, K. GYORGY, *et al.* 1999. The inducible nitric oxide synthase in vascular and cardiac tissue. Eur. J. Pharmacol. **375:** 139–155.
36. TERLUK, M.R., J.E. DA SILVA-SANTOS & J. ASSREUY. 2000. Involvement of soluble guanylate cyclase and calcium-activated potassium channels in the long-lasting hyporesponsiveness to phenylephrine induced by nitric oxide in rat aorta. Naunyn Schmiedebergs Arch. Pharmacol. **361:** 477–483.
37. WU, C.C., C. SZABO, S.J. CHEN, *et al.* 1994. Activation of soluble guanylyl cyclase by a factor other than nitric oxide or carbon monoxide contributes to the vascular hyporeactivity to vasoconstrictor agents in the aorta of rats treated with endotoxin. Biochem. Biophys. Res. Commun. **201:** 436–442.
38. BEASLEY, D. & M. MCGUIGGIN. 1994. Interleukin 1 activates soluble guanylate cyclase in human vascular smooth muscle cells through a novel nitric oxide-independent pathway. J. Exp. Med. **179:** 71–80.

Thioredoxin-Related Regulation of NO/NOS Activities

LI-EN SHAO, TETSUYA TANAKA, RUBY GRIBI, AND JOHN YU

Department of Molecular and Experimental Medicine, The Scripps Research Institute, La Jolla, California, U.S.A.

ABSTRACT: The role of regulation of nitric oxide synthase (NOS) activity in mitigating oxidative stress in neonatal lungs and contributing to pulmonary vasodilation at birth is still unclear. Furthermore, it is known that, depending on interactions between the individual components of the mitogen-activated protein kinase (MAPK) signaling cascades, many biological consequences, including apoptosis, are initiated. Although the importance of nitric oxide (NO) in apoptosis is controversial and likely depends on NO concentrations and cell types, this highly reactive free radical can activate the p38 MAPK signal cascade. Recent studies have suggested that thioredoxin may play an important role as an effector for some of these functions. Thioredoxin is a major redox protein for many enzymes/transcription factors and is involved in cellular functions, such as viability, activation, and proliferation. In addition to its redox regulation, thioredoxin binds directly to the apoptosis signal-regulating kinase 1 (ASK1), thus inhibiting the activation of stress-induced MAPK signaling cascades that lead to apoptosis. Furthermore, NO produced from newly induced neuronal NOS was reported to induce expression of thioredoxin and several other genes for preconditioning-induced neuroprotection. Moreover, although exposure of endothelial cells to NO decreases NOS activity, this inhibition was shown to be reversed by thioredoxin. Finally, the correlation of expression of thioredoxin with endothelial NOS activity seems to suggest an important role played by this protein in perinatal changes of pulmonary artery functions. Therefore, thioredoxin may participate in the regulation of NOS activity and be involved in NO functions via multiple mechanisms.

KEYWORDS: nitric oxide; nitric oxide synthase; thioredoxin; apoptosis; oxygenation

INTRODUCTION

Nitric oxide (NO), a short-lived and highly reactive free radical, plays many important roles as a unique biological messenger in cells.[1] NO is known to have pleiotropic biological activities and has been implicated in many physiologic and pathological functions within the vascular, nervous, and immune systems.[1] The enzyme NOS catalyses L-arginine to NO and L-citrulline.[2] There are three isoforms of NOS: the constitutive and calcium dependent (1) endothelial NOS (eNOS), and (2) neuronal NOS (nNOS), and (3) the calcium independent inducible isoform

Address for correspondence: John Yu, M.D., Ph.D., Department of Molecular and Experimental Medicine (MEM 265), The Scripps Research Institute, 10550 North Torrey Pines Road, La Jolla, CA 92037, U.S.A. Voice: 858-784-7924; fax: 858-784-7977.

(iNOS) that is produced in response to cytokine or bacterial endotoxin stimulation.[3] However, several factors, including oxidative stress, are known to influence basal expression and modulate the production of NO from all three isoforms of NOS.[3] NO pathways may also exhibit negative feedback; previous studies report that exposure of endothelial cells to NO significantly reduces the catalytic activity of constitutive eNOS.[4]

Many studies suggest that thioredoxin, a redox protein,[5] is involved in cell regulation and apoptosis. It has been shown that overexpression of thioredoxin prevents NO-induced reduction of eNOS activity in lung endothelial cells.[6] Although activation of soluble guanylate cyclase is known as the primary physiologic mechanism for NO action, the action of NO has also been reported to be stimulatory or inhibitory to apoptosis.[7–9] Recent studies further indicate that NO is likely to be involved in MAPK signaling cascades.[10,11] Furthermore, stress-related MAPK cascades were shown to be activated by ASK1, which is in turn suppressed by thioredoxin.[12] It was also found that NO produced from newly induced nNOS activates the cGMP pathway that leads to the induction of thioredoxin and several other genes for preconditioning induced neuroprotection.[13]

Thioredoxin is a 12,000-dalton reduction/oxidation (redox) protein that plays an important role in cell viability, activation, and proliferation.[5] As part of a general protein disulfide-reducing system, this protein serves to maintain the redox environment in the cell.[14] It is the major reducing protein for many enzymes/DNA-binding transcription factors and is involved in many steps of gene expression. It is a cofactor for the enzyme ribonucleotide reductase that is essential for DNA synthesis.[15] Thioredoxin is also required for the interaction between transferrin receptor mRNA and the iron responsive element-binding protein, serving to promote cellular proliferation.[16]

Because of a large and rapidly expanding amount of data on NO and NOS, it is difficult to adequately summarize the entire area of research. Therefore, this article addresses only one aspect of the field: thioredoxin-related regulation of NO/NOS activities.

REDOX REGULATION BY THIOREDOXIN

Thioredoxin serves as a general protein disulfide oxidoreductase: it interacts with a broad range of proteins by regulating their activities via thiol-redox control, based on reversible oxidation of two cysteine thiol groups to a disulfide. The redox activity of human thioredoxin arises from a highly conserved catalytic site (-Trp-Cys^{32}-Gly-Pro-Cys^{35}-Lys-). The cysteine residues, Cys^{32} and Cys^{35}, of this active site of thioredoxin undergo reversible oxidation/reduction catalyzed by thioredoxin reductase (see FIGURE 1). The reduced thioredoxin, thioredoxin-$(SH)_2$, has a hydrophobic surface area that is conserved in the thioredoxin-glutaredoxin family. The mechanism for the reducing action of thioredoxin involves the binding of the reduced thioredoxin to a target disulfide protein-S_2 via this hydrophobic region (FIG. 1). The thiolate of Cys^{32} acts as a nucleophile in this hydrophobic environment of the complex, and combines with the protein substrate to form a transient, but covalently linked, mixed disulfide (-Cys^{32}-S-S- protein). Finally, the now deprotonated Cys^{35} reacts with

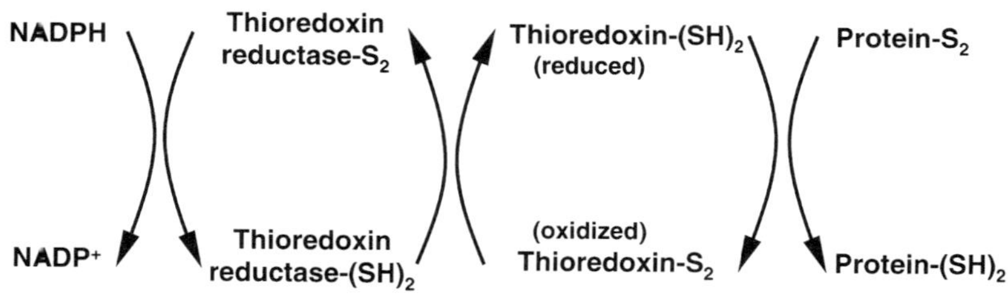

FIGURE 1. Mechanism of NADPH-dependent protein disulfide reduction by the thioredoxin/thioredoxin reductase system. Thioredoxin reductase catalyzes the reduction of oxidized thioredoxin (thioredoxin-S_2) by NADPH using FAD and its redox-active disulfide (-S_2) to form reduced thioredoxin-$(SH)_2$. The reduced thioredoxin [thioredoxin-$(SH)_2$] then directly reduces the disulfide in the subtract protein (Protein-S_2) to Protein-$(SH)_2$ and forms the oxidized thioredoxin-S_2.

the disulfide bond of -Cys^{32}-S-S-protein, releasing the reduced protein substrate, protein-$(SH)_2$, and forming the Cys^{32}-Cys^{35}-disulfide in thioredoxin.[17] This oxidized thioredoxin-S_2 is in turn reduced by thioredoxin reductase with the help of NADPH to form thioredoxin-$(SH)_2$ (FIG. 1). Thus, the reducing power of thioredoxin is regenerated.

MULTIFUNCTIONAL ACTIONS OF THIOREDOXIN

Just as NO is a pleiotropic effector of cells, thioredoxin is a multifunctional ubiquitous protein. It can act as a growth factor, antioxidant, regulator for transcription factors, and is also involved in direct protein binding. In a variety of oxidative stresses and viral infections, thioredoxin is induced to high levels of expression and serves as an endogenous antioxidant to provide cytoprotective activity.[18–20] The direct antioxidant properties of thioredoxin include removal of hydrogen peroxide.[21] However, thioredoxin appears to exert most of its antioxidant properties in cells through thioredoxin peroxidase, which uses thiol groups as reducing equivalents to scavenge oxidants.[22]

Thioredoxin has been shown to exert redox control over the activity of many DNA-binding transcription factors because of the existence of thiol residues located at either allosteric or active sites of these proteins.[23–36] For example, thioredoxin regulates the activity of transcription factors including nuclear factor-κB and the glucocorticoid receptor.[23–26] Furthermore, thioredoxin also modulates activities of many enzymes via its thiol-redox control of the protein disulfide (FIG. 1). The binding of Fos/Jun to the AP-1 site is subject to control by redox factor 1, Ref-1, which is reduced by thioredoxin.[27,28]

Moreover, thioredoxin binds to cellular proteins directly resulting in the formation of a mixed disulfide bond between the cysteine residues of the thioredoxin catalytic site and of the other protein. Such protein binding, however, occurs only with the reduced form of thioredoxin; neither the oxidized nor the mutant redox-inactive forms of thioredoxin participate in direct protein binding. One example of protein to which thioredoxin binds is the apoptosis signal-regulating kinase 1 (ASK1)[12] (see next section). NMR studies suggest additionally that a 13-amino cid peptide of Ref-1 also binds directly to a crescent-shaped groove on the surface of human thioredoxin.[27,28]

Thioredoxin can also be secreted by cells at a high rate through a unique "leaderless" pathway.[30,31] Exogenously added thioredoxin was reported to stimulate cell proliferation by enhancing the sensitivity of cells to cytokines.[32] Furthermore, we previously showed that exogenous thioredoxin stimulates proliferation of clonogenic cells of childhood T-cell leukemia in the absence of PMA, possibly due to ability of thioredoxin to induce the expression of cytokine receptors in the leukemia cells.[33] Of interest, the enhancing effect of exogenous thioredoxin was observed in the subgroup of acute leukemia that had a relatively low level of thioredoxin expression, but not in the cells that express high levels of intracellular thioredoxin.[33] Conceivably, these results indicate that different cell types respond differently to exogenous thioredoxin, based on variations in their intracellular redox potentials.

ANTIAPOPTOTIC ACTION OF THIOREDOXIN

Prior to the discovery that thioredoxin is an inhibitor for the apoptosis-related kinase ASK1, human thioredoxin was originally cloned and recognized as an adult T-cell leukemia-derived factor, produced by human T-cell leukemia virus type-I.[34–36] It was also observed that levels of thioredoxin expression were elevated in primary tumors/cell lines from patients with breast, cervical squamous cell carcinoma, hepatoma, lung, and colon tumors.[19,20,37,38]

In a Pediatric Oncology Group study, we demonstrated that thioredoxin expression is increased in childhood acute lymphoblastic leukemias that have no known viral eiology.[33] We found that in these leukemias, increased expression of thioredoxin is associated with higher WBC counts at diagnosis,[33] one of the important prognostic factors.[39] We also reported that ovarian cancer is yet another type of malignancy with abnormal thioredoxin expression: malignant ovarian cells express high levels of thioredoxin, commencing at the early stages of disease and continuing throughout advancement of the disease and progression to metastasis.[40]

It is now known that overexpression of thioredoxin in cells suppresses apoptosis.[41,42] Stable transfection of thioredoxin gives a twofold increase in thioredoxin levels in the cells and a concomitant threefold increase in resistance to cisplatin.[43] We also observed that there is a difference in thioredoxin expression between ovarian cancer 2008 cells and the cisplatin-resistant isogenic subline, 2008/A and 2008/C13*5.25; levels of thioredoxin expression in these resistant isogenic 2008/A and 2008/C13*5.25 cells were, respectively, 1.5- and 2.3-fold higher than those found in the 2008 cells.[40] Furthermore, increased levels of thioredoxin were shown to be associated with resistance to cisplatin, mitomycin C, doxorubicin, and etoposide.[43,44] Together with the findings of high levels of thioredoxin expression in tumors, these studies suggest that increased thioredoxin expression leads to increased tumor growth resulting from inhibition of apoptosis and a decrease in the sensitivity of cancer cells to drug-induced apoptosis.

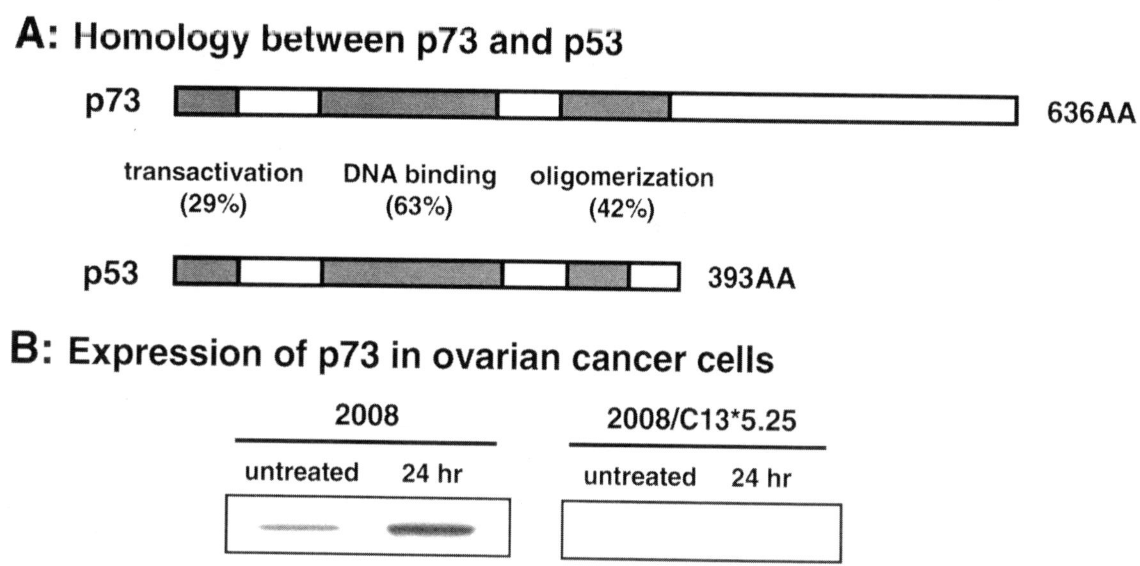

FIGURE 2. Loss of inducibility of the proapoptotic p73 protein in cisplatin-resistant cancer cells during drug treatment. **(A)** Shows the homology of the domains for transactivation, DNA binding, and oligomerization in p73 and p53 proteins. **(B)** Western blot analysis using mAb 789 (Imgenex, La Jolla, CA) indicated that, upon exposure to 1.5 μM of 1-methylpropyl 2-imidazolyl disulfide for 24 h, the expression of the p53-related proapoptotic p73 protein in ovary cancer cells was upregulated in cisplatin-sensitive 2008 cells, but not in cisplatin-resistant 2008/C13*5.25 cells. Similar results were observed when 2008 and 2008/C13*5.25 cells were exposed to cisplatin at 0.25 and 1.5 μM, respectively (corresponding to the IC_{50} of these cell lines toward cisplatin).[40] Conceivably, drug-induced treatment activates apoptosis through a proapoptotic p73 pathway only in drug sensitive subline.[40]

The mechanism by which thioredoxin suppresses apoptosis may also be linked to p73, a p53-independent activator of apoptosis. Recent studies indicate that overproduction of p73 can induce apoptosis in p53-null cells.[45] The p73 gene encodes proteins that are homologous to p53 in the *N*-terminal *trans*-activation domain, the DNA binding domain, and the oligomerization domain (see FIGURE 2A). Therefore, p73 and its isoforms can activate transcription from promoters that contain the p53-binding sequences.[46,47] Although somatic mutation of p73 is rare, hypermethylation and silencing of the p73 gene is noted in many leukemias,[48] suggesting that silencing of p73 contributes to development and/or progression of maligr.ancies.

We found noticeable differences in the expression of proapoptotic p73 (FIG. 2B) and p53 proteins upon drug–induced apoptosis in the ovarian cancer 2008 cells in comparison with the cisplatin-resistant isogenic subline, 2008/C13*5.25.[40] Indeed, many studies have indicated that cisplatin-activated apoptosis[49] can be attributed to the induction and accumulation of p73.[47] Similarly, our studies show that cisplatin-resistant 2008/C13*5.25 cells selectively lost the ability to upregulate p73 protein upon drug treatment (FIG. 2B) while retaining the inducibility of p53 expression (not shown). As noted already, 2008/C13*5.25 cells also exhibit higher levels of thioredoxin expression. Thus, we postulate that thioredoxin may inhibit apoptosis via suppression of p73 in these ovarian cancer cells.

Additional studies of neuroblastoma by Andoh, *et al.* have showed that the anti-apoptotic action of thioredoxin is actually multifactorial, including suppression of cytochrome *c* release, induction of Bcl-2, and inhibition of caspase-9 and -3.[50] Intriguingly, a recent study showed that caspase-3 activity in cell lysates was suppressed by diamide, but restored by thio-reducing agents, including thioredoxin.[51] This observation seems to be at odds with the fact that thioredoxin protects cells from apoptosis since an increase in caspase activity is responsible for the execution of apoptosis. To resolve this dilemma, it was suggested that thioredoxin might reduce thiol groups on other proteins in the cells that negatively regulate caspase activity indirectly.[52]

Recently, thioredoxin was also shown to be the physiologic inhibitor of the apoptosis signal-regulating kinase 1 (ASK1).[12] It was found that overexpression of ASK1 in epithelial cells in low serum conditions induces apoptosis, suggesting that ASK1 is involved in stress-induced apoptosis.[53] Thioredoxin binds directly to the ASK1 amino-terminal non-catalytic domain and blocks activation of ASK1 by tumor necrosis factor.[12] Indeed, it was shown that activation of ASK1 by tumor necrosis factor requires prior dissociation of thioredoxin from the mixed disulfide complex, followed by the oligomerization of ASK1.[54] Therefore, the observations that thioredoxin may be an endogenous inhibitor of apoptosis-related ASK1 provide clues on the connections between induction of thioredoxin during oxidative stress and its cytoprotective actions.

PUTATIVE ROLE OF THIOREDOXIN IN THE SIGNALING OF APOPTOSIS

NO mediates caspase-dependent apoptosis in many cell types.[7] On the other hand, NO can prevent apoptosis by direct inhibition of caspase activity by nitrosation, as well as through cGMP-dependent mechanisms.[8,9] Whereas the importance of NO in apoptosis is controversial, these reactive species are likely to be involved as signaling intermediates in the apoptosis-associated signal transduction mechanism.[10] Lander, *et al.*, showed that NO activates three different MAPK cascades in Jurkat T cells.[55] These include (1) the stress-activated protein kinase/c-Jun *N*-terminal kinase (SAPK/JNK) cascade, (2) the stress-activated p38 MAPK cascade, and (3) the classical growth factor-activated extracellular signal-regulated kinase (ERK) MAPK cascade. These kinases are activated in response to different extracellular stimuli and perform different functions. All three MAP kinase cascades phosphorylate substrates on serine/threonine adjacent to proline residues. In fact, these MAP kinase signaling cascades consist of three distinct members of the protein kinase family: that is, MAP kinase (MAPK), MAPK kinase (MAPKK), and MAPKK kinase (MAPKKK). MAPKKK phosphorylates and thereby activates MAPKK, and the activated form of MAPKK, in turn, phosphorylates and activates MAPK. Finally, after translocation to cell nucleus, the activated MAPK regulates the activities of transcription factors and thereby controls gene expression.[53]

Conceivably, activation of the MAPK cascades by NO may occur either by direct alterations to the kinases themselves, or by modulation of an upstream factor such as GTP-binding protein $p21^{ras}$. More recent studies show that NO rapidly activates the p38 MAPK, but does not activate JNK or ERK pathways in human myeloid leukemia cells, suggesting a pivotal role for this p38 kinase in NO signaling of apoptosis in these cells.[11] As mentioned, thioredoxin has been shown to be the physiclogical inhibitor of a MAPKKK, ASK1. The latter activates the MKK3/6-p38 as well as the SEK1-JNK/SAPK signaling cascades.[53] Thus, by complexing with ASK1 blocking its activation by tumor necrosis factor,[54] thioredoxin presumably suppresses the NO-induced activation of apoptosis signaling cascades.

INDUCTION OF THIOREDOXIN EXPRESSION AND ENOS ACTIVITY

eNOS is constitutively expressed in endothelial cells, cardiac myocytes, and blood platelets. Although the basal expression of eNOS is largely tissue dependent, there are many other factors that influence the basal expression levels.[3] Shear stress in blood vessels has been shown to upregulate expression of eNOS.[56] There is also an increase in eNOS mRNA and protein expression in fetal rat lungs as they are near term.[57,58] Moreover, O_2 ventilation in lungs of fetal lambs is known to induce eNOS gene expression.[59] The production of NO by eNOS in endothelial cells activates guanylate cyclase, increases cGMP, and initiates a cascade that results in smooth muscle relaxation and vasodilation. These changes bring forth a decrease in pulmonary vascular resistance and an increase in pulmonary blood flow. Therefore, increased oxygenation and increased blood flow with increased shear stress induce eNOS gene expression and contribute to pulmonary vasodilation after birth.

In addition to these regulatory mechanisms, there are other controls for eNOS activities at the posttranscriptional and posttranslational levels.[3] Exposure to NO induces *S*-nitrosylation of protein thiols and modulation of enzymatic activities, including a decrease in the catalytic activity of NOS. Exposure to NO of intact pulmonary artery endothelial cells, isolated plasma membranes, and purified eNOS were shown to significantly decrease eNOS activity, even in the absence of changes in the expression of eNOS mRNA.[60] Thus, inhibition of eNOS by NO affects the decrease in pulmonary resistance and the increase in pulmonary blood flow.

Recent studies indicate that thioredoxin may also play a role in the regulation of eNOS activity. Thioredoxin has the ability to modulate the thiol groups of proteins, suggesting a plausible mechanism by which thioredoxin can regulate eNOS. Research has demonstrated that incubation with thioredoxin/thioredoxin reductase, but not glutaredoxin, restores the NO-induced decrease of eNOS activity.[60] This finding of thioredoxin-regulated reversal of NO-induced modulation of NOS protein suggests that an oxidative conformational change may have occurred in vicinal thiols of the eNOS molecule, resulting in the formation of disulfides.[60]

The relationship between thioredoxin expression and eNOS activity is further illustrated by the following observations. It is known that at birth, the sudden exposure of the lung to oxygen poses an acute oxidative stress because the environment in the fetus is relatively anerobic. Das K.C. *et al.* reported that thioredoxin and thioredoxin reductase are specifically upregulated at birth by O_2.[61] Although mRNAs for thioredoxin and thioredoxin reductase were expressed at low levels in fetal lungs constitutively throughout the third trimester, these lung mRNAs were shown to increase rapidly after birth.[61] In fetal lung explant cultures, mRNAs for thioredoxin and thioredoxin reductase were also elevated with exposure to 95% O_2. In addition, the protein and activity of thioredoxin/thioredoxin reductase increased, *in vivo*, in response to hyperoxia. Thus, the rapid increase in gene expression for both thioredoxin and thioredoxin reductase in response to O_2 is correlated to increased expression of eNOS, suggesting an important role for these proteins in perinatal changes in pulmonary vascular resistance.

ACKNOWLEDGMENTS

This work was supported by Cancer Research Fund, under Interagency Agreement #97-12013 (U. C. Davis contract #98-00924V) with the Dept. of Health Services, Cancer Research Section. The views expressed herein represent those of the authors and do not necessarily represent the position of the State of California, Dept. of Health Services. This is publication 14354-MEM from The Scripps Research Institute.

REFERENCES

1. Schmidt, H.H. & U. Walter. 1994. NO at work. Cell **78:** 919–925.
2. Palmer, R.M., D.S. Ashton & S. Moncada. 1988. Vascular endothelial cells synthesize nitric oxide from L-arginine. Nature **333:** 664–666.
3. Govers, R. & T.J. Rabelink. 2001. Cellular regulation of endothelial nitric oxide synthase. Am. J. Physiol. Renal Physiol. **280:** F193–F206.
4. Patel, J.M. & E.R. Block. 1995. Sulfhydryl-disulfide modulation and the role of disulfide oxidoreductases in regulation of the catalytic activity of nitric oxide synthase in pulmonary artery endothelial cells. Am. J. Respir. Cell Mol. Biol. **13:** 352–359.
5. Holmgren, A. 1989. Thioredoxin and glutaredoxin systems. J. Biol. Chem. **264:** 13963–13966.

6. ZHANG, J., L.D. YONG, J.M. PATEL & E.R. BLOCK. 1998. Thioredoxin overexpression prevents NO-induced reduction of NO synthase activity in lung endothelial cells. Am. J. Physiol. **175:** L288–L293.
7. KIM, Y.-M., H.-T. CHUNG, R.L. SIMMONS & T.R. BILLIAR. 2000. Cellular non-heme iron content is a determinant of nitric oxide-mediated apoptosis, necrosis, and caspase inhibition. J. Biol. Chem. **275:** 10954–10961.
8. MANNICK, J.B., A. HAUSLADEN, L. LIU, *et al.* 1999. Fas-induced caspase denitrosylation. Science **284:** 651–654.
9. KIM, Y.-M., H.-T. CHUNG, S.-S. KIM, *et al.* 1999. Nitric oxide protects PC12 cells from serum deprivation-induced apoptosis by cGMP-dependent inhibition of caspase signaling. J. Neurosci. **19:** 6740–6747.
10. JACOBSON, M.D. 1996. Reactive oxygen species and programmed cell death. Trends Biochem. Sci. **21:** 83–86.
11. JUN, C.-D., H.-O. PAE, H.-J. KWAK, *et al.* 1999. Modulation of nitric oxide-induced apoptotic death of HL-60 cells by protein kinase C and protein kinase A through mitogen-activated protein kinases and CPP32-like protease pathways. Cell. Immunol. **194:** 36–46.
12. SAITOH, M., H. NISHITOH, M. FUJII, *et al.* 1998. Mammalian thioredoxin is a direct inhibitor of apoptosis signal-regulating kinase (ASK) 1. EMBO J. **17:** 2596–2606.
13. CHIUEH, C.-C. & T. ANDOH. 2001. *In* Mapping the Progress of Alzheimer's and Parkinson's Disease. Y. Mizuno, Ed.: 1–6. Kluwer Academic/Plenum, New York.
14. HOLMGREN, A. & M. BJORNSTEDT. 1995. Thioredoxin and thioredoxin reductase. Methods Enzymol. **252:** 199–208.
15. LAURENT, T.C., E.C. MOORE & P. REICHARD. 1964. Enzymatic synthesis of deoxyribonucleotides. VI. Isolation and characterization of thioredoxin, the hydrogen donor from *Escherichia coli B*. J. Biol. Chem. **239:** 3436–3444.
16. HENTZE, M.W., T.A. ROUAULT, J.B. HARFORD & R.D. KLAUSNER. 1989. Oxidation–reduction and the molecular mechanism of a regulatory RNA-protein interaction. Science **244:** 357–359.
17. HOLMGREN, A. 1995. Thioredoxin structure and mechanism: conformational changes on oxidation of the active-site sulfhydryls to a disulfide. Structure **3:** 239–243.
18. YODOI, J. & T. TURSZ. 1991. ADF, a growth-promoting factor derived from adult T cell leukemia and homologous to thioredoxin: Involvement in lymphocyte immortalization by HTLV-1 and EBV. Adv. Cancer Res. **57:** 381–411.
19. NAKAMURA, H., H. MASUTANI, Y. TAGAYA, *et al.* 1992. Expression and growth-promoting effect of T-cell leukemia-derived factor. A human thioredoxin homologue in hepatocellular carcinoma. Cancer **69:** 2091–2097.
20. FUJII, S., Y. NANBU, H. NONOGAKI, *et al.* 1991. Coexpression of adult T-cell leukemia-derived factor. A human thioredoxin homologue, and human papillomavirus DNA in neoplastic cervical squamous epithelium. Cancer **68:** 1583–1591.
21. SPECTOR, A., G.Z. YAN, R.R. HUANG, *et al.* 1998. The effect of H_2O_2 upon thioredoxin-enriched lens epithelial cells. J. Biol. Chem. **263:** 4984–4990.
22. CHAE, H.-J., S.J. CHUNG & S.G. RHEE. 1994. Thioredoxin-dependent peroxide reductase from yeast. J. Biol. Chem. **269:** 27670–27678.
23. MATTHEWS, J.R., N. WAKASUGI, J.-L. VIRELIZIER, *et al.* 1992. Thioredoxin regulates the DNA binding activity of NF-KB by reduction of a disulphide bond involving cysteine 62. Nucleic Acids Res. **20:** 3821–3830.
24. CROMLISH, J.A. & R.G. ROEDER. 1989. Human transcription factor IIIC (TFIIIC). Purification, polypeptide structure, and the involvement of thiol groups in specific DNA binding. J. Biol.Chem. **264:** 18100–18109.
25. BANNISTER, A.J., A. COOK & T. KOUZARIDES. 1991. *In vitro* DNA binding activity of Fos/Jun and BZLF1 but not C/EBP is affected by redox changes. Oncogene **6:** 1243–1250.
26. GRIPPO, J.F., W. TIENRUNGROJ, M.K. DAHMER, *et al.* 1983. Evidence that the endogenous heat-stable glucocorticoid receptor-activating factor is thioredoxin. J. Biol. Chem. **258:** 13658–13664.
27. ABATE, C., L. PATEL, F.J. RAUSCHER III & T. CURRAN. 1990. Redox regulation of fos and jun DNA-binding activity *in vitro*. Science **249:** 1157–1161.

28. HIROTA, K., M. MATSUI, S. IWATA, *et al.* 1997. AP-1 transcriptional activity is regulated by a direct association between thioredoxin and Ref-1. Proc. Natl. Acad. Sci. U.S.A. **94:** 3633–3638.
29. QIN, J., G.M. LORE, W.P. KENNEDY, *et al.* 1996. The solution structure of human thioredoxin complexed with its target from Ref-1 reveals peptide chain reversal. Structure **4:** 613–620.
30. RUBARTELLI, A., A. BAJETTO, G. ALLAVENA, *et al.* 1992. Secretion of thioredoxin by normal and neoplastic cells through a leaderless secretory pathway. J. Biol. Chem. **267:** 24161–24164.
31. RUBARTELLI, A., N. BONIFACI & R. SITIA. 1995. High rates of thioredoxin secretion correlate with growth arrest in hepatoma cells. Cancer Res. **55:** 675–680.
32. GASDASKA, J.R., M. BERGGREN & G. POWIS. 1995. Cell growth stimuation by the redox protein thioredoxin occurs by a novel helper mechanism. Cell Growth Different. **6:** 1643–1650.
33. SHAO, L.-E., M.B. DICCIANNI, T. TANAKA, *et al.* 2001. Thioredoxin expression in primary T-cell acute lymphoblastic leukemia and its therapeutic implication. Cancer Res. **61:** 7333–7338.
34. TAGAYA, Y., Y. MAEDA, A. MITSUI, *et al.* 1989. ATL-derived factor (ADF), an IL-2 receptor/Tac inducer homologous to thioredoxin; possible involvement of dithiol-reduction in the IL-2 receptor induction. The EMBO J. **8:** 757–764.
35. ROSEN, A., P. LUNDMAN, M. CARLSSON, *et al.* 1995. A CD4+ T cell line-secreted factor, growth promoting for normal and leukemic B cells, identified as thioredoxin. Int. Immunol. **7:** 625–633.
36. WOLLMAN, E.E., L. D'AURIOL, L. RIMSKY, *et al.* 1988. Cloning and expression of a cDNA for human thioredoxin. J. Biol. Chem. **263:** 15506–15512.
37. GASDASKA, P.Y., J.E. OBLONG, I.A. COTGREAVE & G. POWIS. 1994. The predicted amino acid sequence of human thioredoxin is identical to that of the autocrine growth factor human adult T-cell derived factor (ADF): Thioredoxin mRNA is elevated in some human tumors. Biochim. Biophys. Acta **1218:** 292–296.
38. BERGGREN, M., A. GALLEGOS, J.R. GASDASKA, *et al.* 1996. Thioredoxin and thioredoxin reductase gene expression in human tumors and cell lines, and the effects of serum stimulation and hypoxia. Anticancer Res. **16:** 3459–3466.
39. SMITH, M., D. ARTHUR, B. CAMITTA, *et al.* 1996. Uniform approach to risk classification and treatment assignment for children with acute lymphoblastic leukemia. J. Clin. Oncol. **14:** 18.
40. TANAKA, T., L.-E. SHAO, R. GRIBI, *et al.* 2001. Increased thioredoxin expression and implications in ovarian cancer. Cancer Submitted.
41. BAKER, A., C.M. PAYNE, M.M. BRIEHL & G. POWIS. 1997. Thioredoxin, a gene found overexpressed in human cancer, inhibits apoptosis *in vitro* and *in vivo*. Cancer Res. **57:** 5162–5167.
42. GALLEGOS, A., J.R. GASDASKA, C.W. TAYLOR, *et al.* 1996. Transfection with human thioredoxin increases cell proliferation and a dominant negative mutant thioredoxin reverses the transformed phenotype of breast cancer cells. Cancer Res. **56:** 5765–5770.
43. SASADA, T., S. IWATA, N. SATO, *et al.* 1996. Redox control of resistance to *cis*-diamminedichloroplatium (II) (CDDP): protective effect of human thioredoxin against CDDP-induced cytotoxicity. J. Clin. Invest. **97:** 2268–2276.
44. YOKOMIZO, A., M. ONO, H. NANRI, *et al.* 1995. Cellular levels of thioredoxin associated with drug sensitivity to cisplatin, mitomycin C, doxorubicin, and etoposide. Cancer Res. **55:** 4293–4296.
45. JOST, C.A., M.C. MARIN & W.G. KAELIN. 1997. p73 is a simian [correction of human, see comments 1999] p53-related protein that can induce apoptosis. Nature **389:** 191–194.
46. KAELIN, W.G., JR. 1999. The p53 gene family. Oncogene **18:** 7701–7705.
47. LEVRERO, M., V. DE LAURENZI, A. COSTANZO, *et al.* 2000. The p53/p63/p73 family of transcription factors: overlapping and distinct functions. J. Cell Sci. **113:** 1661–1670.

48. KAWANO, S., C.W. MILLER, A.F. GOMBART, *et al.* 1999. Loss of p73 gene expression in leukemias/lymphomas due to hypermethylation. Blood **94:** 1113–1120.

49. GONG, J.-G., A. COSTANZO, H.-Q. YANG, *et al.* 1999. The tyrosine kinase c-Abl regulates p73 in apoptotic response to cisplatin-induced DNA damage. Nature **399:** 806–809.

50. ANDOH, T., P.B. CHOCK & C.C. CHIUEH. 2001. The roles of thioredoxin in protection against oxidative stress-induced apoptosis in SH-SY5Y cells. J. Biol. Chem. In press.

51. UEDA, S., H. NAKAMURA, H. MASUTANI, *et al.* 1998. Redox regulation of caspase-3 (-like) protease activity: regulatory roles of thioredoxin and cytochrome *c*. J. Immunol. **161:** 6689–6695.

52. BAKER, A., B. DOS SANTOS & G. POWIS. 2000. Redox control of caspase-e activity by thioredoxin and other reduced proteins. Biochem. Biophys. Res. Commun. **268:** 78–81.

53. ICHIJO, H., E. NISHIDA, K. IRIE, *et al.* 1997. Induction of apoptosis by ASK1, a mammalian MAPKKK that activates SAPK/KNK and p38 signaling pathways. Science **275:** 90–94.

54. LIU, H., H. NISHITOH, H. ICHIJO & J.M. KYRIAKIS. 2000. Activation of apoptosis signal-regulating kinase 1 (ASK1) by tumor necrosis factor receptor-associated factor 2 requires prior dissociation of the ASK1 inhibitor thioredoxin. Molec. Cell. Biol. **20:** 2198–2208.

55. LANDER, H.M., A.T. JACOVINA, R.J. DAVIS & J.M. TAURAS. 1996. Differential activation of mitogen-activated protein kinases by nitric oxide-related species. J. Biol. Chem. **271:** 19705–19709.

56. NORIS, M., M. MORIGI, R. DONADELLI, *et al.* 1995. Nitric oxide synthesis by cultured endothelial cells is modulated by flow conditions. Circ. Res. **76:** 536–543.

57. NORTH, A.J., R.A. STAR, T.S. BRANNON, *et al.* 1994. Nitric oxide synthase type I and type III gene expression are developmentally regulated in rat lung. Am. J. Physiol. **266:** L635–L641

58. KAWAI, N., D.B. BLOCH, G. FILIPPOV, *et al.* 1995. Constitutive endothelial nitric oxide synthase gene expression is regulated during lung development. Am. J. Physiol. **268:** L589–L595

59. BLACK, S.M., M.J. JOHENGEN, Z.-D. MA, *et al.* 1997. Ventilation and oxygenation induce endothelial nitric oxicde synthase gene expression in the lungs of fetal lambs. J. Clin. Invest. **100:** 1448–1458.

60. PATEL, J.M., J. ZHANG & E.R. BLOCK. 1996. Nitric oxide-induced inhibition of lung endothelial cell nitric oxide synthase via interaction with allosteric thiols: role of thioredoxin in regulation of catalytic activity. Am. J. Respir. Cell Mol. Biol. **15:** 410–419.

61. DAS, K.C., X.-L. GUO & C.W. WHITE. 1999. Induction of thioredoxin and thioredoxin reductase gene expression in lungs of newborn primates by oxygen. Am. J. Physiol. **276:** L530–L539

Does Nitric Oxide Play a Role in Orofacial Pain Transmission?

JIN-FEI YEO

Department of Oral and Maxillofacial Surgery, Faculty of Dentistry, National University of Singapore, Singapore

ABSTRACT: **Although the synaptology, neural connectivity, and the roles played by nitric oxide (NO) and other neurotransmitters have been extensively studied in spinal pain, such information is rather scanty with respect to orofacial pain transmission. This paper presents the findings of several investigations carried out by the author and his colleagues on the roles of NO in orofacial pain transmission in male Wistar rats, using nicotinamide adenosine dinucleotide phosphate-diaphorase (NADPH-d) histochemistry using light and electron microscopy; and NOS immunohistochemistry and immunofluorescence using both light and confocal laser scanning microscopy. The results revealed that (1) a complicated relation existed between the nitrergic axon terminals and dendrites in the caudal part of the spinal trigeminal nucleus (cSTN); (2) the nitrergic neuronal cells bodies were not projection neurons, but rather, local circuit neurons; (3) although the thalamus projecting neurons in the cSTN did not synthesize NO, they could be modulated by NO diffused from nitrergic neurons; (4) c-fos positive neurons in the superficial laminæ of the cSTN, detected following subcutaneous injection of 0.5 ml of 4% formalin into the left lateral face of the rats, respond to the release of glutamate through activation of *N*-methyl-D-aspartate (NMDA), α-amine-3-hydroxy-5-methyl-4-isoxazole-propionate (AMPA) and metabotropic glutamate (mGlu) receptors expressed by these c-fos neurons; and (5) NO might play a seemingly less important role than glutamate in neural transmission.**

KEYWORDS: **nitric oxide; orofacial pain transmission**

INTRODUCTION

Most painful conditions in the orofacial region may be traced to dental pathology and trauma, and they are acute rather than chronic in nature. The causes of chronic orofacial pain are more complex than the acute counterpart.[1] Orofacial pain is a term defined by the American Academy of Orofacial Pain (AAOP) as "pain conditions that are associated with the hard and soft tissues of the head, face, neck, and all the intraoral structures".[2] Orofacial pain used in this context is, therefore, a broad term encompassing a variety of disease entities within and outside the oral cavity that cause pain, dysfunction, or insidious pathology in the head and neck area. It has been estimated that chronic orofacial pain affects about 7% of the population, a

Address for correspondence: Jin-Fei Yeo, Department of Oral and Maxillofacial Surgery, Faculty of Dentistry, National University of Singapore, 5 Lower Kent Ridge, Singapore 119074. Fax: 65 7785742 / 65 7732600.
omsyeojf@nus.edu.sg

prevalence comparable with that for dental caries or periodontal diseases.[3] Since pain, by definition, is an emergent property of the human brain and requires consciousness,[4] it is not so much the pain than the suffering that brings patients into the dentist's office in the hope of seeking relief. Although relieving orofacial pain has been a major aim of dental professionals, pain in the oral and facial region has also been one of the most elusive complaints to manage.

The research work done by my group in the past six years aimed to gain a better understanding of the mechanism of orofacial pain and the development of a more effective management of such painful conditions.

NITRIC OXIDE—BRIEF INTRODUCTION

Nitric oxide (NO) is a very important inorganic gas in the biological system. It is a paramagnetic species that has an unpaired electron in its outer orbit; that is it is a free radical. Since 1987, when Palmer, Ferrige, and Moncada[5] suggested that the gas might act as a biological mediator, research on this molecule with very simple structure has become a big business and resulted in the publication of no less than 20,000 publications to date.[6] As a result of the concerted efforts of NO scientists, important biological roles of NO have been identified in the cardiovascular, respiratory, nervous, genitourinary, gastrointestinal, immune, hemopoietic, and musculoskeletal systems. In the nervous system NO has been shown to be a novel and important neurotransmitter.[7]

NO is synthesized intracellularly directly from the amino acid, L-arginine, and the enzyme responsible for this conversion is nitric oxide synthase (NOS). The nitrogen is derived from one of the guanidine nitrogen atoms of arginine and an oxygen atom from molecular oxygen. The amino acid, L-citrulline, is a byproduct of the reaction. There are three distinct isoforms of NOS, the endothelial (eNOS or NOS III), the neuronal (nNOS or NOS I) and the inducible (iNOS NOS II) isoform. The last was first identified in macrophages. The three isoforms of NOS are encoded by three different genes in different chromosomes in humans. The nNOS gene is in chromosome 12, the eNOS gene in chromosome 7, and iNOS gene in chromosome 17. The genesis of NO can be blocked by NOS inhibitors like L-N^G-nitroarginine methyl ester (L-NAME) and N^G-monomethy L-arginine (L-NMMA).[6]

The NO produced by nitrergic cells diffuses into adjacent cells and directly activates their guanylate cyclase to produce the secondary messenger cyclic guanosine monophosphate (cGMP), which may activate protein kinase G and other effector systems to produce a cellular response. Because NO has a short half-life (about 5 seconds) it exerts only a local action.[8]

NO AND OROFACIAL PAIN TRANSMISSION

Recently, NO has been implicated in nociception and pain processing. It had also been implicated in hyperalgesia. Intrathecal injection of NO donor had been reported to cause spinally mediated hyperalgesia.[9] It was shown that blockage of NO synthesis eliminated hyperalgesia[10,11] as well as formalin-induced hyperalgesia.[12]

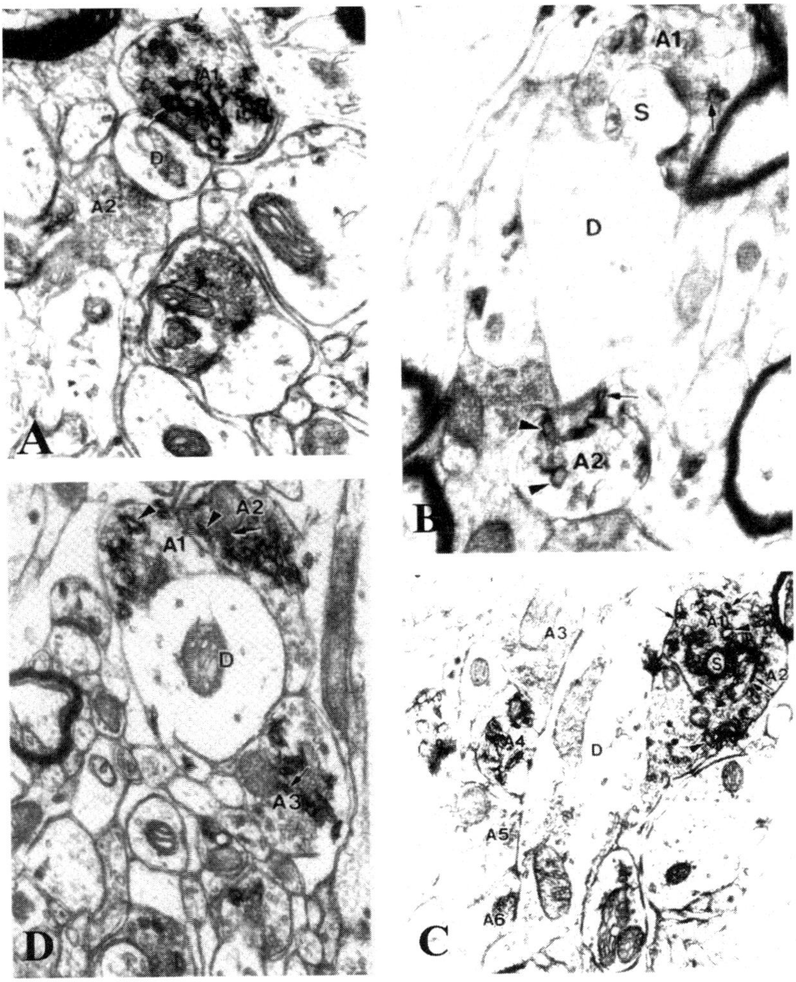

FIGURE 1. **(A)** Electron micrograph showing a NADPH-d negative dendrite (D) postsynaptic to a NADPH-d positive (A1), as well as a NADPH-d negative (A2) axon terminal (×385,000). **(B)** A synapse formed between an NADPH-d positive axon terminal (A1) and the spine (S) of an NADPH-d negative dendrite (D) that is postsynaptic to another NADPH-d positive axon terminal. (×35,000). **(C)** A NADPH-d negative dendritic spine (S) is postsynaptic to a NADPH-d positive axon terminal (A1) that is postsynaptic to another NADPH-d positive axon terminal (A2), thus forming an axoaxonic synapse (indicated by *double arrows*). Note also the NADPH-d negative dendrite (D) that is postsynaptic to a NADPH-d positive axon terminal (A4) and three NADPH-d negative axon terminals (A3, A5, and A6) (×29,750). **(D)** A NADPH-d negative dendrite (D) that is postsynaptic to three NADPH-d positive axon terminals (A1, A2, and A3) (×35,000).

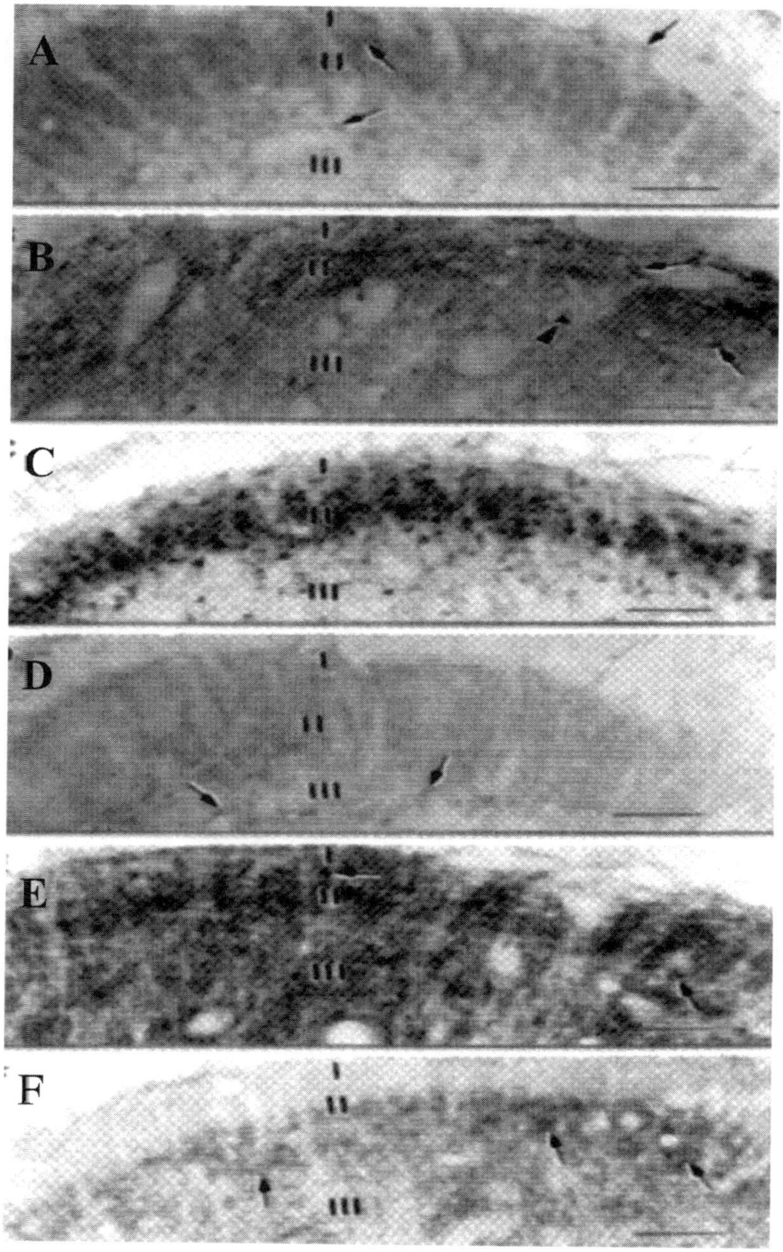
A
I
II
III
B
C
D
E
F

Furthermore, inhibition of spinal nitric oxide synthesis also blocked NMDA-induced hyperalgesia,[9,13] and thermal hyperalgesia resulting from nerve injury.[14] Inhibition of spinal nitric oxide synthesis also produced antinociception in rats injected with formalin at the paw.[15]

NO has been widely reported to play important roles in spinal pain processing. However, very little is known in the case of orofacial nociception, except for the localization of nitrergic neurons in the trigeminal ganglion and in the superficial laminae of the caudal part of the spinal nucleus of the trigeminal nerve (cSTN). The latter is well known to receive the primary afferents of the trigeminal ganglion neurons. The nucleus is also known to be primarily responsible for relaying orofacial nonreflexive nociceptive information.[16,17] From there information is conveyed to the thalamus,[18] and then to the cerebral cortex.

Although NO has been implicated in nociceptive transmission, the neurons involved and their synaptic connections have not been elucidated. A major difficulty in the investigation of the role of NO in nociception is the problem of precise intracellular localization of NO. My research group took an anatomical approach to study the nitrergic neurons present in the cSTN. It was previously shown by Dohrn *et al.*[17] that under light microscopy the staining pattern of NOS is similar to that of nicotinamide adenosine dinucleotide phosphate-diaphorase (NADPH-d). Hence, neurons in the superficial laminæ of the cSTN were stained by the NADPH-d method and examined by conventional electron microscopy to investigate their ultrastructural features and synaptic relationship.

Recent investigations had also demonstrated that noxious stimulation of the tooth pulp, or facial skin, or after tooth extraction,[19–21] or subcutaneous injection of formalin into the lateral face[22] caused *c-fos* expression in the cSTN. The *c-fos* expressing neurons are predominantly situated in lamina I, the outer part of lamina II (IIo), and laminæ V–VI of the spinal cord, all of which correspond to the terminal fields of primary nociceptive afferent fibers and to the distribution of nociresponsive neurons identified by electrophysiological recording.[23–25] Thus, Fos, the nuclear protein encoded by the early gene *c-fos,* as demonstrated by immunohistochemistry, is a suitable indicator for the presence of nociceptive neurons activated by injection of formalin into the lateral face. The formalin test as a means for assessing pain from the hindpaw, the upper lip or vibrissal pad of animals has indeed been widely used.[26–28]

FIGURE 2. (A) NMDAR1 positive neurons (*arrow*) in the three superficial laminæ of the caudal spinal trigeminal nucleus (cSTN). (×110). **(B)** A few lightly to moderately stained GluR1 positive neurons in laminæ I and III of the cSTN (*arrows*). Some moderately to strongly stained GluR1 positive neurons (*arrowheads*) are also present in lamina II of the nucleus (×110). **(C)** Sections stained for GluR2/3 immunohistochemistry. Few weakly (in lamina I) and many strongly (in lamina II) stained neurons in the cSTN. In lamina III, some neurons are weakly and others, strongly stained (×10). **(D)** Weakly or moderately stained GluR4 positive neurons in lamina III of the cSTN (×110). **(E)** Some (in lamina I and outer part of lamina II) and many (in inner part of lamina II and lamina III) mGluR1α immunopositive dots in cSTN. A few weakly stained mGluR1α positive neurons are also seen in the three laminæ (*arrows*) of the nucleus. (×110). **(F)** A few (in lamina I and outer part of lamina II) and some (in inner part of lamina II and lamina III) mGluR2/3 positive dots in the cSTN. No mGluR2/3 immunopositive neuronal cell bodies can be detected (×110).

During the period between 1995 and 2000, my research group made some interesting findings from several studies undertaken on male Wistar rats, using NADPH-d histochemistry under light and electron microscopy, and NOS immunohistochemistry and immunofluoroscence with both light and confocal laser scanning microscopes.[29,30]

Our electron microscopy investigation[29] showed that the NADPH-d labeled neurons in the cSTN measured 20 μm × 11 μm. The reaction product, formazan, was localized in the nuclear envelope, mitochondria, rough endoplasmic reticulum, and multivesicular bodies of the neurons, as well as in the membrane of smooth endoplasmic reticulum at the axon terminals. These neurons were contacted by both axosomatic and axodendritic synapses formed by both NADPH-d positive and NADPH-d negative axon terminals. Two types of NADPH-d positive axon terminal were observed. One, a large terminal containing many stained mitochondria and unstained small round agranular vesicles mixed with some slightly flattened ones. The other type of axon terminal contained pleomorphic synaptic vesicles forming asymmetric synapses with both dendrites and somata. Complicated synaptic configurations demonstrating NADPH-d positive axoaxonic synapses in relation to NADPH-d negative dendritic spines were also observed in which an NADPH-d negative dendritic spine was fully contacted by an NADPH-d positive bouton (a knob-like terminal) that was in turn in contact with another NADPH-d positive bouton. The complicated synaptic relations between axons and dendrites indicated complex synaptic interactions in the cSTN (see FIGURE 1).

Immunohistochemical study of the distribution of NOS and various glutamate receptors in the cSTN of rats showed few strongly stained NOS-like and glutamate (Glu) R1-like, as well as some moderately stained GluR2/3- and NMDAR1-like positive neurons in lamina I. In lamina II, there were many strongly stained NOS and GluR2/3, and a few fairly strongly stained GluR1 immunoreactive neurons. Moderate to densely stained NOS, GluR1, GluR2/3, and weakly stained NMDAR1 positive fibres were seen in laminæ I and II. These results suggested that as a neurotransmitter in the cSTN, glutamate might play a more important role than nitric oxide (see FIGURE 2).

The above investigations were followed by a study of the distribution of neurons in the cSTN that expressed Fos protein in response to subcutaneous formalin injection into the lateral face of the rat. This aimed to determine (1) whether these neurons

FIGURE 3. **(A)** Fos labeling in the caudal medulla of a rat receiving formalin injection into the left lateral face. Note that the nuclei are intensively labeled by Fos immunohistochemistry and are mostly concentrated in lamina I and the outer part of lamina II (IIo) of the left cSTN. A few are also scattered in the deeper laminæ of the nucleus. *Scale bar*, 200 μm. **(B)** Photomicrograph showing double labeling of Fos and NOS in the left cSTN of a rat receiving formalin injection into the left lateral face. Note that neurons containing nuclei labelled by Fos (*light grey*) are more superficially located than those labelled by NOS (*dark grey*). There are two Fos positive nuclei that are surrounded by NOS positive cytoplasm (*arrow*) with insert demonstrating a magnified view of the neuron indicated by the *arrow*. **(C)** Photomicrograph showing an NOS positive neuron with two processes (*dark grey*) in the superficial laminæ of cSTN, one of which extends to the close vicinity of a neuron bearing a Fos positive nucleus (*light grey*). **(D)** Three Fos labeled nuclei (*light grey*) in the superficial laminæ of the cSTN with two of them closely related to NOS positive neurons (*dark grey*).

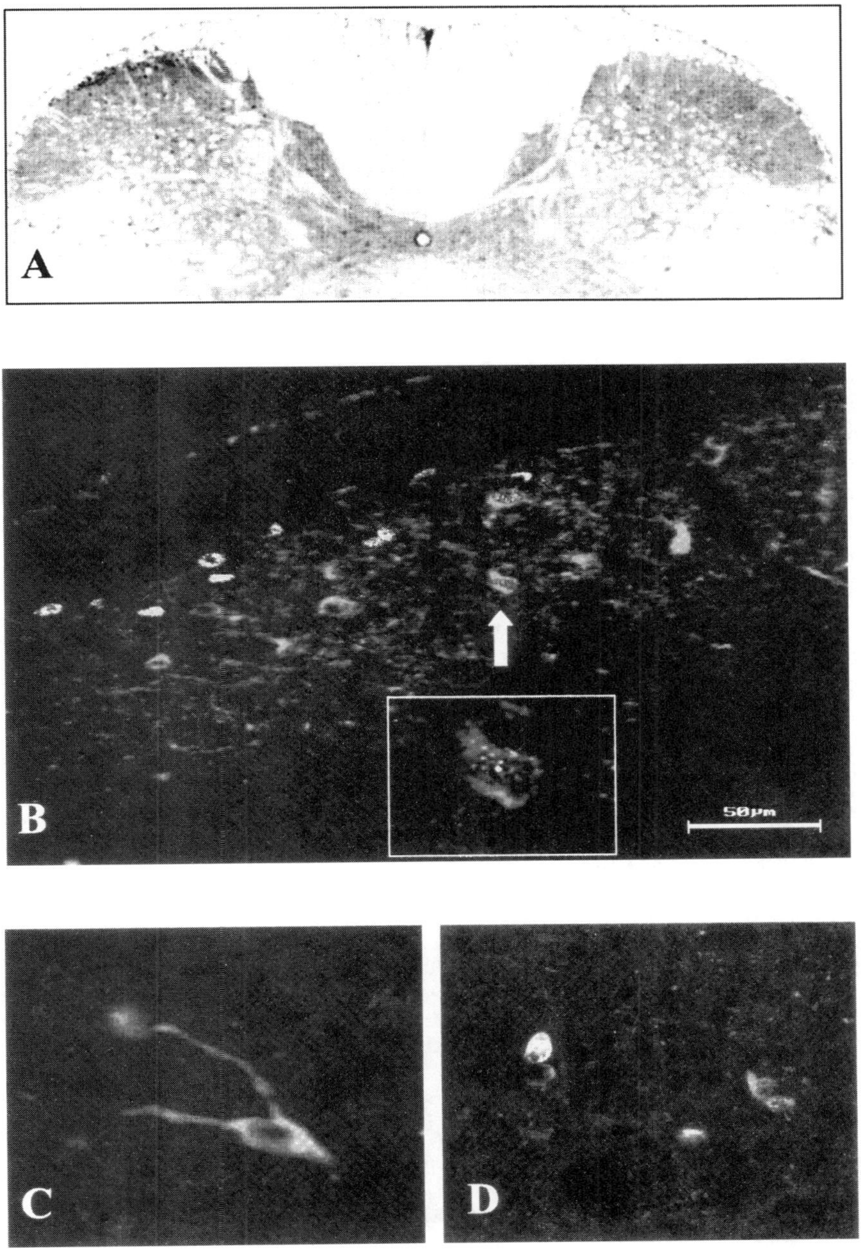
A
B
50µm
C
D

express glutamate receptors or NOS; and (2) if they project to the thalamus. For the latter investigation, the retrograde axonal tracer, tetramethylrhodamine-dextran (TMR-D), was injected into the thalamus or the trigeminothalamic tract to reveal the trigeminothalamic neurons. Since our preliminary study indicated that of all the glutamate receptors, NMDA receptor subunit 1 (NMDAR1) and AMPA receptor subunit 2/3 (GluR2/3) showed the most intense immunostaining in the neurons of the cSTN and should, therefore, play the most prominent role in nociception, we focused on these two receptors. After subcutaneous injection of 0.5 ml of 4% formalin into the left lateral face of the rats, c-fos positive neurons were detected in the superficial laminæ of the cSTN. The same neurons were investigated using immunofluoresence and confocal laser scanning microscopy for possible colocalization with NOS and various glutamate receptors. NOS positive neurons were distributed widely in the superficial laminæ of the cSTN. Some of NMDAR1 and GluR2/3 positive neurons in the cSTN were NOS positive. Most of *c-fos* positive neurons were NMDAR1 and GluR2/3 positive. Some *c-fos* expressing neurons were thalamus projection neurons and most of them were NMDAR1 positive. Only 14% of the *c-fos* expressing neurons were NOS positive, but none of them projected to the thalamus. These neurons were, however, seen in close apposition to NOS positive neurons (see FIGURE 3).

The above results indicated that neurons in the superficial laminæ of the cSTN respond to the release of glutamate through activation of NMDA, AMPA, and mGluR receptors expressed by these c-Fos neurons. Although *c-fos* positive neurons projecting to the thalamus did not synthesize NO, they were modulated by NO diffused from adjacent NOS positive neurons. The NOS positive neurons were most likely local circuit neurons.

CONCLUSION

In conclusion, NO plays important roles in orofacial pain transmission and hyperalgesia. That NADPH-d labeled neurons in the cSTN were contacted by both axosomatic and axodendritic synapses formed by both NADPH-d positive and NADPH-d negative axon terminals demonstrated the involvement of NO in the complicated arrangement of the nitrergic axons and dendrites involved in orofacial nociception. Although most of the neurons involved in orofacial nociception in the cSTN did not synthesize NO, they were modulated by NO that diffused from adjacent NOS positive neurons.

ACKNOWLEDGMENTS

This research was supported by Grants RP940320 and RP960331 from the National University of Singapore.

REFERENCES

1. ZAKRZEWSKA, J.M. & P.J. HAMLYN. 1999. Facial pain. *In* Epidemiology of Pain. I.K. Crombie, Ed.: 171–202. IASP Press, Seattle.

2. OKESON, J.P. 1996. Orofacial Pain-Guidelines for Assessment, Diagnosis, and Management. Quintessence Books, Chicago.
3. TALLEY, R.L., J.R. FRICTION & J.P. OKESON. 2000. Broad support evident for the emerging specialty of orofacial pain. Okla. Dent. Assoc. J. (Summer Issue) 14–17.
4. LOESER, J.D. 2000. Pain and suffering. Clin. J. Pain **16**(2): S2–S6
5. PALMER, R.M., A.G. FERRIGE & S. MONCADA. 1987. Nitric oxide release accounts for the biological activity of endothelial-derived relaxing factor. Nature **327:** 524–526.
6. MONCADA, S. 1999. Nitric oxide: discovery and impact on clinical medicine. J. Roy. Soc. Med. **92:** 164–167.
7. BREDT, D.S. & S.H SNYDER. 1992. Nitric oxide, a novel neuronal messenger. Neuron **8:** 3–11.
8. LUTY, J. & P. HARRISON. 1997. Basic and Clinical Pharmacology, 6. Churchill Livingstone, London.
9. KITTO, K.F., J.E. HALEY & G.W. WILCOX. 1992. Involvement of nitric oxide in spinally mediated hyperalgesia in the mouse. Neurosci. Lett. **148:** 1–5.
10. MELLER, S.T., C. DYKSTRA & G.F. GEBHART. 1996. Acute thermal hyperalgesia in the rat is produced by activation of *N*-methyl-D-aspartate receptors and protein kinase C and prodution of nitric oxide. Neuroscience **71:** 327–335.
11. YAMAMOTO, R., D.S. BREDT, S.H. SYNDER & R.A. STONE. 1993. The localisation of nitric oxide synthase nerve in the rat eye and related cranial ganglia. Neuroscience **54:** 189–200.
12. WIERTELAK, E.P., L.E. FURNESS, R. HORAN, *et al.* 1994. Subcutaneous formalin produces centrifugal hyperalgesia at a non-injected site via the NMDA-nitric oxide cascade. Brain Res. **649:** 19–26.
13. MALMBERG, A.B. & T.L. YAKSH. 1993. Spinal nitric oxide synthesis inhibition blocks NMDA-induced thermal hyperalgesia and produced antinociception in the formalin test in rats. Pain **54:** 291–300.
14. MELLER, S.T., P.S PECHMAN, G.F. GEBHART & T.J. MAVES. 1992. Nitric oxide mediates the thermal hyperalgesia produced in a model of neuropathic pain in the rat. Neuroscience **50:** 7–10.
15. ROCHER, A.K., M. COOK, G.L. WILCOX & K.C.A. KAJANDER. 1996. A nitric oxide synthesis inhibitor (L-NAME) reduces licking behavior and Fos-labelling in the spinal cord of rats during formalin-induced inflammation. Pain **66:** 331–341.
16. BROTON, J.G., J.W. HU & B.J. SESSLE. 1988. Effects of temporo-mandibular joint stimulation on nociception and non-nociceptive neurons of the cat's trigeminal nucleus caudalis (medullary dorsal horn). J. Neurophysiol. **59:** 1575–1589.
17. DOHRN, C.S., M.A. MULLET, R.H. PRICE & A.J. BEITZ. 1994. Distribution of nitric oxide synthase immuno-reactive interneurons in the spinal trigeminal nucleus. J. Comp. Neurol. **346:** 449–460.
18. WAITE, P.M.E. & D.J. TRACEY. 1995. Trigeminal sensory system. *In* The Rat Nervous System. G. Paxinos, Ed.: 705–724. Academic Press, New York.
19. COIMBRA, F. & A. COIMBRA. 1994. Dental noxious input reaches the subnucleus caudalis of the trigeminal complex in the rat, as shown by c-fos expression upon thermal or mechanical stimulation. Neurosci. Lett. **173:** 201–204.
20. STRASSMAN, A.M , B.P. VOS, Y. MINETA, *et al.* 1993. Fos-like immunoreactivity in the superficial medullary dorsal horn induced by noxious and innocuous thermal stimulation of facial skin in the rat. J. Neurophysiol. **70:** 1811–1821.
21. WAKISAKA, S., Y. SASAKI, H. ICHIKAWA & S. MATSUO. 1992. Increase in c-fos-like immunoreactivity in the trigeminal nucleus complex after dental treatment. Proc. Finn. Dent. Soc. **88:** 551–555.
22. HOU, W.Y., B.C. SHYU, T.M. CHEN, *et al.* 1997. Intrathecally administered c-fos antisense oligodeoxynucleotide decreases formalin-induced nociceptive behavior in adult rats. Eur. J. Pharmacol. **329:** 17–26.
23. BULLITT, E. 1991. Somatotopy of spinal nociceptive processing. J. Comp. Neurol. **312:** 279–290.
24. HUNT, S.P., A. PINI & G. EVAN. 1987. Induction of c-fos-like protein in spinal cord neurons following sensory stimulation. Nature **328:** 632–634.

25. PAXINOS, G. & C. WATSON. 1996. The rat brain in stereotaxic coordinates, 3rd edit. Academic Press, San Diego.
26. CLAVELOU, P., J. PAJOT, R. DALLEL & P. RABOISSON. 1989. Application of the formalin test to the study of orofacial pain in the rat. Neurosci. Lett. **103:** 349–353.
27. EISENBERG, E., B.P. VOS & A.M. STRASSMAN. 1993. The NMDA antagonist memantine blocks pain behavior in a rat model of formalin-induced facial pain. Pain **54:** 301–307.
28. WOOLF, C.J. & S.W.N. THOMPSON. 1991. The induction and main-tenance of central sensitization is dependent on *N*-methyl-D-aspartic acid receptor activation: implications for the treatment of post-injury pain hypersensitivity states. Pain **44:** 293–300.
29. YEO, J.F., F.R. TANG & S.K. LEONG. 1997. Ultrastructural study of NADPH-d positive neurons in laminae I and II of the rat caudal spinal trigeminal nucleus. Intl. J. Neurosci. **91:** 29–43.
30. LEONG, S.K., H.P. LIU & J.F.YEO. 2000. Nitric oxide synthase and glutamate receptor immunoreactivity in the rat spinal trigeminal neurons expressing Fos protein after formalin injection. Brain Res. **855:** 107–115.

A Critical Assessment of the Neurodestructive and Neuroprotective Effects of Nitric Oxide

SENG-KEE LEONG,[a] RUN-SHENG RUAN,[b] AND ZHI ZHANG[c]

[a]*Department of Anatomy, National University of Singapore, Singapore*

[b]*Department of Otolaryncology, National University of Singapore, Singapore*

[c]*MRC Collaborative Centre, 1-3 Burtonhole Lane, Mill Hill, London, United Kingdom*

ABSTRACT: Whether nitric oxide is cytodestructive or cytoprotective is of obvious clinical importance. The debate on this subject in the past decade has generated much "heat and light". This paper focuses on the actions of NO on the nervous system and reexamines the controversial issue and the contribution of the authors and their colleagues in the light of recent findings. We also report new findings, critically assesses previous experimental data, and share perspectives on this important subject.

KEYWORDS: critical assessment; neurodestructive effect; neuroprotection effect

INTRODUCTION

Although nitroglycerine has been used since 1867 to relieve chest pain, it is only recently that its action is known to be brought about largely by the vasodilator action of the nitric oxide (NO) it releases. In fact, for many years, NO received poor evaluation, being branded an atmospheric pollutant that contributes to the erosion of the stratospheric ozone layer and causes respiratory distress. The discovery by Furchgott, Murad, and Ignarro and their collaborators, that the gas is released from vascular endothelial cells and is involved in the relaxation of the smooth muscles of the blood vessels,[1–4] helped the gas to rise from the milieu of waste products to the centre stage at Stockholm, where it shared the honour of Murad, Furchgott, and Ignarro when the trio received the Nobel Prize for Medicine in 1998. It should also be mentioned that in the process of identifying the endothelium derived relaxing factor (EDRF) that is capable of vasodilation, it was Palmer *et al.*[5] who first showed, by means of chemiluminescence, that the endothelium released enough of nitric oxide to fully account for the relaxation of the adjacent smooth muscle cells.

Distribution, Production, and Function of NO

NO is widely distributed in the vertebrate and invertebrate animal kingdom and in humans. Its distribution has been reviewed[6–10] and reported in several original papers (e.g., see Refs. 11–28). Not only is NO widely distributed, it is also ubiquitously

Address for correspondence: Seng-Kee Leong, Department of Anatomy, National University of Singapore, Singapore 117597.
antlsk@nus.edu.sg

present in all body systems and is generated by many cell types. Furthermore, plants and microbes also generate NO, using various biochemical pathways.[29]

Endogenous NO is synthesized from the amino acid L-arginine. Its synthesis is catalyzed by the enzyme nitric oxide synthase (NOS), in the presence of molecular oxygen, nicotinamide adenine dinucleotide phosphate (NADPH), and various cofactors like tetrahydrobiopterin. At least three isoforms of NOS have been identified and purified. These are the neuronal NOS (nNOS or NOS I), endothelial NOS (eNOS or NOS III), and inducible NOS (iNOS or NOS II) that were first localized in neurons, vascular endothelium, and activated macrophages, respectively. The first two are constitutively expressed and their activities are calcium and calmodulin dependent. The formation of iNOS requires new protein synthesis, is calcium independent, and is mediated by cytokines, such as IFN-γ and TNF-α, as well as by lypopolysaccharides.[30–32] Each of these isoforms performs a different function and they may, in fact, exert opposite effects in the biological system. The isolation, purification, and structure of these enzymes; the intricate biochemical mechanisms whereby they act on the substrate to generate the simple bioactive inorganic molecule, NO; how NO acts on the soluble guanylyl cyclase to bring about an increased level of cGMP; and the associated biological actions have been extensively studied and eminently reviewed by many authors (see, e.g., Refs. 33–37).

Since NO is generated by diverse cells types, it is expected to have a multiplicity of functions. Thus, when produced by vascular endothelium, it brings about vasodilation and plays a very important role in the maintenance of normal vascular tone, and penile erection.[38] NO is also a neurotransmitter[39] in the peripheral nervous system, being the major non-adrenergic non-cholinergic (NANC) neurotransmitter in the autonomic nervous system. In addition, it plays a very important role in interneuronal communication in the central nervous system (CNS), where neurons stained with NOS or the diaphorase of NADPH (NADPH-d) are widely distributed.[40–42] The action of NO as a neurotransmitter differs from its classical action in that it is not stored in synaptic vesicles and does not bind to membrane associated receptors. Rather, it is synthesized on demand by the catalytic action of NOS on L-arginine. After its formation, it may act on itself or simply diffuse through the cell membrane to act on adjacent cells. Although NO does not have a membrane receptor, it does have its biochemical targets (said to be its intracellular "receptors") the chief of which is the heme moiety of the soluble guanylyl cyclase that it activates, leading to an increased conversion of GTP to cGMP. NO may act as a retrograde messenger to the presynaptic terminals and it is one of the molecules implicated in bringing about long term potentiation. NO is one of the candidate mechanisms for learning and memory storage in the hippocampus.[43] NO also plays an important role in regulating other neurotransmitter release[44] and neuroendocrine function (see, e.g., Refs. 45–50). More detailed descriptions of the actions of NO in the nervous tissue may be found in the review papers by Dawson and Dawson[51] and by Garthwaite and Boulton.[52] Together with its vascular and neural roles, NO also regulates the functions of the renal, respiratory, endocrine, immune (as reviewed by Nathan[34] and by Schmidt and Walter[53]), integumentary,[54] and musculoskeletal[55–60] systems.

Inhibitory Action of NO

Although the detailed mechanisms that cause the effects of NO have not been elucidated, that it has a prominent inhibitory action is clearly portrayed in a number of situations. Thus, its inhibitory actions on smooth muscle contraction, smooth muscle proliferation, adhesion of monocyte and neutrophils to the endothelium, platelet aggregation, and the inotropic responses of cardiac muscles to β-adrenergic agonists are well known. Its inhibition of the respiratory rate of mitochondria in a number of cell types[61] and the ribonucleotide reductase in its antiviral effects[34] are also general knowledge. In addition, it has been shown to exert an inhibitory effect on the carotid chemoreceptors,[62] bone-resorbing activity of the osteoclast,[56] force of contraction of skeletal muscle,[59] and the basal release of dopamine—the putative light adaptive transmitter in the retina.[63,64] Finally, as an important mediator of tumouricidal[65,66] and antimicrobial[67–69] agent, it exerts powerful cytostatic action. However, when examined in a wider context, the action of NO does not always appear to be that simple. For example, in the case of skeletal muscle, although NO inhibits glycolysis, mitochondrial respiration, and phosphocreatine breakdown, it also promotes glucose uptake. Its effects seem to work towards the preservation of intracellular energy stores. Furthermore, although NO generated by iNOS seems to have an overall suppressive effect on bone resorption, since it inhibits both osteoclast activity and precursor recruitment, basal, constitutive NO synthesis within osteoblasts supports proliferation and activity of these cells.[60]

Pathophysiological Action of NO

Other than its important physiological roles in the biological system, NO may be responsible for pathophysiological conditions like septic shock, ischemia, reperfusion injury, and inflammation. It is also said to be the mediator of NMDA neurotoxicity, and of neuronal death in cerebral stroke, as well as the agent responsible for the development of various neurological disorders and abnormalities.[51,70–74] For example, 7-nitroindazole, a relatively selective inhibitor of nNOS, could block 1-methyl-4-phenyl-1,2,3,6-tetrahydropyridine (MPTP) neurotoxicity in mice[75] and protect against profound striatal dopamine depletions and loss of tyrosine hydroxylase-positive neurons in the substantial nigra of MPTP-treated baboons.[76] In the latter case, cognitive and motor functions of 7-nitroindazole treated baboons does not differ significantly from those of the control. At the periphery, elaboration of NO at the distal end of the transected sciatic nerve appears to be toxic to regenerating axons.[77] This so-called "ugly" or "dark side" of NO is responsible for the coining of terms such as "bad guy", "murderer", and "molecule with a killer soul" for NO.

The Controversy

Despite what has been said of the neurodestructive action of NO, many investigators have proposed a neuroprotective function for the molecule. For example, NADPH-d neurons are strikingly spared in Huntington's disease,[14] and are resistant to degeneration following striatal injections of quinolinic acid.[46] Additionally, the NOS-containing neurons are relatively spared in the hippocampal formation in Alzheimer's disease[78] and in the mouse arcuate nucleus following neonatal glutamate treatment.[79] In cortical cultures, NOS-containing neurons have been shown to be relatively resistant to the direct toxicity of NO and its putative metabolite, peroxynitrite.[80,81] The salvaging effect of NO has also been seen in non-neural tissues, such as mesangial cells affected in acute tubular necrosis,[82] and in the hepatic cells in the murine model of hepatic damage.[83] In fact, exogenous nitric oxide can act as hydroxyl radical scavenger and protect nigral neurons from oxidative injury.[84] In addition to these properties, many authors have also expressed their support for the neuroprotective view of NO (see, e.g., Refs. 85–94).

The controversial findings related to the beneficial effects of NO, on the one hand, and its cytotoxic effects, on the other hand, have generated terms like "Jekyll and Hyde", "friend and foe", and "double-faced Janus", to describe the *dual personality* of the free radical gas. That NO can cause cell death has been adequately demonstrated in its lethal cytotoxic action on the invading microorganisms or tumor cells. How a chemical that is normally a friend can turn around to become a foe is far from clear. The common explanation given is that, when NO is inappropriately and excessively produced under pathophysiological conditions, it can become cytotoxic. For example, although necessary to maintain normal vascular tone under physiological conditions, it can, when present in excessive amount in the presence of endotoxin, cause vasodilation, profound hypotension, and loss of normal myocardial function, leading to septic shock.[95] That this explanation is plausible is supported by the fact that excessive amount of essential molecules for life, such as water or oxygen, can bring about deleterious effects on the body system. In the nervous system, several investigators have also demonstrated that it is the amount of NO that determines its physiologic or pathophysiologic action. For example, Dawson *et al.*[81] have shown that the glutamate-stimulated, NO-mediated neurotoxicity in primary cultures of rat foetal cortical, striatal, and hippocampal neurons exhibits a concentration–effect relationship and time course. Koh and Choi[80] and Bonfoco *et al.*[96] have also demonstrated concentration dependent neuronal injury. Where could the excess amount of NO have come from? This could have been made available via the upregulation of nNOS (see, e.g., Ref. 81). In fact, compared with the wildtype controls, nNOS knockout mice showed reduced infarct volumes following permanent middle cerebral artery occlusion,[97] and were more resistant to MPTP neurotoxicity.[98] Greater than normal amount of NO may also be generated through the formation of iNOS in neutrophils 48–96 h after cerebral ischemia,[99] and in microglia or/and astrocytes *in vivo* following induction of bacterial endotoxin and/or lipopolysaccharide,[18,100] fibrillar β-amyloid,[101] or after cerebral ischemia,[99] or injection of ethanol.[102] Since microglia comprise about 12% of brain cells and proliferate in the penumbra zone following focal ischemia they could generate a formidable amount of NO.

OUR FINDINGS REEXAMINED IN THE LIGHT OF RECENT DISCOVERIES

Myriads of investigations have been devoted to exploring the molecular/biochemical aspects of the cytotoxic/cytoprotective functions of NO. However, only a handful have taken the anatomical approach, following a variety of experimental paradigms. Clowry[103] first showed that there was *de novo* expression of NADPH-d in the sciatic motoneurons after transection of the sciatic nerve in the midthigh of neonatal rats. Many of these neurons appeared shrunken and degenerated. Subsequently, many others also reported the dynamic regulation of NOS in response to a variety of neuronal injury paradigms (for a review, see Ref. 104). Yu[105] showed that NADPH-d and NOS reactivity, normally undetectable in the neurons of the facial motor nucleus, dorsal nucleus of the vagus, and hypoglossal nucleus, became obvious after transection of the facial, vagus, and hypoglossal nerve, respectively. Although all the three nuclei displayed NADPH-d/NOS staining, only the dorsal nucleus of the vagus showed some loss of neurons. It was tempting, in these studies, to surmise that the presence of NO, as indicated by the expression of NADPH-d or NOS, was responsible for the death of the axotomized neurons. However, the findings of our group[106] did not support this contention. After sciatic nerve cut at midthigh of one-day-old mice, the motoneurons showed intense NADPH-d staining. Such staining, however, was not present when the nerve was cut at five days of age. There was no significant difference in the number of sciatic motoneurons lost, whether the sciatic nerve was cut at one day or five days. We surmise that the NO generated is not responsible for the death of the axotomized sciatic motoneurons in one-day-old mice. NO could subserve some other functions. It may reflect a failed attempt at recovery, such as occurred with the NADPH-d stained axonal swellings that failed to regenerate and finally collapsed, leaving behind microcysts.[107] It may also act as a neurotrophic factor, since the transport of this factor from the target structure to the cell body is curtailed by the interruption of the axonal continuity, as suggested by Wu *et al.*[108] That NO could be a *friend* to the injured neurons is further supported by our facial nerve compression experiment.[109] After compression of the nerve in the facial canal in the rat, there was a gradual increase in NADPH-d staining in the facial motoneurons. The staining intensity peaked by four weeks after facial nerve compression, and at this time many of the stained neurons displayed prominent processes. Parallel to this was a gradual return of facial function, following which the expression of the enzyme tailed off. All these observations strongly support participation of NO in the regeneration of the damaged neurons and its functional recovery. Subsequent radiometric assay of soluble NOS activities in tissue homogenates by our group[110] also showed an increase in such activities in the facial motor nucleus at postoperative Day 7. This increase was most likely endothelial, as indicated by an increased NADPH-d staining in the vascular endothelium in the facial motor nucleus. Increased activity is expected to lead to vasodilation and increased blood flow, providing metabolic support for the damaged neurons and other cells. This study also demonstrates a second phase of increase in NOS activities that parallels the increase in NADPH-d expression in the facial motoneurons during the period of functional recovery, thus supporting our previous morphological observation.

In these experiments by our group and Yu, downregulation of the NADPH/d and NOS was noted some time after the experimental lesion. What factors could be responsible for this down regulation? Yu[111] suggested two possible factors: (1) target reinnervation by the regenerated axons; and (2) release of glial cell line-derived neurotrophic factor (GDNF) by the proliferating Schwann cells in response to peripheral nerve injury, as previously demonstrated by Hammarberg *et al.*[112]

The above proposition may be of relevance to the interpretation of the observations of Wu and Li[113] and our group.[114] In these studies intense expression of NOS was seen in the motoneurons of the avulsed spinal nerve root[113] and facial nerve.[114] It is possible that the nerve avulsion prevented the retrograde transport of GDNF necessary to bring about the downregulation of NADPH-d/NOS. This could then lead to sustained and excessive generation of NO, resulting in the destruction of the damaged neurons. Our hypothesis is supported by the presence of neurons in the affected facial motor nucleus that are intensely stained with NADPH-d and NOS antiserum, and of fragmented labelled neuronal parts. Furthermore, the administration of L-NAME (N^{ω}-nitro-L-arginine methyl ester) protected 17% of the neurons from death. The administration of NOS inhibitor also reduced the number of dead neurons after spinal nerve root avulsion. These observations support the notion that NO, when present in abundance and for a prolonged period, may exert deleterious cytotoxic effects, as suggested by Dawson *et al.*,[81] Koh and Choi,[80] and Bonfoco *et al.*[96] Of course, our hypothesis needs to be tested by measuring the actual amount of NO released from the motoneurons under various experimental paradigms, such as nerve compression or transection versus avulsion. This would not be difficult since modern development in technology has enabled scientists to detect the gas at nanomolar level. In fact selective detection of NO even in a single cell has been described.[115] It should, however, be noted that even if the above can be proved, what is true in one situation may not be extrapolated to a similar situation with another nerve. For example, after transection of the vagus nerve, the NOS/NADPH-d reactivity in the dorsal nucleus of the vagus nerve continued to rise without abatement.[116] In this case there was also a greater loss in the number of neurons in this nucleus compared to that in the hypoglossal nucleus after transection of the hypoglossal nerve. Is it possible that other molecules besides NO may also be involved in determining the fate of the axotomized neurons?

In the process of exploring the beneficial and destructive effects of NO, we also investigated if NO could exert different effects on neurons projecting peripherally and those projecting within the CNS, and if neurons that normally express NOS reactivity react differently from those that do not. In our literature search, we noted that the nucleus dorsalis (ND), once called Clarke's nucleus, did not normally show NADPH-d staining.[117] The red nucleus (RN) in the normal rat, on the other hand, showed weak nNOS staining in our preliminary investigation. Previous studies had demonstrated the loss of neurons in both the RN[118] and ND[119] after axotomy. In addition, Wu[117] demonstrated NADPH-d staining in ND neurons after ipsilateral spinal cord hemisection, implicating NO as the killer molecule. Does NO also play a role in the death of the RN neurons after axotomy? To answer this question, a right spinal cord hemisection was performed in the rat at the level of T10, cutting the ipsilateral dorsal spinocerebellar tract (i.e., axons of the ND) and the contralateral rubrospinal tract (i.e., axons of the RN). The results have been published in a series of papers.[120–122] They firstly confirmed the findings of previous work on the ND and the results of our preliminary findings in the RN. Weak nNOS staining was detected in the majority of the normal RN neurons, but NADPH-d staining was virtually absent. Following a lower thoracic cord hemisection, only the ipsilateral ND displayed strong NADPH-d and nNOS staining, and they were colocalized in the same neurons. On the other hand increased nNOS reactivity could be demonstrated bilaterally in the red nucleus. NADPH-d staining was hardly detectable in these neurons, as in the normal rat. nNOS mRNA was induced in the ND but upregulated on both sides of the RN, suggesting that the regulation of nNOS in both nuclei occurs at the transcriptional level. To test if the NO generated by the nNOS was responsible for the neuronal death that occurred in the ND and the RN after the cord hemisection, either the NO donor (L-arginine) or inhibitor (L-NAME) was administered following cord hemisection. Neuronal loss in the ipsilateral ND was augmented by L-arginine, but reduced by L-NAME, whereas in the contralateral RN, no significant neuronal loss was found as late as six weeks after L-arginine treatment. The administration of L-NAME had no effect on the number of neurons lost after hemisection. Although our studies did not measure the real time generation of NO and, therefore, could not compare the exact amount of NO generated by the two nuclei, the results clearly indicated that NO generated in the ND, which does not normally express nNOS, should be considered excessive to the nucleus. The same situation occurred in the motoneurons forming the spinal nerve roots.[113] That NO is cytotoxic to the ND neurons is shown by the beneficial effect of L-NAME administration and the deleterious effects of L-arginine administration. NO is, however, beneficial to neurons in the RN, which normally express NOS immunoreactivity. This is evidenced by the observation that L-NAME administration had no effect on the RN neurons, whereas L-arginine administration brought beneficial effects. In fact, after axotomy, there was increased nNOS expression bilaterally in the red nucleus, many of which could not have projected to the hemisection site, and should not, therefore, have been axotomised. Our observation does not, therefore, support the contention that in Huntington's disease, neurons that disappear are probably killed by the surviving NADPH-d expressing neurons.

Our investigations also show contrasting subcellular localization of nNOS in the ND and RN. This might explain the various actions of NO on these two nuclei. The nNOS immunoreactive products are predominantly bound to the membrane of the mitochondria, rough endoplasmic reticulum, Golgi apparatus, and nuclear envelope of the RN; however, they disperse throughout the cytoplasmic matrix of the axotomized neurons in the ND. This finding is significant in the light of two previous findings. First, Borutaité and Brown[123] showed that mitochondria catalyzed a rapid rate of NO breakdown. Liu *et al.*[124] demonstrated an accelerated reaction of nitric oxide with O_2 within the hydrophobic interior of biological membranes and that the reaction of NO with O_2 within the membranes was approximately 300 times more rapid than in the surrounding aqueous medium. They conjectured that the increase in NO disappearance might largely be caused by its autooxidation within the mitochondrial membrane, and this is accelerated by the concentration of reactants within the hydrophobic phase. It is, therefore, possible that the NO produced by membrane-bound nNOS in the RN neurons metabolizes faster and, therefore, is not able to exert a prolonged effect on the particular organelle. On the other hand, when produced by the cytosolic nNOS, it could accumulate excessively in the ND neurons, leading to their death.

It may be argued that the proximity of the hemisection site to the ND contributes to the induction of NADPH-d and NOS expression in the nucleus. Yick *et al.*[125] showed, however, that following hemisection at different distances from L1 spinal segment, the damaged ND neurons at L1 level expressed NADPH-d and NOS reactivity in all cases, and that the majority of neurons in the nucleus degenerated by 40 days post operation, although the time course of neuronal loss varied.

Finally, in exploring the roles of different types of glutamate receptors in the production of NO, we observed that NMDAR1 is the prevailing glutamate receptor subunit in the ND neurons, and that this is colocalized with nNOS expression in the axotomized neurons. That activation of NMDA receptors brought about the death of the axotomized neurons was demonstrated by Sanner *et al.*,[126] who showed that administration of NMDA receptor antagonist, MK801, prevented cell death in the ND. In the RN, in addition to NMDAR1, GluR1-4 immunoreactivities are also present in its neurons, and almost all of the nNOS positive neurons in this nucleus are colocalized with GluR2. The presence of GluR2 could regulate the influx of Ca^{2+} ions and prevent overflow of these ions into neurons, which could lead to their death. In fact, it has been shown that activation of metabotropic receptors can prevent neuronal degeneration during traumatic neuronal injury, anoxia, or nitric oxide exposure.[127–129] Vincent *et al.*[130] showed that this protection might involve the signal transduction pathways of programmed cell death that are dependent upon NO activity. This could possibly account for the greater percentage of neuronal survival and the slower onset of neuronal loss in the RN after cord hemisection, compared to that in the ND. The lesser number of afferents to the RN and the greater number of its target projections compared to the ND[126] may also account for the observed phenomenon. Whatever be the mechanisms that brought about the death of the RN and ND neurons, it appears that over stimulation of the NMDA receptors is the final common pathway to death, since the administration of MK801 for at least two weeks rescued the axotomized neurons from death.[126]

ADDITIONAL EXPERIMENTAL STUDY

Although the NOS enzyme dependent pathway in the genesis of NO is well known, an enzyme-independent formation of nitric oxide has been reported.[131,132] Zhang *et al.*[133] also showed that under hypoxic conditions NO formation by NOS may be impaired and that NO is formed by an oxygen independent mechanism. In a state of hypoxia, nitrite is reduced by xanthine oxidase (XO) to NO. In fact, even under ambient air conditions, XO also reduced nitrite to form NO. This pathway may play a role in redistributing blood flow to ischemic areas as complementary mechanism in which NOS activity is limited. We asked if XO could contribute to the generation of NO, which has been shown to play a role in the death of facial motorneurons observed after facial nerve avulsion.[114] This hypothesis was tested in Sprague Dawley albino rats and nNOS knockout mice. The presence of XO was indicated by the detection of XO antibodies, using conventional immunohistochemistry. NO was detected by the new fluorescent probe, 4,5-diaminofluorescein DAF-2DA, synthesized by the Kojima group.[134] DAF-2DA is the cell permeable diacetate derivative of DAF-2. When taken up by cells, it is hydrolyzed by cytosolic esterases, resulting in the release of DAF-2, which is relatively non-fluorescent at physiologic pH. However, in the presence of NO and O_2, DAF-2 is converted to the fluorescent triazole derivative, DAF-2T, which gives a bright green fluorescence at excitation wave length of 492 nm and emission wavelength of 515 nm. The technique followed that described by the Kojima group and the fluorescent probe was purchased from ALEXIS Corporation, Switzerland. Briefly, the animal was decapitated and the brain

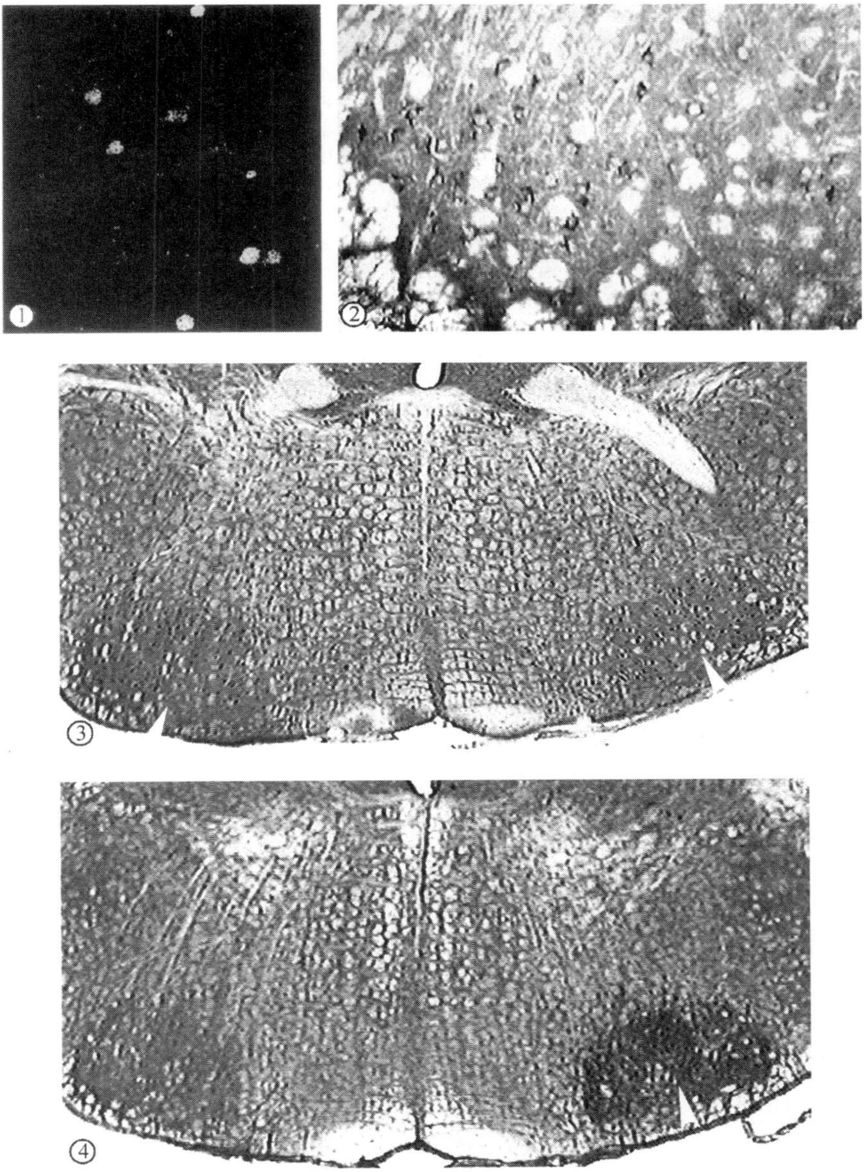

FIGURE 1. Photomicrographs showing some neurons in the ipsilateral facial motor neurons exhibiting bright green fluorescence (①), nNOS immunoreactivity (②), almost all neurons in both facial motor nuclei (*arrows*) labelled with iNOS immunoreactive products bilaterally (③), and almost all neurons in the ipsilateral facial motor nucleus (*arrow*) labelled with xanthine oxidase immunoreactive products (④), one month after facial nerve avulsion in the nNOS knockout mice. Magnification: ① and ②, ×25; ③ and ④, ×10.

stem sectioned at 500 μm thickness at 0°C with a vibratome. The sections were then transferred to artificial CSF at 34°C for preincubation for one hour. The sections were subsequently incubated for 30 minutes with DAF-2DA and viewed in a confocal microscope. The results showed peak expression of nNOS in the facial motor nucleus of the rat at four weeks after avulsion, as shown in our previous study. Surprisingly, the facial motor neurons in the knockout mice also showed expression of nNOS. During this period, intense iNOS and XO immunoreactivities and bright yellow fluorescence were also detected in the facial motor nucleus in both the rat and mice (see FIGURE 1). Whether the fluorescent dyes were coexpressed with the nNOS, iNOS, and XO immunoreactivities requires confirmation. In this experiment the expression of nNOS may be due to the activation of the persisting beta and gamma isoforms of nNOS since the knockouts targeted the first coding exon (second transcribed exon) of the nNOS gene, which eliminates only the alpha isoform.[135]

CRITICAL ASSESSMENT OF PREVIOUS EXPERIMENTAL DATA AND PERSPECTIVES

It is often said that the normal physiological effects of NO are brought about by the low flux of NO reacting rapidly with its biological targets, such as the heme moieties of guanylyl cyclase. Its toxic effects may be attributed to its indirect action. That is, it first reacts with superoxide, ($O_2^{-\cdot}$), or molecular oxygen, O_2, to form reactive nitrogen oxide species and reactive oxygen species. These intermediate compounds then react with a variety of target molecules to bring about deleterious effects. Presumably, the indirect reactions occur only when the local concentration of NO is high (about 1 μm) and maintained for a prolonged period of time, such as occurs when iNOS is formed (for a review, see Ref. 136). Some of the products have been elucidated only in part. The most commonly implicated product is peroxynitrite ($ONOO^-$), formed by the reaction of ·NO with $O_2^{-\cdot}$. This could cause apoptosis of the cortical neurons[81,96] and in various cell lines.[137,138] It can also nitrate or hydroxylate amino acids such as tyrosine,[139–141] forming nitrotyrosine, a toxic product that can disrupt cellular activity. When it oxidizes thiols[142] it decomposes into ·OH. and $NO_2^{-\cdot}$.[141,143] At the cellular level, it could possibly cause DNA damage and activation of poly- (ADP-ribose) synthase with subsequent depletion of NAD and ATP[144] or mitochondrial damage.[145] Whatever be the mechanisms involved, the end results of the neurotoxic, bactericidal, and tumoricidal actions of NO seem to include monoADP-ribosylation and *S*-nitrosylation of glyceraldehyde-3-phosphate dehydrogenase,[146] inhibition of mitochondrial enzymes, such as *cis*-aconitase,[147] inhibition of complexes I and II of the mitochondrial electron transport chain, inhibition of ribonucleotide reductase[148,149] and DNA damage,[150] which activates poly(adenosine 5′-diphosphoribose) synthase (PARS) and the tumor suppressor protein p53.[151,152] In fact Zhang *et al.*[144] showed that NO, the free radical, triggers DNA damage that in turn activates PARS, and this ultimately leads to energy depletion and cell death.

The above hypothesis should not be accepted without reservation. For example, it was pointed out that the antioxidant property of NO manifested in its ability to scavenge peroxyl and other damaging radicals is somewhat overlooked since researchers focus on the toxicity of peroxynitrite. In fact, the cytotoxicity of $ONOO^-$

may have also been over emphasized, for its formation and rapid reaction with antioxidants may provide a mechanism to dispose of excess superoxide radicals, thus maintaining the correct balance between these radicals *in vivo.*[153] Our study[120] also showed normal staining of nitrotyrosine in the red nucleus. In the cardiovascular system, opposite views have also been expressed on whether NO is a friend or a foe in the development of myocardial infarction.[154]

The subject is complex. Many different experimental paradigms have been adopted that could affect diverse cells types. For example, in experimental studies on cerebral ischemia, it is not just the neurons that are affected. The blood vessels and the blood cells streaming through them, the reaction of the surrounding glial cells, and the synaptic connections of the affected neurons should also be considered. The complexity of the subject is further compounded by the observation that NO itself, in smaller amounts, could inhibit iNOS activity (as reviewed by Colasanti and Suzuki.[155]) Finally, in addition to NO, other molecules may also be in operation. Hence, the results of an inflammatory or traumatic event, may not only depend on a balance between the many molecular actions of NO, they could also depend on whether other molecules are involved in bringing about the cytotoxic or beneficial effects. Hence, it is a composite of factors that determine the outcome of the lesion. In addition, the following issues should also be considered.

1. Species differences exist in the patterns of blood supply and the reactions of their cells. Data from lower mammals should not be extrapolated to man without reservation.

2. The same cell types may react differently under various experimental paradigms. For example, although *in vivo* expression of iNOS in microglia and astrocytes has been reported following inflammation and ischemia in the CNS,[99–101] our experiments failed to show the upregulation of the iNOS gene after axotomy of neurons in the red nucleus and nucleus dorsalis.[120]

3. Data from cell cultures do not give the entire truth since cells in culture are frequently hyperoxic and are isolated form other internal milieu that surround them *in vivo.* Adaptation of cells in culture is another issue for consideration.[155] Hence, data obtained from *in vitro* studies may not apply to *in vivo* models. Even in *in vivo* studies, the time frame of a threatening event is a factor that has to be reckoned with. A narrow time window will allow the superoxide generated to be rapidly removed by NO, whereas a longer time frame will allow the excessive amounts of NO formed to react with the concomitantly formed superoxide anions to generate the powerful oxidant, the peroxynitrite.

4. The mere expression of an NOS gene under pathological condition is no indication of a cytotoxic nature of NO. The expression may, in fact, be beneficial. In other words, the genesis of NO may not be the cause of cell death, but rather, the response towards a pathological insult.

5. Although the neurotoxic role of glutamate-stimulated NO release appeared to have been well established, some reports have shown that NO is not involved in glutamate-induced neurotoxicity (for a review, see Ref. 157). The activation of some metabotropic glutamate receptors could prevent neuronal death as supported by our study on the red nucleus. Also controversial is the finding that superoxide dismutase could attenuate the neuronal toxicity induced by gp120, the infectious

agent in AIDS. The problem is, superoxide dismutase is expected to prolong the life of NO and should, therefore, exacerbate the toxic effects of NO. How could it then ameliorate its toxic effects?

6. Much has been said about the close association of nitric oxide with disease production. However, it needs to be mentioned that, although iNOS expression is rapidly induced in the CNS tissues of some rodents after viral or immune mediated insults, the induction of iNOS mRNA did not necessarily correlate with the symptomatology produced by the disease condition.[158] Conversely, the severity of the symptomatology did not correspond to the level of iNOS expression. In seven mice infected with the street rabies virus isolate, all showed paralysis, but iNOS message could only be detected in three animals.

7. Investigators need to take note that NO donors may exert deleterious effects unrelated to the NO it releases. For example, our study[159] showed that the destructive effect of sodium nitroprusside on auditory hair cells is not due to the NO it releases, but rather to the cyanide formed, another end product of the chemical when exposed to light. Hence, results from experiments using nitroprusside as an NO donor need to be reexamined.

8. Numerous NOS inhibitors have been developed to study the functions of NO and these have been extensively reviewed (see, e.g., Refs. 34–36 and 160). The most common are the analogues of the substrate of NOS, L-arginine. These interfere with the binding of the substrate to the enzyme. Although the use of NOS inhibitors is useful in the study of NO functions, the following pitfalls should be noted. (1) Their effects may be different on *in vitro* and *in vivo* models. (2) "Specific" inhibitors may exhibit non-specific or indirect effects. For example, the nNOS blocker, 7-nitroindazole, when inhibiting nNOS in neurons, also affects the tone of vessels innervated by the neurons and reduces blood flow.[161] (3) L-NAME has been shown to reduce infarct size but its action on NO produced by blood vessels can impair recovery of flow. The development of cell selective inhibitors would obviate some of these problems.[160] (4) Although the search for specific NOS inhibitors continues, it should be noted that the application of an NOS inhibitor would not necessarily bring about the suppression of NOS activity. For example, it was observed that application of L-NAME was ineffective in reducing the induction of NOS-immunoreactivity after spinal cord trauma.[162] However, topical application of an antiserum to nitric oxide synthase (NOS) two minutes after spinal cord trauma prevented the trauma-induced up-regulation of NOS immunoreactivity. (5) Some NOS inhibitors like L-NAME reduce basal coronary flow by an NO-independent mechanism.[163] (6) The expression of NOS does not always point to the generation of NO. In fact we observed that nNOS expression in the CNS of the gerbils initially suppressed by L-NAME administration began to express strongly after prolonged period of L-NAME administration (data not included). Future experimentation should aim for the administration of a "NO inhibitor" that would exert a more direct effect. (7) Although the demonstration of NOS expression does not always indicate the genesis of NO, the absence of it does not mean that NO is not being produced since NOS is not the only pathway for the genesis of NO. It may be produced by the xanthine oxidase pathway or non-enzymatically involving hydrogen peroxide and

D- or L-arginine,[132] or by direct reduction of nitrite under the acidic and highly reduced conditions that occur in the ischemic heart.[129]

Some of these problems associated with the use of NOS inhibitors may be avoided by the use of mice without the nNOS,[135] eNOS,[164] or iNOS,[155] or double knockout gene.[74] In fact useful information has been gleaned from these knockouts. However, the limitations of using knockouts that can lead to the wrong interpretation of data cannot be ignored. These include the presence of subtle secondary developmental abnormalities and of compensation,[166] and the substitution of one NOS isoform for another, as shown in the study of Meng *et al.*[167] and in our present study.

(8) The widely used NADPH-d reaction for demonstrating neuronal nitric oxide synthase is not as specific as previously thought, because it visualizes both a nitric oxide synthase-related activity and a nitric oxide synthase-unrelated diaphorase (commented on by Spessert and Claassen[168]). Our studies also clearly indicate that NADPH-d does not always show colocalization with NOS.[114,120]

Research on NO focuses on the *when* (with respect to its production), *what* (its properties and functions), *where* (its distribution), and *how* (its mode of generation and actions); however, the *why* of its action is seldom asked, especially in its killer action. We seek to answer the *why* with another question: What good is it to the nervous system to have a group of neurons isolated from the internal milieu as in the case of ND or having no targets to innervate as in the case of the avulsed spinal motoneurons and facial motoneurons? If neuronal death is truly brought about by NO, then NO should be viewed as an executioner, rather than a killer.

ACKNOWLEDGMENTS

We thank Dr. Paul Hwang for the gift of nNOS knockout mice, and also Mr. Chan Heng Leong, Miss Connie Er Poh Nee, and Miss Zhiyi for technical assistance.

REFERENCES

1. MURAD, F., C.K. MITTAL, W.P. ARNOLD, *et al.* 1978. Guanylate cyclase: activation by azide, nitro compounds, nitric oxide, and hydroxyl radical and inhibition by hemoglobin and myoglobin. Adv. Cyclic. Nucleotide. Res. **9:** 145–158.
2. FURCHGOTT, R.F. & J.V. ZAWADZKI. 1980. The obligatory role of endothelial cells in the relaxation of arterial smooth muscle by acetylcholine. Nature **288:** 373–376.
3. FURCHGOTT, R.F. 1988. Studies on relaxation of rabbit aorta by sodium nitrite: The basis for thr proposal that the acid-activatable factor from bovine retractor penis is inorganic nitrite and the endothelium-derived relaxing factor is nitric oxide. *In* Vasodilation: Vascular Smooth Muscle, Peptides, Autonomic Nerves, and Endothelium. P.M. Vanhoutte, Ed.: 401–414. Raven Press, New York.
4. IGNARRO, L.J., G.M. BUGA, K.S. WOOD, *et al.* 1987. Endothelium-derived relaxing factor produced and released from artery and vein is nitric oxide. Proc. Natl. Acad. Sci. U.S.A. **84:** 9265–9269.
5. PALMER, R.M., A.G. FERRIGE & S. MONCADA. 1987. Nitric oxide release accounts for the biological activity of endothelium-derived relaxing factor. Nature **327:** 524–526.
6. MONCADA, S., R.M. PALMER & E.A. HIGGS. 1991. Nitric oxide: physiology, pathophysiology, and pharmacology. Pharmacol. Rev. **43:** 109–142.
7. MARTINEZ, A. 1995. Nitric oxide synthase in invertebrates. Histochem. J. **27:** 770–776.

8. JIANG, P.J. & S. TERASHIMA. 1996. Distribution of NADPH-diaphorase in the central nervous system of an infrared-sensitive snake, Trimeresurus flavoviridis. Brain Res. **713:** 168–177.
9. JACKLET, J.W. 1997. Nitric oxide signaling in invertebrates. Invert. Neurosci. **3:** 1–14.
10. LEONG, S.K. 1999. Capturing the sublimity of a free radical gas. Free Radic. Res. **31:** 519–629.
11. ALI, S.M., A.S. CHAN & S.K. LEONG. 1996. Histochemical and immunohistochemical localisation of nitrergic neuronal and non-neuronal cells in the bursa of Fabricius of the chicken. Cell Tissue Res. **285:** 273–279.
12. ALI, S.M., A.S. CHAN & S.K. LEONG. 1996. Localization of nitrergic neuronal and non-neuronal cells in the ultimobranchial glands of the chicken. Anat. Embryol. (Berlin) **193:** 161–168.
13. ALONSO, J.R., R. AREVALO, E. GARCIA-OJEDA, *et al.* 1995. NADPH-diaphorase active and calbindin D-28k-immunoreactive neurons and fibers in the olfactory bulb of the hedgehog (*Erinaceus europaeus*). J. Comp. Neurol. **351:** 307–327.
14. FERRANTE, R.J., N.W. KOWALL, M.F. BEAL, *et al.* 1985. Selective sparing of a class of striatal neurons in Huntington's disease. Science **230:** 561–563.
15. GAWRONSKA, B., G. BODEK & A.J. ZIECIK. 2000. Distribution of NADPH-diaphorase and nitric oxide synthase (NOS) in different regions of porcine oviduct during the estrous cycle. J. Histochem. Cytochem. **48:** 867–875.
16. GULATI, P., A.S. CHAN & S.K. LEONG. 1993. NADPH-diaphorase positive cells in the chick and rat thymus. Thymus **22:** 117–124.
17. GULATI, P., A.S. CHAN & S.K. LEONG. 1995. Ultrastructural localisation of NADPH-diaphorase in the chick thymic medulla. Cell Tissue Res. **279:** 405–409.
18. HTAIN, W.W., S.K. LEONG & E.A. LING. 1997. *In vivo* expression of inducible nitric oxide synthase in supraventricular amoeboid microglial cells in neonatal BALB/c and athymic mice. Neurosci. Lett. **223:** 53–56.
19. LEONG, S., H. LIU & J. YEO. 2000. Nitric oxide synthase and glutamate receptor immunoreactivity in the rat spinal trigeminal neurons expressing Fos protein after formalin injection. Brain Res. **855:** 107–115.
20. LIU, H.P., S.K. LEONG & S.S. TAY. 1994. Localization of NADPH-diaphorase positive neurons in the pancreas of the mouse, rat, chick, kitten and monkey. J. Hirnforsch. **35:** 501–510.
21. LIU, H.P., S.S. TAY & S.K. LEONG. 1996. Nitrergic neurons in the pancreas of newborn guinea pig: their distribution and colocalization with various neuropeptides and dopamine-beta-hydroxylase. J. Auton. Nerv. Syst. **61:** 248–256.
22. MUNOZ, M., A. MUNOZ, O. MARIN, *et al.* 1996. Topographical distribution of NADPH-diaphorase activity in the central nervous system of the frog, Rana perezi. J. Comp Neurol. **367:** 54–69.
23. NISBET, A.P., O.J. FOSTER, A. KINGSBURY, *et al.* 1994. Nitric oxide synthase mRNA expression in human subthalamic nucleus, striatum and globus pallidus: implications for basal ganglia function. Brain Res. Mol. Brain Res. **22:** 329–332.
24. OHSHIMA, H., S. OGUCHI, H. ADACHI, *et al.* 1992. Purification of nitric oxide synthase from bovine brain: immunological characterization and tissue distribution. Biochem. Biophys. Res. Commun. **183:** 238–244.
25. RIBERA, J., J. MARSAL, A. CASANOVAS, *et al.* 1998. Nitric oxide synthase in rat neuromuscular junctions and in nerve terminals of Torpedo electric organ: its role as regulator of acetylcholine release. J. Neurosci. Res. **51:** 90–102.
26. SYED, M.A., S.K. LEONG & A.S. CHAN. 1994. Localization of NADPH-diaphorase reactivity in the chick and mouse thyroid gland. Thyroid **4:** 475–478.
27. TIMMERMANS, J.P., M. BARBIERS, D.W. SCHEUERMANN, *et al.* 1994. Nitric oxide synthase immunoreactivity in the enteric nervous system of the developing human digestive tract. Cell Tissue Res. **275:** 235–245.
28. WARD, S.M., C. XUE, C.W. SHUTTLEWORTH, *et al.* 1992. NADPH diaphorase and nitric oxide synthase colocalization in enteric neurons of canine proximal colon. Am. J. Physiol. **263:** G277–G284.
29. PAYNE, W.J., M.Y. LIU, S.A. BURSAKOV & J. LE GALL. 1997. Microbial and plant metabolism of NO. Biofactors **6:** 47–52.

30. DING, A.H., C.F. NATHAN & D.J. STUEHR. 1988. Release of reactive nitrogen intermediates and reactive oxygen intermediates from mouse peritoneal macrophages. Comparison of activating cytokines and evidence for independent production. J. Immunol. **141:** 2407–2412.
31. LYONS, C.R., G.J. ORLOFF & J.M. CUNNINGHAM. 1992. Molecular cloning and functional expression of an inducible nitric oxide synthase from a murine macrophage cell line. J. Biol. Chem. **267:** 6370–6374.
32. XIE, Q.W., H.J. CHO, J. CALAYCAY, *et al.* 1992. Cloning and characterization of inducible nitric oxide synthase from mouse macrophages. Science **256:** 225–228.
33. MONCADA, S., R.M. PALMER & E.A. HIGGS. 1991. Nitric oxide: physiology, pathophysiology, and pharmacology. Pharmacol. Rev. **43:** 109–142.
34. NATHAN, C. 1992. Nitric oxide as a secretory product of mammalian cells. FASEB J. **6:** 3051–3064.
35. KNOWLES, R.G. & S. MONCADA. 1994. Nitric oxide synthases in mammals. Biochem. J. **298**(2): 249–258.
36. HOBBS, A.J., A. HIGGS & S. MONCADA. 1999. Inhibition of nitric oxide synthase as a potential therapeutic target. Annu. Rev. Pharmacol. Toxicol. **39:** 191–220.
37. SANDERS, D.B., T. KELLEY & D. LARSON. 2000. The role of nitric oxide synthase/nitric oxide in vascular smooth muscle control. Perfusion **15:** 97–104.
38. RAJFER, J., W.J. ARONSON, P.A. BUSH, *et al.* 1992. Nitric oxide as a mediator of relaxation of the corpus cavernosum in response to nonadrenergic, noncholinergic neurotransmission. N. Engl. J. Med. **326:** 90–94.
39. SNYDER, S.H. Nitric oxide: first in a new class of neurotransmitters? Science **257:** 494–496.
40. GARTHWAITE, J., S.L. CHARLES & R. CHESS-WILLIAMS. 1988. Endothelium-derived relaxing factor release on activation of NMDA receptors suggests role as intercellular messenger in the brain. Nature **336:** 385–388.
41. BREDT, D.S., C.E. GLATT, P.M. HWANG, *et al.* 1991. Nitric oxide synthase protein and mRNA are discretely localized in neuronal populations of the mammalian CNS together with NADPH diaphorase. Neuron **7:** 615–624.
42. MONCADA, S. & A. HIGGS. 1993. The L-arginine-nitric oxide pathway. N. Engl. J. Med. **329:** 2002–2012.
43. BLISS, T.V. & G.L. COLLINGRIDGE. 1993. A synaptic model of memory: long-term potentiation in the hippocampus. Nature **361:** 31–39.
44. SCHUMAN, E.M. & D.V. MADISON. 1994. Locally distributed synaptic potentiation in the hippocampus. Science **263:** 532–536.
45. VINCENT, S.R. & H. KIMURA. 1992. Histochemical mapping of nitric oxide synthase in the rat brain. Neuroscience **46:** 755–784.
46. DAWSON, T.M., V.L. DAWSON & S.H. SNYDER. 1992. A novel neuronal messenger molecule in brain: the free radical, nitric oxide. Ann. Neurol. **32:** 297–311.
47. KARANTH, S., K. LYSON & S.M. MCCANN. 1993. Role of nitric oxide in interleukin 2-induced corticotropin-releasing factor release from incubated hypothalami. Proc. Natl. Acad. Sci. U.S.A. **90:** 3383–3387.
48. OKAMURA, H., M. YOKOSUKA, B.S. MCEWEN & S. HAYASHI. 1994. Colocalization of NADPH-diaphorase and estrogen receptor immunoreactivity in the rat ventromedial hypothalamic nucleus: stimulatory effect of estrogen on NADPH-diaphorase activity. Endocrinology **135:** 1705–1708.
49. RETTORI, V., A. KAMAT & S.M. MCCANN. 1994. Nitric oxide mediates the stimulation of luteinizing-hormone releasing hormone release induced by glutamic acid *in vitro*. Brain Res. Bull. **33:** 501–503.
50. CECCATELLI, S. 1997. Expression and plasticity of NO synthase in the neuroendocrine system. Brain Res. Bull. **44:** 533–538.
51. DAWSON, V.L. & T.M. DAWSON. 1995. Physiological and toxicological actions of nitric oxide in the central nervous system. Adv. Pharmacol. **34:** 323–342.
52. GARTHWAITE, J. & C.L. BOULTON. 1995. Nitric oxide signaling in the central nervous system. Annu. Rev. Physiol. **57:** 683–706.
53. SCHMIDT, H.H. & U. WALTER. 1994. NO at work. Cell **78:** 919–925.

54. DIPPEL, E., B. MAYER, G. SCHONFELDER, *et al.* 1994. Distribution of constitutive nitric oxide synthase immunoreactivity and NADPH-diaphorase activity in murine telogen and anagen skin. J. Invest. Dermatol. **103:** 112–115.
55. STADLER, J., R.D. CURRAN, J.B. OCHOA, *et al.* 1991. Effect of endogenous nitric oxide on mitochondrial respiration of rat hepatocytes *in vitro* and *in vivo*. Arch. Surg. **126:** 186–191.
56. MACINTYRE, I., M. ZAIDI, A.S. ALAM, *et al.* 1991. Osteoclastic inhibition: an action of nitric oxide not mediated by cyclic GMP. Proc. Natl. Acad. Sci. U.S.A. **88:** 2936–2940.
57. CHARLES, I.G., R.M. PALMER, M.S. HICKERY, *et al.* 1993. Cloning, characterization, and expression of a cDNA encoding an inducible nitric oxide synthase from the human chondrocyte. Proc. Natl. Acad. Sci. U.S.A. **90:** 11419–11423.
58. EVANS, C.H., M. STEFANOVIC-RACIC & J. LANCASTER. 1995. Nitric oxide and its role in orthopaedic disease. Clin. Orthop. **312:** 275–294.
59. REID, M.B. 1998. Role of nitric oxide in skeletal muscle: synthesis, distribution and functional importance. Acta Physiol. Scand. **162:** 401–409.
60. AGUIRRE, J., L. BUTTERY, M. O'SHAUGHNESSY, *et al.* 2001. Endothelial nitric oxide synthase gene-deficient mice demonstrate marked retardation in postnatal bone formation, reduced bone volume, and defects in osteoblast maturation and activity. Am. J. Pathol. **158:** 247–257.
61. BROWN, G.C., J.P. BOLANOS, S.J. HEALES & J.B. CLARK. 1995. Nitric oxide produced by activated astrocytes rapidly and reversibly inhibits cellular respiration. Neurosci. Lett. **193:** 201–204.
62. WANG, Z.Z., L.J. STENSAAS, D.S. BREDT, *et al.* 1994. Localization and actions of nitric oxide in the cat carotid body. Neuroscience **60:** 275–286.
63. BUGNON, O., N.C. SCHAAD & M. SCHORDERET. 1994. Nitric oxide modulates endogenous dopamine release in bovine retina. Neuroreport **5:** 401–404.
64. DJAMGOZ, M.B., J.R. CUNNINGHAM, S.L. DAVENPORT & M.J. NEAL. 1995. Nitric oxide inhibits depolarization-induced release of endogenous dopamine in the rabbit retina. Neurosci. Lett. **198:** 33–36.
65. HIBBS, J.B., JR., Z. VAVRIN & R.R. TAINTOR. 1987. L-Arginine is required for expression of the activated macrophage effector mechanism causing selective metabolic inhibition in target cells. J. Immunol. **138:** 550–565.
66. STUEHR, D.J. & C.F. NATHAN. 1989. Nitric oxide. A macrophage product responsible for cytostasis and respiratory inhibition in tumor target cells. J. Exp. Med. **169:** 1543–1555.
67. GRANGER, D.L., J.B. HIBBS, JR., J.R. PERFECT & D.T. DURACK. 1990. Metabolic fate of L-arginine in relation to microbiostatic capability of murine macrophages. J. Clin. Invest **85:** 264–273.
68. ADAMS, L.B., J.B. HIBBS, JR., R.R. TAINTOR & J.L. KRAHENBUHL. 1990. Microbiostatic effect of murine-activated macrophages for Toxoplasma gondii. Role for synthesis of inorganic nitrogen oxides from L-arginine. J. Immunol. **144:** 2725–2729.
69. LIEW, F.Y., Y. LI & S. MILLOTT. 1990. Tumour necrosis factor (TNF-alpha) in leishmaniasis. II. TNF-alpha-induced macrophage leishmanicidal activity is mediated by nitric oxide from L-arginine. Immunology **71:** 556–559.
70. ANGGARD, E. 1994. Nitric oxide: mediator, murderer, and medicine. Lancet **343:** 1199–1206.
71. BREDT, D.S. 1999. Endogenous nitric oxide synthesis: biological functions and pathophysiology. Free Radic. Res. **31:** 577–596.
72. LIPTON, S.A. 1993. Prospects for clinically tolerated NMDA antagonists: open-channel blockers and alternative redox states of nitric oxide. Trend. Neurosci. **16:** 527–532.
73. MURPHY, S., R.L. MINOR, JR., G. WELK & D.G. HARRISON. 1990. Evidence for an astrocyte-derived vasorelaxing factor with properties similar to nitric oxide. J. Neurochem. **55:** 349–351.
74. HUANG, P.L. 2000. Lessons learned from nitric oxide synthase knockout animals. Semin. Perinatol. **24:** 87–90.
75. SCHULZ, J.B., R.T. MATTHEWS & M.F. BEAL. 1995. Role of nitric oxide in neurodegenerative diseases. Curr. Opin. Neurol. **8:** 480–486.

76. HANTRAYE, P., E. BROUILLET, R. FERRANTE, *et al.* 1996. Inhibition of neuronal nitric oxide synthase prevents MPTP-induced parkinsonism in baboons. Nat. Med. **2:** 1017–1021.
77. ZOCHODNE, D.W., M. MISRA, C. CHENG & H. SUN. 1997. Inhibition of nitric oxide synthase enhances peripheral nerve regeneration in mice. Neurosci. Lett. **228:** 71–74.
78. HYMAN, B.T., K. MARZLOFF, J.J. WENNIGER, *et al.* 1992. Relative sparing of nitric oxide synthase-containing neurons in the hippocampal formation in Alzheimer's disease. Ann. Neurol. **32:** 818–820.
79. XUE, Y.D., P.T. WONG & S.K. LEONG. 1997. Nitric oxide synthase-, *N*-methyl-D-aspartate receptor-, glutamate- and aspartate-immunoreactive neurons in the mouse arcuate nucleus: effects of neonatal treatment with monosodium glutamate. Acta Neuropathol. (Berlin) **94:** 572–582.
80. KOH, J.Y. & D.W. CHOI. 1988. Vulnerability of cultured cortical neurons to damage by excitotoxins: differential susceptibility of neurons containing NADPH-diaphorase. J. Neurosci. **8:** 2153–2163.
81. DAWSON, V.L., T.M. DAWSON, G.R. UHL & S.H. SNYDER. 1993. Human immunodeficiency virus type 1 coat protein neurotoxicity mediated by nitric oxide in primary cortical cultures. Proc. Natl. Acad. Sci. U.S.A. **90:** 3256–3259.
82. MARSDEN, P.A. & B.J. BALLERMANN. 1990. Tumor necrosis factor alpha activates soluble guanylate cyclase in bovine glomerular mesangial cells via an L-arginine-dependent mechanism. J. Exp. Med. **172:** 1843–1852.
83. BILLIAR, T.R., R.D. CURRAN, B.G. HARBRECHT, *et al.* 1990. Modulation of nitrogen oxide synthesis *in vivo*: NG-monomethyl-L-arginine inhibits endotoxin-induced nitrate/nitrate biosynthesis while promoting hepatic damage. J. Leukoc. Biol. **48:** 565–569.
84. MOHANAKUMAR, K.P., I. HANBAUER & C.C. CHIUEH. 1998. Neuroprotection by nitric oxide against hydroxyl radical-induced nigral neurotoxicity. J. Chem. Neuroanat. **14:** 195–205.
85. LEI, S.Z., D. ZHANG, A.E. ABELE & S.A. LIPTON. 1992. Blockade of NMDA receptor-mediated mobilization of intracellular Ca^{2+} prevents neurotoxicity. Brain Res. **598:** 196–202.
86. PAUWELS, P.J. & J.E. LEYSEN. 1992. Blockade of nitric oxide formation does not prevent glutamate-induced neurotoxicity in neuronal cultures from rat hippocampus. Neurosci. Lett. **143:** 27–30.
87. WINK, D.A., I. HANBAUER, M.C. KRISHNA, *et al.* 1993. Nitric oxide protects against cellular damage and cytotoxicity from reactive oxygen species. Proc. Natl. Acad. Sci. U.S.A. **90:** 9813–9817.
88. WINK, D.A., J.A. COOK, R. PACELLI, *et al.* 1996. The effect of various nitric oxide-donor agents on hydrogen peroxide-mediated toxicity: a direct correlation between nitric oxide formation and protection. Arch. Biochem. Biophys. **331:** 241–248.
89. LIPTON, S.A., Y.B. CHOI, Z.H. PAN, *et al.* 1993. A redox-based mechanism for the neuroprotective and neurodestructive effects of nitric oxide and related nitroso-compounds. Nature **364:** 626–632.
90. LIPTON, S.A., D.J. SINGEL & J.S. STAMLER. 1994. Neuroprotective and neurodestructive effects of nitric oxide and redox congeners. Ann. N.Y. Acad. Sci. **738:** 382–387.
91. LIPTON, S.A. 1999. Neuronal protection and destruction by NO Cell Death. Differ. **6:** 943–951.
92. CHIUEH, C.C. 1999. Neuroprotective properties of nitric oxide. Ann. N.Y. Acad. Sci. **890:** 301–311.
93. ANDOH, T., S.Y. LEE & C.C. CHIUEH. 2000. Preconditioning regulation of bcl-2 and p66shc by human NOS1 enhances tolerance to oxidative stress. FASEB J. **14:** 2144–2146.
94. MELINO, G., F. BERNASSOLA, M.V. CATANI, *et al.* 2000. Nitric oxide inhibits apoptosis via AP-1-dependent CD95L transactivation. Cancer Res. **60:** 2377–2383.
95. BRADY, A.J. 1995. Nitric oxide, myocardial failure and septic shock. Int. J. Cardiol. **50:** 269–272.

96. BONFOCO, E., D. KRAINC, M. ANKARCRONA, *et al.* 1995. Apoptosis and necrosis: two distinct events induced, respectively, by mild and intense insults with *N*-methyl-D-aspartate or nitric oxide/superoxide in cortical cell cultures. Proc. Natl. Acad. Sci. U.S.A. **92:** 7162–7166.
97. HUANG, Z., P.L. HUANG, N. PANAHIAN, *et al.* 1994. Effects of cerebral ischemia in mice deficient in neuronal nitric oxide synthase. Science **265:** 1883–1885.
98. PRZEDBORSKI, S., V. JACKSON-LEWIS, R. YOKOYAMA, *et al.* 1996. Role of neuronal nitric oxide in 1-methyl-4-phenyl-1,2,3,6-tetrahydropyridine (MPTP)-induced dopaminergic neurotoxicity. Proc. Natl. Acad. Sci. U.S.A. **93:** 4565–4571.
99. IADECOLA, C., F. ZHANG, S. XU, *et al.* 1995. Inducible nitric oxide synthase gene expression in brain following cerebral ischemia. J. Cereb. Blood Flow Metab. **15:** 378–384.
100. GARCION, E., L. SINDJI, C. MONTERO-MENEI, *et al.* 1998. Expression of inducible nitric oxide synthase during rat brain inflammation: regulation by 1,25-dihydroxy-vitamin D3. Glia **22:** 282–294.
101. WELDON, D.T., S.D. ROGERS, J.R. GHILARDI, *et al.* 1998. Fibrillar beta-amyloid induces microglial phagocytosis, expression of inducible nitric oxide synthase, and loss of a select population of neurons in the rat CNS *in vivo.* J. Neurosci. **18:** 2161–2173.
102. TAKEUCHI, A., K.I. ISOBE, O. MIYAISHI, *et al.* 1998. Microglial NO induces delayed neuronal death following acute injury in the striatum. Eur. J. Neurosci. **10:** 1613–1620.
103. CLOWRY, G.J. 1993. Axotomy induces NADPH diaphorase activity in neonatal but not adult motoneurones. Neuroreport **5:** 361–364.
104. DAWSON, T.M., M. SASAKI, M. GONZALEZ-ZULUETA & V.L. DAWSON. 1998. Regulation of neuronal nitric oxide synthase and identification of novel nitric oxide signaling pathways. Prog. Brain Res. **118:** 3–11.
105. YU, W.H. 1994. Nitric oxide synthase in motor neurons after axotomy. J. Histochem. Cytochem. **42:** 451–457.
106. HE, B.P., S.S. TAY & S.K. LEONG. 1996. Motoneuronal death without expression of NADPH-diaphorase activity after sciatic nerve cut in 5-day-old mice. Brain Res. **733:** 125–128.
107. GUIZAR-SAHAGUN, G., P. GARCIA-LOPEZ, A.L. ESPITIA, *et al.* 1998. Transitory expression of NADPH diaphorase (NOS) in axonal swellings after spinal cord injury. Neuroreport **9:** 2899–2902.
108. WU, Y., Y. LI, H. LIU & W. WU. 1995. Induction of nitric oxide synthase and motoneuron death in newborn and early postnatal rats following spinal root avulsion. Neurosci. Lett. **194:** 109–112.
109. RUAN, R.S., S.K. LEONG & K.H. YEOH. 1994. Expression of NADPH-diaphorase activity in the facial motoneurons after compression of the facial nerve in the albino rat. Brain Res. **652:** 350–352.
110. WONG, P.T., R.S. RUAN, S.K. LEONG & K.H. YEOH. 1995. Compression of the facial nerve caused increased nitric oxide synthase activity in the facial motor nucleus. Neuroscience **67:** 697–702.
111. YU, W.H., A. WALCZEWSKA, S. KARANTH & S.M.MCCANN. 1997. Nitric oxide mediates leptin-induced luteinizing hormone-releasing hormone (LHRH) and LHRH and leptin-induced LH release from the pituitary gland. Endocrinology **138:** 5055–5058.
112. HAMMARBERG, H., F. PIEHL, S. CULLHEIM, *et al.* 1996. GDNF mRNA in Schwann cells and DRG satellite cells after chronic sciatic nerve injury. Neuroreport **7:** 857–860.
113. WU, W. & L. LI. 1993. Inhibition of nitric oxide synthase reduces motoneuron death due to spinal root avulsion. Neurosci. Lett. **153:** 121–124.
114. RUAN, R.S., S.K. LEONG & K.H. YEOH. 1995. The role of nitric oxide in facial motoneuronal death. Brain Res. **698:** 163–168.
115. MALINSKI, T. & Z. TAHA. 1992. Nitric oxide release from a single cell measured in situ by a porphyrinic-based microsensor. Nature **358:** 676–678.
116. YU, W.H. 1997. Regulation of nitric oxide synthase expression in motoneurons following nerve injury. Dev. Neurosci. **19:** 247–254.

117. Wu, W. 1993. Expression of nitric-oxide synthase (NOS) in injured CNS neurons as shown by NADPH diaphorase histochemistry. Exp. Neurol. **120:** 153–159.
118. Egan, D.A., B.A. Flumerfelt & D.G. Gwyn. 1977. Axon reaction in the red nucleus of the rat. Perikaryal volume changes and the time course of chromatolysis following cervical and thoracic lesions. Acta Neuropathol. (Berlin) **37:** 13–19.
119. Sanner, C.A., M. Murray & M.E. Goldberger. 1993. Removal of dorsal root afferents prevents retrograde death of axotomized Clarke's nucleus neurons in the cat. Exp. Neurol. **123:** 81–90.
120. Xu, M., Y.K. Ng & S.K. Leong. 1998. Induction of microglial reaction and expression of nitric oxide synthase I in the nucleus dorsalis and red nucleus following lower thoracic spinal cord hemisection. Brain Res. **808:** 23–30
121. Xu, M., Y.K. Ng & S.K. Leong. 2000. Neuroprotective and neurodestructive functions of nitric oxide after spinal cord hemisection. Exp. Neurol. **161:** 472–480.
122. Xu, M., Y.K. Ng & S.K. Leong. 2000. Distinct subcellular localization and mRNA expression of neuronal nitric oxide synthase in the nucleus dorsalis and red nucleus and their correlation with inducible transcription factors after spinal cord hemisection. Nitric Oxide **4:** 483–495.
123. Borutaite, V. & G.C. Brown. 1996. Rapid reduction of nitric oxide by mitochondria, and reversible inhibition of mitochondrial respiration by nitric oxide. Biochem. J. **315**(1): 295–299.
124. Liu, H.P., S.S. Tay, S. Leong & M. Schemann. 1998. Colocalization of ChAT, DbetaH and NADPH-d in the pancreatic neurons of the newborn guinea pig. Cell Tissue Res. **294:** 227–231.
125. Yick, L.W., W. Wu, K.F. So & S.Y. Wong. 1998. Time course of NOS expression and neuronal death in Clarke's nucleus following traumatic injury in adult rat spinal cord. Neurosci. Lett. **241:** 155–158.
126. Sanner, C.A., T.J. Cunningham & M.E. Goldberger. 1994. NMDA receptor blockade rescues Clarke's and red nucleus neurons after spinal hemisection. J. Neurosci. **14:** 6472–6480.
127. Bruno, V., A. Copani, L. Bonanno, *et al.* 1996. Activation of group III metabotropic glutamate receptors is neuroprotective in cortical cultures. Eur. J. Pharmacol. **310:** 61–66.
128. Maiese, K., J. Wagner & L. Boccone. 1994. Nitric oxide: a downstream mediator of calcium toxicity in the ischemic cascade. Neurosci. Lett. **166:** 43–47.
129. Maiese, K. & L. Boccone. 1995. Neuroprotection by peptide growth factors against anoxia and nitric oxide toxicity requires modulation of protein kinase C. J. Cereb. Blood Flow Metab. **15:** 440–449.
130. Vincent, A.M., Y. Mohammad, I. Ahmad, *et al.* 1997. Metabotropic glutamate receptors prevent nitric oxide-induced programmed cell death. J. Neurosci. Res. **50:** 549–564.
131. Zweier, J.L., P. Wang, A. Samouilov & P. Kuppusamy. 1995. Enzyme-independent formation of nitric oxide in biological tissues. Nat. Med. **1:** 804–809.
132. Nagase, S., K. Takemura, A. Ueda, *et al.* 1997. A novel nonenzymatic pathway for the generation of nitric oxide by the reaction of hydrogen peroxide and D- or L-arginine. Biochem. Biophys. Res. Commun. **233:** 150–153.
133. Zhang, Z., D. Naughton, P.G. Winyard, *et al.* 1998. Generation of nitric oxide by a nitrite reductase activity of xanthine oxidase: a potential pathway for nitric oxide formation in the absence of nitric oxide synthase activity. Biochem. Biophys. Res. Commun. **249:** 767–772.
134. Kojima, H., K. Sakurai, K. Kikuchi, *et al.* 1998. Development of a fluorescent indicator for nitric oxide based on the fluorescein chromophore. Chem. Pharm. Bull. (Tokyo) **46:** 373–375.
135. Huang, P.L., T.M. Dawson, D.S. Bredt, *et al.* 1993. Targeted disruption of the neuronal nitric oxide synthase gene. Cell **75:** 1273–1286.
136. Espey, M.G., K.M. Miranda, M. Feelisch, *et al.* 2000. Mechanisms of cell death governed by the balance between nitrosative and oxidative stress. Ann. N.Y. Acad. Sci. **899:** 209–221.

137. LIN, K.T., J.Y. XUE, M. NOMEN, *et al.* 1995. Peroxynitrite-induced apoptosis in HL-60 cells. J. Biol. Chem. **270:** 16487–16490.
138. TROY, C.M., D. DEROSSI, A. PROCHIANTZ, *et al.* 1996. Downregulation of Cu/Zn superoxide dismutase leads to cell death via the nitric oxide-peroxynitrite pathway. J. Neurosci. **16:** 253–261.
139. ISCHIROPOULOS, H., L. ZHU, J. CHEN, *et al.* 1992. Peroxynitrite-mediated tyrosine nitration catalyzed by superoxide dismutase. Arch. Biochem. Biophys. **298:** 431–437.
140. BECKMANN, J.S., Y.Z. YE, P.G. ANDERSON, *et al.* 1994. Extensive nitration of protein tyrosines in human atherosclerosis detected by immunohistochemistry. Biol. Chem. Hoppe Seyler **375:** 81–88.
141. CROW, J.P. & J.S. BECKMAN. 1995. Reactions between nitric oxide, superoxide, and peroxynitrite: footprints of peroxynitrite *in vivo.* Adv. Pharmacol. **34:** 17–43.
142. RADI, R., J.S. BECKMAN, K.M. BUSH & B.A. FREEMAN. 1991. Peroxynitrite-induced membrane lipid peroxidation: the cytotoxic potential of superoxide and nitric oxide. Arch. Biochem. Biophys. **288:** 481–487.
143. BECKMAN, J.S., T.W. BECKMAN, J. CHEN, *et al.* 1990. Apparent hydroxyl radical production by peroxynitrite: implications for endothelial injury from nitric oxide and superoxide. Proc. Natl. Acad. Sci. U.S.A. **87:** 1620–1624.
144. ZHANG, J., V.L. DAWSON, T.M. DAWSON & S.H. SNYDER. 1994. Nitric oxide activation of poly(ADP-ribose) synthetase in neurotoxicity. Science **263:** 687–689.
145. BOLANOS, J.P., S.J. HEALES, J.M. LAND & J.B. CLARK. 1995. Effect of peroxynitrite on the mitochondrial respiratory chain: differential susceptibility of neurones and astrocytes in primary culture. J. Neurochem. **64:** 1965–1972.
146. ZHANG, J. & S.H. SNYDER. 1992. Nitric oxide stimulates auto-ADP-ribosylation of glyceraldehyde-3-phosphate dehydrogenase. Proc. Natl. Acad. Sci. U.S.A. **89:** 9382–9385.
147. DRAPIER, J.C. & J.B. HIBBS, JR. 1986. Murine cytotoxic activated macrophages inhibit aconitase in tumor cells. Inhibition involves the iron-sulfur prosthetic group and is reversible. J. Clin. Invest. **78:** 790–797.
148. LEPOIVRE, M., B. CHENAIS, A. YAPO, *et al.* 1990. Alterations of ribonucleotide reductase activity following induction of the nitrite-generating pathway in adenocarcinoma cells. J. Biol. Chem. **265:** 14143–14149.
149. KWON, N.S., D.J. STUEHR & C.F. NATHAN. 1991. Inhibition of tumor cell ribonucleotide reductase by macrophage-derived nitric oxide. J. Exp. Med. **174:** 761–767.
150. WINK, D.A., K.S. KASPRZAK, C.M. MARAGOS, *et al.* 1991. DNA deaminating ability and genotoxicity of nitric oxide and its progenitors. Science **254:** 1001–1003.
151. CLEAVER, J.E. & W.F. MORGAN. 1991. Poly(ADP-ribose)polymerase: a perplexing participant in cellular responses to DNA breakage. Mutat. Res. **257:** 1–18.
152. BERGER, M.R., H. BISCHOFF, E. FRITSCHI, *et al.* 1985. Synthesis, toxicity, and therapeutic efficacy of 4-amino-*N*-(2′-aminophenyl)-benzamide: a new compound preferentially active in slowly growing tumors. Cancer Treat. Rep. **69:** 1415–1424.
153. HALLIWELL, B., K. ZHAO & M. WHITEMAN. 1999. Nitric oxide and peroxynitrite. The ugly, the uglier and the not so good: a personal view of recent controversies. Free Radic. Res. **31:** 651–669.
154. CURTIS, M.J. & R. PABLA. 1997. Nitric oxide supplementation or synthesis block—which is the better approach to treatment of heart disease? Trends Pharmacol. Sci. **18:** 239–244.
155. COLASANTI, M. & H. SUZUKI. 2000. The dual personality of NO. Trends Pharmacol. Sci. **21:** 249–252.
156. GUTTERIDGE, J.M. & B. HALLIWELL. 2000. Free radicals and antioxidants in the year 2000. A historical look to the future. Ann. N.Y. Acad. Sci. **899:** 136–147.
157. MADISON, D.V. 1993. Pass the nitric oxide. Proc. Natl. Acad. Sci. U.S.A. **90:** 4329–4331.
158. KOPROWSKI, H., Y.M. ZHENG, E. HEBER-KATZ, *et al.* 1993. *In vivo* expression of inducible nitric oxide synthase in experimentally induced neurologic diseases. Proc. Natl. Acad. Sci. U.S.A. **90:** 3024–3027.

159. RUAN, R.S., S.K. LEONG & K.H. YEOH. 1999. Ototoxicity of sodium nitroprusside is not due to nitric oxide. Exp. Neurol. **158:** 192–201.
160. SOUTHAN, G.J. & C. SZABO. 1996. Selective pharmacological inhibition of distinct nitric oxide synthase isoforms. Biochem. Pharmacol. **51:** 383–394.
161. KELLY, P.A., I.M. RITCHIE & G.W. ARBUTHNOTT. 1995. Inhibition of neuronal nitric oxide synthase by 7-nitroindazole: effects upon local cerebral blood flow and glucose use in the rat. J. Cereb. Blood Flow Metab. **15:** 766–773.
162. SHARMA, H.S., J. WESTMAN, Y. OLSSON & P. ALM. 1996. Involvement of nitric oxide in acute spinal cord injury: an immunocytochemical study using light and electron microscopy in the rat. Neurosci. Res. **24:** 373–384.
163. ELLWOOD, A.J. & M.J. CURTIS. 1996. Mechanism of 5-hydroxytryptamine-induced coronary vasodilation assessed by direct detection of nitric oxide production in guinea-pig isolated heart. Br. J. Pharmacol. **119:** 721–729.
164. HUANG, P.L., Z. HUANG, H. MASHIMO, *et al.* 1995. Hypertension in mice lacking the gene for endothelial nitric oxide synthase. Nature **377:** 239–242.
165. MACMICKING, J.D., C. NATHAN, G. HOM, *et al.* 1995. Altered responses to bacterial infection and endotoxic shock in mice lacking inducible nitric oxide synthase. Cell **81:** 641–650.
166. HUANG, P.L. & E.H. LO. 1998. Genetic analysis of NOS isoforms using nNOS and eNOS knockout animals. Prog. Brain Res. **118:** 13–25.
167. MENG, W., J. MA, C. AYATA, *et al.* 1996. ACh dilates pial arterioles in endothelial and neuronal NOS knockout mice by NO-dependent mechanisms. Am. J. Physiol **271:** H1145–H1150.
168. SPESSERT, R. & M. CLAASSEN. 1998. Histochemical differentiation between nitric oxide synthase-related and -unrelated diaphorase activity in the rat olfactory bulb. Histochem. J. **30:** 41–50.

Hypothesis: Proteasomal Dysfunction

A Primary Event in Neurogeneration that Leads to Nitrative and Oxidative Stress and Subsequent Cell Death

BARRY HALLIWELL

Department of Biochemistry, Faculty of Medicine, National University of Singapore, Singapore

ABSTRACT: It is proposed that a primary mechanism leading to neuronal cell death in common neurodegenerative diseases is interference with proteasome function. This can involve genetic defects, direct inactivation of the proteasome (e.g., by reactive oxygen species), or overloading with proteins. The latter can be caused by excessive production of normal proteins or by the formation of poorly degradable proteins as a result of genetic mutations, faulty posttranslational modification, or protein modification by reactive oxygen or nitrogen species. Blockage of the proteasome leads to increased oxidative and nitrative stress, the latter apparently due to upregulation of nitric oxide synthase. Thus, agents that increase proteasome function may be generally neuroprotective, as may be NOS inhibitors. Proteasome inhibitors should be used with caution as therapeutic agents.

KEYWORDS: proteasome; nitric oxide; neurogeneration; oxidative stress; Alzheimer's disease

ABBREVIATIONS:

AD	Alzheimer's disease.
ALS	Amyotrophic lateral sclerosis.
COX-2	Cyclooxygenase-2.
CuZnSOD	Copper, zinc-containing superoxide dismutase.
FA	Freidreich's ataxia.
GSH	Glutathione, reduced.
HNE	4-Hydroxy-trans-nonenal.
MSA	Multiple system atrophy.
NOS	Nitric oxide synthase.
PD	Parkinson's disease.
SOD	Superoxide dismutase

Address for correspondence: Barry Halliwell, Department of Biochemistry, Faculty of Medicine, National University of Singapore, 10 Kent Ridge Crescent, Singapore 119260. Voice: +65 874 3247; fax: +65 779 1453.
bchbh@nus.edu.sg

GENERAL FEATURES OF NEURODEGENERATIVE DISEASE

Accumulation of Abnormal Proteins

The major neurodegenerative diseases, Alzheimer's disease (AD), Parkinson's disease (PD), amyotrophic lateral sclerosis (ALS), Huntington's disease (HD) and Freidreich's ataxia (FA) are all associated with the presence of abnormal proteins (see TABLE 1). The origins of HD[1] and FA[2] involve specific genetic defects that lead to the production of abnormal proteins, whereas AD, ALS, and PD appear to be mostly sporadic. Nevertheless, familial types of AD,[3] ALS,[4] and PD[5] have been described. Examples are the rare mutations in synuclein or parkin that cause familial PD[5] and the 2% of ALS cases associated with mutations in the gene encoding copper-zinc containing superoxide dismutase (CuZnSOD).[4] Even in the more common sporadic versions of PD, ALS, and AD, abnormal proteins are evident. Thus, the "senile plaques" typical of AD contain not only β-amyloid but a wide range of other proteins.[3,6] Some (or perhaps all) of these proteins are oxidized and nitrated.[7–11]. Similar damage has been described for proteins in the Lewy bodies in sporadic PD.[12,13] Oxidative and nitrative protein damage are also elevated in ALS.[14–16] Protein aggregates are not only detected in the familial forms of ALS, but cytoplasmic inclusion bodies are a hallmark of motor neurone degeneration in sporadic ALS.[17–19] Inclusions from patients with multiple system atrophy (MSA) contain a range of proteins, including α-synuclein, and the same is probably true of all the neurodegenerative diseases.[13,20,21]

In non-dividing cells, such as the great majority of neurones in the adult brain, the protein content of cells is approximately constant. Since protein synthesis is continuous, it must be matched by an equal rate of protein degradation. Cellular proteins can be degraded by the lysosomal system, but a system of equal or greater importance is the proteasome.[22–24] The 20S (named for its sedimentation coefficient) proteasome is a cylindrical structure comprised of multiple protein subunits and containing a narrow channel in which proteolysis occurs. The 20S proteasome can degrade a wide range of proteins, including oxidatively-damaged proteins,[24] but most or all of the 20S proteasome in the cell is associated with a 19S "cap complex" that binds in an ATP-dependent manner and confers specificity for the degradation of polyubiquitinated proteins.[22] Several other proteins are known (e.g., PA28) that can associate with proteasomes and increase (or in some cases decrease) rates of protein clearance.[22,23] Proteasome activity is usually assessed by measuring the

TABLE 1. Processes that generate abnormal proteins

Gene mutations
Aberrant splicing of mRNA
Faulty posttranslational modification
Oxidation of amino acid residues by reactive oxygen species
Nitration of amino acid residues by reactive nitrogen species
Chlorination of amino acid residues by reactive chlorine species
Glycation/glycooxidation/spontaneous deamination

activities of the three active sites that hydrolyse adjacent to different amino acid residues. These are known as the trypsin-like, chymotrypsin-like, and peptidyl-glutamyl-peptidase activities, respectively, and they are usually assayed by employing simple chromogenic substrates.[22]

Oxidative Damage

Increased levels of oxidative damage to DNA, proteins, and lipids (lipid peroxidation) have been detected, using a wide range of biomarkers, in postmortem CNS tissue samples from patients who died with AD, PD, ALS, and in some cases, HD[8–15, 25, 26] (although oxidative damage in human HD appears less prominent than in the other diseases[27]) Oxidative damage also occurs in transgenic animal models of AD, PD, ALS, and HD, and in some cases, oxidative damage has been found to be elevated prior to neuronal cell death.[25,28] Decreases in the level of the antioxidant reduced glutathione (GSH) appear to be early events in the pathogenesis of PD.[29] In animal models of ALS, antioxidants have been shown to prolong lifespan and/or delay disease incidence to some extent and limited evidence suggests that antioxidant agents are beneficial in patients with AD or HD (reviewed in Refs. 1, 14, 26, and 30). Studies on isoprostanes, end products of lipid peroxidation that can be measured in CSF and urine from AD patients, suggest that lipid peroxidation is an early stage in the disease process.[31–33] Another end product of lipid peroxidation is the unsaturated aldehyde 4-hydroxy-*trans*-nonenal (HNE), which is highly neurotoxic.[34–37] HNE binds avidly to proteins,[38] and HNE—protein adducts are demonstrable in senile plaques and tangles in AD, tissues from ALS patients, and Lewy bodies in PD.[37,39–43] The toxicity of HNE to cells in culture is aggravated if the genes that encode mutant synucleins associated with PD, or mutant SODs associated with ALS, are being overexpressed.[44,45]

It is usually assumed that increased levels of oxidative damage are due to the increased generation of such reactive oxygen species as hydroxyl and peroxyl radicals ($\cdot OH$ and $\cdot RO_2$), superoxide ($O_2^{\cdot -}$), and hydrogen peroxide (H_2O_2). This may well be the case. However, increased levels of oxidative damage could equally well be due to decreased clearance of oxidatively-modified biomolecules.[46] Oxidized proteins appear to be largely or entirely removed by the proteasome system.[24,47] Their levels in nervous tissues tend to increase with age,[48,49] consistent with several reports that proteasome activity decreases with age.[50–53] Some animal studies suggest that levels of brain protein carbonyls (a general biomarker of oxidative protein damage[54]) are positively correlated with cognitive impairment.[48,49] Although such a correlation does not prove a cause–consequence relationship, in a study in gerbils the antioxidant *tert*-butyl-α-phenylnitrone decreased carbonyl levels and also improved cognitive function.[54] Caloric restriction had a similar effect in rats.[55] Protein carbonyls can be generated by direct oxidative damage to proteins, by the binding to proteins of cytotoxic aldehydes such as HNE, and by glycoxidation of proteins.[54,56] Indeed, there are suggestions that glycated proteins contribute to the pathology of AD.[57]

Nitrative Damage

Nitric oxide is an essential neurotransmitter and vasodilator in the brain, but only a small percentage of neurons usually express nNOS.[58] Overproduction of ·NO can cause cellular damage in some circumstances, both directly (e.g., by inhibition of cytochrome oxidase, leading to impaired mitochondrial function)[59] and indirectly, by forming more-cytotoxic "reactive nitrogen species". For example, ·NO can react with $O_2^{-\bullet}$.

$$O_2^{-\bullet} + \cdot NO \rightarrow ONOO^- .$$

The resulting peroxynitrite ($ONOO^-$) rapidly protonates,[60]

$$ONOO^- + H^+ \rightarrow ONOOH ,$$

and ONOOH proceeds to generate a range of cytotoxic species that exhibit hydroxylating, oxidizing, and nitrating activity.[60,61] Peroxynitrous acid (ONOOH) rapidly oxidizes and nitrates proteins; for example, cysteine and methionine residues are oxidized, whereas tyrosine and tryptophan can be nitrated.[60] Measurement of 3-nitrotyrosine residues is a widely-used biomarker for ONOOH formation,[60,62] although nitration can also involve other reactive nitrogen species.[61]

Increased levels of nitrated proteins have been observed in nervous tissues from subjects with AD, PD, or ALS.[4,9,14,16,30,62,63] Again, the usual assumption is that this rise in levels is caused by generation of more reactive nitrogen species, but defects in clearance of nitrated proteins could also contribute. How nitrated proteins are removed *in vivo* is uncertain; *denitrases* have been described in rat tissues,[64] but proteasomal degradation of nitrated proteins may also be important.[65] I find it hard to envisage how nitro groups could be removed from an intact protein, and feel it more likely that such proteins are degraded by the proteasome system, presumably releasing free nitrotyrosine. Indeed, elevated free nitrotyrosine levels have been reported in ALS.[66]

A UNIFYING HYPOTHESIS: PROTEASOMAL DYSFUNCTION IS THE PRIMARY DEFECT LEADING TO CELL DEATH IN NEURODEGENERATIVE DISEASES

I propose that interference with proteasomal function is the primary mechanism causing neurodegeneration in the major neurodegenerative diseases, by the following sequence of events,

$$\begin{matrix}\text{abnormal proteins/} \\ \text{oxidative stress}\end{matrix} \rightarrow \begin{matrix}\text{proteasomal} \\ \text{dysfunction}\end{matrix} \rightarrow \begin{matrix}\text{oxidative/} \\ \text{nitrative stress}\end{matrix} \rightarrow \text{cell death} .$$

Of course, all these events are interrelated; FIGURE 1 is an attempt to depict this. Proteasomal dysfunction could be due to any, or any combination of, the following:

1. Direct inactivation of the proteasome. Evidence suggests that HNE may directly damage the proteasome,[67,68] and it is possible that the same is true for ONOOH and other reactive oxygen or nitrogen species. I suggest that the proteasome is one of the primary cellular targets of such species. Indeed, Glockzin *et al.*[69] suggested that ·NO-induced apoptosis in RAW264.7 macrophages involves proteasomal inhibition.

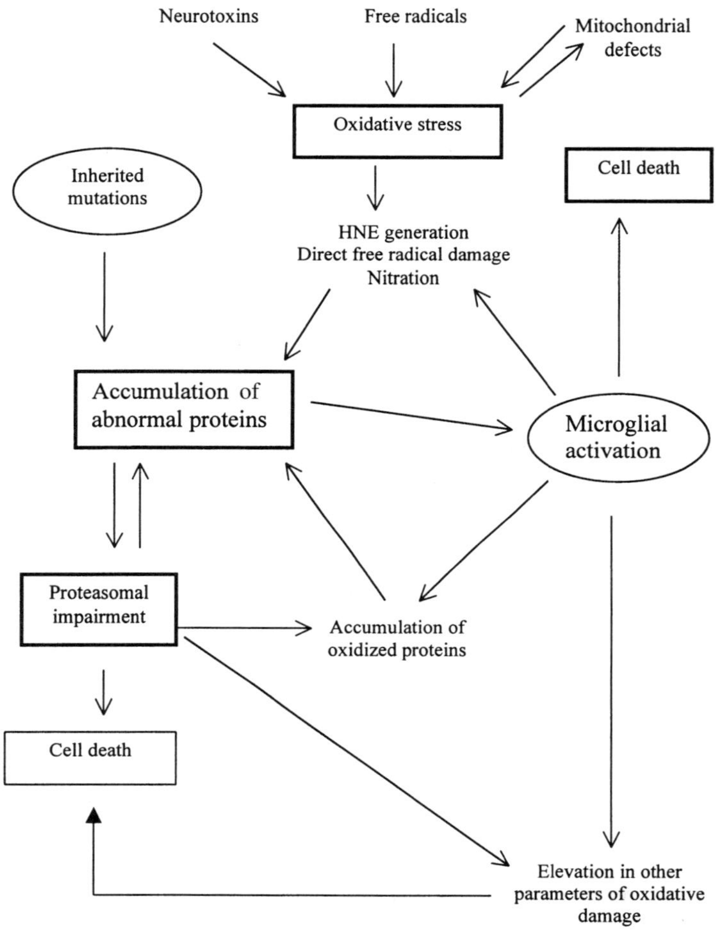

FIGURE 1. The vicious cycle of proteasomal impairment, oxidative damage and cell death.

2. Overload of the proteasome due to the presence of proteins that are difficult to degrade and/or are being synthesized at an abnormally rapid rate, requiring accelerated clearance. For example, the mutant synucleins associated with some cases of familial PD appear to be more slowly degraded than normal synucleins.[70] Their expression in PC12 cells led to decreased proteasomal activity.[71] Parkin is also degraded by the proteasome.[72]

3. Overload of the proteasome due to the presence of increased levels of oxidized and nitrated proteins. For example, HNE-modified proteins, including HNE-modified β-amyloid, can block proteasome function.[73,74]

4. Defects in the genes[75] encoding proteasomal subunits may lead to impaired function. It is possible that this may predispose to disease in later life if proteasomal activity declines with age: at some point a critically-low level will be reached. Evidence for the feasibility of this concept also exists: mutations in a ubiquitin C-terminal hydrolase (L1) have been described in a few cases of familial PD, and the *parkin* gene encodes a ubiquitin conjugating enzyme.[5]

How Does Interference with Proteasome Function Cause Cell Death?

I propose that it does so by causing oxidative and nitrative damage that trigger cell death, often by apoptosis. Apoptosis is strongly suspected to be a major mechanism of cell death in ALS, PD, and AD.[76–78] Both reactive oxygen species and ONOOH can damage mitochondria in a way that leads to cytochrome c release and subsequent apoptosis.[79] In turn, apoptosis is accompanied by antioxidant depletion and increased formation of reactive oxygen species.[80] Lee *et al.*[81] showed that addition of selective proteasomal inhibitors (lactacystin and epoxomicin) to cell lines raised levels of protein carbonyls, oxidative DNA damage, and lipid peroxidation. Furthermore, there was an increase in nNOS levels and increased ·NO formation. The non-selective NOS inhibitor L-NAME delayed apoptosis.[81] In glioma cells, apoptosis induced by proteasome inhibition involved cytochrome c release and caspase activation.[82]

EVIDENCE SUPPORTING THE HYPOTHESIS: TANTALISING BUT INCOMPLETE

The hypothesis stated above and summarized in FIGURE 2 arose from our previous proposals,[26,83] that protein aggregation is an important source of neurotoxicity in all the neurodegenerative diseases. Others have made broadly similar suggestions.[84] In our laboratory, the main evidence supporting the hypothesis comes from the work of Lee *et al.*,[81] who showed that upregulation of nNOS, increased ·NO formation, and increased oxidative damage are early events prior to apoptosis in cells in which the proteasome was almost completely inhibited by selective inhibitors. The molecular mechanisms by which proteasomal blockage leads to induction of nNOS and oxidative damage need to be elucidated, although kinase activation may be involved.[85] The work of Lee *et al.*[81] was performed on NT-2 and SKNMC cell lines, however, and needs to be repeated on neuronal cells. In the same cell lines, Hyun *et al.*[68] found that proteasomal activity was altered after HNE addition and this was followed by mitochondrial dysfunction and apoptosis.

Evidence from postmortem brain samples, showing that proteasomal activity in PD[86] and AD[87] is decreased and that proteasomal components are coprecipitated in ubiquitinated inclusions[88] in both diseases, is consistent with the hypothesis. In the case of AD, no alterations in gene transcription were observed, suggesting loss of proteasome activity due to subsequent inactivation.[87] The data of Hyun *et al.*[68] on the ability of HNE-modified proteins to block the proteasome are also supportive. Apoptosis induced by mutant SOD enzymes in motor neurones involved upregulation of ·NO synthesis and $ONOO^-$ formation.[89] Usually, proteasomal blockage prevents induction of iNOS,[90] but cases where it increases iNOS expression have been

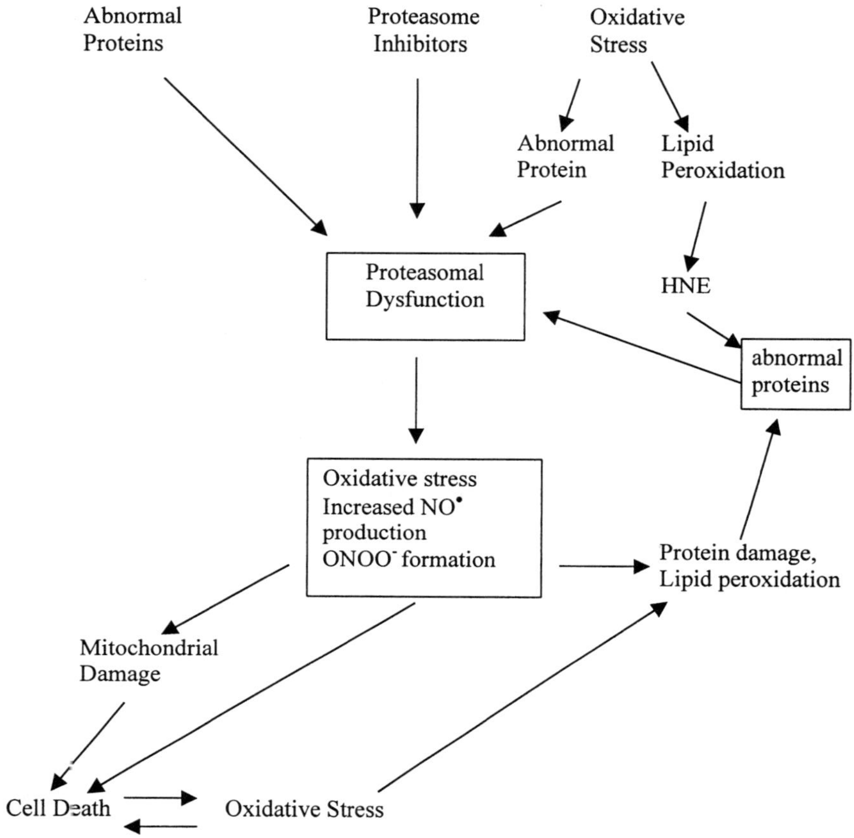

FIGURE 2. Possible sequence of events.

reported.[91] Can dysregulation of the proteasome function prevent cell death? We observed that overexpression of the antiapoptotic protein bcl-2 increases proteasome activity in cells[92] and it also delays cell death associated with the presence of mutant proteins.[45] The latter has been observed *in vivo* and in cell culture.[45,93]

Multiple effects of proteasomal inhibition have been described on a wide range of cells, including causing neurite outgrowth,[94] *protection* of cells against apoptosis (e.g., by preventing upregulation of NF-κB)[91] and interference with cell division.[95,96] These variations may be due to the use of different cell types (e.g., dividing versus non-dividing), inhibitors achieving various degrees of inhibition, or different routes leading to cell death.[95] Nevertheless, proteasomal inhibition caused apoptosis in cultured cerebellar granule cells,[97] and caused cell death and activation of poly (ADP-ribose) polymerase in the PC6 cell line; inhibition of this enzyme decreased cell death.[98] The proteasome modulates the levels of presenilins, and proteasome blockade could further promote AD by causing their accumulation.[99]

It is not clear to what extent the hypothesis might apply to HD and FA. In the former case, evidence for oxidative damage in human patients is weaker when compared to that for AD and PD.[27] Nevertheless, nitrative damage is present and a deleterious role for ·NO has been postulated and is supported by animal experiments.[100] In FA, the defective gene alters mitochondrial oxidative metabolism and leads to oxidative stress,[101,102] but at present we know too little about the mechanisms of neuronal cell death in FA to see whether or not they fit our hypothesis. It is also possible that ·NO is not the only mediator of cell death upregulated as a result of proteasomal inhibition; activations of poly (ADP-ribose) polymerase[98] and cyclooxygenase-2 (COX-2)[103] are alternative or additional mechanisms. It must always be remembered that ·NO is not always neurotoxic; cases exist where extra ·NO is neuroprotective.[104,105]

CONCLUSIONS: THERAPEUTIC IMPLICATIONS

The hypothesis is summarized in FIGURE 2. If it is correct, three conclusions follow

1. Selective inhibitors of nNOS may have a general therapeutic role in the neurodegenerative diseases. Limited data consistent with this view have been provided in the transgenic CuZnSOD mouse model of ALS.[106] Of course, other sources of ·NO, such as increases in iNOS related to the inflammatory component of AD,[107–109] may also be important.

2. The therapeutic use of proteasomal inhibitors; for example, for the treatment of cancer or inflammatory diseases,[110,111] should be approached with caution. Agents that do not cross the blood–brain barrier may be preferable.

3. Agents that upregulate proteasome function, whether by relieving blockage or increasing transcription of the relevant genes, should be neuroprotective. Bcl-2 may act (in part) by this mechanism.[92]

REFERENCES

1. ALEXI, T., C.V. BORLONGAN, R.L.M. FAULL, *et al.* 2000. Neuroprotective strategies for basal ganglia degeneration: Parkinson's and Huntington's diseases. Prog. Neurobiol. **60:** 409–470.
2. PUCCIO, H. & M. KOENIG. 2000. Recent advances in the molecular pathogenesis of Friedreich ataxia. Hum. Mol. Genet. **12:** 887–892.
3. SELKOE, D.J. 1999. Translating cell biology into therapeutic advances in Alzheimer's disease. Nature **399:** A23–A31.
4. GURNEY, M.E., R. LIU, J.S. ALTHAUS, *et al.* 1998. Mutant CuZn superoxide dismutase in motor neuron disease. J. Inherit. Metabol. Disease **21:** 587–597.
5. KAHLE, P.J., U. LEIMER & C. HAASS. 2000. Does failure of parkin-mediated ubiquitination cause juvenile parkinsonism? Trend. Biochem. Sci. **25:** 524–527.
6. SCHMIDT, M.L., A.G. DIDARIO, L. OTVOS, *et al.* 1994. Plaque-associated neuronal proteins: a recurrent motif in neuritic amyloid deposits throughout diverse cortical areas of the Alzheimer's disease brain. Exp. Neurol. **130:** 311–322.
7. MATTSON, M.P. 2000. Free radical-mediated disruption of cellular ion homeostasis, mitochondrial dysfunction and neuronal degeneration in sporadic and inherited Alzheimer's disease. *In* Free Radicals in Brain Pathophysiology. G. Poli, E. Cadenas & L. Packer, Eds.: 323–357. Marcel Dekker, New York.

8. LYRAS, L., N.J. CAIRNS, A. JENNER, *et al.* 1997. An assessment of oxidative damage to proteins, lipids, and DNA in brain from patients with Alzheimer's disease. J. Neurochem. **68:** 2061–2069.
9. SU, J.H., G. DENG & C.W. COTMAN. 1997. Neuronal DNA damage precedes tangle formation and is associated with up-regulation of nitrotyrosine in Alzheimer's disease brain. Brain Res. **774:** 193–199.
10. SAYRE, L.M., G. PERRY, P.L.R. HARRIS, *et al.* 2000. *In situ* oxidative catalysis by neurofibrillary tangles and senile plaques in Alzheimer's disease: a central role for bound transition metals. J. Neurochem. **74:** 270–279.
11. GOOD, P.F., P. WERNER, A. HSU, *et al.* 1996. Evidence of neuronal oxidative damage in Alzheimer's disease. Am. J. Pathol. **149:** 21–28.
12. GOOD, P.F., A. HSU, P. WERNER, *et al.* 1998. Protein nitration in Parkinson's disease. J. Neuropathol. Exp. Neurol. **57:** 338–342.
13. GIASSON, B.I., J.E. DUDA, I.V. MURRAY, *et al.* 2000. Oxidative damage linked to neurodegeneration by selective alpha-synuclein nitration in synucleinopathy lesions. Science **290:** 985–989.
14. COOKSON, M.R. & P.J. SHAW. 1999. Oxidative stress and motor neurone disease. Brain Pathol. **9:** 165–186.
15. FERRANTE, R.J., S.E. BROWNE, L.A. SHINOBU, *et al.* 1997. Evidence of increased oxidative damage in both sporadic and familial amyotrophic lateral sclerosis. J. Neurochem. **69:** 2064–2074.
16. TOHGI, H., T. ABE, K. YAMAZAKI, *et al.* 1999. Remarkable increase in cerebrospinal fluid 3-nitrotyrosine in patients with sporadic amyotrophic lateral sclerosis. Ann. Neurol. **46:** 129–131.
17. JOHNSTON, J.A., M.J. DALTON, M.E. GURNEY, *et al.* 2000. Formation of high molecular weight complexes of mutant Cu, Zn-superoxide dismutase in a mouse model for familial amyotrophic lateral sclerosis. Proc. Natl. Acad. Sci. U.S.A. **97:** 12571–12576.
18. STIEBER, A., J.O. GONATAS & N.K. GONATAS. 2000. Aggregation of ubiquitin and a mutant ALS-linked SOD1 protein correlate with disease progression and fragmentation of the Golgi apparatus. J. Neurol. Sci. **173:** 53–62.
19. OKADO-MATSUMOTO, A., T. MYINT, J. FUJII, *et al.* 2000. Gain in functions of mutant Cu, Zn-superoxide dismutases as a causative factor in familial amyotrophic lateral sclerosis: Less reactive oxidant formation but high spontaneous aggregation and precipitation. Free Rad. Res. **33:** 65–73.
20. GAI, W.P., H.T. POWER, P.C. BLUMBERGS, *et al.* 1999. α-Synuclein immunoisolation of glial inclusions from multiple system atrophy brain tissue reveals multiprotein components. J. Neurochem. **73:** 2093–2100.
21. CHARLES, V., E. MEZEY, P.H. REDDY, *et al.* 2000. Alpha-synuclein immunoreactivity of huntingtin polyglutamine aggregates in striatum and cortex of Huntington's disease patients and transgenic mouse models. Neurosci. Lett. **28:** 29–32.
22. DEMARTINO, G.N. & C.A. SLAUGHTER. 1999. The proteasome, a novel protease regulated by multiple mechanisms. J. Biol. Chem. **274:** 22123–22126.
23. WILK, S., W.E. CHEN & R.P. MAGNUSSON. 2000. Properties of the nuclear proteasome activator PA28gamma (REGgamma). Arch. Biochem. Biophys. **15:** 265–271.
24. GRUNE, T., T. REINHECKEL & K.J. DAVIES. 1997. Degradation of oxidised proteins in mammalian cells. FASEB J. **11:** 526–534.
25. GRUNEWALD, T. & M.F. BEAL. 1999. Bioenergetics in Huntington's disease. Ann. N.Y. Acad. Sci. **893:** 203–213.
26. HALLIWELL, B. 2001. Role of free radicals in the neurodegenerative diseases. Therapeutic implications for antioxidant treatment. Drugs Aging. **18:** 685–716.
27. ALAM, Z.I., B. HALLIWELL & P. JENNER. 2000. No evidence for increased oxidative damage to lipids, proteins, or DNA in Huntington's disease. J. Neurochem. **75:** 840–846.
28. PAPPOLLA, M.A., Y.J. CHYAN, M. SOS, *et al.* 2000. Role of transgenic models for the study of oxidative neurotoxicity in Alzheimer's disease. *In* Free Radicals in Brain Pathophysiology. G. Poli, E. Cadenas & L. Packer, Eds.: 359–381. Marcel Dekker, New York.

29. OWEN, A.D., A.H. SCHAPIRA, P. JENNER, *et al.* 1997. Indices of oxidative stress in Parkinson's disease, Alzheimer's disease and dementia with Lewy bodies. J. Neural Trans. Suppl. **51:** 167–173.

30. MARK, R.J. 1999. Common mechanisms of oxidative damage in chronic neurodegenerative conditions: potential points of therapeutic intervention. Exp. Opin. Ther. Patents **9:** 1339–1346.

31. PRATICO, D., C.M. CLARK, V.M. LEE, *et al.* 2000. Increased 8, 12-iso-iPF2alpha-VI in Alzheimer's disease: correlation of a non-invasive index of lipid peroxidation with disease severity. Ann. Neurol. **48:** 809–812.

32. MONTINE, T.J., M.F. BEAL, M.E. CUDKOWICZ, *et al.* 1999. Increased CSF F2- isoprostane concentration in probable A.D. Neurology **52:** 562–565.

33. MONTINE, T.J., W.R. MARKESBERY, W. ZACKERT, *et al.* 1999. The magnitude of brain lipid peroxidation correlates with the extent of degeneration but not with density of neuritic plaques or neurofibrillary tangles or with APOE genotype in Alzheimer's disease patients. Am. J. Path. 155: 863-868.

34. MARK, R.J., M.A. LOVELL, W.R. MARKESBERY, *et al.* 1997. A role for 4-hydroxynonenal, an aldehyde product of lipid peroxidation, in disruption of ion homeostasis and neuronal death induced by amyloid beta-peptide. J. Neurochem. **68:** 225–264.

35. KRUMAN, I., A.J. BRUCE-KELLER, BREDESEN, *et al.* 1997. Evidence that 4-hydroxynonenal mediates oxidative stress-induced neuronal apoptosis. J. Neurosci. **17:** 5089–5100.

36. BRUCE-KELLER, A.J., Y.J. LI, M.A. LOVELL, *et al.* 1998. 4-Hydroxynonenal, a product of lipid peroxidation, damages cholinergic neurons and impairs visuospatial memory in rats. J. Neuropath. Exp. Neurol. **57:** 257–267.

37. MONTINE, K.S., E. REICH, M.D. NEELY, *et al.* 1998. Distribution of reducible 4-hydroxynonenal adduct immunoreactivity in Alzheimer's disease is associated with APOE genotype. J. Neuropath. Exp. Neurol. **57:** 415–425.

38. WAEG, G., G. DIMSITY & H. ESTERBAUER. 1996. Monoclonal antibodies for detection of 4-hydroxynonenal modified proteins. Free Rad. Res. **25:** 149–159.

39. KELLER, J.N. & M.P. MATTSON. 1998. Roles of lipid peroxidation in modulation of cellular signalling pathways, cell dysfunction, and death in the nervous system. Rev. Neurosci. **9:** 105–116.

40. SAYRE, L.M., D A. ZELASKO, P.L. HARRIS, *et al.* 1997. 4-Hydroxynonenal-derived advanced lipid peroxidation end products are increased in Alzheimer's disease. J. Neurochem. **68:** 2092–2097.

41. TAKEDA, A., M.A. SMITH, J. AVILA, *et al.* 2000. *In* Alzheimer's Disease, Heme Oxygenase is Coincident with Alz50, an epitope of τ induced by 4-hydroxy-2-nonenal modification. J. Neurochem. **75:** 1234-1241.

42. YORITAKA, A., N. HATTORI, K. UCHIDA, *et al.* 1996. Immunohistochemical detection of 4-hydroxynonenal protein adducts in Parkinson's disease. Proc. Natl. Acad. Sci. U.S.A. **93:** 2696–2701.

43. PEDERSEN, W.A., W. FU & J.N. KELLER. 1998. Protein modification by the lipid peroxidation product 4-hydroxynonenal in the spinal cords of amyotrophic lateral sclerosis patients. Ann. Neurol. **44:** 819–824.

44. LEE, M.H., D.H. HYUN, B. HALLIWELL, *et al.* 2001. Effect of the overexpression of wild type or mutant α-synuclein on cell susceptibility to insult. J. Neurochem. **76:** 998–1009.

45. LEE, M.H., D.H. HYUN, B. HALLIWELL, *et al.* 2001.Effect of overexpression of wild type and mutant Cu/Zn-SODs on oxidative stress and cell death induced by H_2O_2, 4-HNE or serum deprivation: potentiation of injury by ALS-related mutant SODs and protection by bcl-2. J. Neurochem. **78:** 209–220.

46. HALLIWELL, B. & J.M.C. GUTTERIDGE. 1999. Free Radicals in Biology and Medicine, Third edit. Chapter 4. Oxford University Press.

47. ULLRICH, O., T. REINHECKEL, N. SITTE, *et al.* 1999. Poly-ADP ribose polymerase activates nuclear proteasome to degrade oxidatively damaged histones. Proc. Natl. Acad. Sci. U.S.A. **96:** 6223–6228.

48. FORSTER, M.J., A. DUBEY, K.M. DAWSON, *et al.* 1996. Age-related losses of cognitive function and motor skills in mice associated with oxidative protein damage in the brain. Proc. Natl. Acad. Sci. U.S.A. **93:** 4765–4769.

49. SMITH, C.D., J.M. CARNEY, P.E. STARKE-REED, *et al.* 1991. Excess brain protein oxidation and enzyme dysfunction in normal aging and in Alzheimer's disease. Proc. Natl. Acad. Sci. U.S.A. **88:** 10540–10543.

50. PONNAPPAN, U., M. ZHONG & G.U. TREBILCOCK. 1999. Decreased proteasome-mediated degradation I T cells from the elderly: A role in immune senescence. Cell Immunol. **192:** 167–174.

51. KELLER, J.N., K.B. HANNI & W.R. MARKESBERY. 2000. Possible involvement of proteasome inhibition in aging: implications for oxidative stress. Mech. Aging Dev. **113:** 61–70.

52. HAYASHI, T. & S. GOTO. 1998. Age-related changes in the 20S and 26S proteasome activities in the liver of male F344 rats. Mech. Aging Dev. **102:** 55–66.

53. KELLER, J.N., F.F. HUANG & W.R. MARKESBERY. 2000. Decreased levels of proteasome activity and proteasome expression in aging spinal cord. Neuroscience **98:** 149–156.

54. CARNEY, J.M., P.E. STARKE-REED, C.N. OLIVER, *et al.* 1991. Reversal of age-related increase in brain protein oxidation , decrease in enzyme activity, and loss in temporal and spatial memory by chronic administration of the spin-trapping compound *N-tert*-butyl-alpha-phenylnitrone. Proc. Natl. Acad. Sci. U.S.A. **88:** 3633–3636.

55. DUBEY, A., M.J. FORSTER, H. LAL, *et al.* 1996. Effects of age and caloric intake on protein oxidation in different brain regions and on behavioural functions of the mouse. Arch. Biochem. Biophys. **333:** 189–197.

56. LIGGINS, J. & A.J. FURTH. 1997. Role of protein-bound carbonyl groups in the formation of advanced glycation endproducts. Biochem. Biophys. Acta. **22:** 123–130.

57. SASAKI, N., R. FUKATSU, K. TSUZUKI, *et al.* 1998. Advanced glycation end products in Alzheimer's disease and other neurodegenerative diseases. Am. J. Pathol. **153:** 1149–1155.

58. BREDT, D.S. 1999. Endogenous nitric oxide synthesis: biological functions and pathphysiology. Free Rad. Res. **31:** 577–596.

59. FORFIA, P.R., T.H. HINTZE, M.S. WOLIN, *et al.* 1999. Role of nitric oxide in the control of mitochondrial function. Adv. Exp. Med. Biol. **471:** 381–388.

60. BECKMAN, J.S. & W.H. KOPPENOL. 1996. Nitric oxide, superoxide, and peroxynitrite: the good, the bad, and the ugly. Am. J. Physiol. **271:** C1424–1437.

61. HALLIWELL, B., K. ZHAO & M. WHITEMAN. 1999. Nitric oxide and peroxynitrite. The ugly, the uglier and the not so good. Free Rad. Res. **31:** 651–669.

62. GREENACRE, S.A.B. & H. ISCHIROPOULOS. 2001. Tyrosine nitration: localization, quantification, consequences for protein function and signal transduction. Free Rad. Res. **34:** 541–581.

63. HENSLEY, K., M.L. MAIDT, Z. YU, *et al.* 1998. Electrochemical analysis of protein nitrotyrosine and dityrosine in the Alzheimer brain indicates region-specific accumulation. J. Neurosci. **18:** 8126–8132.

64. KAMISAKI, Y., K. WADA, K. BIAN, *et al.* 1998. An activity in rat tissues that modifies nitrotyrosine-containing proteins. Proc. Natl. Acad. Sci. U.S.A. **95:** 11584–11589.

65. GRUNE, T., I.E. BLASIG, N. SITTE, *et al.* 1998. Peroxynitrite increases the degradation of aconitase and other cellular proteins by proteasome. J. Biol. Chem. **273:** 10857–10862.

66. BRUIJN, L.I., M.F. BEAL, M.W. BECHER, *et al.* 1997. Elevated free nitrotyrosine levels, but not protein-bound nitrotyrosine or hydroxyl radicals, throughout amyotrophic lateral sclerosis (ALS)-like disease implicate tyrosine nitration as an aberrant *in vivo* property of one familial ALS-linked superoxide dismutase 1 mutant. Proc. Natl. Acad. Sci. U.S.A. **94:** 7606–7611.

67. OKADA, K., C. WANGPOENGTRAKUL, T. OSAWA, *et al.* 1999. 4-hydroxy-2-nonenal-mediated impairment of intracellular proteolysis during oxidative stress. J. Biol. Chem. **274:** 23787–23793.

68. HYUN, D.H., M.H. LEE, B. HALLIWELL, *et al.* 2002. Proteasomal dysfunction induced by 4-hydroxy-2,3-*trans*nonenal, an end-product of lipid peroxidation. J. Neurochem. In press.
69. GLOCKZIN, S., A. VON KNETHEN, M. SCHEFFNER & B. BRUNE. 1999. Activation of the cell death programme by nitric oxide involves inhibition of the proteasome. J. Biol. Chem. **274:** 19581–19586.
70. BENNETT, M.C., J.F. BISHOP, Y. LENG, *et al.* 1999. Degradation of alpha-synuclein by proteasome. J. Biol. Chem. **274:** 33855–33858.
71. TANAKA, Y., S. ENGELENDER, S. IGARASHI, *et al.* 2001. Inducible expression of mutant alpha-synuclein decreases proteasome activity and increases sensitivity to mitochondria-dependent apoptosis. Hum. Mol. Genet. **10:** 919–926.
72. CHOI, P., N. OSTREROVA-GOLTS, D. SPARKMAN, *et al.* 2000. Parkin is metabolised by the ubiquitin/proteasome system. Neuroreport **11:** 2635–2638.
73. FRIGUET, B. & L.I. SZWEDA. 1997. Inhibition of the multicatalytic proteinase (proteasome) by 4-hydroxy-2-nonenal cross-linked protein. FEBS Lett. **405:** 21–25.
74. SHRINGARPURE, R., T. GRUNE, N. SITTE, *et al.* 2000. 4-Hydroxynonenal–modified amyloid-β peptide inhibits the proteasome: possible importance in Alzheimer's disease. Cell Mol. Life Sci. **57:** 1802–1809.
75. TANAKA, K. 1998. Molecular biology of the proteasome. Biochem. Biophys. Res. Commun. **247:** 537–541.
76. PASINELLA, P., M.K. HOUSEWEART, R.H. JR. BROWN, *et al.* 2000. Caspase-1 and -3 are subsequently activated in motor neuron death in Cu, Zn superoxide dismutase-mediated familial amyotrophic lateral sclerosis. Proc. Natl. Acad. Sci. U.S.A. **97:** 13901–13906.
77. TATTON, N.A. 2000. Increased caspase 3 and bax immunoreactivity accompany nuclear GAPDH translocation and neuronal apoptosis in Parkinson's disease. Exp. Neurol. **166:** 29–43.
78. MARTIN, L.J. 2001. Neuronal cell death in nervous system development, disease, and injury (Review). Int. J. Mol. Med. **7:** 455–478.
79. GHAFOURIFAR, P., U. BRINGOLD, S.D. KLEIN, *et al.* 2001. Mitochondrial nitric oxide synthase, oxidative stress and apoptosis. Biol. Signals Recept **10:** 57–65.
80. CHANDRA, J., A. SAMALI & S. ORRENIUS. 2000. Triggering and modulation of apoptosis by oxidative stress. Free Rad. Biol. Med. **29:** 323–333.
81. LEE, M.H., D.H. HYUN, P. JENNER, *et al.* 2001. Effect of proteasome inhibition on cellular oxidative damage and antioxidant defences. J. Neurochem. **78:** 32–41.
82. WAGENKNECHT, B., M. HERMISSON, P. GROSCURTH, *et al.* Proteasome inhibitor-induced apoptosis of glioma cells involves the processing of multiple caspases and cytochrome c release. J. Neurochem. **75:** 2288–2297.
83. HALLIWELL, B. & P. JENNER. 1998. Impaired clearance of oxidised proteins in neurodegenerative disease. Lancet **351:** 1510.
84. KAKIZUKA, A. 1998. Protein precipitation: a common etiology in neurodegenerative disorders? Trend. Genet. **14:** 396–402.
85. MERIIN, A.B., V.L. GABAI, J. YAGLOM, *et al.* 1998. Proteasome inhibitors activate stress kinases and induce Hsp72. J. Biol. Chem. **273:** 6373–6379.
86. MCNAUGHT, K.S.T.P. & P. JENNER. 2001. Proteasomal function is impaired in substantia nigra in Parkinson's disease. Neurosci. Lett. **297:** 191–194.
87. KELLER, J.N., K.B. HANNI & W.R. MARKESBERY. 2000. Impaired proteasome function in Alzheimer's Disease. J. Neurochem. **75:** 436–439.
88. II, K., H. ITO, K. TANAKA, *et al.* 1997. Immunocytochemical co-localization of the proteasome in ubiquitinated structures in neurodegenerative diseases and the elderly. J. Neuropathol. Exp. Neurol. **56:** 125–131.
89. ESTEVEZ, A.G., J.P. CROW, J.B. SAMPSON, *et al.* 1999. Induction of nitric oxide – dependent apoptosis in motor neurons by zinc-deficient superoxide dismutase. Science **286:** 2498–2500.
90. GRISCAVAGE, J.M., S. WILK & L.J. IGNARRO. 1996. Inhibitors of the proteasome pathway interfere with induction of nitric oxide synthase in macrophages by blocking activation of transcription factor NF-kappa B. Proc. Natl. Acad. Sci. U.S.A. **93:** 3308–3312.

91. STASIOLEK, M., V. GAVRILYUK, A. SHARP, *et al.* 2000. Inhibitory and stimulatory effects of lactacystin on expression of nitric oxide synthase type 2 in brain glial cells. The role of Ikappa B-beta. J. Biol. Chem. **275:** 24847–24856.
92. LEE, M.H., D.H. HYUN, K.A. MARSHALL, *et al.* 2001. Effect of overexpression of bcl-2 on cellular oxidative damage, nitric oxide production, antioxidant defenses and the proteasome system. Free Rad. Biol. Med. **31:** 1550–1559.
93. KOSTIC, V., V. JACKSON-LEWIS, F. DE BILBAO, *et al.* 1997. Bcl-2: prolonging life in a transgenic mouse model of familial amyotrophic lateral sclerosis. Science **277:** 559–562.
94. FENTEANY, G. & S.L. SCHREIBER. 1998. Lactacystin, proteasome function and cell fate. J. Biol. Chem. **273:** 8545–8548.
95. LIN, K.I., J.M. BARABAN & R.R. RATAN. 1998. Inhibition versus induction of apoptosis by proteasome inhibitors depends on concentration. Cell Death Differentiation **5:** 577–583.
96. BUCHAN, A.M., H. LI & B. BLACKBURN. 2000. Neuroprotection achieved with a novel proteasome inhibitor which blocks NF-kappaB activation. Neuroreport **11:** 427–430.
97. PASQUINI, L.A., M. BESIO MORENO, A.M. ADAMO, *et al.* 2000. Lactacystin, a specific inhibitor of the proteasome, induces apoptosis and activates caspase-3 in cultured cerebellar granule cells. J. Neurosci. Res. **59:** 601–611.
98. KELLER, J.N. & W.R. MARKESBERY. 2000. Proteasome inhibition results in increased poly-ADP-ribosylation: implications for neuron death. J. Neurosci. Res. **61:** 436–442.
99. CHECLER, F., D.E. ALVES, C. COSTA, *et al.* 2000. Role of proteasome in Alzheimer's disease. Biochim. Biophys. Acta. **1502:** 133–138.
100. SCHULZ, J.B., R.T. MATHEWS, T. KLOCKGETHER, *et al.* 1997. The role of mitochondrial dysfunction and neuronal nitric oxide in animal models of neurodegenerative diseases. Mol. Cell Biochem. **174:** 193–197.
101. BRADLEY, J.L., J.C. BLAKE, S. CHAMBERLAIN, *et al.* 2000. Clinical, biochemical and molecular genetic correlations in Friedreich's ataxia. Hum. Mol. Genet. **9:** 275–282.
102. SCHULZ, J.B., T. DEHMER, L. SCHOLS, *et al.* 2000. Oxidative stress in patients with Friedreich ataxia. Neurology **55:** 1600–1601.
103. ROCKWELL, P., H. YUAN, R. MAGNUSSON, *et al.* 2000. Proteasome inhibition in neuronal cells induces a proinflammatory response manifested by upregulation of cyclooxygenase-2, its accumulation as ubiquitin conjugates, and production of the prostaglandin PGE(2). Arch. Biochem. Biophys. **374:** 325–333.
104. VAN BERGEN, P., P. RAUHALA, C.M. SPOONER, *et al.* 1999. Hemoglobin and iron evoked oxidative stress in the brain: protection by bile pigments, manganese and *S*-nitrosoglutathione. Free Rad. Res. **31:** 631–640.
105. LEONG, S.K. 1999. Capturing the sublimity of a free radical gas. Free Rad. Res. **31:** 619–629.
106. FACCHINETTI, F., M. SASATI, F.B. CUTTING, *et al.* 1999. Lack of involvement of neuronal nitric oxide synthase in the pathogenesis of a transgenic mouse model of familial amyotrophic lateral sclerosis. Neuroscience **90:** 1483–1492.
107. MOLINA, J.A., F.J. JIMENEZ-JIMENEX, M. ORTI-PAREJA, *et al.* 1998. The role of nitric oxide in neurodegeneration. Potential for pharmacological intervention. Drugs and Aging **12:** 251–259.
108. HULL, M., B.L. FIEBICH, G. SCHUMANN, *et al.* 1999. Anti-inflammatory substances—a new therapeutic option in Alzheimer's disease. Drug Discovery Today **4:** 275–282.
109. PARKS, J.K., T.S. SMITH, P.A. TRIMMER, *et al.* Neurotoxic Aβ peptides increase oxidative stress *in vivo* through NMDA-receptor and nitric-oxide-synthase mechanisms, and inhibit complex IV activity and induce a mitochondrial permeability transition *in vitro*. J. Neurochem. **76:** 1050–1056.
110. LUM, R.T., S.S. KERWAR, S.M. MEYER, *et al.* 1998. A new structural class of proteasome inhibitors that prevent NF-kappa B activation. Biochem. Pharmacol. **55:** 1391–1397.
111. DOU, Q.P. & S. NAM. 2000. Pharmacological proteasome inhibitors and their therapeutic potential. Exp. Opin. Ther. Patents **10:** 1263–1272.

A Chemical Perspective on the Interplay Between NO, Reactive Oxygen Species, and Reactive Nitrogen Oxide Species

MICHAEL G. ESPEY,[a] KATRINA M. MIRANDA,[a]
DOUGLAS D. THOMAS,[a] SANDHYA XAVIER,[a] DEBORAH CITRIN,[a]
MICHAEL P. VITEK,[b] AND DAVID A. WINK[a]

[a]*Radiation Biology Branch, National Cancer Institute, National Institutes of Health, Bethesda, Maryland, U.S.A.*

[b]*Department of Neurology, Duke University Medical Center, Durham, North Carolina, U.S.A.*

ABSTRACT: Nitric oxide (nitrogen monoxide, NO) plays a veritable cornucopia of regulatory roles in normal physiology. In contrast, NO has also been implicated in the etiology and sequela of numerous neurodegenerative diseases that involve reactive oxygen species (ROS) and nitrogen oxide species (RNOS). In this setting, NO is often viewed solely as pathogenic; however, the chemistry of NO can also be a significant factor in lessening injury mediated by both ROS and RNOS. The relationship between NO and oxidation, nitrosation, and nitration reactions is summarized. The salient factors that determine whether NO promotes, abates, or interconnects these chemistries are emphasized. From this perspective of NO chemistry, the type, magnitude, location, and duration of either ROS or RNOS reactions may be predicted.

KEYWORDS: oxidative stress; nitrosative stress; nitration; nitrite; peroxidase

INTRODUCTION

Redox biology plays important roles in various aspects of neurobiology, including development, control of cellular function, and neurotoxicity cascades. The reactive intermediates produced from reduction of molecular oxygen (O_2), such as superoxide (O_2^-) and hydrogen peroxide (H_2O_2), are known collectively as reactive oxygen species (ROS). Formation of ROS can result in DNA damage, protein oxidation and lipid peroxidation in a process termed oxidative stress. Oxidative stress has been implicated in the etiology and sequela of numerous neurodegenerative diseases, ranging from acute events, such as stroke, to chronic conditions, such as Alzheimer's disease,[1,2] Parkinson's disease,[3] and amyotrophic lateral sclerosis (ALS).[4,5] The consequence of oxidant and free radical formation in the central nervous system has prompted much investigation into the potential therapeutic benefits of antioxidants and free radical scavengers.

Address for correspondence: Michael Graham Espey, Ph.D., Radiation Biology Branch, National Cancer Institute, NIH Building 10, Room B3-B69, Bethesda, MD 20892, U.S.A. Voice: 301-496-7511; fax: 301-480-2238.
sp@nih.gov

Nitric oxide (nitrogen monoxide, NO) was added to the list of potentially deleterious free radicals soon after the discovery of its synthesis in biological systems. In 1988, Garthwaite and colleagues showed that NO produced by neurons could affect cGMP levels following exposure to the neurotransmitter glutamate.[6] Subsequent studies implicated overproduction of NO and formation of reactive nitrogen oxide species (RNOS) as pivotal features of the cytotoxic cascades leading to neuronal death and neurodegenerative pathology.[7] Peroxynitrite ($ONOO^-$) in particular has garnered much interest in the neuroscience community.[8]

NO, together with ROS, cannot be considered to be merely toxic byproducts; rather, they are integral components of basal metabolism and cellular function. The goal of this review is to emphasize that the formation of NO can ultimately lead to a variety of outcomes for a particular tissue region under either physiologic or pathophysiologic conditions. A multitude of factors may determine whether the result is beneficial or deleterious.[9] Of foremost consideration are the concentrations of NO and other reactants in the system. The effective concentration for each species depends on the balance between the rates of formation, diffusion from their origin, and consumption. Secondary to this is whether the nature of exposure is acute or chronic. Other factors to consider are characteristics of the target substance, including structure, concentration, location, metabolism, and relative roles in maintaining cellular function and health.

The chemistry of NO may be separated into two basic categories, direct effects or indirect effects, chiefly based on the concentration of NO.[10–12] This thesis, termed *the chemical biology of NO,* classifies the reactions of NO in the context of relevance to biological conditions.[12] Direct effects are reactions in which NO reacts directly with the biological target. Reactions of NO with metalloproteins, such oxyhemoglobin, to form a nitrosyl–heme complexes typify direct effects, which predominate at low NO fluxes. Indirect effects are those in which NO initially combines with another reactant, such as O_2 or O_2^-, prior to participating in chemical modification of biological targets. Both direct and indirect effects are components of basal metabolism; however, indirect effects may also result in either oxidative or nitrosative stress, notably under conditions of high and sustained NO flux. These stresses are defined as conditions in which significant numbers of important molecules become abnormally involved in either oxidation or nitrosation chemistry, due to the compromise of cellular defense mechanisms resulting in damage. The involvement of NO in oxidation and nitrosation, as well nitration chemistry, is discussed.

OXIDATION

Oxidation chemistry involves the loss of electrons from a compound and is a process in everyday biology, as well as acute and chronic tissue injury. Moderate to strong oxidants (electron acceptors) can be formed from the NO/O_2, NO/O_2^-, and HNO/O_2 reactions. The reaction between NO and O_2 is termed NO autoxidation. In the gas phase and in hydrophobic layers of cellular membranes, NO autoxidation reaction initially produces NO_2, which then further reacts with an additional NO molecule to form N_2O_3.[13,14]

$$2NO + O_2 \rightarrow 2NO_2 \qquad \text{(gas phase)} \qquad \textbf{(1)}$$

$$NO_2 + NO \leftrightarrow N_2O_3 \qquad (2)$$

With an oxidation potential of 1.2 V versus a standard hydrogen electrode, NO_2 is a strong oxidant that can mediate lipid peroxidation reactions. In aqueous solution, NO_2 is not formed to an appreciable extent during NO autoxidation to N_2O_3 (see Ref. 13 and references therein).

$$4NO + O_2 \rightarrow N_2O_3 \qquad \text{(aqueous phase)} \qquad (3)$$

$$N_2O_3 + H_2O \rightarrow 2HNO_2 \qquad (4)$$

Although N_2O_3 is a relatively mild oxidant with an oxidation potential of 0.7 V, NO autoxidation in either aqueous or gas phase does not cause appreciable DNA strand breaks or oxidize proteins.[15] Conversely, N_2O_3 is the quintessential species in most biological nitrosative conditions (discussed later in this paper).[9–16]

Appreciation of the reaction between NO and O_2^- led to a focused examination of $ONOO^-$ as a chief mediator of oxidative stress in the brain.[3] Decomposition of $ONOO^-$ is suggested to proceed through peroxynitrous acid (ONOOH, pKa 6.8), which results in formation of intermediates *trans*-ONOOH*, $\cdot NO_2$, and $\cdot OH$ that mediate either one- or two-electron oxidations of target molecules.[17–20]

$$NO + O_2^- \rightarrow ONOO^- \qquad (5)$$

$$ONOO^- + H^+ \leftrightarrow cis\text{-}ONOOH \qquad (6)$$

$$cis\text{-}ONOOH \rightarrow trans\text{-}ONOOH^* \qquad (7)$$

$$trans\text{-}ONOOH^* \rightarrow [NO_2^{\cdot} \mid \cdot OH] \rightarrow NO_3^- \qquad (8)$$

$$trans\text{-}ONOOH^* \rightarrow NO_3^- \qquad (9)$$

The full characterization of the intermediates of $ONOO^-$ decomposition and their relative contributions to oxidation *in vivo* persists as an area of debate. The question remains as to the distribution of an activated peroxynitrous acid, such as the *trans*-ONOOH* isomer and the geminal radical pair, $NO_2^{\cdot}$ and $\cdot OH$, **(8)**.

Unsaturated membrane lipids are particularly susceptible to the oxidative process of peroxidation where O_2^- and ferric ions (Fe^{3+}) conspire in a Haber-Weiss redox cycle to form $\cdot OH$ through the intermediacy of ferrous iron (Fe^{2+}) and H_2O_2. An $\cdot OH$ mechanism has also been proposed for lipid peroxidation initiated by exposure to synthetically prepared $ONOO^-$.[21] Damage, such as base modification and strand breaks, has been observed following exposure of purified DNA to synthetic $OONO^-$.[22,23] Amino acids, such as cysteine, methionine, and tryptophan, can be oxidized by $OONO^-$; however, particular attention has been paid to tyrosine (see also the NITRATION section).[17] Oxidation of protein tyrosines to form dityrosyl-bridges can cause structural distortion and pathogenic aggregates. Many studies have been performed with bolus applications of synthetic $OONO^-$/ONOOH to elucidate the potential deleterious effects of the NO/O_2^- reaction. Controlled delivery or *de novo* formation of these radicals, however, can lead to very different results.[24–26] These distinctions are examined in more detail in the NITRATION section.

Nitroxyl (HNO/NO^-) can form oxidizing intermediates following reaction with O_2. Potential sources of HNO are from the decomposition of *S*-nitrosothiols and metal nitrosyl enzyme complexes, as well as the oxidation of *N*-hydroxyarginine (NOHA), an intermediate of the arginine and nitric oxide synthase catalytic cycle (see Ref. 27 and references therein). A convenient method to study HNO behavior in biological conditions is the use of Angeli's salt (AS, $Na_2N_2O_3$).[27–29] This complex decomposes at neutral pH to HNO and NO_2^- with a halflife of approximately 2.5 minutes.[30]

$$N_2O_3^{2-} + H^+ \rightarrow HNO + NO_2^- \tag{10}$$

In vitro, the cytotoxicity of AS is similar to that of alkylhydroperoxides and is both O_2 dependent and metal independent.[31] Furthermore, the reactive intermediate derived from AS exhibited the unique ability to cause double strand breaks in DNA akin to ionizing radiation. The addition of HEPES to buffer can quench the formation of double strand breaks in DNA produced by AS exposure *in vitro,* but does not affect cytotoxicity as determined by clonogenic survival (unpublished results). This observation shows that the reactive intermediate(s) derived from AS act on multiple pathways to produce cell death. The reaction between triplet-state NO^- and O_2 forms $ONOO^-$, suggesting that AS decomposition may result in chemistry similar to that mediated by $ONOO^-$. However, the product of the HNO/O_2 reaction was found to elicit very different chemistry mediating two-electron oxidation reactions and phenolic hydroxylation exclusive of one-electron oxidation.[27] The relevance of this species is still being determined.[32]

Similar to the duality of NO, HNO may conduct seemingly paradoxic chemistries in biological systems. Short-term exposure of HEK cells expressing recombinant NMDA receptor subunits to AS in ærobic solution enhanced glutamatergic whole cell currents and blocked glycine-independent desensitization.[33] Administration of AS to conscious dogs strengthened the ionotropic properties of the heart by increasing contractility without increasing heart rate; an ideal property for treatment of coronary disease.[34] The mechanism may reside in HNO mediated release of the neuropeptide calcitonin gene-related peptide.

NO AS AN ANTIOXIDANT

The previous sections discussed the prooxidant aspects of NO. However, NO also has unique ability to scavenge and abate oxidative stress elicited by ROS and even by NO derived oxidants, such as NO_2 and $ONOO^-$. An antioxidant is a substance that either prevents oxidant formation or scavenges oxidants, thereby forming a less damaging or innocuous species resulting in abatement oxidative stress. Strong oxidants are produced through Fenton-type reactions of H_2O_2 with transition metal complexes.[35,36] Direct reaction of NO with metals form relatively stable metal–nitrosyl complexes thereby inhibiting their availability for reaction with H_2O_2. NO can also scavenge ROS in a competitive manner with potential targets. For example, lipid peroxidation perpetuates oxidative stress through formation of a variety of lipid–oxy and –peroxy adducts. NO can terminate the chain propagation process by forming non-radical end products (e.g., LOONO, see the NITROSYLATION section).[37]

NO can directly react at near diffusion control with O_2^- (**5**), converting a ROS into a RNOS.

Although RNOS, such as $ONOO^-$, NO_2, and HNO are generally less potent oxidants than those formed as a result of Fenton-type reactions, they can also promote oxidative stress.[36] An important concept here is that NO not only promotes RNOS formation, but also limits their oxidation chemistry with biological substrates.

$$trans\text{-}ONOOH^* + NO \rightarrow NO_2 + NO_2^- + H^+ \quad \textbf{(11)}$$

$$NO_2 + NO \rightarrow N_2O_3 \quad \textbf{(12)}$$

In biological systems, CO_2 is present at millimolar concentrations. A similar reaction series for $ONOO^-$ occurs with CO_2, where a nitrosoperoxocarbonate (also referred to as nitrosoperoxocarboxylate) intermediate ($ONOOCO_2^-$) is formed that can further react with NO to ultimately produce N_2O_3, (**12**)–(**14**).[39]

$$ONOO^- + CO_2 \rightarrow ONOOCO_2^- \quad \textbf{(13)}$$

$$ONOOCO_2^- + NO \rightarrow NO_2 + NO_2^- + CO_2 \quad \textbf{(14)}$$

Superoxide also scavenges oxidizing intermediates in the decomposition pathway of $ONOO^-$. However, this interaction simply abates oxidation without conversion to a nitrosating species. The oxidation chemistry of HNO can also be abated by NO.

$$HNO + NO \rightarrow HN_2O_2 \leftrightarrow HN_3O_3 \rightarrow N_2O + NO_2^- + H^+ \quad \textbf{(15)}$$

By these mechanisms, NO serves to abate the oxidation chemistry of both ROS and RNOS. RNOS oxidation chemistry may be strictly limited *in vivo* despite the findings to the contrary from studies employing a single species (e.g., reactions with preformed synthetic $ONOO^-$) under non-biological conditions. The versatile chemistry of NO provides antioxidant mechanisms through ligand–metal, radical–radical and RNOS interactions. Oxidation should be viewed as only a minor component of NO metabolism when one considers the balance of reactions that potentially may occur *in vivo*.

NITROSATION

Nitrosation is at the opposite end of the NO-redox spectrum from oxidation.[9] Although nitrosation is often incorrectly thought to be a form of oxidative stress, nitrosating species are formed at the expense of oxidizing species (e.g., conversion of $ONOO^-$ to N_2O_3, (**11**) and (**12**)). Oxidative stress results in the aberrant relocation of electrons from one molecule to another, forming reactive radical species that in some cases may lead to free radical chain reactions. In contrast, nitrosation results in formation of a covalent bound between a nitrosonium equivalent (NO^+), from N_2O_3 for example, and an appropriate nucleophilic amine, thiol, or hydroxyl residues (**16**)–(**18**).[38]

$$N_2O_3 + RSH \rightarrow RSNO + H^+ + NO_2^- \quad \textbf{(16)}$$

$$N_2O_3 + RR'NH \rightarrow RR'NNO + H^+ + NO_2^- \quad \textbf{(17)}$$

$$N_2O_3 + ROH \rightarrow RONO + H^+ + NO_2^- \quad \textbf{(18)}$$

Nitrosylation is often mistakenly used interchangeably with nitrosation. Nitrosylation is defined as the addition of NO without change in the formal charge of the substrate.

$$NO + M^{n+} \rightarrow M^{n+}\text{-}NO \quad \textbf{(19)}$$

Examples of nitrosylation include addition of NO to heme proteins to form iron nitrosyl complexes. Although these distinctions seem trivial, their ramifications are tremendous with respect to interpretation of biological effects.

Chain termination by NO of reactive radical species, such as those formed in lipid peroxidation, results in nitrosylated non-radicals and may be termed oxidative nitrosylation,

$$ROOH - e^- \rightarrow \cdot ROO \quad \textbf{(20)}$$

$$\cdot ROO + NO \rightarrow ROONO \quad \textbf{(21)}$$

Nitrosylation reactions are part of the direct effects scheme, whereas the primary nitrosating species are isomers of N_2O_3 derived from indirect reactions of NO with either O_2 or O_2^-, **(1)**–**(4)**, **(11)**, and **(12)**.[10,39,40]

The NO mediated conversion of the potential oxidation chemistry of $ONOO^-$ to a nitrosative outcome illustrates how the balances between these reactive intermediates (NO, O_2^-, $ONOO^-$, NO_2, and N_2O_3) is one the most important aspects in understanding the role of NO redox chemistry in biological systems. Under conditions where the ratio of NO to O_2^- is less than one, oxidation begins to increase until the rate of production for each radical is nearly equivalent **(12)**. As the rate of NO formation begins to increase beyond that of O_2^- production, oxidation is diminished as nitrosation concomitantly increases (see FIGURE 1). Therefore, adducts that typify NO autoxidation (e.g., *N*-nitrosamine, *S*-nitrosothiol) may be formed through the intermediacy of $ONOO^-$ provided the NO concentration exceeds that of O_2^-. This scheme suggests that relative concentrations of radicals as well as timing are the most critical determinants for the formation of either oxidizing or nitrosating intermediates.

The activation of activation of $p21^{ras}$ can be used to illustrate the concept of redox balances. Ras is a monomeric G-protein that is coupled with cell surface receptors and is involved in transducing growth and proliferation signals. Activation of these receptors with ligand (or other factors) triggers the GDP/GTP exchange function of ras through the recruitment of guanine exchange factor. The conformation of ras activated in this manner, in turn, permits interaction with a variety of transduction cascades involving mitogen-activated-, extracellular signal-related-, c-Jun-terminal-, and p38 MAP kinases that can lead to induction of proteins, such as COX-2, MCP-1, and VEGF.

Ras may be considered to be a redox sensor. Activation of ras by oxidative stress has been demonstrated in a variety of paradigms. Shear stress on the surface of endothelial cells results in translocation of rac-1 and activation of NAPDH oxidase, which in turn produces O_2^-/H_2O_2.[41] Under these conditions, endothelial ras transduces signals to result in the expression of proteins involved in leukocyte

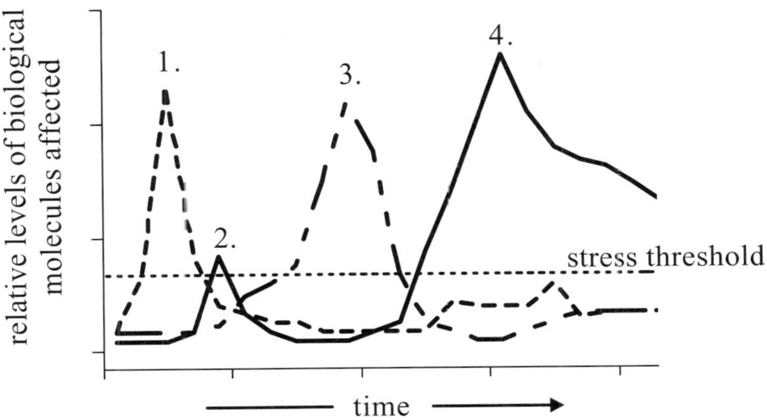

FIGURE 1. Hypothetical model to illustrate the role NO plays in balancing ROS and RNOS-elicited stresses. *1.* The initial insult begins with aberrant production of ROS resulting in oxidation that surpasses the normal defense mechanisms resulting in oxidative stress. *2.* As NO levels begin to increase, both oxidation and nitration are putatively mediated by formation of peroxynitrite. These chemistries subside with a further shift in the balance toward predomination of NO over ROS and peroxynitrite. In this phase, NO abates ROS oxidation and converts peroxynitrite toward a nitrosative end point. *3.* With further increases in NO flux relative to ROS and RNOS, the condition of nitrosative stress occurs via the NO autoxidation pathway. *4.* NO levels begin to decrease relieving nitrosylation-mediated inhibition, for instance, of heme-containing peroxidases. This phase is characterized by nitrative stress due to peroxidase-catalyzed formation of nitrogen dioxide from residual nitrite. KEY: ▪▪▪▪•, oxidative stress; ▪▪▬, nitrosative stress; ▬▬ , nitrative stress.

recruitment, adhesion, and membrane integrity. Low fluxes of NO delivered by synthetic NO donor compounds or by transfected eNOS ameliorate activation of ras mediated by ROS.[41] In contrast, high concentrations of NO donors (millimolar) under ærobic conditions resulted in direct activation of ras via formation of *S*-nitrosothiols.[42,43] These findings illustrate how the balance between NO, ROS, and RNOS can result in either the activation or ablation of signal transduction pathways. The degree to which one will dominate resides in the rates of formation and consumption of each species, as well as spatial and temporal considerations.[9,13,44]

NITRATION

In this section, the fact that NO formation is starting point for most pathways resulting in nitration is emphasized exclusive of nitrite (NO_2^-) derived from bacterial and dietary sources. Nitration involves an electrophilic addition of a NO_2^+ equivalent (nitronium, an electron acceptor) for instance, to an aromatic ring (the site of electron density). Nitration moved beyond the realm of industrial organic synthesis into the biological sciences following development of antibodies that could detect 3-nitrotyrosine (3-NT). These antibodies revealed the extent of nitration in tissue samples derived from a variety of pathological states.[20,45–48] It should be stressed that the degree to which 3-NT represents causal etiopathology or defense and repair processes remains an open question. The challenge for understanding the role of nitration in a particular pathologic state is determination of the predominant pathways that NO travels to elicit this chemistry.

Detection of 3-NT following treatment of free tyrosine, proteins, cells and tissue with bolus applications of synthetic preformed $ONOO^-$ raised the possibility that the NO/O_2^- reaction was the source of nitration observed in biological systems. In general, nitration yields are quite low despite the use of large quantities of synthetic preformed $ONOO^-$.[25,26] Subsequent studies demonstrated that the addition of transition metals (e.g., Fe^{3+}) or the presence of CO_2 to solutions containing free tyrosine enhanced 3-NT formation with synthetic preformed $ONOO^-$, **(13)** and **(14)**.[49–51]

A cornerstone of the argument advocating $ONOO^-$ as the predominant pathway for nitration *in vivo* is the fact that the two radicals NO and O_2^- react with each other at near diffusion control.[52] As mentioned in the NITROSATION section, recapitulation of the chemistry observed with bolus application of preformed synthetic $ONOO^-$ *in vitro* would require that a 1 : 1 ratio between NO and O_2^- be precisely maintained at significantly high level.[53] A tip in the balance from O_2^- to NO would favor the similarly fast reaction between ONOOH and NO, **(11)** and **(12)**; alternatively through decomposition to NO_2, **(8)**, resulting in chemistry of the nitrosation pathway.

Several other pathways in addition to nitration initiated by $ONOO^-$ result in formation of 3-NT *in vitro*. Heme-containing peroxidases are proteins that catalyze the H_2O_2-mediated oxidation of a wide variety of substrates. Initially, H_2O_2 reacts with the heme forming a hypervalent iron–oxo species (e.g., Fe^{4+} or Fe^{5+}) that oxidizes NO_2^- to the nitrating species NO_2.[48,54] This suggests that Fenton-type chemistry may be important in the formation of these adducts by serving as an initiator of the reaction. In addition to its role as an oxidant, HNO may play an important role in nitration through mechanisms that appears to be unique from either $ONOO^-$ or NO_2.[27,55]

Another intriguing metal-mediated mechanism for nitration involves copper–zinc superoxide dismutase (Cu/ZnSOD). Although the chief function of this SOD isoenzyme is to catalyze conversion of O_2^- to H_2O_2 and O_2, SOD can also exhibit peroxidase activity that is most notably enhanced by mutated forms in individuals with ALS.[56] This can result in the putative formation of OH and/or conversion of copper to a hypervalent state, followed by oxidation of NO_2^- to NO_2.[57]

The final nitration mechanism discussed involves a specific type of nitrosation reaction termed *O*-nitrosation. A nitrosating agent, such as N_2O_3, adds a NO^+ equivalent to the hydroxy group of phenolic compounds (e.g., tyrosine) forming an *O*-nitroso complex.[58] This adduct can spontaneously rearrange to form 3-nitrotyrosine.[59] These reaction mechanisms further illustrate that nitrosation and nitration are not mutually exclusive events, but are related in a continuum or circular fashion and can lead to identical products. Several components of the NO autoxidation pathway (NO_2, N_2O_3, and NO_2^-) may ultimately result in nitration of targets preceding, following, or in lieu of either oxidation or nitrosation. As mentioned in the INTRODUCTION, the order, magnitude, and ultimate impact of these chemistries on the individual depends on factors related, not only to NO, but also to the composition of the tissue bed. This may be quite dynamic with respect to the levels of effector and defense molecules for a given disease course.

CONCLUSION

A casual reader of the NO literature can quickly become confused by what often appears to be contradictory actions attributed to this diatomic radical. Indeed, disparity often arises from diverse experimental approaches, reagents, conditions, and interpretation of results. In this brief report, we have attempted to illustrate that the chemistry initiated by NO should not be considered linear toward a particular chemical mode of action whether oxidation, nitrosation, nitrosylation, or nitration. The major challenge that lies before us is to understand and manipulate the various mechanisms in which NO acts to interconnection these chemistries.

ACKNOWLEDGMENT

We dedicate this work to the memory of Dr. Daniel L. Gilbert. He was a source of inspiration and friendship to all who knew him.

REFERENCES

1. COLTON, C.A., O.N. CHERNYSHEV, D.L. GILBERT & M.P. VITEK. 2000. Microglial contribution to oxidative stress in Alzheimer's disease. Ann. N.Y. Acad. Sci. **899:** 292–307.
2. SMITH, M.A., A. NUNOMURA, X. ZHU, *et al.* 2000. Metabolic, metallic, and mitotic sources of oxidative stress in Alzheimer disease. Antioxid. Redox Signal **2:** 413–420.
3. CHIUEH, C.C., R.M. WU, R.K.P. MOHANAKUMA, *et al.* 1994. *In vivo* generation of hydroxyl radicals and MPTP-induced dopaminergic toxicity in the basal ganglia. Ann. N.Y. Acad. Sci. **738:** 25–36.
4. VALENTINE, J.S., P.J. HART & E.B. GRALLA, 1999. Copper-zinc superoxide dismutase and ALS. Adv. Exp. Med. Biol. **448:** 193–203.
5. ESTEVEZ, A.G., N. SPEAR, S.M. MANUEL, *et al.* 1998. Role of endogenous nitric oxide and peroxynitrite formation in the survival and death of motor neurons in culture. Prog. Brain Res. **118:** 269–280.
6. GARTHWAITE, J., S.L. CHARLES & R. CHESS-WILLIAMS. 1988. Endothelium-derived relaxing factor release on activation of NMDA receptors suggests a role as intercellular messenger in the brain. Nature **336:** 385–388.
7. DAWSON, D.A. 1994. Nitric oxide and focal cerebral ischemia: multiplicity of actions and diverse outcome. Cerebrovasc. Brain Metab. Rev. **6:** 299–324.
8. BECKMAN, J.S. 1991. The double-edged role of nitric oxide in brain function and superoxide-mediated injury. J. Dev. Physiol. **15:** 53–59.
9. ESPEY, M.G., K.M. MIRANDA, M. FEELISCH, *et al.* 2000. Mechanisms of cell death governed by the balance between nitrosative and oxidative stress. Ann. N.Y. Acad. Sci. **899:** 209–221.
10. MIRANDA, K.M., M.G. ESPEY, D. JOURD'HEUIL, *et al.* 2000. The chemical biology of nitric oxide. *In* Nitric Oxide. 41–56. Academic Press, San Diego.
11. MIRANDA, K.M., M.G. ESPEY & D.A. WINK. 2000. A discussion of the chemistry of oxidative and nitrosative stress in cytotoxicity. J. Inorg. Biochem. **79:** 237–240.

12. WINK, D.A. & J.B. MITCHELL. 1998. The chemical biology of nitric oxide: insights into regulatory, cytotoxic and cytoprotective mechanisms of nitric oxide. Free Rad. Biol. Med. **25:** 434–456.

13. ESPEY, M.G., K.M. MIRANDA, D.D. THOMAS & D.A. WINK. 2001. Distinction between nitrosating mechanisms within human cells and aqueous solution. J. Biol. Chem. **276:** 30085–30091.

14. WINK, D.A., I. HANBAUER, M.B. GRISHAM, *et al.* 1996. The chemical biology of NO. Insights into regulation, protective and toxic mechanisms of nitric oxide. Cur. Top. Cell. Reg. **34:** 159–187.

15. WINK, D.A., Y. VODOVOTZ, J. LAVAL, *et al.* 1998. The multifaceted roles of nitric oxide in cancer. Carcinogenesis **19:** 711–721.

16. GRISHAM, M.B. & A.M. MILES. 1994. Effects of aminosalicylates and immunosuppressive agents on nitric oxide dependent N-nitrosation reactions. Biochem. Pharmacol. **47:** 1897–1902.

17. PRYOR, W.A. & G.L. SQUADRITO. 1996. The chemistry of peroxtynitrite and peroxynitrous acid: Products from the reaction of nitric oxide with superoxide. Am. J. Phys. **268:** L699–721.

18. CODDINGTON, J.W., S. WHERLAND & J.K. HURST. 2001. Pressure dependence of peroxynitrite reactions. Support for a radical mechanism. Inorg. Chem. **40:** 528–532.

19. KOPPENOL, W.H. 1999. Chemistry of peroxynitrite and its relevance to biological systems. Metal Ions Biol. Syst. **36:** 597–619.

20. RADI, R., A. DENICOLA, B. ALVAREZ, *et al.* 2000. The biological chemistry of peroxynitrite. *In* Nitric Oxide. 57–82. Academic Press, San Diego.

21. RUBBO, H., R. RADI, M. TRUJILLO, *et al.* 1994. Nitric oxide regulation of superoxide and peroxynitrite dependent lipid peroxidation: formation of novel nitrogen containing oxidized lipid derivatives. J. Biol. Chem. **269:** 26066–26075.

22. KING, P.A., V.E. ANDERSON, J.O. EDWARDS, *et al.* 1992. A stable solid that generates hydroxyl radical upon dissolution in aqueous solution: reactions with proteins and nucleic acids. J. Am. Chem. Soc. **114:** 5430.

23. KING, P.A., E. JAMISON, D. STRAHS, *et al.* 1993. Footprinting proteins on DNA with peroxonitrous acid. Nucleic Acids Res. **21:** 2473–2478.

24. PFEIFFER, S. & B. MAYER. 1998. Lack of tyrosine nitration by peroxynitrite generated at physiological pH. J. Biol. Chem. **273:** 27280–2725.

25. REITER, C.D., R.J. TENG & J.S. BECKMAN. 2000. Superoxide reacts with nitric oxide to nitrate tyrosine at physiological pH via peroxynitrite. J. Biol. Chem. **275:** 32460–32466.

26. PFEIFFER, S., K. SCHMIDT & B. MAYER. 2000. Dityrosine formation outcompetes tyrosine nitration at low steady-state concentrations of peroxynitrite. Implications for tyrosine modification by nitric oxide/superoxide *in vivo*. J. Biol. Chem. **275:** 6346–6352.

27. MIRANDA, K.M., M.G. ESPEY, K. YAMADA, *et al.* 2001. Unique oxidative mechanisms for the reactive nitrogen oxide species, nitroxyl anion. J. Biol. Chem. **276:** 1720–1727.

28. FUKUTO, J.M., A.J. HOBBS & L.J. IGNARRO, 1993. Conversion of Nitroxyl (NO^-) in biological systems: the role of physiological and relevance to the biological activity of HNO. Biochem. Biophys. Res. Commun. **196:** 707–713.

29. FEELISCH, M., M. TE POOL, R. ZAMORA, *et al.* 1994. Understanding the controversy over the identity of EDRF. Nature **368:** 62–64.

30. MARAGOS, C.M., D. MORLEY, D.A. WINK, *et al.* 1991. Complexes of NO with nucleophiles as agents for the controlled biological release of nitric oxide. Vasorelaxant effects. J. Med. Chem. **34:** 3242–3247.

31. WINK, D.A., M. FEELISCH, J. FUKUTO, *et al.* 1998. The cytotoxicity of nitroxyl: possible implications for the pathophysiological role of NO. Arch. Biochem. Biophys. **351:** 66–74.

32. BARTBERGER, M.D., J.M. FUKUTO & K.N. HOUK. 2001. On the acidity and reactivity of HNO in aqueous solution and biological systems. Proc. Natl. Acad. Sci. U.S.A. **98:** 2194–2198.

33. COLTON, C.A., M. GBADEGESIN, D.A. WINK, *et al.* 2001. Nitroxyl anion regulates glycine-independent desensitization of the NMDA receptor. J. Neurochem. **78:** 1126–1134.
34. PAOLOCCI, N., W.F. SAAVEDRA, K.M. MIRANDA, *et al.* 2001. Nitroxyl anion exerts redox-sensitive positive cardiac inotropy *in vivo* by calcitonin gene-related peptide signaling. Proc. Natl. Acad. Sci. U.S.A. **98:** 10463–10468.
35. WINK, D.A., Y. VODOVOTZ, M.B. GRISHAM, *et al.* 1999. Antioxidant effects of nitric oxide. Method. Enzymol. **301:** 413–424.
36. WINK, D.A., K.M. MIRANDA, M.G. ESPEY, *et al.* 2001. Mechanisms of the antioxidant effects of nitric oxide. Antioxid. Redox Signal **3:** 203–213.
37. DARLEY-USMAR, V.M., R.P. PATEL, V.V. O'DONNEL & B.A. FREEMAN. 2000. Antioxidant actions of nitric oxide. *In* Nitric Oxide. 265–276. Academic Press, San Diego.
38. WILLIAMS, D.L.H. 1988. *In* Nitrosation. Cambridge Press, Oxford.
39. JOURD'HEUIL, D., K.M. MIRANDA, S.M. KIM, *et al.* 1999. The oxidative and nitrosative chemistry of the nitric oxide/superoxide reaction in the presence of bicarbonate. Arch. Biochem. Biophys. **365:** 92–100.
40. WINK, D.A., J.A. COOK, S. KIM, *et al.* 1997. Superoxide modulates the oxidation and nitrosation of thiols by nitric oxide derived reactive intermediates. J. Biol. Chem. **272:** 11147–11151.
41. CHIU, J.J., B.S. WUNG, H.J. HSIEH, *et al.* 1999. Nitric oxide regulates shear stress-induced early growth response-1. Expression via the extracellular signal-regulated kinase pathway in endothelial cells. Circ. Res. **85:** 238–246.
42. LANDER, H.M., J.S. OGISTE, S.F. PEARCE, *et al.* 1995. Nitric oxide-stimulated guanine nucleotide exchange on p21ras. J. Biol. Chem. **270:** 7017–7020.
43. YUN, H.Y., M. GONZALEZ-ZULUETA, V.L. DAWSON & T.M. DAWSON. 1998. Nitric oxide mediates *N*-methyl-D-aspartate receptor-induced activation of p21ras. Proc. Natl. Acad. Sci. U.S.A. **95:** 5773–5778.
44. ESPEY, M.G., K.M. MIRANDA, R.M. PLUTA & D.A. WINK. 2000. Nitrosative capacity of macrophages is dependent on nitric-oxide synthase induction signals. J. Biol. Chem. **275:** 11341–11347.
45. CROW, J.P. & H. ISCHIROPOULOS. 1996. Detection and quantitation of nitrotyrosine residues in proteins: *in vivo* marker of peroxynitrite. Method. Enzymol. **269:** 185–194.
46. MACMILLAN-CROW, L.A., J.P. CROW, J.D. KIRBY, *et al.* 1996. Nitration and inactivation of manganese superoxide dismutase in chronic rejection of human renal allografts. Proc. Natl. Acad. Sci. U.S.A. **93:** 11853–11858.
47. SOUZA, J.M., E. DAIKHIN, M. YUDKOFF, *et al.* 1999. Factors determining the selectivity of protein tyrosine nitration. Arch. Biochem. Biophys. **371:** 169–178.
48. EISERICH, J.P., M. HRISTOVA, C.E. CROSS, *et al.* 1998. Formation of nitric oxide-derived inflammatory oxidants by myeloperoxidase in neutrophils. Nature **391:** 393–397.
49. BECKMAN, J.S., H. ISCHIROPOULOS, L. ZHU, *et al.* 1992. Kinetics of superoxide dismutase- and iron-catalyzed nitration of phenolics by peroxynitrite. Arch. Biochem. Biophys. **298:** 438–445.
50. LYMAR, S.V., Q. JIANG & J.K. HURST. 1996. Mechanism of carbon dioxide-catalyzed oxidation of tyrosine by peroxynitrite. Biochem. **35:** 7855–7861.
51. GOW, A., D. DURAN, S.R. THOM & H. ISCHIROPOULOS. 1996. Carbon dioxide enhancement of peroxynitrite-mediated protein tyrosine nitration. Arch. Biochem. Biophys. **333:** 42–48.
52. HUIE, R.E. & S. PADMAJA. 1993. The reaction of NO with superoxide. Free Radic. Res. Commun. **18:** 195–199.
53. MILES, A.M., D.S. BOHLE, P.A. GLASSBRENNER, *et al.* 1996. Modulation of superoxide-dependent oxidation and hydroxylation reactions by nitric oxide. J. Biol. Chem. **271:** 40–47.
54. SAMPSON, J.B., Y. YE, H. ROSEN & J.S. BECKMAN, 1998. Myeloperoxidase and horseradish peroxidase catalyze tyrosine nitration in proteins from nitrite and hydrogen peroxide. Arch. Biochem. Biophys. **356:** 207–213.

55. CHAZOTTE-AUBERT, L., S. OIKAWA, I. GILIBERT, *et al.* 1999. Cytotoxicity and site-specific DNA damage Induced by nitroxyl anion (NO^-) in the presence of hydrogen peroxide. Implications for various pathophysiological conditions. J. Biol. Chem. **274:** 20909–20915.
56. WIEDAU-PAZOS, M., J.J. GOTO, S. RABIZADEH, *et al.* 1996. Altered reactivity of superoxide dismutase in familial amyotrophic lateral sclerosis. Science **271:** 515–518.
57. SINGH, R.J., S.P. GOSS, J. JOSEPH & B. KALYANARAMAN. 1998. Nitration of gamma-tocopherol and oxidation of alpha-tocopherol by copper-zinc superoxide dismutase/H_2O_2/NO_2^-: role of nitrogen dioxide free radical. Proc. Natl. Acad. Sci. U.S.A. **95:** 12912–12917.
58. WINK, D.A., R.W. NIMS, J.F. DARBYSHIRE, *et al.* 1994. Reaction kinetics for nitrosation of cysteine and glutathione in aerobic nitric oxide solutions at neutral pH. Insights into the fate and physiological effects of intermediates generated in the NO/O_2 reaction. Chem. Res. Toxicol. **7:** 519–525.
59. SIMON, D.I., M.E. MULLINS, L. JIA, *et al.* 1996. Polynitrosylated proteins: characterization, bioactivity, and functional consequences. Proc. Natl. Acad. Sci. U.S.A. **93:** 4736–4741.

Nitric Oxide and Superoxide, a Deadly Cocktail

ALVARO G. ESTÉVEZ[a] AND JOAQUÍN JORDÁN[b]

[a]*Department of Physiology and Biophysics, University of Alabama at Birmingham, Birmingham, Alabama, U.S.A*

[b]*Departmento Ciencias Médicas, Facultad de Medicina, Universidad de Castilla–La Mancha, Albacete, Spain*

Abstract: Free radicals and oxidants may have contrasting effects on cells. Nitric oxide is a cellular messenger that acts by several mechanisms, including activation of soluble guanylate cyclase, nitrosylation of thiols, and formation of peroxynitrite. The action of nitric oxide depends on oxidative conditions in the cell. In motor neurons, nitric oxide enhances brain-derived neurotrophic factor effects by stimulation of cGMP production. In the absence of trophic factors, nitric oxide induces motor neuron apoptosis by reacting with superoxide to form peroxynitrite. The mechanism of induction of motor neuron apoptosis by the formation of peroxynitrite is more complicated than simply the formation of a strong oxidant and it also involves decreased formation of cGMP.

Keywords: nitric oxide; peroxynitrite superoxide

The biology of free radicals and oxidants with short halflives and diffusion-limiting reactions is better described by fluxes in production and rates of reaction than by concentrations or affinity constants. The rate of reaction gives information about the relative contribution of each possible reaction to the final outcome for a specific radical species. The rate of reaction of superoxide with superoxide dismutase (SOD) is 2×10^9 $M^{-1}sec^{-1}$, and the rate of reaction of superoxide with nitric oxide is $6–10\times10^9$ $M^{-1}sec^{-1}$. These numbers indicate that if the concentration of nitric oxide (NO) is five times lower than the concentration of superoxide dismutase, 50% of the superoxide produced will react with NO to form peroxynitrite. An additional requirement for this competition to occur is that the superoxide dismutase must be present in the same compartment in which, and when, the production of superoxide takes place. Furthermore, to determine how relevant a specific reaction may be in the cell, the rates of reaction as well as the diffusion properties of the molecules involved must be considered. In addition, very low flux of an oxidant could damage a cell over hours or days, since the products of oxidative damage can accumulate in cells. For these reasons, free radicals and oxidants traditionally have been considered byproducts of oxidative metabolism with toxic properties that can be lethal to cells by causing *non*-specific damage to all macromolecules.

However, not all reactions involving free radicals are damaging to cells. The identification of NO as an endothelial-derived relaxing factor (EDRF) was the first example

Address for correspondence: Alvaro G. Estévez, Department of Physiology and Biophysics, The University of Alabama at Birmingham, MCLM 850, 1918 University Boulevard, Birmingham, Alabama 35294-0005, U.S.A. Voice: 205–934–4271; fax: 205-975-7679.
A.G. Estévez: estevez@physiology.uab.edu
J. Jordán: jjb@med-ab.uclm.es

of a physiological non-toxic biological activity of a free radical. More recent reports show that reactive oxygen species (ROS) such as superoxide and hydrogen peroxide are used as signaling molecules by cells, adding a new group of compounds to the second messenger list.

Nitric oxide is a free radical synthesized from the conversion of arginine to citrulline by a family of nitric oxide synthases (NOS). Three isoforms of this flavoprotein have been purified and cloned from mammalian cells: neuronal, inducible, and endothelial NOS, also known as NOS I, II, and III, respectively. Multiple isoforms of the enzyme may be expressed in the same tissue and even in a single cell. Neuronal and endothelial NOS isoforms are regulated by a variety of different stimuli in different cell types via calcium/calmodulin activation, as well as by *de novo* synthesis. In contrast, the activity of inducible NOS is regulated solely by the *de novo* synthesis of the enzyme in a Ca^{2+}-independent manner. It has been known for some time that NO can have opposite effects on various biological systems. Nitric oxide is used as a cellular messenger by the vascular and nervous systems but as a toxin used by the immune system to kill bacteria and in viral infections. The first signaling pathway described for NO involved the activation of soluble guanylate cyclase in the smooth muscle of circulatory vessels. Shortly afterward, it was found that glutamate stimulates the production of cyclic GMP via nitric oxide production in the nervous system. At the same time, Dawson and collaborators found that the toxicity of glutamate was mediated by NO.[1] These and other experiments showed that, depending on the conditions of its production, NO could become cytotoxic and mediate damage in pathological conditions.[2] Since the original studies, growing evidence has confirmed that NO may act as a neuroprotective agent by stimulating cGMP synthesis or nitrosylation. Alternatively, NO may mediate neuronal injury through the formation of secondary oxidants or by inhibition of mitochondrial respiration.[3]

However, NO is a relatively stable and non-reactive free radical that under physiologically, and even pathophysiologically, relevant conditions does not exert toxic effects on cells by itself.[4] It can react with superoxide, producing the potent oxidant peroxynitrite,[5] which can induce apoptosis or necrosis, depending on the concentration.[6,7] Additional evidence that NO toxicity is mediated, at least in part, by peroxynitrite comes from experiments performed in PC12 cells and sympathetic neurons. Trophic factor deprivation induces apoptosis in both of these cell types, and can be prevented by addition of NO donors by a mechanism involving cyclic GMP production.[8] However, in the presence of trophic factors, downregulation of superoxide dismutase induces PC12 apoptosis by a mechanism requiring NO and peroxynitrite formation.[9]

Superoxide radicals are formed by the one electron reduction of molecular oxygen. Its sources include NOS, NADPH oxidases, cyclo/oxygenase, lipoxygenase, xanthine oxidase, and the mitochondrial electron transport chain.[10] Neuronal superoxide production increases after activation of excitatory amino acid receptors.[11–13] Superoxide may oxidize and reduce other compounds and potentially forms more reactive species via secondary reactions. The rapid reaction of superoxide with SOD forms H_2O_2 and, potentially, secondary radicals such as $O_2^{-\cdot}$ are generated via the Haber-Weiss reaction.[10] However, the Fenton chemistry is too slow to compete with other possible reactions of superoxide and hydrogen peroxide.[4,14,15] Every cell contains antioxidant defenses to control reactive oxygen species (ROS) levels, including superoxide dismutase (SOD), catalase, and glutathione. When the equilibrium of oxidant/antioxidant is disturbed, ROS may modify proteins, lipids, and DNA, leading to catalytic activity changes, deregulation of metabolic function, lipoperoxidation, and activation of cell death pathways.[16,17] More recently, increased production of ROS has been reported during normal physiological processes, such as tyrosine phosphorylation of certain proteins induced by platelet-derived growth factor or epidermal growth factor (for a review see Ref. 18).

Production of ROS is also deleterious to the cell, since treatment with antioxidants prevents cell death in different models of injury. Rat superior cervical ganglion neurons require the presence of NGF to develop and survive in culture. NGF-deprivation of sympathetic neurons in culture leads to apoptosis, which is prevented by overexpression of human wildtype copper/zinc-SOD.[19,20] Similarly, trophic factor deprivation induces apoptosis in cultured motor neurons, which can be prevented by liposome-delivered Cu, Zn SOD, or by a membrane-permeant superoxide mimetic, but not by extracellular addition of Cu, Zn SOD.[21,22] However, inhibition of nitric oxide production also prevents motor neuron apoptosis induced by trophic factor deprivation without affecting superoxide production,[21,22] suggesting that neither NO nor superoxide by itself is sufficient to stimulate motor neuron apoptosis. Indeed, in motor neuron cultures, NO functions as a survival signal, thus NOS inhibition in brain-derived neurotrophic factor (BDNF)-treated cultures induces cell death, which is overcome by exogenous NO production.[23] This apparent paradox has been attributed to alternate redox states of nitrogen species, whereby, depending on the redox milieu, they may be neuroprotective via nitrolsylation chemistry.[24]

Motor neuron cultures offer a good model in which to study these opposite effects of NO because, under various conditions, NO participates in the induction of apoptosis and enhances survival.[21,23] As already described, trophic factor deprivation induces motor neuron apoptosis by a mechanism that involves induction of neuronal NOS and the increased production of superoxide and NO, which leads to peroxynitrite formation.[21] Based on these results, we conclude that formation of peroxynitrite is responsible for the induction of apoptosis. This interpretation agrees with previous studies showing that peroxynitrite induces apoptosis in various cell types.[6,7] However, peroxynitrite formation may be "scavenging" NO from other possible protective actions, such as nitrosylation of cysteine residues or activation of soluble guanylate cyclase to produce cGMP.

In summary, the actions of NO depend on the oxidative state of the cell. Some effects may be explained by the formation of more reactive derivatives, such as peroxynitrite or nitrogen dioxide, whereas others may be the consequence of simple competition for NO. This competition may result in the inhibition of cellular pathways that are necessary for cell survival. In conclusion, the mechanisms of NO-mediated toxicity are more complicated than previously suspected, and the direct actions of strong oxidants may not be as relevant as the formation of the oxidant itself.

REFERENCES

1. DAWSON, V.L., T.M. DAWSON, E.D. LONDON, *et al.* 1991. Nitric oxide mediates glutamate neurotoxicity in primary cortical cultures. Proc. Natl. Acad. Sci. U.S.A. **88:** 6368–6371.
2. MURPHY, M.P. 1999. Nitric oxide and cell death. Biochem. Biophys. Acta **1441:** 401–414.
3. LIPTON, S.A. 1999. Neuronal protection and destruction by NO. Cell Death Differ. **6:** 943–951.
4. BECKMAN, J.S. & W.H. KOPPENOL. 1996. Nitric oxide, superoxide, and peroxynitrite—the good, the bad, and the ugly. Am. J. Physiol. **271** (Cell Physiol. 40): C1424–C1437.
5. BECKMAN, J.S., T.W. BECKMAN, J. CHEN, *et al.* 1990. Apparent hydroxyl radical production by peroxynitrite: Implications for endothelial injury by nitric oxide and superoxide. Proc. Natl. Acad. Sci. U.S.A. **87:** 1620–1624.
6. BONFOCO, E., D. KRAINC, M. ANKARCRONA, *et al.* 1995. Apoptosis and necrosis: two distinct events induced, respectively, by mild and intense insults with *N*-methyl-D-aspartate or nitric oxide/superoxide in cortical cell cultures. Proc. Natl. Acad. Sci. U.S.A. **92:** 7162–7166.
7. ESTÉVEZ, A.G., R. RADI, L. BARBEITO, *et al.* 1995. Peroxynitrite-induced cytotoxicity in PC12 cells: evidence for an apoptotic mechanism differentially modulated by neurotrophic factors. J. Neurochem. **65:** 1543–1550.
8. FARINELLI, S.E., D.S. PARK & L.A. GREENE. 1996. Nitric oxide delays the death of trophic factor-deprived PC12 cells and sympathetic neurons by a cGMP-mediated mechanism. J. Neurosci. **16:** 2325–2334.
9. TROY, C.M., D. DEROSSI, A. PROCHIANTZ, *et al.* 1996. Downregulation of Cu/Zn superoxide dismutase leads to cell death via the nitric oxide-peroxynitrite pathway. J. Neurosci. **16:** 253–261.
10. KEHRER, J.P. 2000. The Haber-Weis reaction and mechanism of toxicity. Toxicology **149:** 43–50.
11. LAFON-CAZAL, M., S. PIETRI, M. CULCASI & J. BOCKAERT. 1993. NMDA-dependent superoxide production and neurotoxicity. Nature **364:** 535–537.
12. LAFON-CAZAL, M., M. CULCASI, F. GAVEN, *et al.* 1993. Nitric oxide, superoxide and peroxynitrite: putative mediator of NMDA-induced cell death in cerebellar granule cells. Neuropharmacology **32:** 1259–1266.
13. GONZALEZ-ZULUETA, M., L.M. ENSZ, G. MUKHINA, *et al.* 1998. Manganese superoxide dismutase protects nNOS neurons from NMDA and nitric oxide-mediated neurotoxicity. J. Neurosci. **18:** 2040–2055.
14. BECKMAN, J.S. 1994. *In* The Neurobiology of ·NO and ·OH. C.C. Chiueh, D.L. Gilbert & C.A. Colton, Eds.: 69–75. New York Academy of Sciences, New York.
15. BECKMAN, J.S. & J.P. CROW. 1993. Pathological implications of nitric oxide, superoxide and peroxynitrite formation. Biochem. Soc. Trans. **21:** 330–334.
16. JORDAN, J., M.F. GALINDO, J.H. PREHN, *et al.* 1997. p53 Expression induces apoptosis in hippocampal pyramidal neuron cultures. J. Neurosci. **17:** 1397–1405.
17. JORDAN, J.J., M.F. GALINDO, S. CALVO, *et al.* 2000. Veratridine induces apoptotic death in bovine chromaffin cells through superoxide production. Br. J. Pharmacol **130**(7): 1496–1504.
18. FINKEL, T. 2000. Redox-dependent signal transduction. FEBS Lett. **476:** 52–54.
19. JORDAN, J., G.D. GHADGE, J.H.M. PREHN, *et al.* 1995. Expression of human copper/zinc-superoxide dismutase inhibits the death of rat sympathetic neurons caused by withdrawal of nerve growth factor. Mol. Pharmacol. **47:** 1095–1100.
20. GREENLUND, L.J., T.L. DECKWERTH & E.M. JOHNSON. 1995. Superoxide dismutase delays neuronal apoptosis: a role for reactive oxygen species in programmed neuronal death. Neuron **14:** 303–315.
21. ESTÉVEZ, A.G., N. SPEAR, S.M. MANUEL, *et al.* 1998. Nitric oxide and superoxide contribute to motor neuron apoptosis induced by trophic factor deprivation. J. Neurosci. **18:** 923–931.

22. ESTÉVEZ, A.G., J.B. SAMPSON, Y.-X. ZHUANG, *et al.* 2000. Liposome-delivered superoxide dismutase prevents nitric oxide-dependent motor neuron death induced by trophic factor withdrawal. Free Rad. Biol. Med. **28:** 437–446.
23. ESTÉVEZ, A.G., N. SPEAR, J.A. THOMPSON, *et al.* 1998. Nitric oxide dependent production of cGMP supports the survival of rat embryonic motor neurons cultured with brain derived neurotrophic factor. J. Neurosci. **18:** 3708–3714.
24. CHIUEH, C.C. 1999. Neuroprotective properties of nitric oxide. Ann. N.Y. Acad. Sci. **890:** 301–311.

Apolipoprotein-E Allele-Specific Regulation of Nitric Oxide Production

CAROL A. COLTON, CANDICE M. BROWN,
MEGGAN CZAPIGA, AND MICHAEL P. VITEK

Division of Neurology, Duke University Medical Center, Durham North Carolina, U.S.A.

ABSTRACT: Cognitive decline and dementia are key features of Alzheimer's disease (AD) that result from failure of neuronal function. Affected neurons demonstrate indices of nitrosative stress resulting from changes in nitric oxide (NO) mediated redox balance. Neurofibrillary tangles, a characteristic neuropathologic feature of AD, and dysfunctional neurons frequently display 3-nitrotyrosine or other markers of nitrosative stress and immunoreactive nitric oxide synthase (NOS), suggesting that NOS-containing neurons are affected in AD. Our previous studies showed that apolipoprotein E treatment of macrophages increased NO production. Using transgenic mouse models expressing human apoE2, apoE3, or apoE4 protein isoforms and no mouse apoE, we now report an isoform specific difference in microglial NO production. Mice expressing the apoE4 protein isoform have a greater NO production than mice expressing the apoE3 protein isoform. The supply of arginine, the sole substrate for NOS, is dependent on cationic amino acid transporters (CATs) that also demonstrate a similar pattern of apoE isoform dependency. Although arginine transport is greater in APOE4 microglia, this effect is not limited to tissue macrophages. Cortical neurons in primary culture from APOE4 transgenic mice exhibit a similar increase in arginine uptake over neurons cultured from APOE3 mice. The inappropriate levels of arginine transport and of NO in the presence of the APOE4 compared to the APOE3 gene and its products are likely to have significant impact in the CNS.

KEYWORDS: **apolipoprotein-E; arginine transport; nitric oxide production**

INTRODUCTION

The progressive dysfunction and death of neurons in Alzheimer's dementia is enhanced in patients carrying one or more APOE4 alleles. Although the exact pathway conferring this enhancement is still a matter of active investigation, our new data support a mechanism in which nitric oxide levels are enhanced in APOE4 when compared with APOE3 cells. This increase in NO production appears to be intimately linked to the increased transport of arginine into APOE4 cells as compared to APOE3 cells. By connecting these observations with increased basal levels of NO production in APOE4 cells compared to APOE3 cells, a chronic nitrosative stress condition is established in nNOS neurons. Amyloid plaques and neurofibrillary

Address for correspondence: Carol A. Colton, Ph.D., Division of Neurology, Box 2900, Bryan Research Bldg., Duke University Medical Center, Durham, NC 27710. Voice: 919-668-2758; fax: 919-684-6514.
glia01@aol.com

tangle densities increase with the presence of apoE4 protein isoforms in comparison with apoE3 protein isoforms. Since the amyloid-β peptide of amyloid plaques is able to stimulate microglia and astrocytes to produce reactive nitrogen and reactive oxygen species, then their presence surrounding plaques may create a pathological oxidative-stress environment that leads to neuronal death. In contrast, neurons containing neurofibrillary tangles typically possess immunoreactive nNOS, are not typically surrounded by activated microglia and astrocytes, and display footprints of a pathologic nitrosative-stress environment that also leads to neuronal death. Even though both oxidative and nitrosative stress environments can kill neurons, transgenic mice displaying amyloid plaque pathologies rarely suffer neuronal death, in contrast to transgenic mice displaying neurofibrillary tangle pathologies and that exhibit significant neuronal death and death of the young animal. These finding suggest that nitric oxide, a product essential to the generation of nitrosative stress, whose production is regulated by apolipoprotein E in an isoform specific fashion, may be the smoking gun underlying neuronal death in Alzheimer's and other neurodegenerative diseases.

A key feature of Alzheimer's disease (AD) is loss of synapses and neurons in the CNS, resulting in failure of normal neuronal function and, ultimately, cognitive decline and dementia. The characteristics of the neurodegenerative process and the apparent progression from region to region in the brain have been well described.[1–3] The most susceptible region in the CNS is the entorhinal cortex. From there, the neurodegeneration spreads to the hippocampus, to the neocortex, and then to subcortical regions, leaving some regions, such as the cerebellum, relatively untouched.[1] The associated dementia caused by damage to these key processing areas of the brain has been linked to the formation of neurofibrillary tangles (NFTs) and senile (amyloid) plaques. NFTs are intraneuronal accumulations of insoluble paired helical filaments of tau, a microtubule-associated protein that helps to stabilize the cytoskeleton. Extracellular amyloid plaques are composed of fibrillar deposits of amyloid-β (Aβ) peptides, the proteolytic cleavage products of amyloid precursor protein (APP), dystrophic neurites, and activated glia. The formation of tangled neurons in the entorhinal cortex and the perforant path of the hippocampus is significantly associated with the cognitive decline of individuals with AD.[3,4]

The neurons that contain NFTs in AD have been examined in a number of studies. One of the characteristic features of these neurons is the presence of "footprints" of reactive nitrogen species that are not observed in aged matched, unaffected individuals. For example, Smith *et al.*[5] observed increased levels of 3-nitrotyrosine (NT) immunoreactivity in neurons from AD brain when compared to aged matched controls. The most intense staining was found in neurons that also contained intraneuronal NFTs.[6] Higher levels of nitrotyrosine were also seen in AD lysates by using Western blot analysis.[7] Nitrotyrosine is formed when the aromatic ring of tyrosine interacts with reactive nitrogen–oxygen species (RNOs), forming nitrated moieties that become part of the protein structure. Nitrotyrosine formation can, in turn, be detected by an immunoreaction with antibodies specific for the nitrotyrosine moieties[8] or as 3-nitrotyrosine in protein digests of tissue using HPLC.[9] Hensley *et al.*[9] demonstrated region specific elevated levels of 3-nitrotyrosine and dityrosine in AD. Consistently high levels were observed in the hippocampus with low levels in the cerebellum. Furthermore, CSF levels of 3-nitrotyrosine are increased by about sixfold in AD as compared to non-demented, age matched control individuals.[10] The expression of nitric oxide regulatory proteins, such as dimethylargininase (also known as N^g, N^g-dimethyl-L-arginine dimethylaminohydrolase, DDAH) and heme oxygenase-1 are also observed in NFT-containing neurons.[11,12] Dimethylargininase catalyzes the hydrolysis of free, methylated arginine derivatives, whereas heme oxygenase 1 (HO-1) catalyzes the decomposition of heme moieties. Both enzymes participate in the regulation of nitric oxide synthase (NOS), the enzyme that catalyzes the production of NO.[13] Finally, double labeling studies have demonstrated that neurons containing NFTs are neurons that coexpress NOS.[4] Collectively, these data indicate an important role for NO in those neurons undergoing changes that lead to neurofibrillary tangle formation.

THE SOURCE OF NO IN THE BRAIN IN AD

Nitric oxide is produced by the enzymatic 5-electron oxidation of arginine to citrulline in the presence of specific cofactors (Ca^{2+}-calmodulin, H_4-biopterin, NADPH, and FAD/FMN). Three isoforms of nitric oxide synthase have been described and are termed, inducible NOS (iNOS), endothelial NOS (eNOS), and neuronal NOS (nNOS). The normal adult brain contains numerous NOS-positive neurons (see <http:@nadph.anatomy.lsumc.edu> for a comprehensive view of NOS immunoreactive neurons in the normal adult mouse brain).[14] Generally, a regional distribution pattern for NOS is seen in the CNS, where the highest levels of expression and the highest activity for NOS (commonly measured as NADPH-diaphorase activity) are observed in the substantia innominata, nucleus accumbens, and the cerebellar cortex. The lowest levels are observed in the corpous callosum, thalamus, dentate gyrus, and occipital cortex. The intensity of NOS immunoreactivity also varies, with the highest level seen in cortical layers 2 to 4 and a lighter, granular staining in the medial septum and hippocampus.[14,15] In normal adult brain, nNOS is the predominant form of NOS, with some neuronal staining for eNOS (primarily in the CA-1 pyramidal neurons of the hippocampus) and sparse but measurable neuronal staining for iNOS, particularly in cerebellar granule cells when stimulated with immune factors.[16] Recently published data indicate that the percentage of NOS neurons in the CNS of adult humans is generally larger than that observed in the CNS of adult rodent.[17]

Each of the NOS protein isoforms has been observed in AD. Immunoreactivity for iNOS was detected in neurofibrillary tangle-bearing neurons in AD brain sections but not age-matched control sections.[18] Ordinarily, iNOS expression is associated with immune activation and immunocompetent cells[19] and, thus, robust iNOS expression is predicted within the reactive astrocytes and microglia surrounding amyloid deposits. However, iNOS mRNA or protein is not readily observed in glia in AD or in aged human brain although expression in astrocytes has been observed.[17,18,20] Transgenic mice expressing the Swedish double mutations of human APP that develop extensive amyloid plaques, do not express iNOS in either microglia or astrocytes. Microglia from these mice demonstrate other characteristics of activation and are stimulated to express iNOS by treatment of the mice with LPS.[21] These data are surprising, since microglia are the CNS specific macrophage and vigorously produce NO *in vitro* when activated by injury or inflammatory mediators.[22] Astrocytes also directly participate in the immune response by secreting factors, such as cytokines, and indirectly participate by modulating microglial function. Furthermore, microglia and astrocytes respond similarly to activating agents and both are stimulated *in vitro* by Aβ peptides, a component of amyloid plaques.[23] Ishii *et al.*[24] have demonstrated that injection of Aβ peptides into living rodent brain induces NO production that is reversed by inhibition of iNOS. Thus, microglia and/or astrocytes in the rodent brain are able to generate NO in response to plaque components. The differences between *in vitro* and *in vivo* conditions remain puzzling.

Dorheim *et al.*[25] have observed an increase in eNOS immunoreactivity around affected blood vessels in AD patients and in association with sprouting neurites within areas of degeneration. Recently, a polymorphism in the NOS3 gene (encoding eNOS) has been correlated with late onset AD, although this finding has been disputed.[26] Data also support a role for nNOS neurons in AD. Double labeling studies demonstrated a strong association between nNOS immunoreactivity and the presence of neurofibrillary pathology.[4,6] Consistent with the proposed toxicity of NO, significant loss of nNOS-expressing neurons are observed in those areas of the brain that are susceptible to AD pathology; that is, the frontal cortex, the entorhinal cortex, and the hippocampus.[4,6] In contrast, Hyman *et al.*[27] demonstrated that NADPH-diaphorase positive neurons are selectively spared in regions of the frontal and temporal cortex and in the hippocampus in AD. Part of the confusion over the number and survival of nNOS neurons stems from the use of NADPH diaphorase histochemical staining as an indicator of nNOS. Blottner *et al.*[28] demonstrated that NADPH-diaphorase positive are not necessarily immunoreactive for NOS. Even with this caveat, our overall interpretation is that NOS containing neurons appear to be "at risk" in AD.

A LINK BETWEEN NO AND A GENETIC RISK FACTOR FOR AD

We have examined the potential interconnections between NO, oxidative and nitrosative stress, and apolipoprotein E4, a major risk factor for AD. The APOE gene is polymorphic and the three alleles for APOE, namely, APOE2, APOE3, and APOE4, are found in about 3, 85, and 12% of the human population, respectively.[29] Individuals who express the APOE 4 gene and apoE4 protein isoform demonstrate an increased *risk* for AD.[29,30] The presence of APOE4/4 shifts the age of onset of the sporadic form of AD from about 80–85 to 60 years.[30] Cognitive and memory losses occur earlier in the APOE 4 population, and the presence of APOE4 is associated with an increased density of amyloid plaques and neurofibrillary tangles when compared to four non-APOE positive individuals also diagnosed as AD.[2,31] The modulation of NO by apoE provides a means by which apoE can regulate redox balance in the CNS. We have shown that treatment of immune-activated peripheral macrophages and microglia with apolipoprotein E (apoE) enhanced NO production.[32,33] This effect of apoE is consistent with its potential role in the innate immune response.[34,35] The action of ApoE depends partly on the costimulatory presence of agents that induce NOS2, such as gamma interferon (γIFN), lipopolysaccharide (LPS), a bacterial coat component, or polyinosinic:polycytidylic acid (PIC), a viral mimetic.

TABLE 1. Isoform specific effects of APOE on microglial NO Production

	APOE3 Nitrite (nmoles/μg protein)	Percent of control γIFN + PIC /γIFN	APOE4 Nitrite (nmoles/μg protein)	Percent of control γIFN + PIC /γIFN	Percent increase APOE4 /APOE3
γIFN only (100 U/ml)	4.73 ± 0.70 (36)		10.2 ± 1.07* (39)		215
γIFN + PIC (100 U/ml + 50 μg/ml)	5.30 ± 1.0 (36)	112	18.1 ± 2.3* (37)	177	341

NOTE: Microglia were cultured in a standard fashion and enriched to greater than 95% microglia by using serum deprivation.[22] To activate iNOS and induce NO production, the microglia were treated for 24 hours in serum free media containing 100 U/ml recombinant murine γIFN alone or in combination with 50 μg/ml polyinosinic:polycytidylic acid (PIC). Supernatant nitrite levels were determined using the Seivers nitric oxide analyzer standardized against known sodium nitrite concentrations. The protein content of each well was measured using albumin as the standard and data were normalized to nmole/μg protein. $*p < 0.001$ as determined by the unpaired Student's t test; n, number of wells.

Although apoE alters NO production under conditions where iNOS is stimulated, apoE does not directly increase the expression of iNOS mRNA or protein.[32] Interestingly, the effect of apoE on NO is found in other cell types that express NOS. NO production in platelets is stimulated by apoE.[36]

Our studies have employed two mouse models; (1) a mouse made transgenic for human APOE isoforms on a mouse $APOE^{-/-}$ background, denoted Tg[60] and (2) a "targeted replacement mouse" in which Exons 2, 3, and 4 of the mouse APOE gene were replaced with their counterparts from the human APOE2, APOE3, and APOE4 genes, denoted TR[37]. These targeted replacements allow the promoter and control regions of the mouse-APOE gene to regulate the expression of human-apoE protein translated from Exon-2, Exon-3, and Exon-4 of the human-APOE gene. Both mouse models demonstrated a significant isoform-specific effect of APOE, such that NO production in APOE4 mice is greater than that in APOE3 mice. TABLE 1 provides a summary of the results for the TR APOE3 and the TR APOE4 mice. APOE4 mice demonstrated a 215% increase in NO compared with APOE3 when treated with γIFN alone and a 340% increase when treated with γIFN plus PIC. Interestingly, when PIC + γIFN treated cells were compared with γIFN treated cells within each genotype, PIC exerted a minimal effect in the APOE3 TR microglia but was associated with an approximate twofold increase in nitrite in APOE4 TR mice.

REGULATION OF NO—HOW IS THE BALANCE SHIFTED IN AD?

The AD-specific changes in NOS levels and the presence of *footprints* of NO in AD brain indicate that NOS regulation may be significantly different in AD as compared to unaffected individuals. Furthermore, our data demonstrate a connection between the APOE4 genotype, a well described risk factor for AD, and the dysregulation of NO levels. The mechanism(s) for these changes are not known and may involve multiple pathways. In general, the production of nNOS protein and NO is influenced by a variety of factors at the mRNA level and at the protein level. For example, the level of NOS messenger RNA can be altered by changes in mRNA transcription, mRNA stability, and mRNA translation. NOS enzymatic activity can be altered by cofactor availability (Ca^{2+}-calmodulin, tetrahydrobiopterin, FAD, and NADPH), by inhibitory factors, by phosphorylation of the enzyme, by the presence of metals (Zn^{2+}) and by the availability of L-arginine, the substrate for NOS.[13,38] For iNOS-containing cells, the presence and the amount of "inducing" agents modulates NO production.[9] Although each of these factors is important, NO production is critically dependent on the intracellular supply of substrate, L-arginine. L-Arginine is the only amino acid used by NOS for the generation of citrulline and NO, and no other physiological amino acid or guanidino-containing molecule can substitute for arginine.[39]

FIGURE 1 provides a general scheme for arginine utilization in a typical neuron, astrocyte or microglia. A number of intracellular pathways rely on a supply of arginine, including protein synthesis, diverse signaling processes, and polyamine production, and hence, cell proliferation. Two outstanding recently published reviews provide the details of cellular arginine metabolism in general.[13,39] In NOS containing cells, however, the arginine supply is important for NO production and activation of the NOS enzyme. The relationship between extracellular arginine concentration and NO production has been well described and is readily demonstrated in macrophages. Nicholson *et al.*[40] recently showed that reduction of arginine transport using a mouse arginine transporter knockout model completely blocks induced NO production in macrophages. Furthermore, the absolute dependency on extracellular arginine in macrophages is evident despite the presence of an intracellular pool of arginine, the so called "arginine paradox".[41] In cells expressing either eNOS or nNOS, a separate intracellular pool of arginine is accessible to NOS and is used for NO production.[39,41] Part of the intracellular arginine pool can be resupplied by a

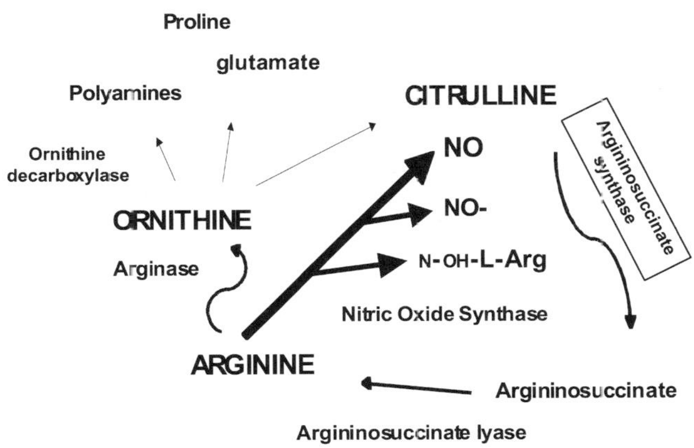

FIGURE 1. Arginine utilization pathways in NOS-containing cells (adapted from Wu and Morris[39]).

recycling pathway; that is, the conversion of citrulline to arginine through the citrulline–NO cycle (also known as the citrulline–arginine cycle). Two cytosolic enzymes are important for this process; argininosuccinate synthetase (AS) and argininosuccinate lyase (AL).[13,39] In normal adult brain the rate limiting enzyme, AS, has been localized primarily to neurons, but is not tightly colocalized with its companion enzyme, AL.[13] The citrulline–arginine cycle enzymes also do not necessarily colocalize with NOS-containing neurons, suggesting that certain populations of NOS neurons do not directly use this recycling pathway to provide arginine. Although originally disputed, recent evidence indicates that astrocytes and microglia also express AS when immune stimulated.[13] Again, there is a disparate location of the arginine supply enzymes, AS and AL, and the arginine utilization enzyme, NOS. In immune activated primary mixed glial cultures from rodent brain, astrocytes express AS and AL whereas microglia express NOS. Treatment of mouse astrocyte cultures with oxidative stressors, such as peroxynitrite, induced the release of arginine from the astrocytes.[42] Collectively, these data have prompted the hypothesis that arginine in the CNS is supplied by the astrocytes and is used by neurons and microglia.[13] Whether this is true in human CNS or in human neurodegenerative disease is not yet clear. The production and regulation of NOS in humans is significantly different from that in rodents, thus arginine supply and utilization may also be different.[19] For example, in AD, astrocytes express iNOS and AS, whereas microglia are immunoreactive for AS but not iNOS.[16] It is clear, however, that a better understanding of arginine regulation in human CNS is important to understanding NO function in normal and diseased human brain.

ARGININE UPTAKE BY CATIONIC AMINO ACID TRANSPORTERS—INFLUENCE OF APOE

We have examined [^{3}H] L- arginine uptake in microglia from APOE3 Tg, APOE4 Tg, and wildtype mice. Similarly to NO, an isoform specific difference (APOE4 Tg > APOE3 Tg > wildtype mice) in microglial arginine transport was observed.[43] The effect is not limited to microglia, however, and arginine uptake in cortical neurons also demonstrates apoE isoform specificity. As shown in FIGURE 2, uptake in neurons from APOE4 Tg mice was significantly greater than uptake in neurons from either APOE3 Tg or APOE knockout mice. The close similarity of this data to that observed in macrophages suggests the involvement of common pathways. A logical common target is the arginine transport system located in cell membranes. Arginine uptake in most cells, including macrophages and neurons, is mediated by several transport systems. Of these, the Y^+ system, now know as cationic amino acid transporters (CATs), is the most selective for arginine and is critical for transport of arginine into most cells.[13,39,41,44] Four homologous cationic transport proteins, known as CAT1, CAT2A, CAT2B, and CAT3, have been identified as system y+ transporters. This transport family is cationic amino acid selective, Na^+-independent, temperature dependent, and is inhibited by lysine.[13,39,41,44] In particular, CAT2B is associated with cells containing iNOS, can be induced by inflammatory signals, such as LPS, and is a likely candidate for the primary arginine transporter in activated

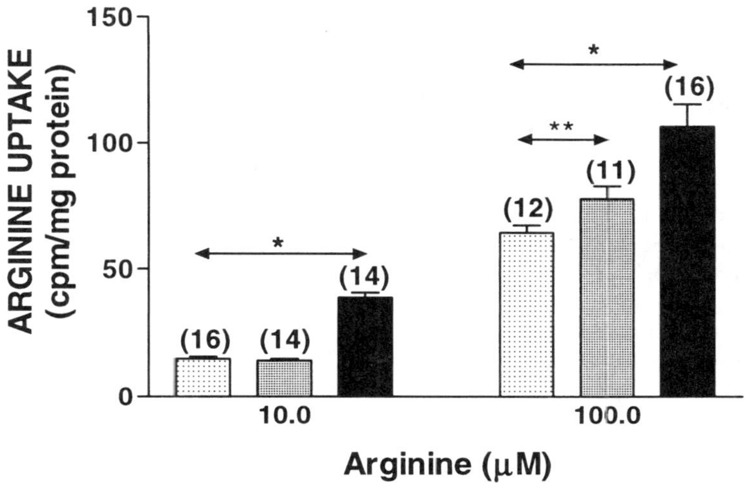

FIGURE 2. Neuronal arginine uptake is greater in APOE4 Tg mice compared to APOE3 Tg or to APOE knock out mice. [^{3}H]-arginine uptake was determined for mixed cortical neurons in primary culture from APOE knockout (▒), human APOE3 Tg (▒), and human APOE4 Tg mice (■). Primary mixed cortical cultures were prepared using standard techniques.[61] Arginine uptake was determined essentially as described by Kakuda *et al.*,[62] using uptake buffer (in mM: 137 NaCl; 2.7 KCl; 1.5 $CaCl_2$; 1.2 KH_2PO_4; 1.0 $MgSO_4$; 20 HEPES, pH 7.4) containing a defined arginine concentration (10 or 100 μM) plus [^{3}H]-arginine (1 μCi/ml). Uptake was followed for 10 min at 37°C and then stopped by washing the cells with ice cold uptake buffer containing 10 mM unlabeled arginine. Cells were lysed in 1.0% Triton X-100 and radioactivity of lysates was determined using liquid scintillation counting. Non-specific uptake/binding values were determined at 4°C and subtracted from the total cell-associated radioactivity observed following uptake at 37°C. Data were then normalized to protein content and are presented as the average cpm/mg protein ± SEM for a minimum of 11 wells (n is the number of wells) for three different culture groups at 10 or 100 μM arginine in the media. Significance was determined using an unpaired Student's t test. *$p < 0.001$; **$p < 0.02$.

macrophages and microglia.[40] CAT1 is constitutively expressed in most, if not all, cells and is downregulated, in general, by the same signals that upregulate CAT2.[44] CAT3 has been localized to brain, although its strict specificity for neurons has not been fully determined.[45]

The regulation of arginine transport by apoE protein isoforms is novel and the mechanism is unknown. Our data on RAW cells, however, indicate that apoE regulates the expression of CATs and the affinity of the transporter for arginine.[32] Whether the presence of APOE4 and apoE4 protein alters CAT expression or function in a different manner from APOE3 and apoE3 protein remains unknown. It is likely, however, that these changes are directly involved in the increased NO production associated with the action of apoE protein isoforms.

IS THE INCREASED RISK OF APOE4 IN AD DUE TO NO?

Nitric oxide is a complex mediator of a number of biological effects, many of which have important implications to the normal physiology of the CNS. For example, NO frequently acts as an intracellular or extracellular messenger, regulating key pathways. Many of the NO-mediated events are associated with the regulation of neurotransmitter release, synaptic efficacy, and synaptic modeling.[46,47] Synaptic processes, such as long term potentiation, are linked to NMDA receptor activation and to the subsequent Ca^{2+}-mediated stimulation of nNOS.[47] NO also plays a role in long term depression and in the activity-dependent refinement of synaptic connections.[46] During development, NO has both positive and negative effects on growth cone extension, axonal bifurcation, and dendritic spine number, clearly demonstrating a complex role in synaptic formation and remodeling.[46,48] Interestingly, although NO plays a key role in synaptic plasticity, there are redundant systems that can compensate for the loss of NO. Thus, removal of NO during development is not lethal to mice but has more subtle effects on certain regions of the CNS.[46] Excess NO, however, can result in cell injury and death. Over activation of the NMDA receptor, and hence, increased NO production induces *excitotoxic death* pathways in neurons.[46] NO-mediated cytotoxicity is also associated with the immune defense of the tissue and macrophages that are well known to kill invading organisms via a NO-mediated mechanism or other cells including neurons.[49]

The apparent dichotomy of action, the so-called "good" and "bad" effects of NO, may be explained by the type of reactive nitrogen species generated by NOS, the wide variety of potential targets, the types of biochemical interactions with those targets, and the level of NO generated by the tissue. Wink *et al.*[50,51] separated NO-mediated effects into two broad categories, direct and indirect effects. Direct effects of NO can be defined as those effects in which NO reacts directly with a target molecule. One example is the activation of guanylate cyclase by interaction of NO with the heme moiety. The consequent increased production of cGMP regulates a number of normal, intracellular events, such as the presynaptic release of glutamate. Thus, direct effects of NO are generally associated with normal physiology and low levels of NO. Indirect effects of NO are most commonly observed during pathological events and are associated with higher tissue production of NO. In this case, NO forms an intermediate with oxygen forming N_2O_3 or, more generally, reactive nitrogen oxygen species (RNOS). These molecules can initiate oxidative stress via reactants, such as peroxynitrite, or nitrosative stress via reactants, such as nitrosated thiols or amines.[50,52,53] Both oxidative and nitrosative stress can cause damage to target molecules, such as proteins, lipids, or DNA. Oxidative stress is more likely when the flux of oxygen free radicals (e.g., superoxide anion) is greater than the NO flux, whereas nitrosative stress occurs when the NO flux is greater than the superoxide flux.[54]

Since the level of NO can be a key factor in determining the balance between direct and indirect NO effects, an abnormal shift in the production of NO is likely to have long term consequences in the CNS. Our data suggest that the presence of the APOE4 gene and the apoE4 protein isoform is a factor in the regulation of NO level and may influence the overall redox balance of the tissue. Furthermore, the local environment surrounding the NOS-containing cells affects the determination of an oxidative or nitrosative outcome. Cells in the vicinity of an amyloid plaque; that is, activated microglia and reactive astrocytes, exhibit footprints of both reactive nitrogen and reactive oxygen species. Although not robust in humans, NOS is localized to the cellular components of plaques. Furthermore, there is evidence for an activated NADPH oxidase system, which generates superoxide anion as part of the respiratory burst.[55] Although the reasons for microglial activation are not fully evident, we and others[59] have observed an effect of Aβ peptides on both resting and activated superoxide anion production. NO production is also stimulated by Aβ peptide in rodent and human macrophages.[33,49] Thus, an oxidative environment is likely to predominate within the vicinity of an active plaque. Indices of oxidative damage are commonly found both in AD and in transgenic mice that develop dense amyloid deposits.[56] Interestingly, lipid peroxidative products are frequently found in areas distant to plaques.[5] However, free iron and reactive glycation end-products are observed in close proximity to the amyloid deposits.[56] Enhanced microglial NO production by the presence of APOE4 might alter this oxidative environment. Stimulation of the immune response under these conditions can be expected to provide additional insults that could be repetitive and additive in nature. In contrast, microglia are not commonly found in the vicinity of neurons containing neurofibrillary tangles. Instead, the presence of nNOS and footprints of NO within the NFT-containing neurons are common features indicative of a nitrosative environment. Chronically increased NO levels produced by the NOS neurons under these conditions could further increase the nitrosative damage leading to a greater susceptibility of this neuronal population to dysfunction and death.

IF NOS NEURONS ARE AT RISK, WHY IS THE CEREBELLUM SPARED IN AD?

One of the major confounding issues in AD is the regional specificity of the disease process. As mentioned previously, the entorhinal cortex and hippocampus are among the first areas to demonstrate neuropathology in AD, whereas the cerebellum is rarely affected. The cerebellum, however, has numerous nNOS expressing neurons that do not seem to be a risk in AD. Although this appears to conflict with the loss of nNOS neurons and the nitrosative stress observed in the entorhinal cortex and the hippocampus, there are potential explanations for this striking difference in susceptibility. The cerebellum in the normal adult brain exhibits high concentrations of methylarginines that serve as endogenous NOS inhibitors and competitively reduce NO production.[13] Interestingly, dimethylargininase, the enzyme that catalyzes the degradation of methylarginines is expressed in the hippocampus and increased expression is seen in AD.[12] Thus, the hippocampus may lack the endogenous "stops" of NOS found in the cerebellum. The profile of activation of astrocytes and microglia is also clearly different in the cerebellum. Both of these cells are active participants in the formation of AD neuropathology in the hippocampus and cortex, particularly in the progression of diffuse amyloid deposits to compact plaques. Sheng *et al.*[13] demonstrated the lack of IL-1 or S100 β expression by astrocytes or microglia in the cerebellum in sharp contrast to the robust immunoreactivity found in the hippocampus.[58] Stryen *et al.*[59] confirmed the general non-responsive nature of these normally reactive cells, indicating that the cerebellar environment suppresses, or does not foster, successful immune activation. Lack of an inflammatory response has been a common explanation for the presence of only diffuse amyloid deposits in the cerebellum.[57,58] Regional variation in apoE expression may also be a factor. Localization of apoE immunoreactivity to plaques is commonly found in the hippocampus, but is less frequent in the cerebellum.[59] Xu *et al.*[60] observed APOE mRNA in hippocampal neurons, but not in cerebellar neurons, supporting the idea of a regional variation in APOE expression levels. Finally, it is possible that cerebellar NOS neurons employ an inherently different pathway for use and scavenging of NO. During development, NO is a slight, but significant, survival factor for cerebellar neurons and withdrawal of NO leads to increased apoptosis.[46] Collectively, these data point to unique mechanisms for NO in the cerebellum and that may dictate different susceptibility to damage.

REFERENCES

1. NAGY, Z., N. HINDLEY, H. BRAAK, *et al.* 1999. The progression of Alzheimer's disease from limbic regions to the neocortex: clinical,radiological and pathological relationships. Dement. Geriatr. Cogn. Disord. **10:** 115–120.
2. OHM, T.G., H. SCHARNAGL, W. MARZ & J. BOHL. 1999. Apolipoprotein E isoforms and the development of low and high Braak stages of Alzheimer's disease-related lesions. Acta Neuropathol. (Berlin) **98:** 273–280.
3. THAL, D.R., M. HOLZER, U. RUB, *et al.* 2000. Alzheimer-related tau-pathology in the perforant path target zone and in the hippocampal stratum oriens and radiatum correlates with onset and degree of dementia. Exp. Neurol. **163:** 98–110.
4. THORNS, V., L. HANSEN & E. MASLIAH. 1998. nNOS expressing neurons in the entorhinal cortex and hippocampus are affected in patients with Alzheimer's disease. Exp. Neurol. **150:**14–20.
5. SMITH, M., P.L. RICHEY HARRIS, L. SAYRE, *et al.* 1997. Widespread peroxynitrite-mediated damage in Alzheimer's disease. J. Neurosci. **17:** 2653–2657.
6. GOOD, P., P. WERNER, A. HSU, *et al.* 1996. Evidence for neuronal oxidative damage in Alzheimer's disease. Amer. J. Pathol. **149:** 21–28.
7. SU, J.H., G. DENG & C. COTMAN. 1997. Neuronal DNA damage precedes tangle formation and is associated with up-regulation of nitrotyrosine in Alzheimer's disease brain. Brain Res. **774:** 193–199.
8. YE, Y., M. STRONG, Z. HUANG & J.S. BECKMAN. 1996. Antibodies that recognize nitrotyrosine. Methods Enz. **269:** 201–209.
9. HENSLEY, K., M. MAIDT, Z. YU, *et al.* 1998. Electrochemical analysis of protein nitrotyrosine and dityrosine in the Alzheimer brain indicates region-specific accumulation. J. Neurosci. **18:** 8126–8132.
10. TOHGI, R., T. ABE, K. YAMAZAKI, *et al.* 1999. Alterations of 3-nitrotyrosine concentrations in the cerebrospinal fluid during aging and in patients with Alzheimer's disease. Neurosci. Lett. **269:** 52–54.
11. SMITH, M., K. KUTTY, P. RICHEY, *et al.* 1994. Heme Oxygenase-1 is associated with the neurofibrillary pathology of Alzheimer disease. Amer. J. Pathol. **145:** 42–47.
12. SMITH, M., M. VASAK, M. KNIPP, *et al.* 1998. Dimethylarginase, a nitric oxide regulatory protein, in Alzheimer disease. Free Rad. Biol. Med. **25:** 898–902.
13. WIESINGER, H. 2001. Arginine metabolism and the synthesis of nitric oxide in the nervous system. Prog. Neurobiol. **64:** 365–391.
14. CORK, R.J., M.L. PERRONE, D. BRIDGES, *et al.* 1998. A web-accessible digital atlas of the distribution of nitric oxide synthase in the mouse brain. Prog. Brain Res. **118:** 37–50.

15. NORRIS, P., R.L. FAULL & P.C. EMSON. 1996. Neuronal nitric oxide synthase (nNOS) mRNA expression and NADPH-diaphorase staining in the frontal cortex, visual cortex and hippocampus of control and Alzheimer's disease brains. Brain Res. Mol. Brain Res. **41:** 36–49.
16. HENEKA, M.T. & D.L. FEINSTEIN. 2001. Expression and function of inducible nitric oxide synthase in neurons. J. Neuroimmunol. **114:** 8–18.
17. BLUM-DEGEN, D., T. HEINEMANN, J. LAN, *et al.* 1999. Characterization and regional distribution of nitric oxide synthase in the human brain during normal ageing. Brain Res. **834:** 128–135.
18. VODOVOTZ, Y., S. LUCIA, K. FLANDERS, *et al.* 1996. Inducible nitric oxide synthase in tangle bearing neurons of patients with Alzheimer's disease. J. Exp. Med. **184:** 1425–1433.
19. WEINBERG, J.B., M.A. MISUKONIS, P.J. SHAMI, *et al.* 1995. Human mononuclear phagocyte inducible nitric oxide synthase (iNOS): analysis of iNOS mRNA, iNOS protein, biopterin, and nitric oxide production by blood monocytes and peritoneal macrophages. Blood **86**(3): 1184–1195.
20. WALLACE, M., J. GEDDES, D. FARQUHAR & M. MASSON. 1997. Nitric oxide synthase in reactive astrocytes adjacent to beta-amyloid plaques. Exp. Neurol. **144:** 266–272.
21. HARTLAGE-RUBSAMEN, M., J. APELT & R. SCHLIEBS. 2001. Fibrillary beta-amyloid deposits are closely associated with atrophic nitric oxide synthase (NOS)-expressing neurons but do not upregulate the inducible NOS in transgenic Tg2576 mouse brain with Alzheimer pathology. Neurosci. Lett. **302:** 73–76.
22. COLTON, C., J. SNELL, O. CHERYNSHEV & D. GILBERT. 1994. Induction of superoxide anion and nitric oxide production in cultured microglia. *In* The Neurobiology of NO and OH, C. Chieuh, D. Gilbert & C. Colton, Eds.: 54–63. New York Academy of Sciences, New York.
23. MEDA, L., M. CASSATELLA, G. SZENDREL, *et al.* 1995. Activation of microglial cells by b-amyloid protein and interferon-g. Nature **374:** 647–650.
24. ISHII, K., F. MUCIHAUSER, U. LICHT, *et al.* 2000. Subacute NO generation by Alzheimer's beta amyloid in the living brain: reversal by inhibition of the inducible NO synthase. FASEB J. **14:** 1405–1409.
25. DORHEIM, M. 1994. Nitric oxide synthase activity is elevated in brain microvessels in Alzheimer's disease. Biochem. Biophys. Res. Com. **205:** 659–665.
26. CRAWFORD, F., M. FREEMAN, L. ABDULLAH, *et al.* 2000. No association between the NOS3 codon 298 polymorphism and Alzheimer's disease in a sample from the United States. Ann. Neurol. **47:** 687.
27. HYMAN, B., K. MARZLOFF, J. WENNIGER, *et al.* 1992. Relative sparing of nitric oxide synthase-containing neurons in the hippocampal formation in Alzheimer's disease. Ann. Neurol. **32:** 818–820.
28. BLOTTNER, D., Z. GROZDANOVIC & R. GOSSRAU. 1995. Histochemistry of nitric oxide synthase in the nervous system. Histochem. J. **27:** 785–811.
29. BREITNER, J.C., G. JARVIK, B.L. PLASSMAN, *et al.* 1998. Risk of Alzheimer disease with the epsilon 4 allele for apolipoprotein E in a population based study of men aged 62–73 years. Alzheimer Dis. Assoc. Disord. **12:** 40–44.
30. ROSES, A.D. 1997. Apolipoprotein E, a gene with complex biological interactions in the aging brain. Neurobiol. Dis. **4:** 170–185.
31. SCHMECHEL, D., A. SAUNDERS, W. STRITTMATTER, *et al.* 1996. Increased amyloid b-peptide deposition in cerebral cortex as a consequence of apolipoprotein E genotype in late-onset Alzheimer disease. Proc. Natl. Acad. Sci. U.S.A. **90:** 9649–9653.
32. COLTON, C.A., M. CZAPIGA, J. SNELL-CALLANAN, *et al.* 2001. Apolipoprotein E acts to increase NO production in macrophages by stimulating arginine transport. Biochem. Biophys. Acta **1535:** 134–144.
33. VITEK, M., J. SNELL, O. CHERNYSHEV & C. COLTON. 1997. Modulation of nitric oxide production in human macrophages by apolipoprotein E and amyloid beta peptide. Biophys. Biochem Res. Comm. **240:** 391–394.
34. BARGER, S. & A. HARMON. 1997. Microglial activation by Alzheimer amyloid precursor protein and modulation by apolipoprotein E. Nature **388:** 878–881.

35. EGENSPERGER, R., S. KOSEL, U. VON EITZEN & M.B. GRAEBER. 1999. Microglial activation in Alzheimer's disease: association with APOE genotype. Mol. Neuropathol. **3:** 439–447.
36. RIDDELL, D., A. GRAHAM & J.S. OWENS. 1997. Apolipoprotein E inhibits platelet aggregration through the L-arginine: nitric oxide pathway. J. Biol. Chem. **272:** 89–95.
37. SULLIVAN, P., H. MEZOUR, S.H. QUARFORDT & N. MAEDA. 1998. Type III hyperlipoproteinemia and spontaneous athersclerosis in mice resulting from gene replacement of mouse Apoe with human Apoe2. J. Clin. Invest. **102:** 130–135.
38. BOUCHER, J.L., C. MOALI & J.P. TENU. 1999. Nitric oxide biosynthesis, nitric oxide synthase inhibitors and arginase competition for L-arginine utilization. Cell. Mol. Life Sci. **55:** 1015–1028.
39. WU, G. & S.M. MORRIS, JR. 1998. Arginine metabolism: nitric oxide and beyond. Biochem. J. **336:** 1–17.
40. NICHOLSON, B., C. MANNER, J. KLEEMAN & C. MACLEOD. 2001. Sustained NO production in macrophages requires the arginine transporter. J. Biol. Chem. **273:** 15881–15883.
41. CLOSS, E., J. SCHELD, M. SHARAFI & U. FORSTERMANN. 2000. Substrate supply for nitric oxide synthase in macrophages and endothelial cells: role of cationic amino acid transporters. Mol. Pharmacol. **57:** 68–74.
42. VEGA-AGAPITO, V., A. ALMEIDA, S. HEALES, *et al.* 1999. Peroxynitrite anion stimulates arginine release from cultured rat astrocytes. J. Neurochem. **73:** 1446–1452.
43. COLTON, C., D. COOK, C. BROWN, *et al.* 2002. Nitric oxide, arginine transport and apolipoprotein E: a link between oxidative stress and the genetics of AD. J. Neurobiol. Aging. In press.
44. DEVES, R. & C.A. BOYD. 1998. Transporters for cationic amino acids in animal cells: Discovery, structure and function. Physiol. Rev. **78:** 487–545.
45. HOSOKAWA, H., T. SAWAMURA, S. KOBAYASHI, *et al.* 1997. Cloning and characterization of a brain specific cationic amino acid transporter. J. Biol. Chem. **272:** 8717–8722.
46. CONTESTABILE, A. 2000. Roles of NMDA receptor activity and nitric oxide production in brain development. Brain Res. Rev. **32:** 476–509.
47. BREDT, D. & S. SNYDER. 1992. Nitric oxide: a novel neuronal messenger molecule. Neuron **8:** 3–11.
48. INGLIS, F., F. FURIA, S. ZUCKERMAN, *et al.* 1998. The role of nitric oxide and NMDA receptors in the development of motor neuron dendrites. J. Neurosci. **19:** 10493–10501.
49. II, M., M. SUNAMOTO, K. OHNISHI & Y. ICHIMORI. 1996. B-amyloid protein dependent nitric oxide production from microglial cells and neurotoxicity. Brain Res. **720:** 93–100.
50. WINK, D., I. HANBAUER, M. GRISHAM, *et al.* 1996. Chemical biology of nitric oxide: Regulation and protective and toxic mechanisms. Current Topic. Cell. Reg. **34:** 159–186.
51. WINK, D.A., M. FEELISCH, Y. VODOVOTZ, *et al.* 1999. The chemical biology of nitric oxide. *In* Reactive Oxygen Species in Biological Systems: An Interdisciplinary Approach. D.L. Gilbert & C.A. Colton, Eds.: 245–291. Kluwer Academic, New York.
52. WINK, D., K. MIRANDA & M.G. ESPEY. 2001. Cytotoxicity related to oxidative and nitrosatice stress by nitric oxide. Exp. Biol. Med. **226:** 621–623.
53. ESPEY, M.G., K. MIRANDA, R. PLUTA & D. WINK. 2000. Nitrosative capacity of macrophages is dependent on nitric-oxide synthase induction signals. J. Biol. Chem. **275:** 11341–11347.
54. MILES, A., S. BOHLE, P. GLASSBRENNER, *et al.* 1996. Modulation of superoxide-dependent oxidation and hydroxylation reactions by nitric oxide. J. Biol. Chem. **271:** 40–47.
55. KLEGERIS, A. & P. MCGEER. 1997. Beta-amyloid protein enhances macrophage production of oxygen free radicals and glutamate. J. Neurosci. Res. **49:** 229–235.
56. SMITH, M., K. HIRAI, K. HSIAO, *et al.* 2001. Amyloid-b deposition in Alzheimer transgenic mice is associated with oxidative stress. J. Neurochem. **70:** 2212–2215.

57. SHENG, J., R. MRAK & W.S.T. GRIFFIN. 1997. Neuritic plaque evolution in Alzheimer's disease is accompanied by transition of activated microglia from primed to enlarged to phagocytic. Acta Neuropathol. **94:** 1–5.
58. GRIFFIN, W.S.T., L. STANLEY, C. LING, *et al.* 1989. Brain interleukin 1 and S-100 immunoreactivity are elevated in Down syndrome and Alzheimer disease. Proc. Natl. Acad. Sci. U.S.A. **86:** 7611–7615.
59. STYREN, S., M.I. KAMBOH & S. DEKOSKY. 1998. Expression of differential immune factors in temportal cortex and cerebellum: The role of a-1-antichymotrypsin, apolipoprotein E and reactive glia in the progression of Alzheimer's disease. J. Comp. Neurol. **396:** 511–520.
60. XU, P., D. SCHMECHEL, T. ROTHROCK-CHRISTIAN, *et al.* 1996. Human apolipoprotein E2, E3 and E4 isoform specific transgenic mice: Human like pattern of glial and neuronal imunoreactivity in central nervous system not observed in wildtype mice. Neurobiol. Dis. **3:** 229–245.
61. COLTON, C., F. PAGAN, J. SNELL, *et al.* 1995. Protection from oxidation enhances the survival of cultured mesencephalic neurons. Exp. Neurol. **32:** 54–61.
62. KAKUDA, D., M.J. SWEET, C. MACLEOD, *et al.* 1999. CAT2-mediated L-arginine transport and nitric oxide production in activated macrophages. Biochem. J. **340:** 549–553.

The Association Between Neuronal Nitric Oxide Synthase and Neuronal Sensitivity in the Brain After Brain Injury

PHILIP K. LIU,[a,b] CLAUDIA S. ROBERTSON,[a] AND ALEX VALADKA[a]

[a]*Department of Neurosurgery, Baylor College of Medicine, Houston Texas, U.S.A.*

[b]*Graduate Program in Cardiovascular Sciences and Molecular and Cell Biology, Baylor College of Medicine, Houston Texas, U.S.A.*

ABSTRACT: Injury to the central nervous system is the leading cause of disability in the United States. Neuronal death is one of the causes of disability. Among patients who survive this type of injury, various degrees of recovery in brain function are observed. The molecular basis of functional recovery is poorly understood. Clinical observations and research using experimental injury models have implicated several metabolites in the cascade of events that lead to neuronal degeneration. The levels of intracellular ATP (energy source) and pH are decreased, whereas levels of extracellular glutamate, intracellular calcium ions, and oxidative damage to RNA/DNA, protein, and lipid are increased. These initiating events can be associated with energy failure and mitochondrial dysfunction, resulting in functional or structural brain damage. The injured brain is known to express immediate early genes. Recent studies show that reactive oxygen species (ROS) cause lesions in genes from which mRNA is transcribed as part of the endogenous neuroprotective response. Although degenerating proteins and lipids may contribute to necrosis significantly after severe injury, abnormalities in genetic material, if not repaired, disturb cellular function at every level by affecting replication, transcription, and translation. These lesions include abnormal nucleic acids, known as oxidative lesions of DNA (ODLs) or of RNA (ORLs). In this review, we focus on our current understanding of the various effects of neuronal nitric oxide synthase on the formation of modified bases in DNA and RNA that are induced in the brain after injury, and how ODLs and ORLs affect cell function.

KEYWORDS: oxidative stress; immediate early gene; gene repair; 8-oxoguanosine glycosylase/AP lyase; Fos/AP-1; stroke; neuroprotection; nNOS; head injury; cerebral ischemia-reperfusion injury

INTRODUCTION

Reactive oxygen species (ROS) are normal byproducts of oxygen metabolism in mitochondria, but they can also be generated in the cytoplasm under certain conditions during an activation of inflammatory cells.[1–8] Therefore, mitochondria are constantly under attack from ROS. Nevertheless, mitochondria have effective repair

Address for correspondence: Philip K. Liu, Ph.D., Department of Neurosurgery, Baylor College of Medicine, 6560 Fannin, Suite 944 Houston, TX 77030, U.S.A. Voice: 713-798-5177; fax: 713-798-4124.

philipl@bcm.tmc.edu

mechanisms to remove the damage caused by ROS and to avoid the consequence of genetic damage.[9,10] Pathologically, as observed in brain injury after the transient ischemia followed by reperfusion, the reduction in oxygen supply perturbs the energy metabolism of the brain and causes ROS levels to increase. ROS (such as nitric oxide [NO], superoxide anion [$O_2^{-\bullet}$] and hydroxyl radicals) are highly reactive chemicals that attack and damage DNA. Nitric oxide interacts with oxygen, superoxide anion, and thiol compounds to form three different groups of chemicals: reactive nitrogen species, peroxynitrite, and *S*-nitrosothiol (*S*-nitroso glutathione), respectively.[11–15] Specifically, reactive NO appears to increase after cerebral ischemia,[16] and it can lead to a reduction in energy production.[17] The superoxide anion is formed under normal conditions as a consequence of mitochondrial respiration. A slight increase in reactive NO, which occurs during ischemia, favors the formation of peroxynitrite, especially in the ischemic core.[18] Peroxynitrite attacks proteins, DNA/RNA, and lipids. The most notorious products of peroxynitrite interaction with these cellular components are 3-nitrotyrosine in proteins and base modifications in DNA and RNA.[8,19–22] Peroxynitrite has a halflife of 1–2 seconds and a diffusion distance of about 100 micrometers. Therefore, it can diffuse into the nuclei and into cells other than that in which it is formed. Peroxynitrite undergoes a cleavage reaction to form a hydroxyl radical or singlet oxygen,[23] both of which are highly reactive and can cause damage to all cellular components. Modified RNA and DNA can change their coding properties and block chain elongation during transcription and replication. Therefore, NO directly or indirectly, through the reaction with other ROS or molecules, can exert adverse effects at every level of cellular function.

Our hypothesis is that acute brain injury causes DNA damage in nuclear genes of the brain, resulting in transient interruptions in endogenous neuroprotective responses, such as expression of repair function. This delay leaves the injured brain vulnerable to secondary insults as a result of an altered homeostatic state. We have chosen to study a specific endogenous neuroprotective pathway (see FIGURE 1) that is initiated by the expression of the c-fos gene. Expression of this gene after ischemic injury has been studied in a number of laboratories. However, little attention has been paid to whether the mRNA produced by the gene is functional. We review whether or not the c-fos mRNA transcripts that are produced after ischemic injury are defective and the consequences of defective c-fos mRNA on the ability of the brain to repair or to recover. We present evidence to suggest that c-fos transcripts produced after ischemic injury contain lesions, at least for a transient period of time. Furthermore, recent studies indicate that ischemia-induced damage to the c-fos gene can be repaired. As background to this review, we also discuss the following topics: DNA damage after brain injury, the role of NO and superoxide anion (peroxynitrite) in oxidative damage to DNA after brain injury, the early gene response to brain injury, and the neuroprotective effects of Fos/AP-1 and NGF after ischemic injury. Because oxidative stress that is produced in traumatic brain injury models may not be easily distinguishable from that induced in cerebral ischemia models, we present results that are observed in both types of injury, but comment on differences between certain studies.

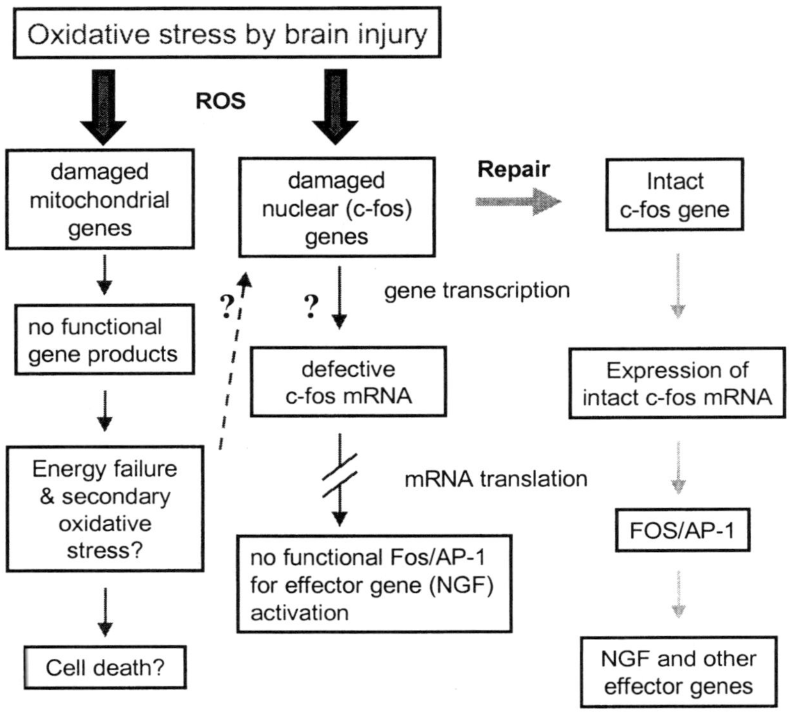

FIGURE 1. Diagram of the proposed mechanism of cerebral ischemia/reperfusion-induced sensitivity to oxidative stress as discussed in the text. Brain injury, via oxidative damage induced by cerebral ischemia/reperfusion, results in damaged immediate early genes (IEGs) and the production of defective mRNA transcripts. This damage delays the activation of downstream neuroprotective effector genes, leaving the brain transiently more vulnerable to oxidative stress. Conversely, a reduction in the level of damaged genes reduces oxidative stress (shown by treatments with 7-nitroindazole) via repair. The specific pathway that is present involves the product of c-fos (Fos/AP-1) and subsequent activation of nerve growth factor (NGF). Least is known about the effect of oxidative stress on gene expression in mitochondria.

DNA DAMAGE AFTER BRAIN INJURY CAUSED BY OXIDATIVE STRESS

Oxidative stress induced by brain injury causes two types of DNA damage. The first that is discovered, DNA strand breaks with fragments of finite length, occurs as early as 2–6 hours after injury.[24–26] This type of DNA damage is a signal that shows the cell is undergoing the process of self-destruction, or apoptosis, after injury.[27–32] Evidence of apoptotic cell death has been found in human brain tissue at autopsy and in contused tissue removed at surgery.[33,34] Evidence for enzymes that repair apoptotic DNA damage is lacking and, therefore, this type of DNA injury is believed to be irreversible once the process starts.[35] Consequently, efforts to develop therapeutic interventions have concentrated on inhibiting those steps that initiate apoptosis.[30,36–43] This type of damage to DNA is referred to as DNA fragmentation.[44]

The second type of DNA injury is the result of direct attacks on free nucleic acids, RNA and DNA, by excess electrons during oxidative stress.[44–47] Recent evidence supports the suggestion that this type of DNA lesion occurs before apoptotic DNA fragmentation.[20,24] DNA lesions caused by oxidative stress generally take the form of modified base lesions, DNA strand breaks,[48–50] and DNA–DNA or protein–DNA crosslinks.[51] There could be as many as 70 different types of lesions if we include those that occur in RNA, DNA, and free nucleotides; 20 of these are in DNA.[52] ODLs in nucleic acids can be detected by using spin traps and high-pressure liquid chromatography (HPLC),[6] gas chromatography/mass spectrometry (GC/MS),[53] HPLC with electrochemical detection (HPLC-EC),[8,54] and immunochemistry with monoclonal antibodies.[19,20,45,55,56] In addition, base lesions can be detected in DNA using lesion-specific repair enzymes.[8,19,57–60] DNA strand breaks and other ODLs can be detected by the inability of the DNA to serve as a template for PCR.[61] Recently, a nucleic acid probe has been generated that binds Oh8dG in oligo DNA and generates a fragment shift when subjected to polyacrylamide gel electrophoresis (PAGE).[62] We have also developed an *in situ* detection system in which DNA lesions are identified in tissue sections.[20,21] FIGURE 2 shows seven ODLs that have been identified in the brain after injury. The activity of 8-oxo-guanosine-glycosylase/lyase (OGG), which removes almost all of the ODLs except the strand breaks shown in FIGURE 2, has been identified and shown to increase after cerebral oxidative stress.[63]

When single strand breaks (SSBs) in both strands of DNA occur in close proximity to each other, they can result in a double-strand break (DSB). This is more likely to occur in the presence of high levels of ROS. DSBs can become a precursor of DNA fragmentation if there is an excess of them or if the breakpoint is within a gene that is essential for the viability of the cell. Although few analytical methods can easily decipher the cause of SSBs, these breaks can be readily demonstrated using the comet assay.[64,65] SSBs and DSBs induced by ROS, like all other forms of ODLs, are of unequal length and are randomly distributed. Unlike DNA fragmentation during apoptosis, SSBs and DSBs resulting from oxidative stress appear as a smear upon gel electrophoresis. In addition, SSBs with 3′-OH ends can be generated under four different conditions: (1) ROS attack on the phosphate backbone (oxidative DNA damage), (2) glycosylase/endonuclease removal of ODLs (an intermediate during DNA repair), (3) nascent DNA in a replication fork of dividing cells (an intermediate in DNA replication), and (4) as products of digestion of DNA by non-specific nucleases (apoptosis). It is not an easy task to distinguish SSBs among these four mechanisms. SSBs with 3′-OH ends as intermediates in DNA repair and replication are part of normal cellular existence and generally are not considered to be DNA damage. Therefore, SSBs with 3′-OH ends can not be generalized as DNA damage. Very few methods can definitively separate SSBs formed during normal cellular functions from those formed under pathophysiologic conditions. To avoid confusion, we discuss only abnormal bases that result from oxidative stress.

If only abnormal bases are considered, Oh8dG or Oh8G are the most reliably produced lesions by oxidative stress (FIG. 2). Oh8dG not only promotes mismatch pairing with deoxyadenosine triphosphate, but also inhibits the methylation of adjacent cytosine bases in CpG sequences.[66] Apyrimidinic/apurinic (AP) sites allow the

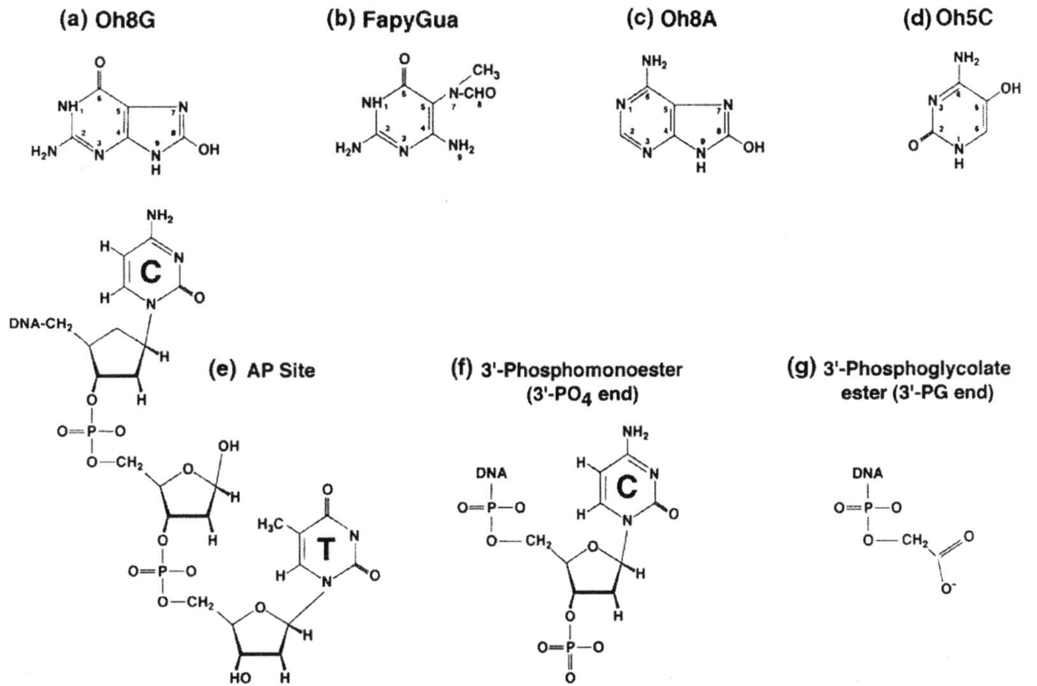

FIGURE 2. Modified nucleic acids found in the brain after cerebral oxidative stress: 8-hydroxyl-guanosine (Oh8G); 2,6 diamino-4-hydroxy-5-*N*-methylformamidopyrimidine (FapyGua): 8-hydroxyadenosine (Oh8A); 5-hydroxycytosine (Oh5C); apurinic/apyrimidinic (AP) sites; and single strand breaks with 3′ blocking termini of 3′-phosphomonoester (3′PO4) and 3′-phosglycolate ester (3′PG).

random insertion of nucleotides during *trans*-lesion synthesis[67,68] and have been implicated in neuronal death that occurs after brain injury and with aging.[21,28,69,70] Furthermore, Oh8dGTP, a lesion of a free nucleotide triphosphate induced by ROS, can be incorporated and paired with adenosine in the template, thereby creating an AT to CG transversion. We have demonstrated an elevated level of ODLs in three nuclear genes (actin, c-fos, and DNA polymerase-β genes) using two models of cerebral ischemia/reperfusion. Elevation occurs immediately after the release of vessel occlusion and remains elevated for at least the first 30 minutes of the reperfusion period. Although more than 80% of ODLs are repaired within two hours of reperfusion,[19–21] several genes are transcribed in response to brain injury during this time interval. Moreover, because there are only two copies of each gene in a cell, only two lesions (one in each copy of the gene) are required to inactivate the function of the gene. However, the continued replenishment of protein, lipid, and RNA depends on an intact copy of the gene. Thus, genetic repair mechanisms are needed to ensure the replacement of other cellular components. Gene damage is generally reversible via DNA repair processes. Without proper repair, a damaged gene could be associated with either the initiation of various diseases by changing the expression of genes or with subsequent cell death. Numerous studies on brain injury suggest a role for oxidative damage in neuronal death after ischemia. However, no work has concentrated specifically on the mechanism of oxidative DNA damage formation, or on the role this damage may play in causing the production of aberrant mRNA, thereby delaying the induction of the endogenous defense response by the brain.

ROLE OF NITRIC OXIDE IN CAUSING OXIDATIVE DNA DAMAGE AFTER BRAIN INJURY

Singlet oxygen and hydroxyl radicals are the most potent forms of ROS. Hydroxyl radicals have a very short halflife, and attack only nearby molecules, as seen in the formation of Oh8G in RNA. Several studies have shown an elevation of ROS following brain injury, whereas other studies have demonstrated oxidative gene damage following brain injury. Taken together, these data suggest an interaction between hydroxyl radicals and nuclear genes. One way that this might be accomplished is from a homolytic cleavage of peroxynitrite molecules. Peroxynitrite is formed from superoxide anions and NO and can be cleaved to form hydroxyl radical and NO_2.[11] Singlet oxygen can also be generated from peroxynitrite.[23] The precursor, NO, is synthesized from at least three isoforms of NO synthase (NOS): endothelial (eNOS), neuronal (nNOS), and inducible (iNOS) forms. In the central nervous system, cells that express nNOS are detected by the presence of NADPH diaphorase activity.[71–74] The eNOS and nNOS isoforms are constitutively expressed, and their activities are stimulated by increased intracellular calcium levels within the central and peripheral nervous systems. NO can exist as a nitrosonium ion (NO^+) or as a radical NO (·NO), depending on the redox state of the cell.[75] Furthermore, NO is an important cellular messenger, transmitting biological signals for a variety of physiological functions, such as vasodilation, neurotransmission, and the immune response.[76]

The detrimental effects of NO, especially NO produced from nNOS, are being increasingly recognized.[19,21,56,77–79] The exact function of nNOS under normal physiological conditions is not clear. However, NO that is produced from NOS-positive cells can be transmitted to neighboring NOS-negative cells and combine with ischemia induced superoxide anions to form peroxynitrite.[80] Moreover, the halflife of peroxynitrite is longer than that of hydroxyl radicals, and peroxynitrite can diffuse beyond neighboring cells, thereby amplifying the detrimental effects of NO. Superoxide dismutase (SOD) reduces superoxide anion and enhances the interaction of NO with glutathione, producing *S*-nitrosoglutathione, which has been shown to be neuroprotective.[15] Additionally, an isoform of SOD in the NOS-positive cells may act as a scavenger, reducing the level of superoxide anions, and thereby reducing the neurotoxicity of NO by limiting its ability to produce peroxynitrite. Therefore, the neurotoxic effect of NO is also dependent on the availability of superoxide anions.[80,81]

Pharmacologically, the neurotoxic effect of NO produced by nNOS can be largely eliminated by blocking the formation of NO using a specific nNOS inhibitor, 7-nitroindazole,[56,77,82] or its derivative, 3-bromo-7-nitroindazole,[19,21] thereby greatly reducing the neurotoxic effect of NO and its byproducts. Indeed, treatment with 7-nitroindazole significantly reduces the cellular Oh8dG levels in an experimental model,[21,56] as well as necrotic cell death in brain injury models of the rat,[82] the mouse,[21,81,83] and the gerbil.[84] nNOS- or iNOS-knockout animals show a consistent decrease in cell death after cerebral ischemia,[18,79,83,85–87] whereas eNOS-knockout animals show an increase in cell death following experimental stroke.[88] Reducing endothelial NO in eNOS-knockout mice may increase the ischemic effect by preventing vessel dilation. A mutation in the eNOS gene in humans is associated with coronary spasm and hypertension.[89–97] Two other groups, however, found no such association.[98,99] Extracellular glutamate increases the influx of calcium, which elevates nitric oxide (NO) levels by activating nNOS. All factors or pathophysiologic conditions that increase extracellular glutamate and, thereby, calcium influx, may indirectly cause an increase in ODLs.[46]

The consequences of inhibiting NOS may have variable results in different models of brain injury. The elevation of NO lasts for hours in cerebral ischemia models. However, during the first few minutes after brain injury, the activity of the constitutive forms of NOS is increased, and NO concentrations are elevated.[100,101] Subsequently, the activity of constitutive NOS decreases to below preinjury levels and may remain low for several days. Later, inducible NOS activity is increased in the brain.[102] Inhibition of the early increase in NOS activity has been found to have varying effects. Some studies show that *N*(omega)-nitro-L-arginine methyl ester (L-NAME), a nonspecific inhibitor of NOS, is protective,[79,103] whereas other studies show a detrimental or no effect.[100,104] The use of 7-nitroindazole in the early postinjury period has consistently demonstrated a protective effect on both contusion volume and behavioral outcomes.[100,103] Later, when inducible NOS activity is increased, both protective and detrimental effects have been observed with inhibition.[102,105]

EARLY GENE RESPONSE TO BRAIN INJURY

The injured brain also expresses immediate early genes (IEGs), and a large number of other genes, such as heat shock protein, neurotrophin genes (including brain-derived neurotrophic factor), β-amyloid precursor protein, and antiapoptosis genes (bcl-2 and bcl-x-long).[5,7,106–110] The expression of IEGs, such as the *c-fos* gene, precedes the expression of cell–cycle checkpoint genes and is required for the subsequent expression of nerve growth factor (NGF), which is protective after brain injury.[111–114] The *c-fos* gene is selectively expressed in ipsilateral cortex and hippocampus.[110] In a transient ischemia/reperfusion injury model, the *c-fos* gene was induced immediately after the release of vessel occlusion, with a diminished c-fos signal in the core during the first 30 minutes of reperfusion.[5,19,111] Expression of the *c-fos* gene is limited to the ipsilateral cortex during the first few hours of reperfusion in the focal cerebral ischemia/reperfusion model of the rat.[5,115] In both the fluid percussion and the cortical impact injury models of brain injury, the expression of IEGs has been observed as early as five minutes after injury and persists for as long as 5–6 hours after injury.[116–125] The expression of IEGs is related to the severity of the injury, being more widespread throughout the brain and for a longer duration with increasingly severe injuries.[119] The temporal pattern of expression of IEGs also depends on the location within the brain. Raghupathi and McIntosh[119] found a more prolonged expression of IEGs at the site of impact than in the ipsilateral hippocampus in a brain injury model. Additionally, a secondary insult, such as hypoxia, has been observed to potentiate the expression of c-fos in the injured brain.[126] Expression of IEGs has also been observed in human brain tissue removed as a part of the treatment of a traumatic brain injury.[107,124] Whitfield and Pickard found a direct correlation between increasing expression of c-fos and c-jun mRNA in contused brain tissue and a worsening outcome following brain injury.[107] The initial elevation of *c-fos* mRNA after brain injury temporally coincides with the presence of ODLs in the c-fos gene. *In vitro,* RNA polymerase and DNA polymerase can perform transcription and replication "bypassing" the ODL in a given gene, and adenosine is mostly likely to be inserted opposite the Oh8dG in the template (reviewed in Ref. 44). It is very likely that induced transcription of the damaged *c-fos* gene immediately after brain injury results in an increase of faulty fos mRNA. Unless the faulty mRNA contains only silent mutations, the mRNA may not be able to support translation that gives rise to normal peptides. A reduction of ODLs in the nuclear genes using a specific nNOS inhibitor, 7-nitroindazole, allows intact c-fos mRNA to be transactivated at an earlier time following ischemic injury.[5,106] Thus, the presence of ODLs could delay the expression of the endogenous response by delaying the expression of functional protein. Theoretically, 7-nitroindazole could help prevent cell death by facilitating a more timely expression of intact c-fos mRNA for activator protein-1 (AP-1) peptide.

NEUROPROTECTIVE EFFECTS OF FOS AND NERVE GROWTH FACTOR (NGF) FOLLOWING ISCHEMIC BRAIN INJURY

Transcription of these IEGs, in turn, triggers expression of a variety of target genes (or late effector genes). AP-1 is a dimer of Fos with itself or with products of other IEGs. Fos/AP-1 is thought to play an important role as a gene regulator in brain response to injury by binding to the promoter regions of the late effector genes. In the cortical impact injury model, Fos/AP-1 transcription factor has been found to be increased in the injured brain between 1 and 24 h postinjury,[110,117] and the expression of NGF mRNA is induced after the significant expression of Fos/AP-1.[118] Fos/AP-1 activates many genes in response to injury, some of which have neuroprotective properties, whereas others are neurotoxic.[127] The overall effect (protective versus toxic) of c-fos expression and Fos/AP-1 activity is likely to vary with the circumstances and the conditions under which other gene activators are present. Several important neuroprotective genes that may be regulated by Fos/AP-1 are NGF, metallothionein, and genes mediating DNA synthesis. Several studies report that NGF expression is associated with c-fos expression.[111,128–133] Therefore, the expression of Fos/AP-1 in the brain could be an endogenous neuroprotective response to injury. This conclusion is supported by a study demonstrating that an inhibition of Fos using antisense DNA increases infarct volume after stroke.[112] Conversely, transgenic mice overexpressing NGF under Fos/AP-1 control exhibit reduced necrosis after experimental stroke.[113]

The expression of c-fos mRNA increased significantly at five minutes and peaked at one hour after traumatic brain injury compared to animals undergoing sham operation. However, Fos/AP-1 activity and NGF immunoreactivity were not significantly increased until one hour postinjury.[118] The data support the hypothesis that NGF expression is positively correlated with the expression of Fos/AP-1 activity. NGF has a neuroprotective effect after neuronal injury[134] and after stroke.[112,113] Some studies have demonstrated a beneficial effect of NGF following brain injury, especially on cognitive function.[114,135] The mechanisms by which cells escape (or are unable to escape) the effect of oxidative damage may depend on IEGs and subsequent gene expression. The expression of Fos/AP-1 immediately after brain injury could be important for neuronal survival. The presence of intact mRNA during early reperfusion may be of great importance for functional recovery.

CONCLUSION AND FUTURE DIRECTIONS

Several methods that have been used to minimize neurodegeneration are to prevent (1) energy failure, (2) the formation of peroxynitrite, and/or (3) ODLs. The isoforms of NOS that are expressed in various models of brain injury may have important implications for neuronal viability. NOS may play a role in preventing energy failure by allowing vessel dilation. Conversely, expression of nNOS in the presence of elevated levels of superoxide ions is neurotoxic. A reduction of ROS by inhibiting nNOS using 7-nitroindazole can increase neuroprotection following ischemic brain injury by reducing peroxynitrite formation and, secondarily, the amount of ODLs/ORLs. The attenuation of ODLs in nuclear genes increases the functionality of the mRNA that is transcribed. One example shown here is the expression of the c-fos gene and its activation of NGF. Studies have shown that the neuroprotective drug 7-nitroindazole, or its derivative 3Br7NI, increases expression of intact c-fos mRNA earlier than placebo.[5,106] Future studies should be directed towards elucidating the mechanism of action of intact c-fos mRNA, the development of antioxidants that target NO, superoxide anion, or both, and the development of other nNOS inhibitors.

ACKNOWLEDGEMENTS

We thank Mr. T. Arora for assisting in the preparation of this manuscript. This research is supported by grants from NINDS (NS34810, PKL) (NS38660, CSR) and American Heart Association (9640202N). PKL is an Established Investigator of the American Heart Association.

REFERENCES

1. LIPTON, P. 1999. Ischemic cell death in brain neurons. Physiol. Rev. **79:** 1431–1568.
2. LEWEN, A., *et al.* 2000. Free radical pathways in CNS injury. J. Neurotrauma **17:** 871–890.
3. ELIASSON, M.J., *et al.* 1997. Poly(ADP-ribose) polymerase gene disruption renders mice resistant to cerebral ischemia. Nat. Med. **3:** 1089–1095.
4. CHEN, J., *et al.* 1998. Induction of caspase-3-like protease may mediate delayed neuronal death in the hippocampus after transient cerebral ischemia. J. Neurosci. **18:** 4914–4928.
5. CUI, J.K. & P.K. LIU. 2001. Neuronal NOS inhibitor that reduces oxidative DNA lesions and neuronal sensitivity increases the expression of intact c-fos transcripts after brain injure. J. Biomed. Sci. **8:** 336–341.
6. FLOYD, R.A. & J.M. CARNEY. 1992. Free radical damage to protein and DNA: mechanisms involved and relevant observations on brain undergoing oxidative stress. Ann. Neurol. **32:** 22–27.
7. SALMINEN, A., P.K. LIU & C.Y. HSU. 1995. Alteration of transcription factor binding activities in the ischemic rat brain. Biochem. Biophys. Res. Commun. **212:** 939–944.
8. LIU, P.K., *et al.* 1996. Damage, repair and mutagenesis in nuclear genes after mouse forebrain ischemia-reperfusion. J. Neurosci. **16:** 6795–6806.
9. ENGLANDER, E.W., *et al.* 1999. Hypoxia-induced mitochondrial and nuclear DNA damage in the rat brain. J. Neurosci. Res. **58:** 262–269.
10. CHEN, D., J. LAN & W. PEI. 2000. Detection of DNA base-excision repair activity for oxidative lesions in adult rat brain mitochondria. J. Neurosci. Res. **61:** 225–236.
11. BECKMAN, J.S., *et al.* 1990. Apparent hydroxyl radical production by peroxynitrite: implications or endothelial injury from nitric oxide and superoxide. Proc. Natl. Acad. Sci. U.S.A. **87:** 1620–1624.
12. WINK, D.A., *et al.* 1994. Reaction kinetics for nitrosation of cysteine and glutathione in aerobic nitric oxide solutions at neutral pH. Insights into the fate and physiological effects of intermediates generated in the NO/O_2 reaction. Chem. Res. Toxicol. **7:** 519–525.
13. HOGG, N., R.J. SINGH & B. KALYANARAMAN. 1996. The role of glutathione in the transport and catabolism of nitric oxide. FEBS Lett. **382:** 223–228.
14. GOW, A.J., D.G. BUERK & H. ISCHIROPOULOS. 1997. A novel reaction mechanism for the formation of *S*-nitrosothiol *in vivo*. J. Biol. Chem. **272:** 2841–2845.
15. RAUHALA, P. A.M.-Y. LIN & C.C. CHIUEH. 1998. Neuroprotection by *S*-nitrosoglutathione of brain dopamine neurons from oxidative stress. FASEB J. **12:** 165–173.

16. KUMURA, E.T., *et al.* 1996. Generation of nitric oxide and superoxide during reperfusion after focal cerebral ischemia in rats. Am. J. Physiol. **270**(Cell Physiol. 39): C748–752.
17. BRORSON, J.R., P.T. SCHUMACKER & H. ZHANG. 1999. Nitric oxide acutely inhibits neuronal energy production. J. Neurosci. **19:** 147–158.
18. ELIASSON, M.L., *et al.* 1999. Neuronal nitric oxide synthase activation and peroxynitrite formation in ischemic stroke linked to neural damage. J. Neurosci. **19:** 5910–5918.
19. CUI. J.K., E.H. HOLMES & P.K. LIU. 1999. Oxidative damage to the c-fos gene and reduction of its transcription after focal cerebral ischemia. J. Neurochem. **73:** 1164–1174.
20. CUI, J.K., *et al.* 2000. Oxidative DNA damage precedes DNA fragmentation after experimental stroke in rat brain. FASEB J. **14:** 955–967.
21. HUANG, D., *et al.* 2000. Nitric oxide mediates oxidative DNA damage in cell nuclei and neuronal death in cerebral cortex after experimental forebrain ischemia. FASEB J. **14:** 407–417.
22. NAGAYAMA, T., *et al.* 2000. Induction of oxidative DNA damage in the peri-infarct region after permanent focal cerebral ischemia. J. Neurochem. **75:** 1716–1728.
23. KHAN, A.U., *et al.* 2000. The decomposition of peroxynitrite to nitroxyl anion (NO^-) and singlet oxygen in aqueous solution. Proc. Natl. Acad. Sci. U.S.A. **97:** 2984–2989.
24. CLARK, R.S., *et al.* 2001. Detection of single- and double-strand DNA breaks after traumatic brain injury in rats: comparison of in situ labeling techniques using DNA polymerase I, and terminal deoxynucleotidyl transferase. J. Neurotruma **18:** 675–689.
25. DU, C., R. HU & C.A. CSERNANSKY. 1996. Very delayed infarction after mild focal cerebral ischemia: a role for apoptosis? J. Cereb. Blood Flow Metab. **16:** 195–201.
26. CHEN, J., K. JIN & M. CHEN. 1997. Early detection of DNA strand breaks in the brain after transient focal ischemia: Implications for the role of DNA damage in apoptosis and neuronal cell death. J. Neurochem. **69:** 232–245.
27. RINK, A., *et al.* 1995. Evidence of apoptotic cell death after experimental traumatic brain injury in the rat. Am. J. Pathol. **147:** 1575–1583.
28. NEWCOMB, J.K., et al. 1999, Temporal profile of apoptotic-like changes in neurons and astrocytes following controlled cortical impact injury in the rat. Exp. Neurol. **157:** 76–88.
29. FUJIMURA, M., *et al.* 1999. Early decrease of apurinic/apyrimidinic endonuclease expression after transient focal cerebral ischemia in mice. J. Cereb. Blood Flow Metab. **19:** 495–501.
30. MORITA-FUJIMURA, Y., *et al.* 1999. Early decrease in apurinic/apyrimidinic endonuclease is followed by DNA fragmentation after cold injury-induced brain trauma in mice. Neurosci. **93:** 1465–1473.
31. CONTI, A.C., *et al.* 1998. Experimental brain injury induces regionally distinct apoptosis during the acute and delayed post-traumatic period. J. Neurosci. **18:** 5663–5672.
32. COLICOS, M.A., C.E. DIXON & P.K. DASH. 1996. Delayed, selective neuronal death following experimental cortical impact injury in rats: possible role in memory deficits. Brain Res. **739:** 111–119.
33. SMITH, F.M., *et al.* 2000. TUNEL-positive staining of surface contusions after fatal head injury in man. Acta. Neuropathol. (Berlin) **100:** 537–545.
34. NG, I., *et al.* 2000. Apoptosis occurs after cerebral contusions in humans. Neurosurgery **46:** 949–956.
35. GAVRIELI, Y.Y. SHERMAN & S.A. BEN-SASSON. 1992. Identification of programmed cell death *in situ* via specific labeling of nuclear DNA fragmentation. J. Cell Biol. **119:** 493–501.
36. LU, J., *et al.* 2000. Changes in apoptosis-related protein (p53, Bax, Bcl-2 and Fos) expression with DNA fragmentation in the central nervous system in rats after closed head injury. Neurosci. Lett. **290:** 89–92.

37. O'DELL, D.M., *et al.* 2000. Traumatic brain injury alters the molecular fingerprint of TUNEL-positive cortical neurons *in vivo*: a single-cell analysis. J. Neurosci. **20:** 4821–4828.
38. BUKI, A., *et al.* 2000. Cytochrome c release and caspase activation in traumatic axonal injury. J. Neurosci. **20:** 2825–2834.
39. ALLEN, J.W., S.M. KNOBLACH & A.I. FADEN. 1999. Combined mechanical trauma and metabolic impairment *in vitro* induces NMDA receptor-dependent neuronal cell death and caspase-3-dependent apoptosis. FASEB J. **13:** 1875–1882.
40. MORITA-FUJIMURA, Y., *et al.* 1999. Inhibition of interleukin-1beta converting enzyme family proteases (caspases) reduces cold injury-induced brain trauma and DNA fragmentation in mice. J. Cereb. Blood Flow Metab. **19:** 634–642.
41. CLARK, R.S., *et al.* 2000. Caspase-3 mediated neuronal death after traumatic brain injury in rats. J. Neurochem. **74:** 740–753.
42. WHALEN. M.J., *et al.* 1999. Reduction of cognitive and motor deficits after traumatic brain injury in mice deficient in poly(ADP-ribose) polymerase. J. Cereb. Blood Flow Metab. **19:** 835–842.
43. LAPLACA, M.C., *et al.* 1999. Temporal patterns of poly(ADP-ribose) polymerase activation in the cortex following experimental brain injury in the rat. J. Neurochem. **73:** 205–213.
44. LIU, P.K., *et al.* 2001. Ischemic injury and faulty gene transcripts in the brain. Trend. Neurosci. **24:** 579–586.
45. NUNOMURA, A., G. PERRY & M.A. PAPPOLLA. 1999. RNA oxidation is a prominent feature of vulnerable neurons in Alzheimer's disease. J. Neurosci. **19:** 1959–1964.
46. LAN, J., *et al.* 2000. Formation of the base modification 8-hydroxyl-2′-deoxyguanosine and DNA fragmentation following seizures induced by systemic kainic acid in the rat. J. Neurochem. **74:** 302–309.
47. TOBITA, M., *et al.* 1995. DNA single-strand breaks in postischemic gerbil brain detected by *in situ* nick translation procedure. Neurosci. Lett. **200:** 129–132.
48. HENNER, W.D., S.M. GRUNBERG & W.A. HASELTINE. 1982. Sites and structure of gamma radiation-induced DNA strand breaks. J. Biol. Chem. **257:** 11750–11754.
49. DEMPLE, B. & L. HARRISON. 1994. Repair of oxidative damage to DNA: enzymology and biology. Annu. Rev. Biochem. **63:** 915–948.
50. WARD, J.F. 1998. Nature of lesions formed by ionizing radiation. *In* DNA Damage and Repair, Vol. 2. J.A. Nickoloff & M.F. Hoekstra, Eds.: 65–84. Humana Press, Totowa.
51. OLINSKI, R., Z. NACKERDIEN & M. DIZDAROGLU. 1992. DNA-protein cross-linking between thymine and tyrosine in chromatin of gamma-irradiated or H_2O_2-treated cultured human cells. Arch. Biochem. Biophys. **297:** 139–143.
52. KASPRZAK, K.S., B A. DIWAN & J.M. RICE. 1992. Nickel(II)-mediated oxidative DNA base damage in renal and hepatic chromatin of pregnant rats and their fetuses. Possible relevance to carcinogenesis. Chem. Res. Toxicol. **5:** 809–815.
53. DIZDAROGLU, M. 1992, Oxidative damage to DNA in mammalian chromatin. Mut. Res. **275:** 331–342.
54. MECOCCI, P., *et al.* 1993. Oxidative damage to mitochondrial DNA shows marked age-dependent increases in human brain. Ann. Neurol. **34:** 609–616.
55. PARK, E.M., *et al.* 1992. Assay of excised oxidative DNA lesions: isolation of 8-oxoguanine and its nucleoside derivatives from biological fluids with a monoclonal antibody column. Proc. Natl. Acad. Sci. U.S.A. **89:** 3375–3379.
56. SCHULZ, J.B., *et al.* 1995. Blockade of neuronal nitric oxide synthase protects against excitotoxicity *in vivo*. J. Neurosci. **15:** 8419–8429.
57. BOHR, V.A., *et al.* 1985. DNA repair in an active gene: removal of pyrimidine dimers from the DHFR gene of CHO cells is much more efficient than in the genome overall. Cell **40:** 359–369.
58. MELLON, I., G. SPIVAK & P.C. HANAWALT. 1987. Selective removal of transcription-blocking DNA damage from the transcribed strand of the mammalian DHFR gene. Cell **51:** 241–249.
59. DRIGGERS, W.J., S.P. LEDOUX & G.L. WILSON. 1993. Repair of oxidative damage within the mitochondrial DNA of RINr38 cells. J. Biol. Chem. **268:** 22042–22045.

60. TAFFE, B.G., *et al.* 1996. Gene-specific nuclear and mitochondrial repair of formamidopyrimidine DNA glycosylase-sensitive sites in Chinese hamster ovary cells. Mut. Res. **364:** 183–192.
61. YAKES, F.M. & B. VAN HOUTON. 1997. Mitochondrial DNA damage is more extensive and persists longer than nuclear DNA damage in human cells following oxidative stress. Proc. Natl. Acad. Sci. U.S.A. **94:** 514–519.
62. RINK, S.M., J.C. SHEN & L.A. LOEB. 1998. Creation of RNA molecules that recognize the oxidative lesion 7,8-dihydro-8-hydroxy-2′-deoxyguanosine (8-oxodG) in DNA. Proc. Natl. Acad. Sci. U.S.A. **95:** 11619–11624.
63. LIN, L., *et al.* 2000. Upregulation of base excision repair activity for Oh8dG in the mouse brain after forebrain ischemia-reperfusion. J. Neurochem. **74:** 1098–1105.
64. OLIVE, P.L., J.P. BANATH & C.D. FJELL. 1994. DNA strand breakage and DNA structure influence staining with propidium iodide using the alkaline comet assay. Cytometry **16:** 305–312.
65. GOBBEL, G.T. & M. BELLINZONA. 1998. Response of postmitotic neurons to X-irradiation: implications for the role of DNA damage in neuronal apoptosis. J. Neurosci. **18:** 147–155.
66. ENDRES, M., *et al.* 2000. DNA methyltransferase contributes to delayed ischemic brain injury. J. Neurosci. **20:** 3175–3181.
67. LOEB, L.A. & B.D. PRESTON. 1986. Mutagenesis by apurinic/apyrimidinic sites. Annu. Rev. Genet. **20:** 201–230.
68. ZOU, W. & P.W. DOETSCH. 1993. Effects of abasic sites and DNA single-strand breaks on prokaryotic RNA polymerases. Proc. Natl. Acad. Sci. U.S.A. **90:** 6601–6605.
69. ATAMNA, H., I. CHEUNG & B. AMES. 2000. A method for detecting abasic sites in living cells: age-dependent changes in base excision repair. Proc. Natl. Acad. Sci. USA **97:** 686–691.
70. LIU, P.K., *et al.* 2002. The *in situ* detection of apyrimidinic/apurinic sites and DNA breaks bearing extension blocking termini. *In* Detection of DNA Damage: Methods and Protocols. V.V. Didenko, Ed. Humana Press, New Jersey. In press.
71. BREDT, D.S., *et al.* 1991. Nitric oxide synthase protein and mRNA are discretely localized in neuronal populations of the mammalian CNS together with NADPH diaphorase. Neuron **7:** 615–624.
72. CORK, R.J., *et al.* 1998. A web-accessible digital atlas of the distribution of nitric oxide synthase in the mouse brain. Prog. Brain Res. **118:** 37–50.
73. DAWSON, T.M., *et al.* 1991. Nitric oxide synthase and neuronal NADPH diaphorase are identical in brain and peripheral tissues. Proc. Natl. Acad. Sci. U.S.A. **8:** 7797–7801.
74. VINCENT, S.R. & H. KIMURA. 1992. Histochemical mapping of nitric oxide synthase in the rat brain. Neurosci. **46:** 755–784.
75. LIPTON, S.A., *et al.* 1993. A redox-based mechanism for the neuroprotective and neurodestructive effects of nitric oxide and related nitroso-compounds. Nature **364:** 626–632.
76. DAWSON, T.M., *et al.* 1998. Nitric oxide: Diverse actions in the central and peripheral systems. Neuroscientist **4:** 96–112.
77. AYATA, C., *et al.* 1997. Mechanisms of reduced striatal NMDA excitotoxicity in type I nitric oxide synthase knock-out mice. J. Neurosci. **17:** 6908–6917.
78. ESTEVEZ, A.G., *et al.* 1998. Nitric oxide and superoxide contribute to motor neuron apoptosis induced by trophic factor deprivation. J. Neurosci. **18:** 923–931.
79. HUANG, Z., *et al.* 1994. Effects of cerebral ischemia in mice deficient in neuronal nitric oxide synthase. Science **265:** 1883–1885.
80. GONZALEZ-ZULUETA, M., *et al.* 1998. Manganese superoxide dismutase protects nNOS neurons from NMDA and nitric oxide-mediated neurotoxicity. J. Neurosci. **18:** 2040–2055.
81. KAMII, H., *et al.* 1996. Effects of nitric oxide synthase inhibition on brain infarction in SOD-1-transgenic mice following transient focal cerebral ischemia. J. Cereb. Blood Flow Metab. **16:** 1153–1157.
82. YOSHIDA, T., *et al.* 1994. The NOS inhibitor, 7-nitroindazole, decreases focal infarct volume but not the response to topical acetylcholine in pial vessels. J. Cereb. Blood Flow Metab. **14:** 924–929.

83. IADECOLA, C., *et al.* 1997. Delayed reduction of ischemic brain injury and neurological deficits in mice lacking the inducible nitric oxide synthase gene. J. Neurosci. **17:** 9157–9164.
84. O'NEILL, M.J., C. HICKS & M. WARD. 1996, Neuroprotective effects of 7-nitroindazole in the gerbil model of global cerebral ischemia. Eur. J. Pharm. **310:** 115–122.
85. ZAHARCHUK, G., *et al.* 1997. Neuronal nitric oxide synthase mutant mice show smaller infarcts and attenuated apparent diffusion coefficient changes in the peri-infarct zone during focal cerebral ischemia. Magn. Reson. Med. **37:** 170–175.
86. HARA, H., *et al.* 1996. Reduced brain edema and infarction volume in mice lacking the neuronal isoform of nitric oxide synthase after transient MCA occlusion. J. Cereb. Blood Flow Metab. **16:** 605–611.
87. PANAHIAN, N., *et al.* 1996. Attenuated hippocampal damage after global cerebral ischemia in mice mutant in neuronal nitric oxide synthase. Neurosci. **72:** 343–354.
88. HUANG, Z., *et al.* 1996. Enlarged infarcts in endothelial nitric oxide synthase knockout mice are attenuated by nitro-L-arginine. J. Cereb. Blood Flow Metab. **16:** 981–987.
89. GUZIK, T.J., *et al.* 2001. Relationship between the G894T polymorphism (Glu298Asp variant) in endothelial nitric oxide synthase and nitric oxide-mediated endothelial function in human atherosclerosis. Am. J. Med. Genet. **100:** 130–137.
90. SHIMASAKI, Y., *et al.* 1998. Association of the missense Glu298Asp variant of the endothelial nitric oxide synthase gene with myocardial infarction. J. Am. Coll. Cardiol. **31:** 1506–1510.
91. YOSHIMURA, M., *et al.* 1998. A missense Glu298Asp variant in the endothelial nitric oxide synthase gene is associated with coronary spasm in the Japanese. Hum. Genet. **103:** 65–69.
92. YOSHIMURA, M., *et al.* 2000. Genetic risk factors for coronary artery spasm: significance of endothelial nitric oxide synthase gene T-786 → C and missense Glu298Asp variants. Investig. Med. **48:** 367–374.
93. HINGORANI, A.D., *et al.* 1999. A common variant of the endothelial nitric oxide synthase (Glu298 → Asp) is a major risk factor for coronary artery disease in the UK. Circulation **100:** 1515–1520.
94. SHOJI, M., *et al.* 2000. Positive association of endothelial nitric oxide synthase gene polymorphism with hypertension in northern Japan. Life Sci. **66:** 2557–2562.
95. MIYAMOTO, Y., *et al.* 1998. Endothelial nitric oxide synthase gene is positively associated with essential hypertension. Hypertension **32:** 3–8.
96. BENJAFIELD, A.V. & B.J. MORRIS. 2000. Association analyses of endothelial nitric oxide synthase gene polymorphisms in essential hypertension. Am. J. Hypertens. **13:** 994–998.
97. NAKAYAMA, M., *et al.* 1999. T-786 → C mutation in the 5′-flanking region of the endothelial nitric oxide synthase gene is associated with coronary spasm. Circulation **99:** 2864–2870.
98. SCHNEIDER, M.P., *et al.* 2000. Functional gene testing of the Glu298Asp polymorphism of the endothelial NO synthase. J. Hypertens. **18:** 1767–1773.
99. KATO, N., *et al.* 1999. Lack of evidence for association between the endothelial nitric oxide synthase gene and hypertension. Hypertension **33:** 933–936.
100. WADA, K., *et al.* 1998. Role of nitric oxide in traumatic brain injury in the rat. J. Neurosurg. **89:** 807–818.
101. CHERIAN, L., J.C. GOODMAN & C.S. ROBERTSON. 2000. Brain nitric oxide changes after controlled cortical impact injury in rats. J. Neurophysiol. **83:** 2171–2178.
102. WADA, K., *et al.* 1998. Inducible nitric oxide synthase expression after traumatic brain injury and neuroprotection with aminoguanidine treatment in rats. Neurosurgery **43:** 1436.
103. WADA, K., *et al.* 1999. Effects of L-NAME and 7-NI on NOS catalytic activity and behavioral outcome after traumatic brain injury in the rat. J. Neurotrauma **16:** 203–212.
104. LU,Y.C., *et al.* 1997. Inhibition of nitric oxide synthase potentiates hypertension and increases mortality in traumatically brain-injured rats. Mol. Chem. Neuropathol. **30:** 125–137.
105. SINZ, E.H., *et al.* 1999. Inducible nitric oxide synthase is an endogenous neuroprotectant after traumatic brain injury in rats and mice. J. Clin. Invest. **104:** 647–656.

106. CUI, J.K. & P.K. LIU. 2002. Expression of aberrant mRNA during oxidative stress in the mouse brain: an implication of transcription from nuclear genes that contained abasic sites. Submitted.
107. WHITFIELD, P.C. & J.D. PICKARD. 2000. Expression of the immediate early genes c-Fos and c-Jun after head injury in man. Neurol. Res. **22:** 138–144.
108. GILLARDON, F., *et al.* 1996. Activation of c-Fos contributes to amyloid b-peptide-induced neurotoxicity. Brain Res. **706:** 169–172.
109. SHARP, F.R., *et al.* 1999. Heat–shock protein protection. Trend. Neurosci. **22:** 97–99.
110. AN, G., *et al.* 1993. Expression of c-fos and c-jun family genes after focal cerebral ischemia. Ann. Neurol. **33:** 457–464.
111. CUI, J.K., C.Y. HSU & P.K. LIU. 1999. Suppression of postischemic hippocampal nerve growth factor expression by a c-fos antisense oligodeoxynucleotide. J. Neurosci. **19:** 1335–1344.
112. ZHANG, Y.J., *et al.* 1999. Suppression of post-ischemic-induced fos protein expression by an antisense oligonucleotide to c-fos mRNA leads to increased tissue damage. Brain Res. **19:** 112–117.
113. GUEGAN, C., *et al.* 1998. Reduction of cortical infarction and impairment of apoptosis in NGF-transgenic mice subjected to permanent focal ischemia. Mol. Brain Res. **55:** 133–140.
114. DIXON, C.E., *et al.* 1997. Nerve growth factor attenuates cholinergic deficits following traumatic brain injury in rats. Exp. Neurol. **146:** 479–490.
115. LIU, J.S., *et al.* 1995. Delayed transhemishperic c-fos gene expression after focal cerebral ischemia-reperfusion injury in rats. J. Formos. Med. Assoc. **94:** 649–654.
116. HAYES, R.L., *et al.* 1995. Changes in gene expression following traumatic brain injury in the rat. J. Neurotrauma **12:** 779–790.
117. YANG, K., *et al.* 1994. Increased expression of c-fos mRNA and AP-1 transcription factor after cortical impact injury in rats. Brain Res. **664:** 141–147.
118. YANG, K. 1995. Regional and temporal profiles of c-fos and nerve growth factor mRNA expression in rat brain after lateral cortical impact injury. J. Neurosci. Res. **42:** 571–578.
119. RAGHUPATHI, R. & T.K. MCINTOSH. 1996. Regionally and temporally distinct patterns of induction of c-fos, c-jun and junB mRNAs following experimental brain injury in the rat. Brain Res. Mol. Brain Res. **37:** 134–144.
120. RAGHUPATHI, R., T.K. MCINTOSH & D.H. SMITH. 1995. Cellular response to experimental brain injury. Brain Pathol. **5:** 437–442.
121. RAGHUPATHI, R., *et al.* 1995. Regional induction of c-fos and heat shock protein-72 mRNA following fluid-percussion brain injury in the rat. J. Cereb. Blood Flow. Metab. **15:** 467–473.
122. DUNN-MEYNELL, A.A., E. GOVEK & B.E. LEVIN. 1997. Intracarotid glucose selectively increases Fos-like immunoreactivity in paraventricular, ventromedial and dorsomedial nuclei neurons. Brain Res. **748:** 100–106.
123. KATANO, H., *et al.* 1998. Differential induction of immediate early gene mRNAs following cryogenic and impact trauma with/without craniotomy in rats. Brain Res. **800:** 69–77.
124. DUTCHER, S.A., *et al.* 1999. Patterns of immediate early gene mRNA expression following rodent and human traumatic brain injury. Neurol. Res. **21:** 234–242.
125. PHILLIPS, L.L. & E.T. BELARDO. 1992. Expression of c-fos in the hippocampus following mild and moderate fluid percussion brain injury. J. Neurotrauma **9:** 323–333.
126. DAVE, J.R., R.A. BAUMAN & J.B. LONG. 1997. Hypoxia potentiates traumatic brain injury-induced expression of c-fos in rats. Neuroreport **8:** 395–398.
127. CAMADOLA, S., G. POLI & M.P. MATTSON. 2000. The lipid peroxidation product 4-hydroxy-2,3-nonenal increases AP-1-binding activity through caspase activation in neurons. J. Neurochem. **74:** 159–168.
128. KRUIJER, W., D. SCHUBERT & I.M. VERMA. 1985. Induction of the proto-oncogene fos by nerve growth factor. Proc. Natl. Acad. Sci. U.S.A. **82:** 7330–7334.
129. HENGERER, B., *et al.* 1990. Lesion-induced increase in nerve growth factor mRNA is mediated by c-fos. Proc. Natl. Acad. Sci. U.S.A. **87:** 3899–3903.

130. Ip, N.Y., *et al.* 1993. Cultured hippocampal neurons show responses to BDNF, NT-3, and NT-4, but not NGF. J. Neurosci. **13:** 3394–3405.
131. Ghosh, A., J. Carnahan & M.E. Greenberg. 1994. Requirement for BDNF in activity-dependent survival of cortical neurons. Science **263:** 1618–1623.
132. Johnson, E.M., Jr., *et al.* 1995. Neuronal apoptosis: current understanding of molecular mechanisms and potential role in ischemic brain injury. J. Neurotrauma **12:** 843–852.
133. Liu, P.K., *et al.* 1994. Suppression of ischemia-induced Fos expression and AP-1 activity by an antisense oligodeoxynucleotide to c-fos mRNA. Ann. Neurol. **36:** 566–576.
134. Kromer, L.F. 1987. Nerve growth factor treatment after brain injury prevents neuronal death. Science **235:** 214–216.
135. Sinson, G., *et al.* 1997. Improvement of cognitive deficits and decreased cholinergic neuronal cell loss and apoptotic cell death following neurotrophin infusion after experimental traumatic brain injury. J. Neurosurg. **86:** 511–518.

A Reassessment of the Peroxynitrite Scavenging Activity of Uric Acid

M. WHITEMAN, U. KETSAWATSAKUL, AND B. HALLIWELL

Department of Biochemistry, Faculty of Medicine, National University of Singapore, Singapore

ABSTRACT: Peroxynitrite is implicated in numerous human diseases. Hence, there is considerable interest in potential therapeutic peroxynitrite scavengers. It has been claimed that uric acid is a powerful peroxynitrite scavenger. We previously observed that uric acid is a powerful inhibitor of tyrosine nitration induced by peroxynitrite, but fails to prevent α_1-antiproteinase (α_1-AP) inactivation induced by peroxynitrite. However, the reactivity of peroxynitrite is significantly modified by bicarbonate and this has not been considered in evaluating the scavenging activity of uric acid and other endogenous antioxidant compounds. In the presence of bicarbonate (25 mM), the ability of uric acid, ascorbate, Trolox, and GSH to inhibit peroxynitrite-mediated tyrosine and guanine nitration is decreased. Protection against peroxynitrite-mediated α_1-AP inactivation is also decreased by ascorbate, Trolox, and GSH, but it is enhanced by uric acid. Bicarbonate also inhibits the ability of these compounds to prevent peroxynitrite-mediated ABTS radical cation formation. However, the abilities of these antioxidants to prevent peroxynitrite-mediated bleaching of pyrogallol red are enhanced by bicarbonate. These results show that physiologic concentrations of bicarbonate substantially modify the ability of uric acid to prevent peroxynitrite-mediated reactions. This study highlights the need to use several different assays in the presence of physiologically relevant concentrations of bicarbonate when assessing compounds for peroxynitrite scavenging, in order to avoid misleading results.

KEYWORDS: nitric oxide; nitrosative stress; 3-nitrotyrosine; oxidative stress; peroxynitrite; uric acid

ABBREVIATIONS:

ABTS	2,2′-azinobis-(3-ethylbenzthiazoline-6-sulfonic acid diammonium salt.
$ABTS^{+\cdot}$	2,2′azinobis(3-ethylbenzthiazoline-6-sulfonic acid radical cation.
α_1-AP	α_1-antiproteinase.
HCO_3^-	bicarbonate anion.
$CO_3^{-\cdot}$	carbonate radical anion.
GSH	reduced glutathione.
H_2O_2	hydrogen peroxide.
·OH	hydroxyl radical.
LDL	low density lipoprotein.
$\cdot NO_2$	nitrogen dioxide.
·NO	nitrogen monoxide.

Address for correspondence: Department of Biochemistry, Faculty of Medicine, National University of Singapore, 10 Kent Ridge Crescent, Singapore 119260. Voice: +65-874-8891; fax: +65-779-1453.

bchwml@nus.edu.sg

8-NO_2 guanine	8-nitroguanine.
3-NO_2 tyrosine	3-nitrotyrosine.
NO_2^+	nitryl cation.
$ONOO^-$	oxoperoxonitrate (1-).
$ONOOCO_2$	nitrosoperoxycarbonate anion.
PR	pyrogallol red.
SANA	*N*-succinyl $(ala)_3$ *p*-nitroanilide.
$NaHCO_3$	sodium bicarbonate.
$O_2^{-\cdot}$	superoxide.

INTRODUCTION

Peroxynitrite (oxoperoxonitrate [1-]; $ONOO^-$) is a reactive nitrogen species formed from the near diffusion-controlled reaction of nitrogen monoxide (·NO) with superoxide ($O_2^{-\cdot}$), see Equation **(1)**.[1] Its formation *in vivo,* usually inferred from the measurement of a stable biomarker for RNS activity, 3-nitrotyrosine, has been implicated in numerous human disease conditions (reviewed in Refs. 2 and 3).

$$\cdot NO + O_2^{-\cdot} \xrightarrow{k\ =\ 6.7 \times 10^9\ M^{-1} sec^{-1}} ONOO^- \qquad \textbf{(1)}$$

Under physiologic conditions $ONOO^-$ has a halflife of less than one second and it is converted to its protonated form peroxynitrous acid, ONOOH, which in turn decays to generate multiple toxic products with reactivities similar to those of the nitryl cation (NO_2^+), nitrogen dioxide ($\cdot NO_2$) and hydroxyl radical (·OH). Peroxynitrite and species derived from it can also oxidize and nitrate lipids,[4] proteins,[5] DNA,[6,7] and carbohydrates.[8] Addition of $ONOO^-$ to biological fluids leads to the rapid depletion of endogenous antioxidants, such as uric acid, ascorbate, GSH, and other thiols.[9] Cells exposed to $ONOO^-$ show DNA damage,[6,10] DNA strand breakage,[6] and activation of poly(ADP-ribosyl)polymerase[11–13] and caspase-3,[14,15] as well as activation of several MAP kinases, including ERKI/2, p38, JNK, and Src.[16–20] Furthermore, $ONOO^-$ reacts with the amino acid tyrosine to form a nitrated biomarker, 3-nitrotyrosine,[2,3] the formation of which has been measured in numerous human diseases, including neurodegenerative and inflammatory diseases, as well as cancer, diabetes, arthritis, and asthma (reviewed in Refs. 2 and 3).

The chemistry of peroxynitrite is complex. It has been shown that $ONOO^-$ reacts with CO_2 at a rate that is significantly higher than the reaction of $ONOO^-$ with other biomolecules, including endogenous and synthetic antioxidants, to form nitrosoperoxycarbonate anion. The bicarbonate anion (HCO_3^-) is one of the most abundant constituents of the extracellular fluids. It is present at a concentration of approximately 25 mM in blood plasma and 12 mM in intracellular fluids. Most of the HCO_3^- dissolved in plasma and cytosol is present as carbon dioxide (CO_2) with a normal plasma concentration of approximately 1.3 mM and this HCO_3^-–CO_2 system is one of the major pH buffering systems in the blood. It is the peroxynitrite anion ($ONOO^-$), rather than peroxynitrous acid (ONOOH), that reacts with bicarbonate via the formation of a novel adduct; the nitrosoperoxycarbonate anion ($ONOOCO_2^-$).[21] More recently this adduct was shown to be formed from the reaction between $ONOO^-$ and CO_2, present in equilibrium with HCO_3^-, with a rate constant of 3–5.8 × 10^4 M^{-1} sec^{-1} at 37°C **(2)**, one of the fastest reactions known for $ONOO^-$.[21,22]

$$ONOO^- + CO_2 \rightarrow ONOOCO_2^- \quad (2)$$

High, but physiologic concentrations of HCO_3^- (5–30 mM) diminish the bactericidal and parasiticidal activity of $ONOO^-$,[22,23] inhibit $ONOO^-$-dependent hydroxylation and nitration of phenylalanine[24] and benzoate,[25] and inhibit methionine and GSH oxidation.[22,26–28] However, peroxynitrite-dependent nitration of tyrosine, guanine, *p*-hydroxyphenylacetate, phenol, and proteins are increased when HCO_3^- is added to the reaction mixture.[22,29–32] Thus, measurement of tyrosine nitration may reflect $ONOOCO_2^-$-mediated tyrosine nitration rather than purely $ONOO^-$. Interestingly HCO_3^- inhibits LDL and protein oxidation induced by $ONOO^-$, but enhances LDL and protein oxidation induced by the $ONOO^-$ donor, SIN-1.[33] Therefore, the HCO_3^-|CO_2 system significantly influences the reactivity and chemistry of $ONOO^-$ toward various biomolecules.

Due to the cytotoxicity of $ONOO^-$ and its apparent formation at sites of tissue injury,[2,3] there has been considerable interest in the ability of natural and synthetic antioxidants to diminish $ONOO^-$ damage.[34–37] Uric acid, an end product of human purine metabolism, has recently been proposed to fulfil this role as a $ONOO^-$ scavenger and neuroprotective agent.[38–42] Uric acid is present in human blood plasma at concentrations in the range 150–500 μM.[43] In addition to its $ONOO^-$-scavenging, uric acid is a scavenger of singlet oxygen, peroxyl, and ·OH radicals,[44,45] ozone,[46] and hypochlorous acid,[47] and it can chelate iron and copper ions to limit ·OH generation.[43] Indeed, allantoin, an oxidation product of uric acid has been detected in increased amounts under conditions of oxidative stress.[43,48]

It has been suggested that uric acid acts as a peroxynitrite "scavenger" in multiple sclerosis (MS), since uric acid levels are reported to be lower in patients with multiple sclerosis (MS)[38,49] and low levels of uric acid in sera correlate with a higher degree of MS disease activity.[49] Uric acid administration to animals with experimental allergic encephalomyelitis as a model for human multiple sclerosis has been shown to prevent disease progression and CNS inflammation, to preserve blood–brain barrier integrity and to reduce oxidative stress, reportedly by scavenging $ONOO^-$,[38–42] and to inhibit 3-nitrotyrosine formation. Therefore, the role of uric acid as a natural protective agent in human diseases where $ONOO^-$ formation is implicated[2,3] needs to be investigated.

Recently, our laboratory reported that the $ONOO^-$ scavenging activity of certain plant phenolic compounds could be substantially modified by the presence of physiologic concentrations of added HCO_3^- in the reaction media.[50] Since uric acid has been proposed as a natural $ONOO^-$ scavenger in neurodegenerative diseases, and since previous investigations on the $ONOO^-$-scavenging activity of uric acid were only performed in the absence of added bicarbonate,[34] it was logical to reevaluate the $ONOO^-$-scavenging activities of uric acid. For that study we examined five model systems $ONOO^-$-mediated tyrosine[51] and guanine nitration,[7] α_1-antiproteinase (α_1-AP) inactivation,[34] pyrogallol red bleaching,[52] and $ABTS^{+\cdot}$ radical formation.[53,54] Nitration of tyrosine residues in proteins has been observed in numerous human diseases[2,3] whereas α_1-AP inactivation[55] is observed at sites of inflammation; $ONOO^-$ formation provides a plausible mechanism to explain these events.[56,57]

MATERIALS AND METHODS

Materials

Ascorbic acid was obtained from Calbiochem (Singapore). HPLC grade methanol and acetonitrile were purchased from BDH (Singapore). Trolox, uric acid, calf thymus DNA (type I), and all other reagents, including α_1-antiproteinase (A9024) and porcine elastase (E0258), were obtained from SigmaAldrich (Singapore) and were of the highest quality available commercially. Distilled water, passed through an Elga water purification system was used for all purposes. 8-Nitroguanine was synthesized as described elsewhere.[58]

Synthesis of Peroxynitrite

Synthesis of peroxynitrite was essentially as described in Reference 59. Briefly, an acidic solution (0.6 M HCl) of H_2O_2 (0.7 M) was mixed with KNO_2 (0.6 M) on ice for one second and the reaction quenched with ice-cold NaOH (1.2 M). Residual H_2O_2 was removed by mixing with granular MnO_2 prewashed with NaOH (1.2 M). The stock solution was filtered and then frozen overnight (−20°C) and the top layer of the solution collected for the experiment. The concentration of stock $ONOO^-$ was redetermined before each experiment at 302 nm using a molar absorption coefficient of 1,670 $M^{-1}cm^{-1}$.[60] Concentration values of 200–250 mM were usually obtained. Once thawed, $ONOO^-$ solutions were kept on ice for no longer than 30 min before use.

Prevention of α_1-Antiproteinase Inactivation

Elastase and α_1-antiproteinase (α_1-AP) were measured essentially as described in Reference 36. Briefly, 40 µl of α_1-AP stock solution (4 mg/ml) were incubated in 100 mM K_2HPO_4–KH_2PO_4 buffer, pH 7.4, containing 25 mM $NaHCO_3$ and various antioxidants, in a water bath for 15 minutes at 37°C. After that time, $ONOO^-$ (500 µM) was added, the samples incubated for five minutes, and 2.0 ml phosphate buffer added. Residual elastase activity was measured by the addition of 100 µl elastase substrate (Succinyl $(Ala)_3$ *p*-nitroanilide, SANA; 4mg/ml) and the change in absorbance at 410 nm measured for 30 seconds. To assess the role of $ONOO^-$ decomposition products on α_1-AP activity, "decomposed" $ONOO^-$ was prepared by adding 500 µM $ONOO^-$ to 100 mM phosphate buffer, pH 7.4, at 37°C, five minutes prior to the addition of α_1-antiproteinase or antioxidants.[36]

Measurement of Tyrosine Nitration

Tyrosine nitration was measured as described in Reference 34. Briefly, 1 mM tyrosine was incubated in a solution containing 100 mM K_2HPO_4–KH_2PO_4 buffer, pH 7.4, with 25 mM $NaHCO_3$ and various antioxidants, in a water bath for 15 minutes at 37°C. After that time, 1 mm $ONOO^-$ was added and the solution further incubated for 15 minutes. Reverse phase HPLC analysis was performed using a Spherisorb 5ODS2 C_{18} column with a mobile phase consisting of 100 mM K_2HPO_4–KOH buffer, pH 3.01, with 20% (v/v) methanol. Peak area (retention time of 3.8

min) was measured and concentrations calculated from a standard curve. The detection limit for 3-NO_2 tyrosine was 0.1 μm.

Measurement of Guanine Nitration

Guanine nitration was measured as described in Reference 58. Briefly, 1 mM guanine was incubated in a solution containing 100 mM K_2HPO_4–KH_2PO_4 buffer, pH 7.4, with or without 25 mM $NaHCO_3$ and various antioxidants, in a water bath for 15 minutes at 37°C. After that time, 1 mm $ONOO^-$ was added and the solution further incubated for 15 minutes. HPLC analysis was performed as already described except that a mobile phase consisting of 100 mM ammonium phosphate, pH 6.00 was used. The peak area (retention time of 10 min) was measured and concentrations were calculated from a standard curve. The detection limit for 8-NO_2 guanine was 0.1 μM.

Assessment of Pyrogallol Red Bleaching by Peroxynitrite

Pyrogallol red[52] (pyrogallosulphonephthalein; 100 μM) was dissolved in 100 mM K_2HPO_4–KH_2PO_4 buffer, pH 7.4, with or without 25 mM $NaHCO_3$ and various antioxidants, and left at room temperature for 10 minutes. After that time, 20 μM $ONOO^-$ were added (as determined from preliminary experiments) and the samples were diluted 1:5 before analysis by UV-visible spectroscopy (Beckman model DU640B). The decreases in absorbance at 542 nm were recorded.[52]

Inhibition of $ABTS^{+\cdot}$ Formation

The one electron oxidation of ABTS by $ONOO^-$ yields the radical cation $ABTS^{-\cdot}$, which can be quantified spectrophotometrically at 734 nm.[53] To examine the effects of urate and other antioxidants on this process, 2 mM ABTS in 100 mM phosphate buffer, pH 7.4, were mixed, with or without various concentrations of antioxidants, with 25 mM bicarbonate for five minutes at room temperature. After that time, 150 μM $ONOO^-$ were added (as determined from preliminary experiments). Increases in absorbance at 734 nm were immediately recorded and yields of $ABTS^+$ calculated using an extinction coefficient of 16,000 M^{-1} cm^{-1}.[54]

Data Analysis

Graphs were plotted showing mean ± standard error of the mean (SEM). In all cases the mean values were calculated from data taken from at least four separate experiments performed on different days using freshly prepared reagents. Where significance was tested, an independent *t*-test (Student's; two populations) was used; $*p \leq 0.1$ $**p \leq 0.05$; $***p \leq 0.01$.

RESULTS

Inhibition of Tyrosine Nitration

When the amino acid tyrosine is exposed to $ONOO^-$ at pH 7.4, 3-nitrotyrosine is formed.[59] As expected,[34,36] the addition of 1 mM $ONOO^-$ to a buffered solution of 1 mM tyrosine led to the formation of 70–80 μM 3-nitrotyrosine. FIGURE 1 shows the effect of the presence of uric acid and other antioxidants in the reaction mixture prior to $ONOO^-$ addition. Uric acid significantly decreased the formation of 3-nitrotyrosine at all concentrations tested. At 60 μM uric acid was significantly more effective at protecting tyrosine from $ONOO^-$-mediated nitration than ascorbate ($p < 0.01$) and GSH ($p < 0.1$). However, when comparing the effectiveness of uric acid against Trolox, no significant difference was observed below 250 μM ($p < 0.05$).

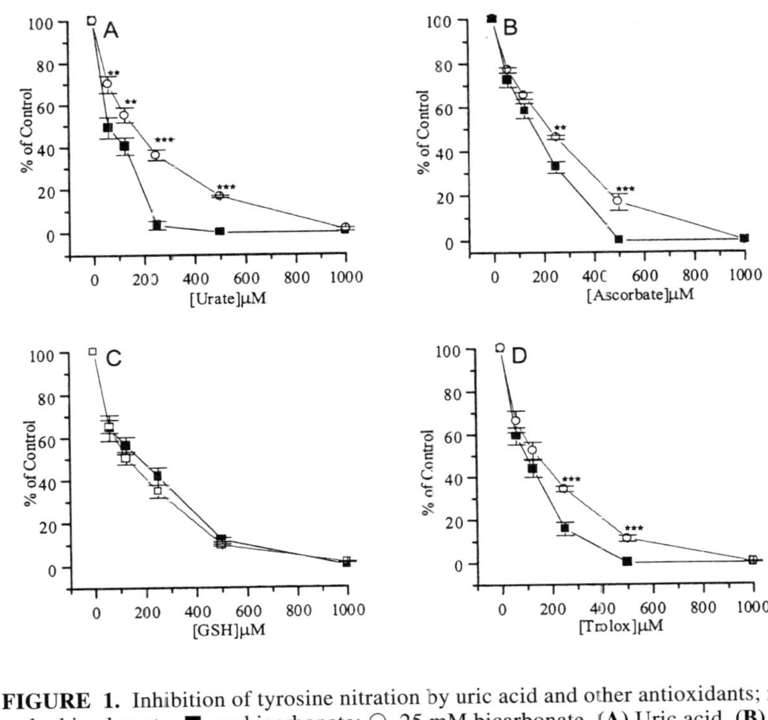

FIGURE 1. Inhibition of tyrosine nitration by uric acid and other antioxidants; modulation by bicarbonate: ■, no bicarbonate; ○, 25 mM bicarbonate. **(A)** Uric acid, **(B)** ascorbate, **(C)** GSH, and **(D)** trolox. Tyrosine (1 mM) incubated with or without HCO_3^- (25 mM) in phosphate buffer, pH 7.4, at 37°C for 15 mins with or without antioxidants (0–1 mM) and $ONOO^-$ (1 mM) added. The 3-NO_2 tyrosine formed was measured by HPLC, as described in MATERIALS AND METHODS. The data shown are mean ± SEM of six or more independent experiments. Significance testing was performed using an independent t-test (Student's; two populations); *$p \leq 0.1$, **$p \leq 0.05$, ***$p \leq 0.01$, comparing no HCO_3^- addition with HCO_3^- addition (25 mM). The control ($ONOO^-$ with tyrosine in the presence or absence of HCO_3^-) was taken as a 100% value.

In the absence of added bicarbonate, uric acid had the lowest IC_{50} (64 ± 14 μM), followed by Trolox (98 ± 8 μM), GSH (194 ± 18 μM), and ascorbate (212 ± 23 μM).

Inclusion of 25 mM bicarbonate in the reaction mixture led to a significantly higher yield of 3-nitrotyrosine, as expected[50] (95–110 μM; $p < 0.01$). The presence of HCO_3 significantly decreased the ability of uric acid, ascorbate, and Trolox, but not that of GSH, to inhibit tyrosine nitration (FIG. 1). Higher concentrations of these antioxidants were needed in the presence of 25 mM bicarbonate to achieve a degree of inhibition similar to that in the absence of bicarbonate. The effectiveness of uric acid at inhibiting $ONOO^-$-mediated tyrosine nitration was substantially reduced, as indicated by the fact that the IC_{50} value for uric acid increased twofold (158 ± 14 μM). In contrast, the IC_{50} values for ascorbate and Trolox were modestly increased (223 ± 14 μM and 111 ± 9 μM, respectively) whereas the IC_{50} value for GSH decreased (132 ± 18 μM).

The addition of decomposed $ONOO^-$[59] to a solution of tyrosine and bicarbonate did not result in the formation of detectable 3-nitrotyrosine (data not shown). Control experiments showed that the addition of HCO_3^- did not change the pH of the reaction mixture (data not shown).

Inhibition of Guanine Nitration

Guanine is also nitrated by $ONOO^-$ at physiologic pH.[6,7,58] As expected, the addition of 1 mM $ONOO^-$ to 1 mM guanine led to the formation of 4 ± 1 μM 8-NO_2 guanine. Uric acid significantly inhibited $ONOO^-$-mediated guanine nitration at all concentrations tested (see FIGURE 2), but in contrast to the tyrosine nitration assay, uric acid was not the most effective agent tested. The IC_{50} value for uric acid was 174 ± 23 μM, whereas the IC_{50} values for ascorbate, GSH, and Trolox were 43 ± 4 μM, 50 ± 5 μM, and 263 ± 8 μM, respectively. As observed with tyrosine nitration assay, inclusion of 25 mM bicarbonate in the reaction mixture significantly increased the yield of 8-NO_2 guanine (5 ± 1 μM). However, FIGURE 2 also shows that the ability of uric acid to inhibit 8-NO_2 guanine formation was greatly reduced in the presence of bicarbonate (IC_{50} 282 ± 39 μM; $p < 0.0$ 1 compared to absence of bicarbonate) The IC_{50} values for ascorbate (170 ± 10 μM), GSH (IC_{50} 184 ± 32 μM), and Trolox (IC_{50} 312 ± 19 μM) were also significantly increased ($p < 0.01$).

The addition of decomposed $ONOO^-$[59] to a solution of guanine and bicarbonate did not result in the formation of detectable 8-NO_2 guanine (data not shown). Control experiments showed that the addition of HCO_3^- did not change the pH of the reaction mixture (data not shown).

Inactivation of α_1-Antiproteinase (α_1-AP)

In agreement with previous studies[34,36,56] the addition of $ONOO^-$ led to substantial loss of the ability of α_1-AP to inhibit the proteolytic enzyme elastase. The inclusion of 25 mM HCO_3^- in the reaction mixture resulted in a small (−10%) but significant ($p < 0.05$) decrease in the ability of $ONOO^-$ to inactivate α_1-AP. Addition of ascorbate, GSH, and Trolox to the reaction mixtures significantly protected (α_1-AP from $ONOO^-$-mediated inactivation, whereas uric acid offered only slight protection even when tested at high (1 mM) concentrations ($p < 0.5$), in accord with previous results[34] (see FIGURE 3). However, in contrast to the observations with the tyrosine and guanine nitration assays, in the presence of bicarbonate, uric acid was able to significantly protect α_1-AP against $ONOO^-$-mediated inactivation, at low concentrations and to a much greater extent in the absence of bicarbonate (FIG. 3A; at all concentrations tested, $p < 0.01$). However, it did not result in further protection at high concentrations.

In contrast to uric acid, the ability of the other antioxidants tested (ascorbate, GSH, and Trolox) to protect α_1-AP from $ONOO^-$-mediated inactivation was decreased to a limited extent by the presence of bicarbonate (FIG. 3B–D). Control experiments showed that none of the compounds tested interfered with elastase or the ability of α_1-AP to inhibit elastase.

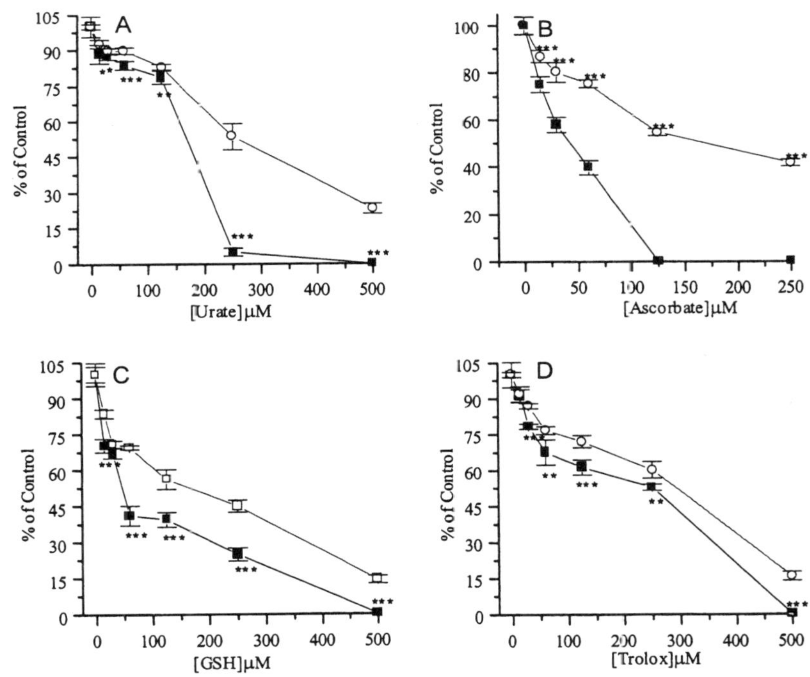

FIGURE 2. Inhibition of $ONOO^-$-mediated guanine nitration by uric acid and other antioxidants; modulation by bicarbonate: ■, no bicarbonate; ○, 25 mM bicarbonate. **(A)** Uric acid, **(B)** ascorbate, **(C)** GSH, and **(D)** trolox. Guanine (1 mM) incubated with or without HCO_3^- (25 mM) in phosphate buffer, pH 7.4, at 37°C for 15 mins with or without antioxidants (0–1 mM) and $ONOO^-$ (1 mM) added. The 8-NO_2 guanine formed was measured by HPLC, as described in MATERIALS AND METHODS. The data shown are mean ± SEM of six or more independent experiments. Significance testing was performed using an independent *t*-test (Student's; two populations); $*p \leq 0.1$, $**p \leq 0.05$, $***p \leq 0.01$, comparing no HCO_3^- addition with HCO_3^- addition (25 mM). The control ($ONOO^-$ with guanine in the presence or absence of HCO_3^-) was taken as a 100% value.

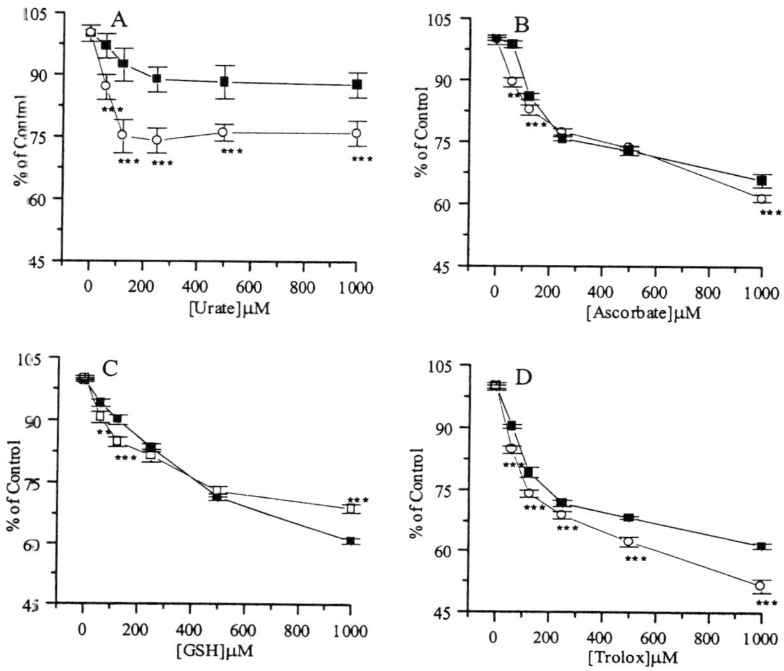

FIGURE 3. Inhibition of $ONOO^-$-mediated α_1-antiproteinase (α_1-AP) inactivation by uric acid and other antioxidants; modulation by bicarbonate: ■, no bicarbonate; ○, 25 mM bicarbonate. **(A)** Uric acid, **(B)** ascorbate, **(C)** GSH, and **(D)** trolox. α_1-AP was incubated with or without HCO_3^- (25 mM) in phosphate buffer, pH 7.4, at 37°C for 15 mins with or without antioxidants (0–1 mM), and $ONOO^-$ (500 µM) added. Residual elastase activity was measured as described in MATERIALS AND METHODS. The data shown are mean ± SEM of six or more independent experiments. Significance testing was performed using an independent *t*-test (Student's; two populations); $*p \leq 0.1$, $**p \leq 0.05$, $***p \leq 0.01$, comparing no HCO_3^- addition with HCO_3^- addition (25 mM). The control ($ONOO^-$-treated α_1-AP in the presence or absence of HCO_3^-) was taken as a 100% value.

Inhibition of Pyrogallol Red Bleaching

An additional assay for assessing $ONOO^-$ scavenging *in vitro* is pyrogallol red (PR) bleaching.[52] $ONOO^-$ and PR concentrations of 200 µM and 100 µM, respectively, were chosen from the results of initial experiments (data not shown), since at these concentrations the effects of $ONOO^-$ scavengers could be accurately assessed. TABLE 1 shows the concentration of uric acid and other antioxidants required to inhibit bleaching of PR by $ONOO^-$ (200 µM) by 30% (IC_{30}). Uric acid was unable to prevent pyrogallol red bleaching at any concentration tested (up to 1 mM) in the absence of bicarbonate. However, when bicarbonate (25 mM) was added to the reaction mixture an IC_{30} value of 114 ± 9 µM was obtained. This value is substantially higher than that for the other antioxidants tested (TABLE 1). The effectiveness of

TABLE 1. Inhibition of $ONOO^-$-mediated bleaching of pyrogallol red (PR) by uric acid: a comparison with other antioxidants and modulation by bicarbonate

	IC_{30} (µM)	
Compound	$ONOO^-$ (200 µM) no bicarbonate	$ONOO^-$ (200 µM) + 25 mM bicarbonate
Urate	> 1 µM	113 ± 9
Trolox	119 ± 5	39 ± 1
Ascorbate	26 ± 3	19 ± 2
GSH	39 ± 2	16 ± 3

NOTES: IC_{30} values for uric acid, trolox, ascorbate, and GSH were obtained as follows; PR (100 µM) was incubated with or without HCO_3^- (25 mM) in phosphate buffer, pH 7.4, at 37°C for 15 mins with or without antioxidants (0–1 mM) and increasing concentrations of $ONOO^-$ (20–200 µM) added. PR bleaching was then measured by a loss in absorbance at 542 nm, as described in MATERIALS AND METHODS. Data are mean (µM) ± SEM of six or more independent experiments. Significance testing was performed using an independent *t*-test (Student's; two populations); $p \leq 0.01$, comparing no HCO_3^- addition with HCO_3^- addition (25 mM). Control, $ONOO^-$-treated PR in the presence or absence of HCO_3^-.

Trolox, ascorbate, and GSH in preventing $ONOO^-$-mediated PR bleaching was significantly enhanced in the presence of bicarbonate (TABLE 1).

Addition of $ONOO^-$ to phosphate buffer (to generate decomposed $ONOO^-$)[58] and subsequent addition of PR, did not result in PR bleaching (data not shown),

Inhibition of $ABTS^{+\cdot}$ Formation

The addition of $ONOO^-$ (150 µM) to a 2 mM solution of ABTS generated $ABTS^{+\cdot}$, leading to the formation of 26.56 ± 1.48 µM $ABTS^{+\cdot}$, the identity of which was confirmed by spectrophotometry (data not shown).[53] Inclusion of uric acid, ascorbate, GSH, or Trolox (see FIGURE 4) in the reaction media significantly decreased the level of $ABTS^{+\cdot}$ in a concentration dependent manner with IC_{50} values of 9.9 ± 0.1 µM, 88.2 ± 3.6 µM, 19.0 ± 2.8 µM, and 14.9 ± 0.5 µM, respectively. The addition of 25 mM bicarbonate to the reaction media increased the yield of $ABTS^{+\cdot}$ by about 10% to 31.56 ± 1.88 µM ($p < 0.5$ as compared to no bicarbonate addition). However, the ability of urate, ascorbate, and to a lesser extent GSH and Trolox, to inhibit $ONOO^-$ mediated ABTS oxidation was significantly inhibited (FIG. 4) with the most pronounced effects observed with ascorbate. Consequently, IC_{50} values for uric acid (12.2 ± 1 µM) and ascorbate (114.4 ± 1.5 µM) were significantly increased ($p < 0.05$ and 0.01, respectively). In contrast, the ability of GSH (IC_{50} 21.6 ± 0.3 µM) and Trolox (IC_{50} 15.8 ± 0.2 µM) to decrease the levels of $ABTS^{+\cdot}$ were not significantly different in the presence of added bicarbonate. However, it should be noted that this assay cannot discriminate between inhibition of $ONOO^-$-mediated $ABTS^{+\cdot}$ formation by uric acid, ascorbate, GSH, or Trolox and antioxidant-mediated $ABTS^{+\cdot}$ scavenging. Nevertheless, the different effects of uric acid and ascorbate in decreasing $ONOO^-$-mediated $ABTS^{+\cdot}$ formation in the presence and absence of added bicarbonate are still significant (FIG. 4A and 4B).

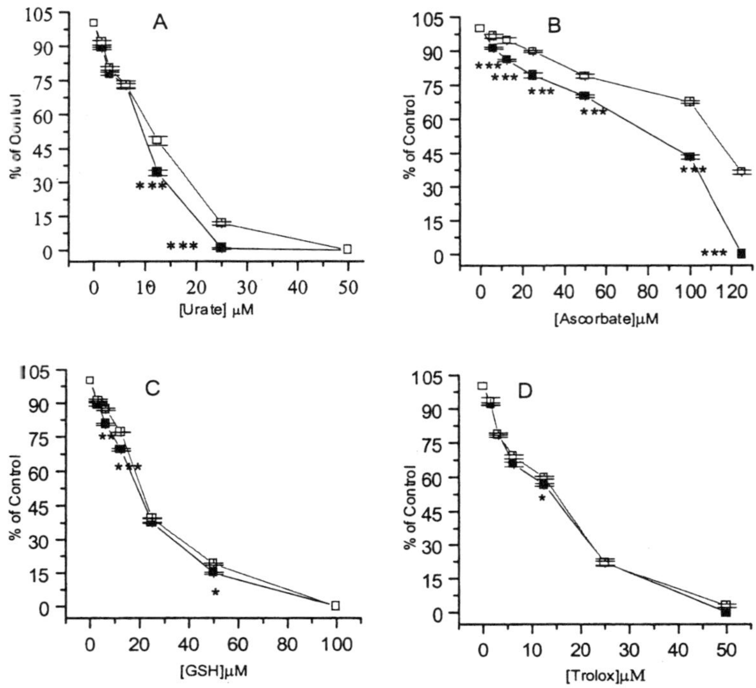

FIGURE 4. Inhibition of $ONOO^-$-mediated ABTS oxidation by uric acid and other antioxidants; modulation by bicarbonate: ■, no bicarbonate; □, 25 mM bicarbonate. **(A)** Uric acid, **(B)** ascorbate, **(C)** GSH, and **(D)** trolox. ABTS (2 mM) was incubated with or without HCO_3^- (25 mM) in phosphate buffer, pH 7.4, at roon tenperature for five minutes with or without antioxidants (0–1 mM), and $ONOO^-$ (500 μM) added. $ABTS^{+\cdot}$ formation was then measured by increase in absorbance at 734 nm as described in MATERIALS AND METHODS. The data shown are mean ± SEM of six or more independent experiments. Significance testing was performed using an independent *t*-test (Student's; two populations); $*p \le 0.1$, $**p \le 0.05$, $***p \le 0.01$, comparing no HCO_3^- addition with HCO_3^- addition (25 mM). The control ($ONOO^-$-treated ABTS in the presence or absence of HCO_3^-) was taken as a 100% value.

Addition of $ONOO^-$ to phosphate buffer (to generate decomposed $ONOO^-$ [58]) and subsequent addition of ABTS did not result in the formation of $ABTS^{+\cdot}$ (data not shown).

DISCUSSION

Peroxynitrite-mediated oxidations are complex and involve one and two electron oxidation pathways, involving $ONOO^-$, ONOOH, and its oxidative decomposition products.[23–32] The oxidative and nitrosative chemistry of $ONOO^-$ is significantly modified by the presence of HCO_3^-/CO_2. This rapid rate of reaction between CO_2

and $ONOO^-$ (one of the fastest known for $ONOO^-$; $k_2 = 3–5.8 \times 10^4\ M^{-1}\ sec^{-1}$)[21,22] and the high normal plasma concentrations of bicarbonate and carbon dioxide *in vivo* (25 mM and 1.3 mM, respectively) make the reaction of $ONOO^-$ with HCO_3^-/CO_2 highly plausible *in vivo*. Higher levels of HCO_3^- and CO_2 could be achieved during pathologic events, such as ischaemia reperfusion injury[60] and probably other inflammatory conditions in which $ONOO^-$ formation has been shown to occur, such as Alzheimer's disease.[2,3] The results of several investigators have suggested that HCO_3^-/CO_2 substantially modifies the reactivity of $ONOO^-$. Bicarbonate modifies both the oxidative and nitrosative properties of $ONOO^-$. For example, bicarbonate inhibits $ONOO^-$-mediated hydroxylation and nitration of phenylalanine, and inhibits the oxidation of GSH and benzoate.[22,25–27] HCO_3^- also decreases $ONOO^-$-mediated bactericidal actions.[23,24] However, the effects of HCO_3^-/CO_2 are complex, since enhancement of tyrosine, guanine, *p*-hydroxyphenylacetate, and protein nitration and increases in the rate of dihydrorhodamine and ABTS oxidation are observed.[26,28–30] Uric acid has been proposed as a neuroprotective agent in human neurodegenerative diseases, such as multiple sclerosis (MS),[38–42] by virtue of its $ONOO^-$ scavenging actions. Administration of uric acid to experimental animals with MS limits the extent of protein tyrosine nitration and ameliorates disease progression. Low levels of uric acid are observed in MS patients[39] and correlate with disease activity.[43] Indeed, epidemiologic studies have suggested that gout (hyperuricaemia) and MS are almost mutually exclusive.[39] Whether the decreased levels of uric acid are due to impaired purine metabolism or due to its conversion of oxidized products, such as allantoin, oxaluric acid, or parabanic acid, that do not scavenge $ONOO^-$ and/or intermediates[34] is currently unknown. Therefore, examining the effects of uric acid, a proposed natural scavenger of $ONOO^-$ *in vivo*, and other endogenous antioxidants on the ability to inhibit $ONOO^-/HCO_3^-$-mediated processes was pertinent, especially since the effectiveness of proposed $ONOO^-$ scavengers is compromised when HCO_3^- is added to the reaction media.[53]

In the absence of bicarbonate, uric acid substantially inhibited $ONOO^-$-mediated nitration reactions of tyrosine and guanine (FIGS. 1A and 2A) and the formation of $ABTS^{+\cdot}$ from ABTS (FIG. 4A), but in contrast to ascorbate, GSH, and trolox, uric acid was unable to protect α_1-AP from $ONOO^-$-mediated inactivation (FIG. 3A) or PR from $ONOO^-$-mediated bleaching (TABLE 1) unless bicarbonate (25 mM) was added to the reaction media. Uric acid has previously been shown to react with $ONOO^-$ to form secondary species capable of inactivating α_1-AP[34] and, recently, to form free radical species, such as the aminocarbonyl radical, that are capable of accelerating $ONOO^-$-induced liposome and LDL oxidation ($k_2 = 4.8 \times 10^2\ M^{-1}\ sec^{-1}$, pH 7.3, 37°C).[62] Uric acid also reacts directly with $ONOO^-$ to form a nitrated/nitrosated derivative, 2-nitrito-4-amino-5-hydroxyimidzaoline, a vasodilatory compound capable of releasing ·NO.[62] However, if this compound is formed by $ONOO^-$ in the presence of HCO_3^-/CO_2 and its significance in human neurodegenerative diseases are currently unknown. Identification of this potential biomarker for uric acid *in vivo* should increase our understanding of the role of uric acid and $ONOO^-/ONOOCO_2^-$ in human diseases and the role of uric acid-mediated neuroprotection under conditions in which it has been examined *in vivo*. It is possible that the protective effects observed with uric acid in experimental animals on neurodegenerative diseases may be due to its reacting with nitrating or oxidizing species derived from $ONOO^-/ONOOH$ rather than with $ONOO^-/ONOOH$ themselves.[64]

The reaction of uric acid with ONOOH at pH 7.4 is slow (k = 155 M^{-1} sec^{-1})[64] and the high concentrations of bicarbonate/CO_2 found *in vivo* may effectively outcompete uric acid for $ONOO^-$. Indeed, ONOOH has been reported to react with CO_2 nearly 920 times faster than with uric acid.[64] These data highlight the fact that $ONOO^-$-scavenging activity of any compound is strongly dependent on the assay employed and that a variety of assays should be performed in the presence of bicarbonate in order to determine $ONOO^-$ scavenging activity. Thus, the data attributing the neuroprotective effects of uric acid *in vivo* to scavenging purely $ONOO^-$ alone should be reassessed.

What mechanisms could account for the discrepancies between the assays used? The reaction of $ONOO^-$ with CO_2 yields the $ONOOCO_2$ adduct.[21,22] It has been proposed that this adduct undergoes either a heterolytic or homolytic cleavage at the weak peroxo O–O bond to give the strong oxidizing and nitrating species $\cdot NO_2$ and $CO_3^{-\cdot}$ radicals, or to give CO_3^{2-} and NO_2^- ions.[65,66] $CO_3^{-\cdot}$ is a strong one-electron oxidant that may, itself, inactivate α_1-AP, since it is known to inactivate enzymes, oxidize amino acids, and proteins.[67] The bactericidal potency of $\cdot OH$ generated extracellularly is enhanced when HCO_3^- is present in the reaction medium.[21,22] $\cdot OH$ has also been shown to inactivate α_1-AP and to oxidize uric acid into species capable of inactivating α_1-AP.[68] Formation of $\cdot OH$ directly from $ONOO^-$ is also possible to some extent.[69,70] $CO_3^{-\cdot}$ reacts with various biomolecules 10^2–10^7 times slower than $\cdot OH$, but as such it may diffuse for longer to selectively reach and oxidize critical targets, such as α_1-AP *in vivo*.[67] However, if $CO_3^{-\cdot}$ is responsible for α_1-AP, inactivation needs to be investigated.

α_1-AP can be inactivated by $ONOO^-$ via the oxidation of its reactive site methionine.[54,55] However, $ONOO^-$-mediated protein methionine oxidation is inhibited by CO_2.[29,30] Addition of HCO_3^- to the reaction media resulted in only about 10% reduction in the ability of $ONOO^-$ to inactivate α_1-AP. An additional mechanism may be via the oxidation of the protein's sole cysteine residue.[71] Indeed, $ONOO^-$-mediated protein thiol oxidation is increased in the presence of HCO_3^-[72] and certain thiol compounds at low concentrations have been shown to aggravate $ONOO^-$-mediated inactivation of α_1-AP,[73] a process enhanced by HCO_3^-.[74] An alternative mechanism of α_1-AP inactivation may be via the nitration of one or more of the six tyrosine residues in the protein,[75] a process enhanced by HCO_3^-. However, the precise mechanism of $ONOO^-$-mediated inactivation of α_1-AP in the presence of HCO_3^- is unknown and is currently being investigated in our laboratory. We have found that $ONOO^-$ not only damages tyrosine, methionine, and cysteine residues in α_1-AP, but also in many other amino acids.[76]

Although local concentrations of bicarbonate (and CO_2) ultimately determine the pathway of $ONOO^-$ decomposition formed *in vivo,* up to about 35% of the $ONOOCO_2^-$ intermediate produces reactive free radicals, such as $\cdot NO_2$ and $CO_3^{-\cdot}$[65,66,77,78] that mediate nitration and oxidation reactions. Tyrosyl radicals ($\cdot$Tye) are more effectively generated by the reaction of $ONOO^-$-derived $CO_3^{-\cdot}$ ($k = 4.5 \times 10^7$ M^{-1} sec^{-1}) than by the reaction of $ONOO^-$-derived $\cdot NO_2$ ($k = 3.2 \times 10^5$ M^{-1} sec^{-1}).[79,80] Uric acid and the other antioxidants tested here may not react fast enough with the $ONOO^-$/ HCO_3^- adduct to prevent nitration and oxidation reactions to critical target molecules, such as α_1-AP, tyrosine, and guanine. This study highlights the need to conduct $ONOO^-$ scavenging experiments in the presence of added bicarbonate and to use more than one scavenging assay to avoid misleading conclusions.[50]

In summary, uric acid inhibits $ONOO^-$-mediated processes *in vitro* and these effects are significantly altered by the presence of physiologically relevant concentrations of bicarbonate. Although uric acid is depleted in MS patients and substantial 3-nitrotyrosine formation is observed,[38,49] the precise mechanism for protection may involve the scavenging of $ONOOCO_2^-$-derived intermediates, such as $CO_3^{-\cdot}$ and $\cdot NO_2$ rather than only $ONOO^-$. Trapping secondary radical species from the rapid reaction of $ONOO^-$ with CO_2 by uric acid may represent a new and valuable approach for ameliorating the adverse effects of $ONOO^-$ in human diseases.[64] The data presented in this paper highlight the fact that the apparent $ONOO^-$-scavenging activity of particular compounds strongly depends on the assay employed.

ACKNOWLEDGMENTS

We are grateful to the National Medical Research Council of Singapore (NMRC) (RI-83000057213) and the National University of Singapore Academic Research Fund (R-183000053214) for their generous research support.

REFERENCES

1. HUIE, R.E. & S. PADMAJA. 1993. The reaction of NO with superoxide. Free Rad. Res. Comm. **18:** 195–199.
2. HALLIWELL, B., K. ZHAO & M. WHITEMAN. 1999. Nitric oxide and peroxynitrite. The ugly, the uglier and the not so good. Free Rad. Res. **31:** 651–669.
3. GREENACRE, S.A.B. & H. ISCHIROPOULOS. 2001. Tyrosine nitration: localisation, quantification, consequences for protein function and signal transduction. Free Rad. Res. **34:** 541–581.
4. RADI, R., J.S. BECKMAN, K.M. BUSH, *et al.* 1991. Peroxynitrite-induced membrane lipidperoxidation—the cytotoxic potential of superoxide and nitric-oxide. Arch. Biochem. Biophys. **288:** 481–487.
5. ISCHIROPOULOS, H. & A.B. AL-MEHDI. 1995. Peroxynitrite-mediated oxidative protein modifications. FEBS Lett. **364:** 279–282.
6. SPENCER, J.P.E, J. WONG, A. JENNER, *et al.* 1996. Base modification and strand breakage in isolated calf thymus DNA and in DNA from human skin epidermal keratinocytes exposed to peroxynitrite or 3-morpholinosydnonimine. Chem. Res. Toxicol. **9:** 1152–1158.
7. YERMILOV, V., J. RUBIO, M. BECCHI, *et al.* 1995. Formation of 8-nitroguanine by the reaction of guanine with peroxynitrite *in vitro*. Carcinogenesis **16:** 2045–2050.
8. BECKMAN, J.S., T.W. BECKMAN, J. CHEN, *et al.* 1990. Apparent hydroxyl radical production by peroxynitrite: implications for endothelial injury from nitric oxide and superoxide. Proc. Natl. Acad. Sci. U.S.A. **87:** 1620–1624.
9. VAN DER VLIET A., D. SMITH, C.A. O'NEILL, *et al.* 1994. Interactions of peroxynitrite with human plasma and its constituents: oxidative damage and antioxidant depletion, Biochem. J. **303:** 296–301.
10. WHITEMAN, M., J.P.E. SPENCER, D.L. MATTEY, *et al.* 2000. Mechanisms of peroxynitrite toxicity to cultured human synovial fibroblasts: a comparison with hypochlorous acid. Brit. J. Pharmacol. **129:** U98.

11. SZABO, C., B. ZINGARELLI, M. O'CONNOR, *et al.* 1996. DNA strand breakage, activation of poly(ADP)ribosyl synthetase, and cellular energy depletion are involved in the cytotoxicity in macrophages and smooth muscle cells exposed to peroxynitrite. Proc. Natl. Acad. Sci. U.S.A. **93:**1753–1758.
12. COOKSON, M.R., P.G. INCE & P.J. SHAW. 1998. Peroxynitrite and hydrogen peroxide induced cell death in the NSC34 neuroblastoma × spinal cord cell line: role of poly(ADP-ribose) polymerase. J. Neurochem. **70:** 501–508.
13. SZABO, C. & V.L. DAWSON. 1998. Role of poly(ADP-ribose) synthetase in inflammation and ischaemia-reperfusion. Trend. Pharmacol. Sci. **19:** 287–298.
14. LIN, K., J. XUE, M.C. LIN, *et al.* 1998. Peroxynitrite induced apoptosis in HL-60 cells by activation of a caspase-3 protease. Am. J. Physiol. (Cell Physiol. 43) **274:** C855–C860.
15. ZHUANG, S.G. & G. SIMON. 2000. Peroxynitrite-induced apoptosis involves activation of multiple caspases in HL-60 cells. Am. J. Cell Physiol. **279:** C341–C351.
16. SAEKI, M., Y. KAMISAKI & S. MAEDA. 2000. S involvement of mitogen-activated protein kinase in peroxynitrite-induced cell death of human neuroblastoma SH-SY5Y cells. Neurosci. Res. **38:** 213–216.
17. JOPE, R.S., L. ZHANG & L. SONG. 2000 Peroxynitrite modulates the activation of p38 and extracellular regulated kinases in PC 12 cells. Arch. Biochem. Biophys. **376:** 355–370.
18. OH-HASHIJ, K.W., H. MARUYAMA, T. YI, *et al.* 1999. Mitogen-activated protein kinase pathway mediates peroxynitrite-induced apoptosis in human dopaminergic neuroblastoma SH-SY5Y cells. Biochem. Biophys. Res. Commun. **263:** 504–509.
19. SCHIEKE, S.M., K. BRIVIBA, L.O. KLOTZ, *et al.* 1999. Activation pattern of mitogen-activated protein kiinases elicited by peroxynitrite: attenuation by selenite supplementation FEBS Lett. **448:** 301–303.
20. KIRDYCHEVA, M., F. AHMED, G.M. ANTHONY, *et al.* 2000. Mitogen-activated protein kinase phosphorylation in kidneys of beta(S) sickle cell mice. J. Am. Soc. Nephrol. **11:** 1026–1032.
21. LYMAR, S.V. & J.K. HURST. 1995. Rapid reaction between peroxynitrite ion and carbon dioxide: implications for biological activity, J. Am. Chem. Soc. **117:** 867–868.
22. DENICOLA, A., B.A. FREEMAN, M. TRUJILLO, *et al.* 1996. Peroxynitrite reaction with carbon dioxide/bicarbonate: kinetics and influence on peroxynitrite-mediated oxidations. Arch. Biochem. Biophys. **333:** 49–58.
23. ZHU L., C. GUNN & J.S. BECKMAN. 1992. Bactericidal activity of peroxynitrite. Arch. Biochem. Biophys. **298:** 452–457.
24. HURST, J.K. & S.V. LYMAR. 1997. Toxicity of peroxynitrite and related reactive nitrogen species toward *Escherichia coli.* Chem. Res. Toxicol. **10:** 802–810.
25. VAN DER VLIET, A., C.A. O'NEILL, B. HALLIWELL, *et al.* 1994. Aromatic hydroxylation of phenylalnine and tyrosine by peroxynitrite. FEBS Lett. **339:** 89–92.
26. BARLETT, B.S., R.L. LEVINE & E.R. STADTMAN. 1998. Carbon dioxide stimulates peroxynitritemediated nitration of tyrosine residues and inhibits oxidation of methionine residues of glutamine synthetase: both modifications mimic effects of adenylylation. Proc. Natl. Acad. Sci. U.S.A. **95:** 2784–2789.
27. ZHANG, H., G.L. SQUADRITO, R.M. UPPU, *et al.* 1997. Inhibition of peroxynitrite-mediated oxidation of glutathione by carbon dioxide. Arch. Biochem. Biophys. **339:** 193–189.
28. TIEN, M., B. BERLETT, R.L. LEVINE, *et al.* 1999. Peroxynitrite-mediated modification of proteins at physiological carbon dioxide concentration: pH dependence of carbonyl formation, tyrosine nitration, and methionine oxidation. Proc. Natl. Acad. Sci. U.S A. **96:** 7809–7814.
29. LEMERCIER, J.N., S. PADMAJA, R. CUETO, *et al.* 1997. Carbon dioxide modulation of hydroxylation and nitration of phenol by peroxynitrite. Arch. Biochem. Biophys. **345:** 160–170.
30. UPPU, R.M., G.L. SQUADRITO & W.A. PRYOR. 1996. Acceleration of peroxynitrite oxidations by carbon dioxide. Arch. Biochem. Biophys. **327:** 335–343.
31. GOW, A., D. DURAN, S.R. THOM, *et al.* 1996. Effects of peroxynitrite-induced protein modifications on tyrosine phosphorylation and degradation. FEBS Lett. **385:** 63–66.

32. YERMILOV, V., Y. YOSHIE, I. RUBIO, *et al.* 1996. Effects of carbon dioxide/bicarbonate on induction of DNA single-strand breaks and formation of 8-nitroguanine, 8-oxoguanine and basepropenal mediated by peroxynitrite. FEBS Lett. **399:** 67–70.
33. THOMAS, S.R., M.J. DAVIES & R. STOCKER. 1998. Oxidation and antioxidation of human lowdensity lipoprotein and plasma exposed to 3-morpholinosydnonimine and reagent peroxynitrite. Chem. Res. Toxicol. **11:** 484–494.
34. WHITEMAN, M. & B. HALLIWELL. 1996. Protection against peroxynitrite-dependent tyrosine nitration and α_1-antiproteinase inactivation by ascorbic acid: a comparison with other biological antioxidants. Free Rad. Res. **25:** 275–283.
35. ARTEEL, G.E., K. BRIVIBA & H. SIES. 1998. Protection against peroxynitrite. FEBS Lett. **445:** 226–230.
36. WHITEMAN, M., H. TRITSCHLER & B. HALLIWELL. 1996. Protection against peroxynitrite-dependent tyrosine nitration and α_1-antiproteinase inactivation by oxidized and reduced lipoic acid. FEBS Lett. **379:** 74–76.
37. PANNALA, A.S., C.A. RICE-EVANS, B. HALLIWELL, *et al.* 1997. Inhibition of peroxynitritemediated tyrosine nitration by catechin polyphenols. Biochem. Biophys. Res. Commun. **232:** 164–168.
38. HOOPER, D.C., S. SPITSIN, R.B. KEAN, *et al.* 1998. Uric acid, a natural scavenger of peroxynitrite, in experimental allergic encephalomyelitis and multiple sclerosis. Proc. Natl. Acad. Sci. U.S.A. **95:** 675–680.
39. HOOPER, D.C., O. BAGASRA, J.C. MARINI, *et al.* 1997. Prevention of experimental allergic encephalomyelitis by targeting nitric oxide and peroxynitrite: implications for the treatment of multiple sclerosis. Proc. Natl. Acad. Sci. U.S.A. **94:** 2528–2533.
40. HOOPER, D.C., G.C. SCOTT, A. ZBOREK, *et al.* 2000. Uric acid, a peroxynitrite scavenger, inhibits CNS inflammation, blood-CNS barrier permeability, and tissue damage in a mouse model of multiple sclerosis. FASEB J. **14:** 691–698.
41. SCOTT, G.S. & D.C. HOOPER. 2000. The role of uric acid in protection against peroxynitritemediated pathology. Med. Hypoth. **56:** 95–100.
42. KEAN R.B., S.V. SPITSIN, T. MIKHEEVA, *et al.* 2000. The peroxynitrite scavenger uric acid prevents inflammatory cell invasion into the central nervous system in experimental allergic encephalomyelitis through maintenance of blood-central nervous system barrier integrity. J. Immunol. **165:** 6511–6518.
43. BECKER, B.F. 1993. Towards the physiological function of uric acid. Free Rad. Biol. Med. **14:** 615–631.
44. AMES, B.A., G. COHEN & P. HOCHSTEIN. 1981. Uric acid provides an antioxidant defence in humans against oxidants and radicals caused by aging and cancer: a hypothesis. Proc. Natl. Acad. Sci. U.S.A. **78:** 6858–6862.
45. FUJII, T.T. YOSHIKAWA, Y. NAITO, *et al.* 1999. Peroxyl radical-scavenging activity of plasma determined by electron paramagnetic resonance J. Clin. Biochem. Nutr. **27:** 103–112.
46. CROSS, C.E., P.A. MOTCHNIK, B.A. BRUENER, *et al.* 1992. Oxidative damage to plasma constiuents by ozone. FEBS Lett. **298:** 269–272.
47. VAN DER VLIET, A., M.L. HU., C.A. O'NEILL, *et al.* 1994. Interactions of human blood plasma with hydrogen peroxide and hypochlorous acid. J. Lab. Clin. Med. **124:** 701–707.
48. GROOTVELD, M. & B. HALLIWELL. 1987. Measurements of allantoin and uric acid in human body fluids. A potential index of free radicals *in vivo*? Biochem. J. **243:** 803–808.
49. DRULOVIC, J., I. DUJMOVIC, N. STOJSAVLJEVIC, *et al.* 2001. Uric acid levels in sera from patients with multiple sclerosis. J. Neurol. **248:** 121–126.
50. KETSAWATSAKUL, U., M. WHITEMAN & B. HALLIWELL. 2000. A reevaluation of the peroxynitrite scavenging activity of some dietary phenolics. Biochem. Biophys. Res. Commun. **279:** 692–699.
51. WHITEMAN, M., H. TRITSCHLER & B. HALLIWELL. 1996. Protection against peroxynitrite-dependent tyrosine nitration and α_1-antiproteinase inactivation by oxidized and reduced lipoic acid. FEBS Lett. **379:** 74–46.
52. BALAVOINE, G.G.A. & Y.V. GELETII. 1999. Peroxynitrite scavenging by different antioxidants. Part 1: convenient assay. Nitric Oxide Biol. Chem. **3:** 40–54.

53. PANNALA, A.S., T.S. CHAN, P.J. O'BRIEN, *et al.* 2001. Flavonoid B-ring chemistry and antioxidant activity: fast reaction kinetics. Biochem. Biophys. Res. Commun. **282:** 1161–1168.
54. GOLDSTEIN, S. & G. CZAPSKI. 1997. The effect of bicarbonate on oxidation by peroxynitrite: implications for its biological activity. Inorg. Chem. **36:** 5113–5117.
55. WONG, P.S. & J. TRAVIS. 1980. Isolation and properties of oxidized α_1-proteinase inhibitor from human rheumatoid synovial fluid. Biochem. Biophys. Res. Commun. **93:** 1449–1454.
56. MORENO, J.J. & W.A. PRYOR. 1992. Inactivation of α_1-proteinase inhibitor by peroxynitrite. Chem. Res. Toxicol. **5:** 425–431.
57. GOLE, M.D., J.M. SOUZA, I. CHOI, *et al.* 2000. Plasma proteins modified by tyrosine nitration in acute respiratory distress syndrome Am. J. Physiol. Lung Cell Mol. Physiol. **278:** L961–L967.
58. ZHAO, K., M. WHITEMAN, J.P. SPENCER, *et al.* 2001. DNA damage by nitrite and peroxynitrite: protection by dietary phenols. Meth. Enzymol. **335:** 296–308.
59. BECKMAN, J.S., J. CHEN, H. ISCHIROPOULOS, *et al.* 1994. Oxidative chemistry of peroxynitrite. Meth. Enzymol. **223:** 229–240.
60. HUGHES, M.N. & H.G. NICKLIN. 1968. The chemistry of pernitrites. Part 1. Kinetics and decomposition of pemitrous acid. J. Chem. Soc. (A) **64:** 450–452.
61. LEVINSKI, N.G. 1994. *In* Harrison's Principles of Internal Medicine. T.R. Harrison, K.J. Isselbacher, E. Braunwald, *et al*, Eds.: 253–262. McGraw-Hill, New York.
62. SANTOS, C.X.C., E.I. EDSON & O. AUGUSTO. 1999. Uric acid oxidation by peroxynitrite: multiple reactions, free radical formation, and amplification of lipid oxidation. Arch. Biochem. Biophys. **372:** 396–294.
63. SKINNER, K.A., C.R. WHITE, R. PATEL, *et al.* 1998. Nitrosation of uric acid by peroxynitrite—formation of a vasoactive nitric oxide donor. J. Biol. Chem. **273:** 24491–24497.
64. SQUADRITO, G.L., R. CUETO, A.E. SPLENSER, *et al.* 2000. Reaction of uric acid with peroxynitrite and implications for the mechanism of neuroprotection by uric acid. Arch. Biochem. Biophys. **376:** 333–337.
65. BONINI, M.G., R. RADI, G. FERRER-SUETA, *et al.* 1999. Direct EPR detection of the carbonate radical anion produced from peroxynitrite and carbon dioxide. J. Biol. Chem. **274:** 10802–1080.
66. MELI, R., T. NAUSER & W.H. KOPPENOL. 1999. Direct observation of intermediates in the reaction of peroxynitrite with carbon dioxide. Helv. Chim. Acta. **82:** 772–725.
67. CHEN, S.N. & M.Z. HOFFMAN. 1973. Rate constants for the reaction of the carbonate radical with compounds of biochemical interest in neutral aqueous solution. Radiat. Res. **56:** 40–47.
68. ARUOMA, O.I. & B. HALLIWELL. 1989. Inactivation of alpha-1-antiproteinase by hydroxyl radicals: the effect of uric acid. FEBS Lett. **244:** 76–80.
69. RICHESON, C.E., P. MULDER, V.W. BOWRY, *et al.* 1998. The complex chemistry of peroxynitrite decomposition: new insights. J. Am. Chem. Soc. **120:** 7211–7219.
70. CODDINGTON, J.W., J.K. HURST & S.V. LYMAR. 1998. Hydroxyl radical formation during peroxynitrous acid decomposition. J. Am. Chem. Soc. **121:** 2438–2443.
71. TYAGI, S.C. 1991. Reversal of inhibition of neutrophil elastase by thiol modified α_1-proteinase inhibitor, J. Biol. Chem. **266:** 5279–5285.
72. SCORZA, G. & M. MINETTE. 1998. One-electron oxidation pathway of thiols by peroxynitrite in biological fluids: bicarbonate and ascorbate promote the formation of albumin disulphide dimers in human blood plasma. Biochem. J. **329:** 405–413.
73. WHITMAN, M. & B. HALLIWELL. 1997. Thiols and disulphides can aggravate peroxynitritedependent inactivation of α_1-antiproteinase. FEBS Lett. **414:** 497–500.
74. WHITEMAN, M., U. KETSAWATASKUL & B. HALLIWELL. 2000. Enhancement of thiol and disulphidemediated peroxynitrite-dependent α_1-antiproteinase inactivation by bicarbonate. Brit. J. Pharmacol. **129**(Suppl. S.): U124.
75. MIEZEWA, M. & S.K. CHAN. 1987. Chemical modification of human α_1-proteinase inhibitor by tetranitromethane. Biochem. J. **246:** 37–42.
76. WHITEMAN, M. 1998. Mechanisms of Oxidative Damage to Biomolecules in Rheumatoid Arthritis. 164–172. Ph.D. Thesis. University of London.

77. LYMAR, S.V. & J.K. HURST. 1998. CO_2-catalyzed one-electron oxidations peroxynitrite: properties of the reactive intermediate. Inorg. Chem. **37:** 867–868.
78. GOLDSTEIN, S. & G. CZAPSKI. 1998. Formation of peroxynitrate from the reaction of peroxynitrite with CO_2: evidence for carbonate radical production. J. Am. Chem. Soc. **120:** 3458–3463.
79. GOLDSTEIN, S., G. CZAPSKI, I. LIND, *et al.* 2000. Tyrosine nitration by simultaneous generation of ·NO and $O_2^{-\bullet}$ under physiological conditions. How the radicals do the job. J. Biol. Chem. **275:** 3031–3036.
80. HODGES, G.R., J. MARWAHA, *et al.* 2000. A novel procedure for generating both nitric oxide and superoxide *in situ* from chemical sources at any chosen mole ratio. First application: tyrosine oxidation and a comparison with preformed peroxynitrite. Chem. Res. Toxicol. **13:** 1287–1293.

Possible Roles of Nitric Oxide in the Physiology and Pathophysiology of the Mammalian Cochlea

RUN-SHENG RUAN

Department of Otolaryngology, National University of Singapore, Singapore

ABSTRACT: Nitric oxide (NO) has been implicated as a mediator of vasodilation and neurotransmission in the mammalian cochlea. This is demonstrated by the presence of nitric oxide synthase (NOS) and nitric oxide (NO) in the blood vessels and the organ of Corti in the cochlea. It is not certain if the neurons in the spiral ganglion produce NO since no fluorescent signal could be detected by 4,5-diaminofluorescein diacetate (DAF-2DA), a fluorescent indicator of NO. To determine if NO/peroxynitrite plays any role in neurodestruction observed in ischemic cochlea of the guinea pig, the effects of NO donors, such as *S*-nitrosocysteine (*S*-NC) and nitroglycerine (NTG); peroxynitrite generators, such as 3-morpholinosydnonimine (SIN-1); peroxynitrite inhibitors, such as superoxide dismutase plus catalase (SOD/Cat); and NOS inhibitors, such as N^G-nitro-L-arginine methyl ether (L-NAME) were tested on normal and ischemic cochleae. The level of NO in the cochlea after 20 to 120 minutes of ischemia was indicated by measurement of nitrites/nitrates in the perilymph. The evidence gathered from these experiments indicates that NO or peroxynitrite is not necessarily destructive to auditory hair cells, and in fact, exogenous NO may protect neural structures in the cochlea from damage under ischemic conditions.

KEYWORDS: role of nitric oxide; physiology; pathophysiology; mammalian cochlea

INTRODUCTION

NO is formed together with citrulline when L-arginine is catalyzed by nitric oxide synthase (NOS) in the presence of molecular oxygen, nicotinamide adenine dinucleotide phosphate (NADPH), and calmodulin, to which Ca^{2+} binds.[1] A number of NOS isoforms have been purified from various tissues.[1–3] Both sequence analysis and biochemical studies show that these isoforms form a quite closely related family of proteins. In general, the synthases are classified into two broad categories: (1) a constitutive isoform purified from neurons (NOS I) and from the vascular endothelium (NOS III), both of which are calcium/calmodulin dependent; and (2) an inducible isoform (NOS II) characterized in macrophages and other cells, and not typically calcium/calmodulin dependent.

Address for correspondence: Run-Sheng Ruan, Department of Otolargncology, National University of Singapore, Lower Kent Ridge Road, Singapore 119074. Fax: (65) 7753820. entrrs@nus.edu.sg

NO has been implicated as a mediator of vasodilation and neurotransmission in the mammalian cochlea.[4] Since it is a labile gas, NADPH-diaphorase (NADPH-d) histochemistry and/or NOS immunohistochemistry have been used in the past few years to indicate the presence of NO.

NOS IMMUNOHISTOCHEMISTRY AND NADPH-DIAPHORASE HISTOCHEMISTRY IN THE COCHLEA

All three isoforms of NOS have been studied independently in the mammalian cochlea. NOS II is not found in the structures of normal cochlea.[5,6] The presence of NOS I and NOS III have been demonstrated histochemically and immunohistochemically in mammalian cochlea by using light and electron microscopy. There are, however, controversies in the distribution of the enzymes in the cochlea (see TABLES 1 and 2). The studies of my group showed that NOS and NADPH-d were present not only in the guinea pig, but also in human cochlea.[7] With both positive and negative controls, NOS I was localized in type I neuronal cell bodies and processes of the spiral ganglion. At subcellular level, NOS I immunoreactivity was scattered throughout the cytoplasm and axoplasm of the type I neurons, with no clear relationship to any of the subcellular organelles. Immunopositive inner radial and inner spiral fibers were evident at the habenular perforata and below the inner hair cells. The nerve endings below the immunopositive outer hair cells, however, were not labeled. The cells of Deiters, tympanic lamina cells, and pillar cells were labeled.

The conflicting findings in the localization of NOS, as shown in TABLES 1 and 2, may be attributed to the different methodologies employed, in particular the conditions of tissue fixation and the various types of antibodies used. It has been demonstrated that commercial antibodies against NOS I exhibit markedly different labeling patterns in guinea pig cochlea.[4] Therefore, to resolve the disparity, both positive and negative controls should be included in the immunohistochemical localization of NOS in the cochlea.

Interestingly, like others,[5,6,8] our electron microscope study revealed that NOS I and NOS III immunoreactivities were almost identical for all cell types (unpublished observation). Why the two isoforms are present in the same cell types remains to be clarified by further study.

Production of NO Detected by a DAF-2DA Technique

One of the major limitations in understanding the diverse roles of NO has been the difficulty in measuring NO, since it is a free radical with a half-life of only a few seconds. Recent development of the fluorescent indicator of NO, 4,5-diaminofluorescein diacetate (DAF-2DA), offers the possibility of a direct visualization of NO in living cells.[9] By using this technique, we detected *in vivo* the NO-containing cells in guinea pig cochlea. A solution of 100 μl of 1 or 10 μM DAF-2DA (Alexis Biochemicals, Switzerland) in artificial perilymph (APL), adjusted to pH 7.3, was infused into the scala tympani of the cochlea. The solution remained for a period of 0.5–2 hours. The cochlea receiving only APL was used as a control. Before decapitation, the cochlea of the animal was fixed by the introduction of 4% paraformaldehyde into the perilymphatic space. The cochlea was removed and kept

TABLE 1. Expression of NOS I in the cochlea

Publication	IHC	OHC	NE	CN	GC	PC	DC	HC	EC	SL	SV	LE	ES
Michel *et al.* (1999)					+	+	+	+		+			+
Riemann and Reuss (1999)			+										
Fessenden *et al.* (1999)					+								
Hess *et al.* (1998)	+	+		+	+								+
Heinrich *et al.* (1998)	+	+				+	+						
Gosepath *et al.* (1997)	+	+			+					+	+		
Ruan *et al.* (1997)		+	+	+	+	+	+						
Franz *et al.* (1996)		+	+	+	+					+		+	

KEY: Inner hair cell (IHC); outer hair cell (OHC); nerve ending (NE); cochlear nerve (CN); ganglion cell (GC); pillar cell (PC); cell of Deiters (DC); cell of Hensen (HC); epithelial cell (EC); spiral ligment (SL); spiral vascularis (SV); limbus epithelium (LE); endosteum (ES).

TABLE 2. Expression of NOS III in the cochlea

Publication	IHC	OHC	NE	CN	GC	PC	DC	HC	EC	SL	SV	LE	ES
Michel *et al.* (1999)	+	+	+	+					+	+	+	+	
Hess *et al.* (1998)	+	+		+	+					+		+	
Heinrich *et al.* (1998)	+	+	+			+	+		+				
Gosepath *et al.* (1997)					+				+				
Yamane *et al.* (1997)									+				
Franz *et al.* (1996)	+	+			+				+			+	

KEY: Inner hair cell (IHC); outer hair cell (OHC); nerve ending (NE); cochlear nerve (CN); ganglion cell (GC); pillar cell (PC); cell of Deiters (DC); cell of Hensen (HC); epithelial cell (EC); spiral ligment (SL); spiral vascularis (SV); limbus epithelium (LE); endosteum (ES).

in the same fixative at 4°C for two hours, before being transferred to 10% EDTA for decalcification. Five days after EDTA treatment in the dark, cryosections of the cochlea were cut at 30 μm, mounted on a slide, coverslipped, and viewed in a confocal microscope.

In the cochlea infused with DAF-2DA, a fluorescent signal was detected by confocal microscopy in the cells of the endosteum, spiral ligament, spiral vascularis, spiral limbus, inner and outer sulcus, bone, and vestibular membrane (see FIGURE 1b and d). The endothelial cells of the blood vessels throughout the cochlea were labeled (FIG. 1c). The signal was also detected in the organ of Corti, including the inner and outer hair cells, the cells of Hensens, and at the base of the inner and outer hair cells (FIG. 1b). Neither the spiral ganglion neurons nor their processes showed fluorescence in Rosenthal's canal or in the modiolus (FIG. 1e). In the cochlea treated with APL alone, no fluorescent signal was detectable in any cell type (FIG. 1a).

A recent report, using DAF-2DA to detect NO production in isolated lateral wall tissue and the organ of Corti of guinea pig cochlea, revealed that the fluorescent signal was located in the efferent nerves and their endings near the inner and outer hair cells, the inner and outer hair cells themselves, the endothelial cells of the blood vessels, and the stria vascularis.[10] These investigators also found that NO signal increased when L-arginine was applied to the medium, and decreased when the tissue was exposed to N^{G}-nitro-L-arginine methyl ether (L-NAME), an NOS inhibitor. Interestingly, our experiment showed an enhancement of the fluorescent signal with the introduction of L-NAME into the cochlea two hours before incubation with DAF-2DA (FIG. 1f). This difference in results may be due to the different experimental conditions. It is possible that the suppression of NOS expression could have brought about the upregulation of xanthine oxidase, which has been shown to reduce nitrite to form NO.[11] Also, NO could be generated by non-enzymatic pathways.[12] Additional studies need to be made in order to elucidate the difference in results.

LOCALIZATION OF SOLUBLE GUANYLYL CYCLASE (SGC) IN THE MAMMALIAN COCHLEA

The NO/cyclic GMP pathway in the auditory system has been studied in recent years. sGC has been localized in the cells of Hensen, cells of Deiters, inner and outer hair cells and their nerve endings, the limbus and pillar cells in the organ of Corti, vascular pericytes, the spiral ligament in the lateral wall, spiral ganglion neurons and their processes, the cochlea nerve, and smooth muscle layer of blood vessels in the modiolus.[5,6,13–15] These, together with previous studies that showed the formation of cGMP in the cochlea,[16,17] indicate the existence of the NO/sGC signaling pathway in the auditory system.

POSSIBLE ROLE OF NO IN AUDITORY PHYSIOLOGY

It has been shown that topical application of sodium nitroprusside (SNP), an NO donor, at the round window of the cochlea results in a striking increase in cochlear blood flow. In contrast, the application of a competitive inhibitor of NOS, L-NAME,

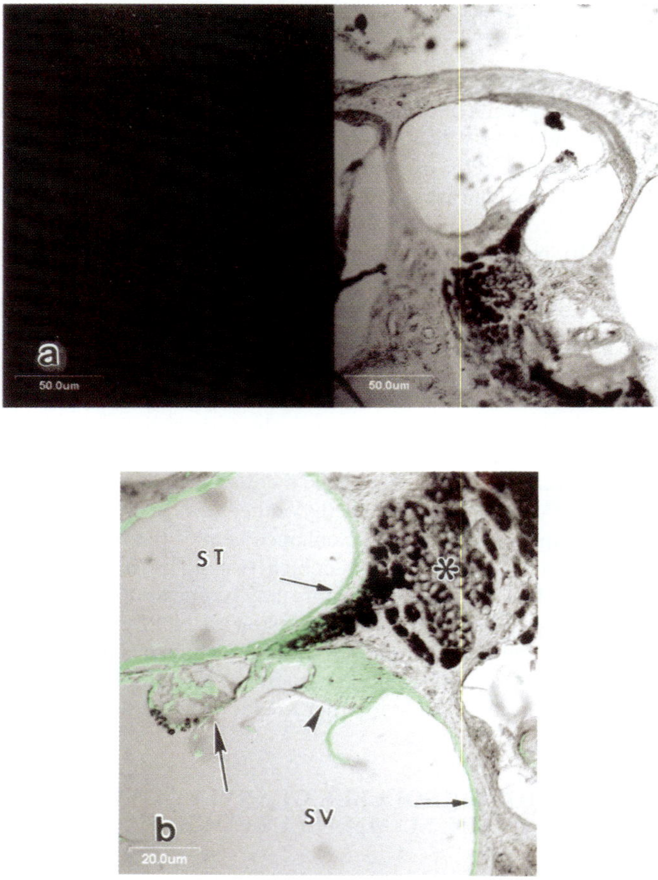

FIGURE 1. Confocal micrographs of the guinea pig cochlea after incubation with artificial perilymph alone **(a)**; DAF-2DA **(b)**–**(e)**; and DAF-2DA following L-NAME treatment **(f)**. Photographs **a**, **e**, and **f** show fluorescence images and the corresponding transmission images; and **b**, **c**, and **d** show an overlay of fluorescence and transmission images. The *bright green* colors indicate the presence of NO. The *small arrows* in **b** indicate the endosteum; the *arrow head* indicates the spiral limbus; and the *larger arrow* indicates the organ of Corti. The *arrows* in **c** indicate vessels in the modiolus of the cochlea. The *arrow* in **d** indicates the spiral vascularis, and the *arrow head* indicates the outer sulcus. The *arrow* in **f** indicates the lateral wall of the cochlea and the *arrow head* indicates the organ of Corti. * Indicates the spiral ganglion. Note that no signal was detected in the spiral ganglion neurons after incubation with DAF-2DA **(e)**. KEY: ST, scala tympani; SV, scala vestibuli; M, modiolus of the cochlea; SL, spiral ligment.

FIGURE 1 c–f, see opposite page.

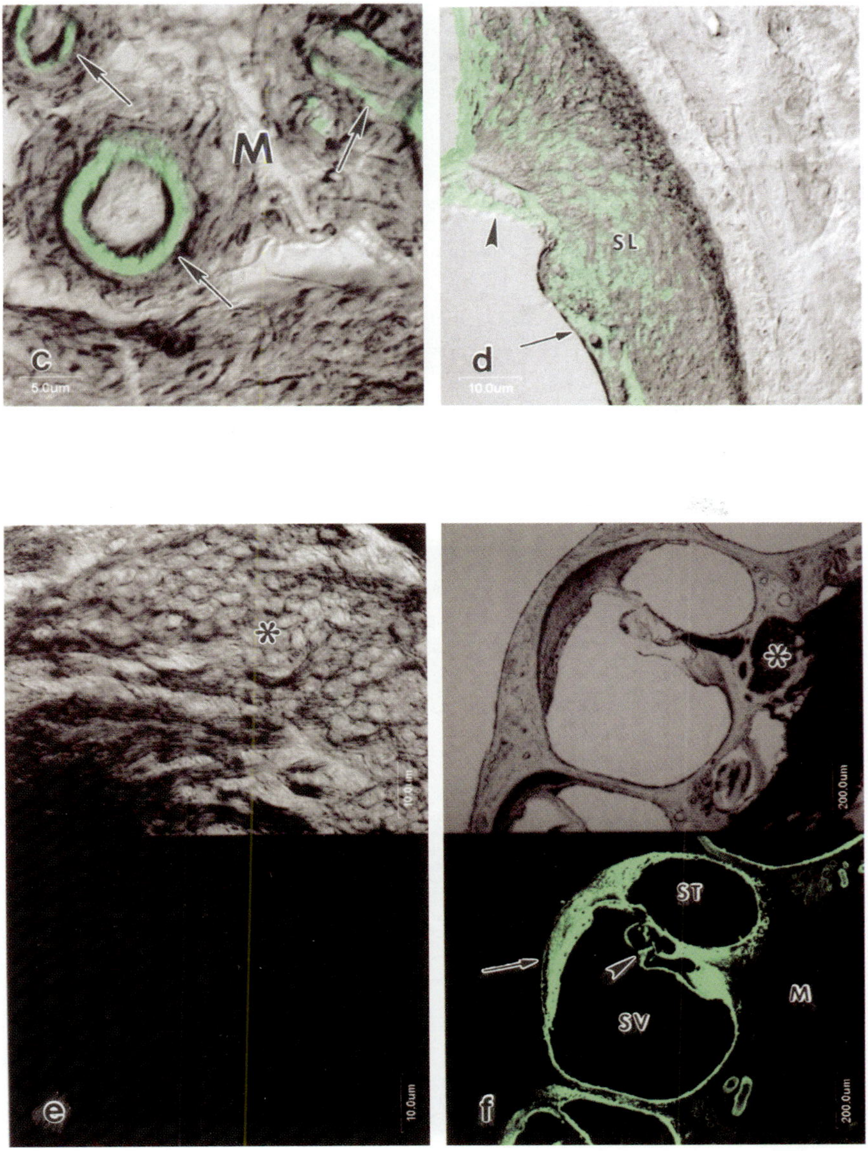
M
c
5.0um
SL
d
10.0um
10.0um
e
10.0um
200.0um
ST
M
SV
f
200.0um

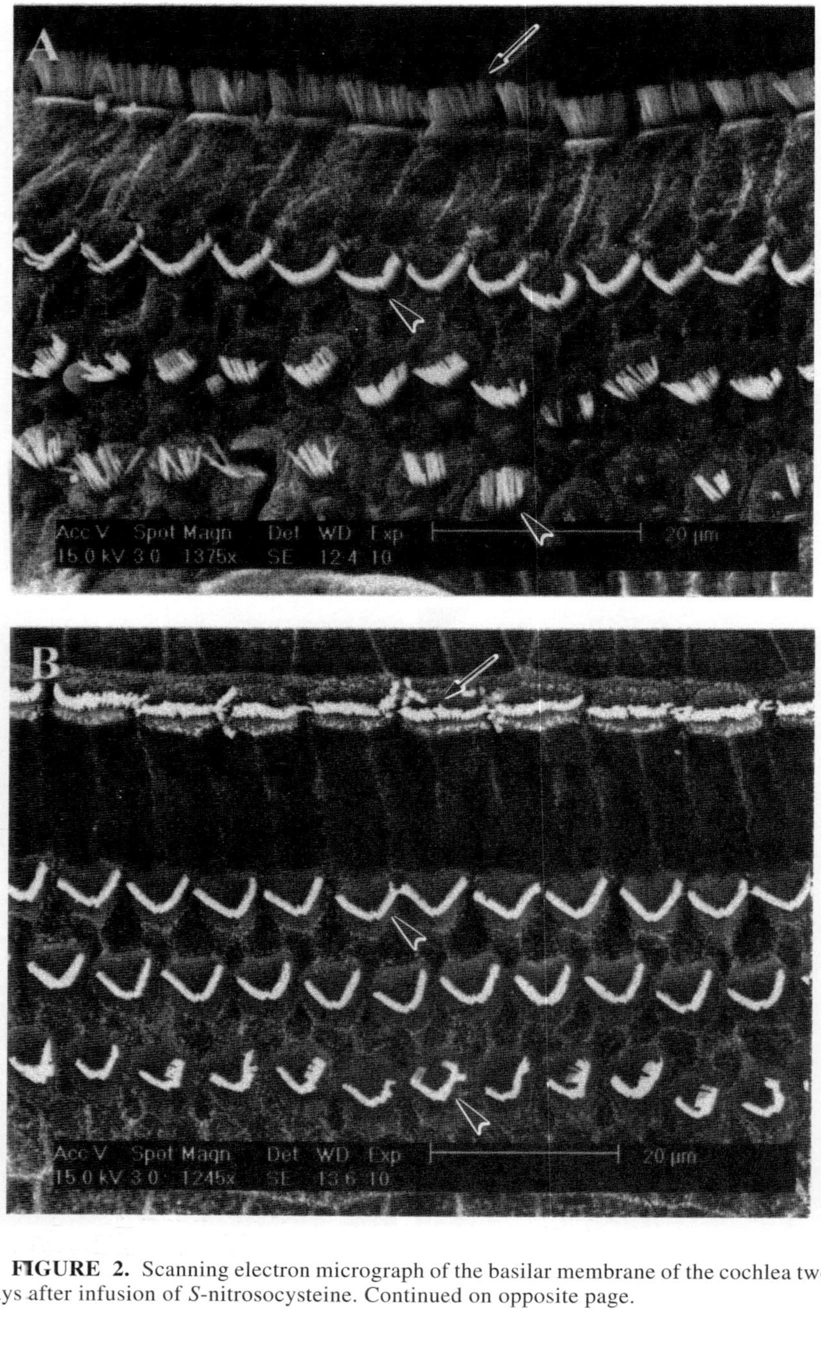

FIGURE 2. Scanning electron micrograph of the basilar membrane of the cochlea two days after infusion of *S*-nitrosocysteine. Continued on opposite page.

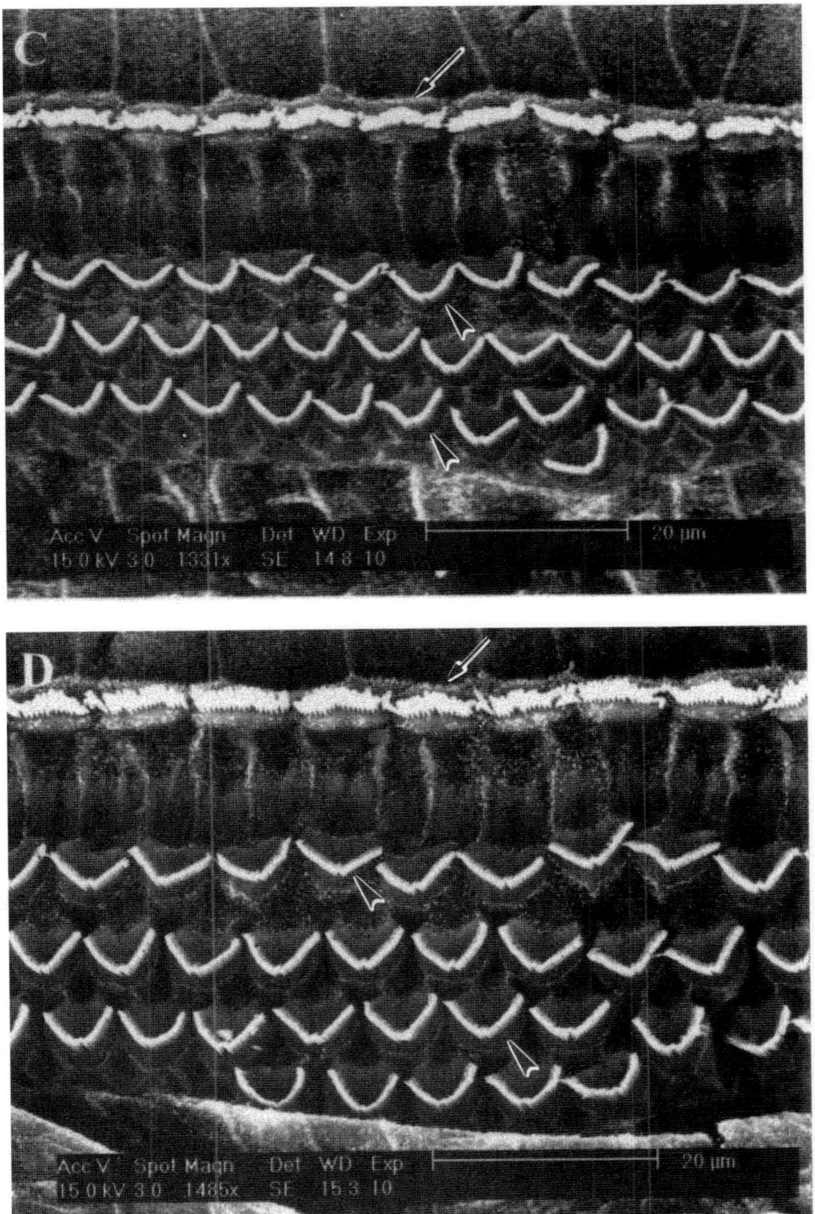

FIGURE 2/continued. The apical, upper middle, lower middle, and basal turn of the cochlea are shown in **A**, **B**, **C**, and **D**, respectively. *Arrows* indicate the inner hair cell bundles and *arrowheads* indicate outer hair cell bundles.

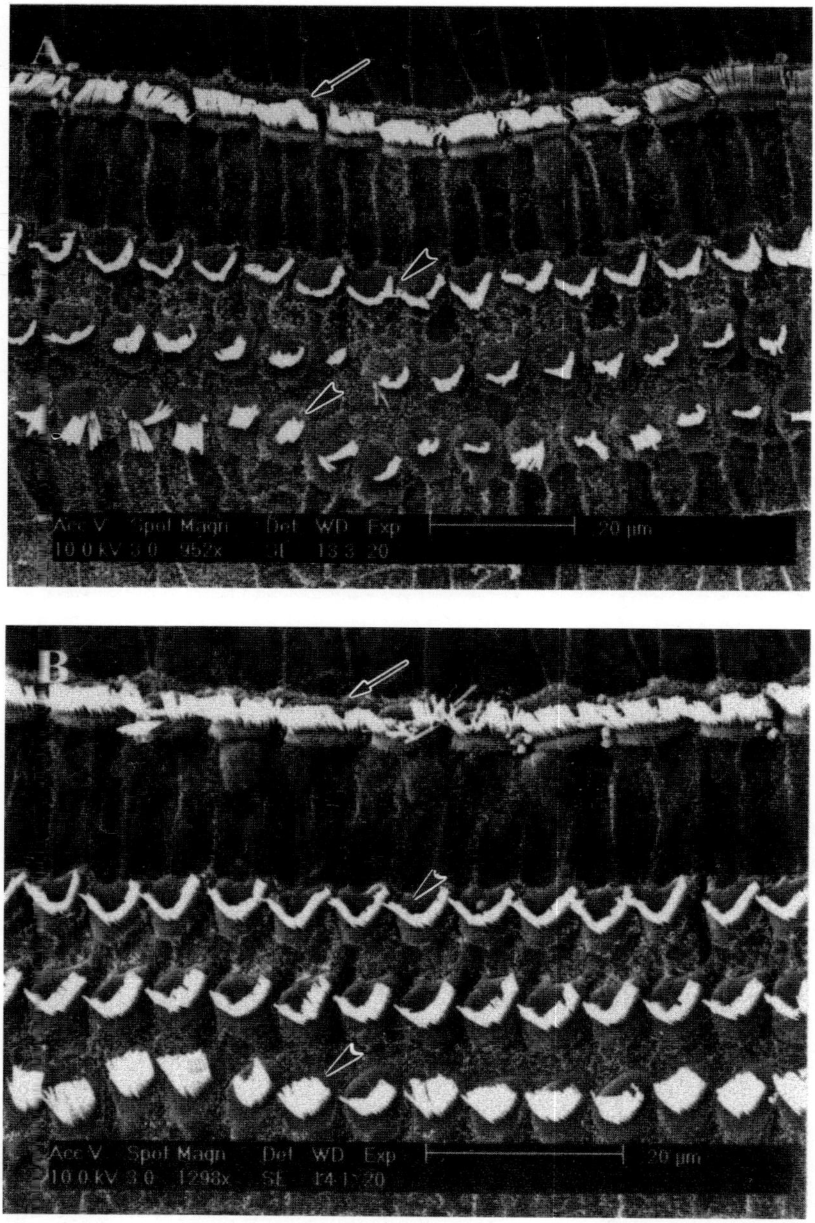

FIGURE 3. Scanning electron micrograph of the basilar membrane of the cochlea two days after infusion of 3-morpholinosydnonimine. Continued on opposite page.

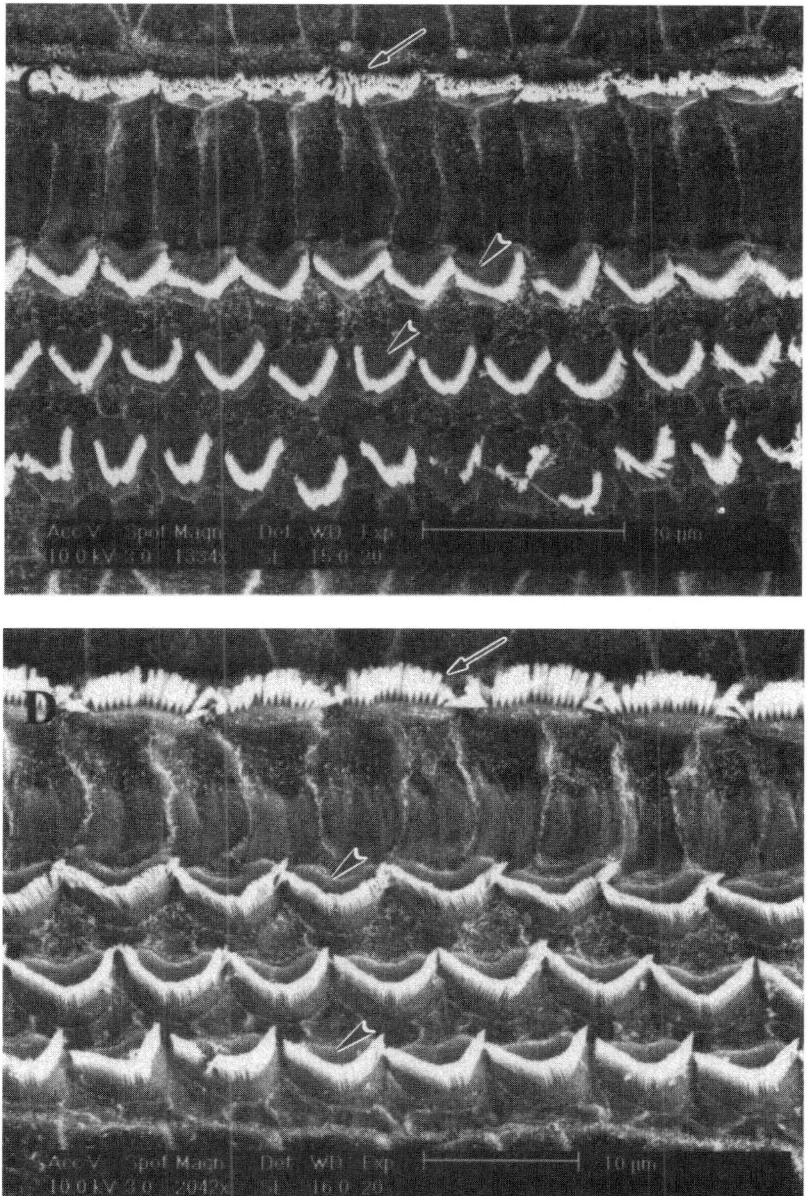

FIGURE 3/continued. The apical, upper middle, lower middle, and basal turn of the cochlea are shown in **A**, **B**, **C**, and **D**, respectively. *Arrows* indicate the inner hair cell bundles and *arrowheads* indicate outer hair cell bundles.

either intravenously or at the round window membrane, causes a reduction in cochlear blood flow.[18–20] These findings suggest an important role for NO in regulating cochlear blood flow, and are supported by the presence of NADPH-d, NOS, sGC, and DAF-DA fluorescence in the endothelial cells of the blood vessels throughout the cochlea. As a signaling molecule, NO released from the endothelium most likely activates the NO/sGC pathway of adjacent smooth muscle cells to bring about vasodilation.[5,13,14,21]

NO has also been implicated in the signaling processes of the auditory system.[4–6,13,15] The current findings from DAF-2DA studies indicate a broad cellular origin of NO in the cochlea. The distribution of NO-containing cells does not completely match that of NOS-positive cells, as demonstrated by NOS immunohistochemisty. Surprisingly, although NOS, citrulline, and sGC immunohistochemistry have been demonstrated in the neuronal cell bodies and processes of the spiral ganglion,[5,7,22] no signal was detected in these neurons with DAF-2DA. Why NOS localization is not matched with NO production remains to be studied. The fluorescent signal for NO was detected at the base of the inner hair cells, where the inner radial and spiral nerve fibers were demonstrated to be NOS I positive in our previous study.[7] This may indicate a role for NO in auditory neurotransmission.

EFFECTS OF NO ON PATHOPHYSIOLOGIC STATES IN THE COCHLEA

As well as its physiologic role, NO has also been implicated in neuronal damage observed in stroke and neurodegenerative diseases through over-activation of the *N*-methyl-D-aspartate (NMDA) receptors.[23,24] It has been postulated that, in the case of ischemic cochlea, over-activation of NMDA receptors triggers a massive influx of Ca^{2+} into the neuron to activate NOS, resulting in excessive NO formation, which leads to swelling of nerve endings at the base of the inner hair cells. This notion has strong support from demonstrations that NMDA receptor antagonists or NOS I inhibitor provide marked protection against neural damage.[25,26] However, the results from our experiments show a protective effect of NO on auditory hair cells under ischemic condition.[27]

Effects of NO and Peroxynitrite on Normal Cochlea

To test the possible ototoxicity of NO and peroxynitrite, my group performed a quantitative study of the inner and outer hair cells with a scanning electron microscope after delivery of high doses of various NO donors into the guinea pig cochlea. The donors delivered were 3-morpholinosydnonimine (SIN-1; donor of peroxynitrite) and *S*-nitrosocysteine (*S*-NC; NO donor). There was no statistical difference in the number of outer and inner hair cells at each turn of the cochlea, when compared with the number in control animals treated with normal saline (see FIGURES 2 and 3). To rule out the toxic effects of other chemicals formed when SNP breaks down, SNP exposed to room light for 48 hours to release NO was also tested. Administration of NO-depleted SNP caused significant hair cell loss, which was comparable with that observed in the animal receiving fresh SNP. Moreover, the hair cells were protected from damage when treated with freshly prepared SNP mixed with acetylcysteine to neutralize cyanide, a byproduct of SNP when it breaks down to produce NO.[28] It appears from these observations that the auditory hair cells are highly resistant to NO and peroxynitrite. Hearing impairment after application of SNP, as observed by independent laboratories, may be due to the toxic effects of cyanide on the cochlea instead of NO.[18,29–31]

Effects of NO on Ischemic Cochlea

To determine if there is any increase in the level of NO in the cochlea during cochlear ischemia, the amount of nitrates/nitrites was measured in the perilymph at 20, 30, 60, and 120 minutes after decapitation. The results showed that the level of nitrites/nitrates in the perilymph declined with an increasing period of ischemia (see TABLE 3).

The effects of NO on ischemic cochlea were tested by the infusion of superoxide dismutase plus catalase (inhibitor of peroxynitrite), L-NAME (NOS inhibitor), *S*-NC (NO donor), or nitroglycerin (NO donor) into the scala tympani before decapitation. Fifteen minutes after decapitation, the cochlea was processed for transmission electron microscopy. Swollen nerve terminals were counted at the base of at least two inner hair cells of each cochlea. The number of affected nerve endings taken from the cochlea treated with the respective chemicals was statistically compared with that for control cochleae treated with saline. The results showed that nitroglycerin and *S*-NC significantly protected most of the dendrites from swelling. On the other hand, there was no significant protection on the nerve endings after the infusion of SOD/Cat or L-NAME (see TABLE 4).

To summarize: (1) *S*-NC and SIN-1 do not cause hair cell loss in normal cochlea; (2) there is a decrease in nitrates/nitrites in the ischemic cochlea; and (3) inhibition of NO/peroxynitrite formation does not prevent neuropathologic development in ischemic cochlea. It is, therefore, reasonable to postulate that NO or peroxynitrite may not be responsible for the neurodestruction in the ischemic cochlea. In fact, exogenous NO may even have protective effects on the neural structures of the cochlea under ischemic condition, since, after infusion of NTG and S-NC into the cochlea, most of the nerve endings at the base of the inner hair cells of the ischemic cochlea were seen to be normal.[23]

How exogenous nitric oxide can be neuroprotective in the cochlea under ischemic condition is not clear. Lipton *et al.*[32] presented evidence that NO in the form of nitrosonium ion (NO^+) reacts with the thiol group of redox modulatory sites of the

TABLE 3. Values of nitrates/nitrites in the perilymph of three cochlea at 20, 30, 60, and 120 minutes after decapitation of guinea pigs

Time after death (minutes)	Nitrates/nitrites (μM)
0	28.5
20	8
30	6
60	2.8
120	3.6

TABLE 4. Mean and standard error for the number of swollen nerve endings at the base of inner hair cells in the cochlea treated with various chemicals at 15 minutes after decapitation

Chemicals	Number of swollen nerve endings
Saline ($n = 7$)	9.43 ± 1.51
SOD/catalase ($n = 9$)	9.78 ± 1.09
0.1 mM L-NAME ($n = 7$)	9.14 ± 1.35
1 mM L-NAME ($n = 8$)	9.13 ± 1.25
0.1 mM *S*-NC ($n = 8$)	9.03 ± 1.25
1 mM *S*-NC ($n = 9$)	4.56 ± 1.88*
0.44 mM NTG ($n = 6$)	1.33 ± 1.21*
4.4 mM NTG ($n = 9$)	1.11 ± 1.05*

NOTES: SOD, superoxide dismutase; *S*-NC, *S*-nitrosocysteine; NTG, nitroglycerin; *n*, number of cells observed. The symbol * indicates those groups of cochlea that showed significant difference ($p < 0.001$) from the saline-treated cochlea, as analyzed by a Student's *t*-test and one-way ANOVA with post hoc *t*-test.

NMDA receptors to inhibit excessive Ca^{2+} influx, a prerequisite for NMDA receptor-mediated neurotoxicity. Specifically, NTG does not spontaneously liberate ·NO to any significant extent, but does react readily with thiol groups, forming derivative thionitrites or thionitrates, an NO^+ equivalent.[32–37]

As mentioned, NO/cGMP signaling pathway is present in the cochlea. It has been reported that increase in intracellular cGMP may be neuroprotective, since this is the case with nonmotor neurons after exposure of spinal cord to glutamate toxicity.[38] In addition to the guanylate cyclase/cGMP pathway, NO may also exert its action through the Ras protein signaling pathway.[39] The Ras pathway is supposed to be a critical mediator of ischemic tolerance through changes in neuronal vulnerability by regulating gene expression and protein synthesis.[40–42] However, whether these endogenous protective mechanisms are relevant to our findings on the ischemic cochlea model remains to be elucidated.

REFERENCES

1. MARLETTA, M.A. 1993. Nitric oxide synthase structure and mechanism. J. Biol. Chem. **268:** 122231–12234.
2. POLLOCK, J.S., U. FORSTERMANN, J.A. MITCHELL, *et al.* 1991. Purification and characterization of particulate endothelium-derived relaxing factor synthase from cultured and native bovine aortic endothelial cells. Proc. Natl. Acad. Sci. U.S.A. **88:** 10480–10484.
3. NATHAN, C. 1992. Nitric oxide as a secretary product of mammalian cells. FASEB J. **6:** 3051–3064.
4. FESSENDEN, J.D. & J. SCHACHT. 1998. The nitric oxide/cyclic GMP pathway: a potential major regulator of cochlear physiology. Hear. Res. **118:** 168–176.
5. MICHEL, O., A. HESS, W. BLOCH, *et al.* 1999. Localization of the NO/cGMP-pathway in the cochlea of guinea pig. Hear. Res. **133:** 1–9.

6. HESS, A., W. BLOCH, S. ARNHOLD, *et al.* 1998. Nitric oxide synthase in the vestibulocochlear system of mice. Brain Res. **813:** 97–102.
7. RUAN, R.S., S.K. LEONG & K.H. YEOH. 1997. Localization of nitric oxide synthase and NADPH-diaphorase in guinea pig and human cochlea. J. Brain Res. **38:** 433–441.
8. HEINRICH, U.-R., J. MAURER, K. GOSEPATH & W. MANN. 1998. Immunoelectron microscopic localization of nitric oxide synthase III in the guinea pig organ of Corti. Eur. Arch. Otorhinolaryngol. **255:** 483–490.
9. KOJIMA, H., K. SAKURAI, K. KIKUCHI, *et al.* 1998. Development of a fluorescent indicator for nitric oxide based on the fluorescein chromophore. Chem. Pharm. Bull. **46:** 373–375.
10. SHI, X., T. REN & A.L. NUTTALL. 2001. Nitric oxide distribution and production in the guinea pig cochlea. Hear. Res. **153:** 23–31.
11. MASSEY, V. & C.M. HARRIS. 1997. Milk xanthine oxidoreductase: the first one hundred years. Biochem. Soc. Trans. **25:** 750–755.
12. ZWEIER, J.L., A. SAMOUILOV & P. KUPPUSAMY. 1999. Non-enzymatic nitric oxide synthesis in biological system. Biochem. Biophys. Acta **1411:** 250–262.
13. FESSENDEN, J.D. & J. SCHACHT. 1997. Localization of soluble guanylate chclase activity in the guinea pig cochlea suggests involvement in regulation of blood flow and supporting cell physiology. J. Histochem. Cytochem. **45:** 1401–1408.
14. HEINRICH, U.-R., J. MAURER, D. KOESLING, *et al.* 2000. Immuno-electron microscopic localization of the α_1 and β_1-subunits of soluble guanylyl cyclase in the guinea pig organ of Corti. Brain Res. **885:** 6–13.
15. TAKUMIDA, M., M. ANNIKO, R. POPA & D.M. ZHANG. 2000. Localization of soluble guanylate cyclase activity in the guinea pig inner ear. Acta Otolaryngol. **120:** 28–33.
16. GUTH, P.S. & M. STOCKWELL. 1977. Guanylate cyclase and cyclic guanosine monophosphate in the guinea pig cochlea. J. Neurochem. **28:** 263–265.
17. THALMANN, R., S. PALOHEIMO & I. THALMANN. 1979. Distribution of cyclic nucleotides in the organ of Corti. Acta Otolaryngol. **87:** 375–380.
18. OHLSÉN, A., E. HULTCRANTZ & B. ENGSTROM. 1993. The effect of topical application of vasodilation agents on cochlear electrophysiolgy. Acta Otolaryngol. **113:** 55–61.
19. BRECHTELSBAUER, P.B., A.L. NUTTALL & J.M. MILLER. 1995. Basal nitric oxide production in the cochlear vasculature. Otolaryngol. Head Neck Surg. **111:** 70.
20. VASS, Z., P.B. BRECHTELSBAUER, A.L. NUTTAL & J.M. MILLER. 1996. Nitric oxide mediates capsicain-induced increase in cochlear blood flow. Hear. Res. **100:** 114–119.
21. FESSENDEN, J.D., R.A. ALTSCHULER, A.F. SEASHOLTZ & J. SCHACHT. 1999. Nitric oxide/cyclic guanosine monophosphate pathway in the peripheral and central auditory system of the rat. J. Comp. Neurol. **404:** 52–63.
22. ZDANSKI, C.J., J. PRAZMA, P. PETRUSZ, *et al.* 1994. Nitric oxide synthase is an active enzyme in the spiral ganglion cells of the rat cochlea. Hear. Res. **79:** 39–41.
23. CHOI, D.W. 1992. Bench to bedside: the glutamate connection. Science **258:** 241–243.
24. DAWSON, V.L., T.M. DAWSON, E.D. LONDON, *et al.* 1991. Nitric oxide mediates glutamate neurotoxicity in primary cortical cultures. Proc. Natl. Acad. Sci. U.S.A. **88:** 6368–6371.
25. PUEL, J.-L., R. PUJOL, F. TRIBILLAC, *et al.* 1994. Excitatory amino acid antagonists protect cochlear auditory neurons from excitoxicity. J. Comp. Neurol. **341:** 241–256.
26. TABUCHI, K., Z. ITO, T. WADA, *et al.* 2000. Effect of 7-nitroindazole upon cochlear dysfunction induced by transient local anoxia. Ann. Otol. Rhinol. Laryngol. **109:** 715–719.
27. RUAN, R.S., S.K. LEONG & K.H. YEOH. 2001. Effects of nitric oxide on normal and ischemic cochlea of the guinea pig. Exp. Neurol. **169:** 200–207.
28. RUAN, R.S., S.K. LEONG & K.H. YEOH. 1999. Ototoxicity of sodium nitroprusside is not due to nitric oxide. Exp. Neurol. **158:** 192–201.
29. CHEN, C., A. NENOV, R. SKELLETT, *et al.* 1995. Nitroprusside suppresses cochlear potentials and outer hair cell responses. Hear. Res. **87:** 1–8.
30. DAIS, C.G., J. PRAZMA, S.S. BALL, *et al.* 1996. Effect of sodium nitroprusside on compound action potential thresholds in the gerbil cochlea. Hear. Res. **99:** 1–6.

31. KONG, W.J., T. REN & A.L. NUTALL. 1996. Electrophysiological and morphological evaluation of the acute ototoxicity of sodium nitroprusside. Hear. Res. **99:** 22–30.
32. LIPTON, S.A., Y.-B. CHOI, Z.-H. PAN, *et al.* 1993. A redox-based mechanism for the neuroprotective and neurodestructive effects of nitric oxide and related nitroso-compounds. Nature **364:** 626–632.
33. BATES, J.N., M.T. BADER, R. GUERRA, JR. & D.G. HARRISON. 1991. Nitric oxide generation from nitroprusside by vascular tissue. Evidence that reduction of the nitroprusside anion and cyanide loss are required. Biochem. Pharmacol. **42**(Suppl.): S157–S165.
34. HOGG, N., V.M. DARLEY-USMAR, M.T. WILSON & S. MONCADA. 1992. Production of hydroxyl radicals from the simultaneous generation of superoxide and nitric oxide. J. Biochem. **281:** 419–424.
35. IGNARRO, L.J., H. LIPTON, J.C. EDWARDS, *et al.* 1981. Mechanism of vascular smooth muscle relaxation by organic nitrates, nitrites, nitroprusside and nitric oxide: evidence for the involvement of *S*-nitrosothiols as active intermediates. J. Pharmacol. Exp. Ther. **218:** 739–749.
36. LEI, S.Z., Z.H. PAN, S.K. AGGARWAL, *et al.* 1992. Effects of nitric oxide production on the redox modulatory site of NMDA receptor channel complex. Neuron **8:** 1087–1099.
37. MARKS, G.S., B.E. MCLAUGHLIN, K. NAKATSU & J.F. BRIEN. 1992. Direct evidence for nitric oxide formation from glyceryl trinitrite during incubation with intact bovine pulmonary artery. Can. J. Physiol. Pharmacol. **70:** 308–311.
38. URUSHITANI, M., S. SHIMOHAMA, T. KIHARA, *et al.* 1998. Mechanism of selective motor neuronal death after exposure of spinal cord to glutamate: involvement of glutamate-induced nitric oxide in motor neuron toxicity and nonmotor neuron protection. Ann. Neurol. **44:** 796–807.
39. GONZALEZ-ZULUETA, M., A.B. FELDMAN, L.J. KLESSE, *et al.* 2000. Requirement for nitric oxide activation of p21^{ras}/extracellular regulated kinase in neuronal ischemic preconditioning. Proc. Natl. Acad. Sci. U.S.A. **97:** 436–441.
40. BARONE, F.C., R.F. WHITE, P.A. SPERA, *et al.* 1998. Ischemic preconditioning and brain tolerance: Temporal histological and functional outcomes, protein synthesis requirement, and interleukin-1 receptor antagonist and early gene expression. Stroke **29:** 1937–1951.
41. BERGMANN, A., J. AGAPITE, K. MCCALL & H. STELLER. 1998. The Drosophila gene *hid* is a direct molecular target of Ras-dependent survival signalling. Cell **95:** 331–341.
42. GOTOH, Y. & E. NISHIDA. 1995. Activation mechanism and function of the MAP kinase cascade. Mol. Reprod. Dev. **42:** 486–492.
43. RIEMANN, R. & S. REUSS. 1999. Nitric oxide synthase in identified olivocochlear projection neurons in rat and guinea pig. Hear. Res. **135:** 181–189.
44. GOSEPATH, K., I. GATH, J. MAURER, *et al.* 1997. Characterisation of nitric oxide synthase isoforms expressed in different structures of the guinea pig cochlea. Brain Res. **747:** 26–33.
45. FRANZ, P., C. HAUSER-KRONBERGER, P. BÖCK, *et al.* 1996. Localization of nitric oxide synthase I and III in the cochlea. Acta Otolaryngol. (Stockh.) **116:** 726–731.
46. YAMANE, H., M. TAKAYAMA, K. KONISHI, *et al.* 1997. Nitric oxide synthase and contractile protein in the rat cochlear lateral wall: possible role of nitric oxide in regulation of strial blood flow. Hear. Res. **108:** 65–73.

Nitric Oxide in Liver Diseases

Friend, Foe, or Just Passerby?

WEI MIN HON, KANG HOE LEE, AND HOON ENG KHOO

Departments of Medicine and Biochemistry, National University of Singapore, Singapore

ABSTRACT: Research on the free radical gas, nitric oxide (NO), during the past twenty years is one of the most rapid growing areas in biology. NO seems to play a part in almost every organ and tissue. However, there is considerable controversy and confusion in understanding its role. The liver is one organ that is clearly influenced by NO. Acute versus chronic exposure to NO has been associated with distinct patterns of liver disease. In this paper we review and discuss the involvement of NO in various liver diseases collated from observations by various researchers. Overall, the important factors in determining the beneficial versus harmful effects of NO are the amount, duration, and site of NO production. A low dose of NO serves to maximize blood perfusion, prevent platelet aggregation and thrombosis, and neutralize toxic oxygen radicals in the liver during acute sepsis and reperfusion events. NO also demonstrates antimicrobial and antiapoptosis properties during acute hepatitis infection and other inflammatory processes. However, in the setting of chronic liver inflammation, when a large sustained amount of NO is present, NO might become genotoxic and lead to the development of liver cancer. Additionally, during prolonged ischemia, high levels of NO may have cytotoxic effects leading to severe liver injury. In view of the various possible roles that NO plays, the pharmacologic modulation of NO synthesis is promising in the future treatment of liver diseases, especially with the emergence of selective NO synthase inhibitors and cell-specific NO donors.

KEYWORDS: nitric oxide; liver diseases; friend or foe

INTRODUCTION

During the past decade, particular attention has been paid to the small, diffusible, unique molecule—nitric oxide (NO). NO, the end product of the enzyme NO synthase (NOS), influences various physiological processes in essentially every organ and tissue. It has a remarkably broad spectrum of such functions as regulation of vascular tone, neurotransmission, antimicrobial defence mechanisms, and immunomodulation.[1] The role of NO in regulating an organ function is always complex, resulting in sometimes conflicting experimental data. In biological systems, NO has a very short half-life of less than five seconds and, therefore, direct measurement of NO is not easy and not commonly practiced. NO is rapidly inactivated in tissues by oxidation to its metabolite nitrites and nitrates, which readily diffuse into the

Address for correspondence: Wei Min Hon, Ph.D., Department of Medicine, Faculty of Medicine, National University of Singapore, 10 Kent Ridge Crescent, Singapore, 119260. Voice: (65) 779 5555 ext 2368; fax: (65) 779-4112.
mdchonwm@nus.edu.sg

circulation. Thus, concentrations of nitrite and nitrate are often used as indicators of the presence of NO. NO elicits many of its actions by signal transduction through activation of the soluble guanylate cyclase, binding to the iron in the heme center, resulting in an intracellular increase of cGMP levels.[2] Hence, increase of NO normally leads to an increase in cGMP concentration. Thus, cGMP is also used as an indicator of the presence of NO. NO can bind to non-heme iron, particularly to the iron in the iron-sulfur centers of numerous enzymes, thus altering their biological activity.[3] These enzymes include those in the respiratory cycle, the tricarboxylic acid (TCA) cycle, and the DNA synthesis pathway. Indirectly, NO can combine with superoxide anion to form the highly reactive peroxynitrite that has been linked to lipid peroxidation and nitrosation of tyrosine residues in proteins.[4] At different redox states, NO may appear in various chemical forms such as NO^+, $\cdot NO$, and NO^-. These NO-related species have distinct chemical reactivities and they can all influence the fate of cells in response to various insults. This review focuses on the possible roles of NO in liver diseases where the basic chemistry or biology of NO has been extensively discussed and reviewed.[1,5,6]

The liver is one of the most important body organs in that it performs so many different functions at the same time. The liver makes proteins, eliminates waste material from the body, produces cholesterol, stores and releases glucose and metabolizes many drugs used in medicine. Liver diseases, such as viral hepatitis, liver cirrhosis, and liver cancer appear to be on the increase and each year more than 43,000 Americans die of liver disease. There are few effective treatments for most life-threatening liver diseases other than liver transplants. It has been postulated that NO is involved with the etiology or progression of liver diseases because NO plays a major role in homeostasis regulation as well as in immunology defence mechanisms. However, the exact role of NO in most of these hepatic diseases is still unknown. Some results indicate that NO is beneficial whereas others postulate a detrimental effect. At the same time, some results even suggest that the presence of NO in certain situations is merely a consequence of other more important disease mechanisms.

There are three isoforms of NOS all of which are expressed in the liver. Of these, the inducible NOS (iNOS) and endothelial NOS (eNOS) are the most important. Neuronal NOS (nNOS) appears to be restricted to nerve endings found in large blood vessels. Under physiologic conditions only the constitutive eNOS is believed to be present in the liver and the low level of NO produced regulates hepatic perfusion, preventing platelet adhesion, thrombosis, and polymorphoneuclear cell (PMN) accumulation.[7] iNOS has been found in almost every cell type of the liver, such as hepatocytes,[8,9] Kupffer cells,[10] hepatic endothelial cells,[11] and Ito cells.[12] However, the degree and duration of NO production in various situations are probably determined by factors such as the type of stimulus, as well as the cell types that are being stimulated to express iNOS. Although most human cell types require a combination of cytokines to activate iNOS expression, interleukin-1β (IL-1β) alone at high doses can induce iNOS mRNA in primary human hepatocytes.[13] Hence, iNOS seems to be more readily induced in hepatocytes than other human cell types. In pathologic conditions, such as endotoxaemia, hemorrhagic shock, ischemia-reperfusion, sepsis, infection, hepatitis, liver cancer, and liver regeneration, iNOS can be induced and produce a significant sustained amount of NO. This NO serves as an important regulator and effector. Study of the actions of NO has been facilitated by the availability of various NOS inhibitors, the most common of which are N^G-nitro-L-arginine methyl ester (L-NAME), N^G-monomethyl-L-arginine (L-NMMA), and N^G-nitro-L-arginine (L-NNA).

The role of NO in various liver pathological conditions has yet to be clearly defined, which may be due to the differences among animal models used as well as the variations in experimental setup. Furthermore, the activity of NO produced by human cells is much lower than that produced by the rodent cells. Hence, the results observed in animal models may not be directly relevant to human pathophysiologic conditions. Therefore, in this article, out discussion is mainly based on the evidence observed in NOS knockout animal models and human patients.

LIVER DAMAGE DURING INFLAMMATION

Cytokines, such as tumor necrosis factor-α (TNF-α), are produced in the liver in endotoxaemia, systemic inflammation, and fulminant hepatic failure. They are responsible for the extensive hepatocellular injury under these conditions. Even though the mechanism of hepatocellular dysfunction has been widely studied, there are still many unanswered questions. Basically, the injured cells either survive or die through apoptosis or necrosis.[14] Because cytokines are well-known stimuli of iNOS, various experiments have been carried out to study the role of NO in these situations. However, results to date concerning the role of NO in liver damage during inflammatory conditions remains controversial. Both beneficial and detrimental NO-mediated properties have been described.

Non-specific inhibition of NOS in endotoxaemia models results in increased liver damage, suggesting a beneficial role of NO in liver during sepsis. Suppression of endogenous NO might aggravate hepatic injury, partly caused by decrease in hepatic blood flow accompanied with increase oxidative stress.[18,19] Recently, Takemura *et al.*[20] further demonstrated that a potent inhibitor of NOS markedly deteriorates the lipopolysacharride (LPS)-induced liver injury by disturbing hepatic microvascular blood flow. Zhu and Fung[21] showed that NO protects liver injury through scavenging lipid radical and, thus, inhibiting lipid peroxidation chain reaction.

Another possible protective mechanism of NO during liver injury might be through its antiapoptotic action. Increasing evidence shows that NO is a potent antiapoptotic molecule in hepatocytes *in vitro*[22] and *in vivo*[23] despite the fact that NO induces apoptosis in various other cell types.[24–26] The antiapoptotic actions of NO in hepatocytes were reviewed by Li and Billiar.[27]

NO is able to prevent TNF-mediated activation of the proapoptotic protease caspase 3 and to protect hepatocytes from cytokine-mediated death, conferring protection against TNF-induced liver injury.[23,28,29] NO acts via the *S*-nitrosylation of procaspases and active caspase enzymes.[30] Moreover, NO stimulation of the cGMP/protein kinase G pathway also appears to contribute to the protective effect of NO.[31] The antiapoptotic role of NO is further illustrated in two reports. First, Mojena *et al.*[32] studied the effect of preexistent hepatic NO synthesis on liver injury induced by LPS in animals carrying an iNOS transgene. These animals expressed iNOS in liver cells under fasting conditions. LPS-induced liver injury was impaired in animals expressing iNOS. Their results indicated that NO protects the liver by inhibiting the synthesis of proinflammatory mediators, such as TNF-α, and prevents apoptosis of the injured cells by inhibiting caspase 3. In the second report, Morio *et al.* analyzed the hypothesis by using knockout mice lacking the gene for the TNF-α receptor, TNF-α cytokine or iNOS. They found that there is significantly less carbon tetrachloride (CCl_4)-induced hepatic injury in mice lacking TNF-α receptor-1 or the TNF-α cytokine. In contrast, liver injury is increased in knockout mice lacking the gene for iNOS. These data further confirm the hepatoprotective effects of NO, which may be due in part to inhibition of TNF-α.

In contrast to the documented protective effects of NO, its detrimental role in liver injury has also been reported. Administration of aminoguanidine (iNOS selective inhibitor) markedly diminished the severity of the liver injury induced by various methods of induction.[34–36] Apparently, hydroxyl radicals produced by the reaction of NO and superoxide anion via peroxynitrite may be involved in the pathogenesis of hepatotoxicity. Recently, Sass *et al.* demonstrated that iNOS knockout mice are protected from liver damage following concanavalin A (Con A) treatment. Furthermore, they found that the amount of plasma TNF-α and intrahepatic TNF-α mRNA and protein is significantly reduced in iNOS knockout mice. These data imply that iNOS-derived NO regulates proinflammatory genes *in vivo,* thereby contributing to inflammatory liver injury.

The contradictory findings about the roles of NO during liver injury might be due to the differences in the redox status of the liver and the dosage and type of the insult. At the beginning of hepatic injury, when only low amount of NO is being produced, NO might protect the liver through its vasodilation, antioxidative, and antiapoptotic effects. However, in the event of massive injury (high dosage of inducers and elevated oxidative stress) large amounts of NO produced might favor the cells to move to the non-returnable channel—necrosis and cell death. As hepatocellular injury is not the major determinant of survival for patients with multiorgan failure due to sepsis or other reasons, the role of NO in hepatic injury has not been well studied. Most of the time, NO in liver injury was studied in conjunction with other liver diseases, such as hepatitis, liver cirrhosis, and ischemic-reperfusion, which we discuss in detail in a later section.

VIRAL HEPATITIS INFECTION

Exposure to hepatitis viruses, such as hepatitis virus B (HBV) and hepatitis virus C (HCV), can give rise to several outcomes: no obvious clinical infection, fulminant hepatitis, a self-limiting acute hepatitis, or a chronic infection that may progress to cirrhosis and hepatocellular carcinoma. Infection with the hepatitis virus is very common worldwide especially in Asia and Africa. The factors that determine the development of chronic viral hepatitis have not been fully identified. It is characterized by a parenchymal infiltration of activated cytotoxic T lymphocytes—the main cause of hepatic injury.

Antiviral activity of NO is well documented.[38,39] It has been shown to be able to inhibit the growth of DNA and RNA viruses, such as herpes simplex virus,[40] Coxsackie B3 virus,[41] and Japanese encephalitis virus.[42] The exact role of NO in hepatitis viral infection is not yet known. However, increased hepatocellular iNOS expression could be part of a nonspecific host response to viral infection inhibiting viral replication. This possible antiviral effect of NO may partly explain the increase in viral titers observed when corticosteroids, which are also iNOS inhibitors, were used in patients with chronic viral hepatitis during the 1980s.[43] Recently, the ability of NO to inhibit the replication of HBV virus was demonstrated in the livers of NO-knockout mice.[44] This result may indicate a role for NO as an antiviral agent for hepatitis infection since NO may reduce the expression of viral antigens in the cells, thereby diminishing the severity of the immune-mediated liver disease.

Initial experimental evidences for a connection between hepatitis infection and NO came from several reports of Liu *et al.*[45–47] First, these authors demonstrated that woodchucks chronically infected with woodchuck hepatitis virus produce high nitrate and *N*-nitroso carcinogenic compounds.[45] Subsequently, they showed that hepatocytes, isolated from woodchucks that were chronic carriers of woodchuck hepatitis virus, formed twice as much nitrite as hepatocytes from noninfected animals.[46] Moreover, woodchuck hepatitis virus surface antigen alone was sufficient to induce high levels of NO in culture hepatocytes.[47]

In 1997, Kane *et al.*[48] showed that 60% of liver biopsies of HCV-positive patients expressed iNOS by RT-PCR. Using immunohistochemistry, 100% of the HCV-positive patients expressed iNOS compared to 12.5% of controls. p53 was not detected in either group, but there was upregulation of p21 over baseline expression in a number of the HCV-positive patients. Kane *et al.* concluded that chronic expression of NO in HCV hepatitis might play a role in DNA mutagenesis and the development of HCC. In the same year, Mihm *et al.*[48] also demonstrated, using quantitative, competitive RT-PCR, that iNOS mRNA expression was increased in the liver tissue from chronically HCV-infected patients. Moreover, it was positively correlated with interferon-γ (IFN-γ) expression, as well as hepatic HCV RNA content in these patients. In contrast, other researchers demonstrated unchanged or decreased NO in patients with chronic hepatitis infection.[49–51] However, in these studies, the researchers measured only the plasma nitrite/nitrate level as an index of NO generation. The methods employed by these researchers for the measurement of serum nitrite/nitrate might not be sensitive enough to detect small changes. Besides this, other possible explanations for the discrepancy include differences in the duration of disease, severity of inflammation, and the degree of fibrosis of various patient groups chosen.

Subsequently, other researchers have found other links between NO and chronic hepatitis viral infection. Majano *et al.*[52] demonstrated intense expression of iNOS in liver biopsies of patients with chronic viral hepatitis by immunohistochemistry and *in situ* hybridization. Immunohistochemistry localized iNOS protein only to hepatocytes whereas some mononuclear cell infiltrate and vascular endothelium were also positive for the mRNA by *in situ* hybridization. Thereafter, they demonstrated that transfection with either the HBV genome or HBV X gene into HepG2 cells resulted in induction of iNOS mRNA expression. Maximal induction of this transcript and NO production was observed in cytokine-stimulated HBV-transfected cells. Subsequently, by applying a non-radioactive *in situ* hybridization method, Schweyer *et al.*[53] also found increased IFN-γ and iNOS gene expression in liver biopsy specimens from patients with chronic HCV infection. However, in contrast, IFN-γ and iNOS mRNA were observed in CD3+ lymphocytes infiltrating portal tracts and hepatic lobules, but not in hepatocytes. The differences might be due to the use of different hybridization probes as well as the hybridization conditions. More work needs to be done to confirm these results.

Recently, Garcia-Monzon *et al.*[54] demonstrated, by means of double immunostaining, that NO-mediated nitration of hepatocellular proteins is markedly induced in the inflamed liver tissue from patients with chronic viral hepatitis, and appears to be associated with the histological severity of viral chronic liver disease. The capability of HBV to increase the transcription of human hepatic iNOS by transactivating its promoter has also been studied. Amaro *et al.*[55] demonstrated that HBV X protein transactivates the iNOS promoter.

All these results demonstrated that NO was indeed increased during viral hepatitis. Increased production of NO during viral hepatitis has two implications. First, NO may serve as an antiviral agent that aids the host to fight against the viral infection. Second, when considered in the light of the known genotoxicity of NO,[56] long-term elevated production of NO free radicals in chronic hepatitis may directly cause reactions with cellular DNA leading to mutagenesis, as well as the formation of hepatocarcinogenic *N*-nitroso compounds. The well known relationship between viral hepatitis infection with increased risk of liver cancer may be the result of increased production of NO in addition to the antiapoptotic properties (apoptosis represents the paramount mechanism for eliminating virus-infected hepatocytes) of certain hepatitis viral proteins.[57]

LIVER CIRRHOSIS

Persistent hepatic injury leads to cirrhosis—a scarring process in the liver that includes both increased fibrogenesis and wound contraction. Cirrhotic patients often present with several systemic hemodynamic disturbances, characterized by symptoms such as hypotension, low systemic vascular resistance, and a reduced sensitivity to vasoconstrictors. During the progression of cirrhosis, vascular resistance continues to decrease and the low arterial pressure may lead to secondary disturbances in renal and hepatic blood flow, ascites, and portal hypertension.

Since Vallance and Moncada[58] first proposed that NO could be responsible for the hyperdynamic circulation in cirrhosis, there has been growing evidence that implicates role of NO in the pathophysiology of liver cirrhosis. According to their hypothesis, NO overproduction may be due to increased incidence of endotoxemia during liver cirrhosis, and endotoxaemia may induce NO overproduction directly or indirectly through cytokines. Their hypothesis was supported by data provided by other experiments: (1) NOS is induced in the endothelium and smooth muscle when vascular tissue is exposed to endotoxin or cytokines *in vitro*;[59,60] (2) infusion of endotoxin in humans leads to the gradual appearance of peripheral vasodilatation;[61,62] (3) inhibitors of NOS increases blood pressure of patients with septic shock;[63] and (4) high circulating levels of endotoxin are found in cirrhotic patients with or without clinical evidence of infection.[64,65]

Evidence supporting a role for NO has been obtained from various experimental models of cirrhosis.[66,67] For example, NO-dependent vasodilation is increased in the aortic rings of cirrhotic rats, and administration of NOS inhibitors to cirrhotic rats increases systemic vascular resistance, thus modulating hyperdynamic circulation.[68] Moreover, the production of NO is observed to increase in cirrhotic patients by using various parameters. One of the earliest reports was that urinary cGMP levels are elevated in cirrhotic patients, directly correlating with the severity of hemodynamic changes.[69] Subsequently, a case report described an event where a cirrhotic patient with severe hypotension and resistance to other pressor agent demonstrated an elevation in blood pressure in response to methylene blue infusion.[70] Methylene blue blocks the action of NO through the inhibition of guanylate cyclase. This provides clinical evidence that NO is responsible, at least in part, for the hypotension associated with liver failure. Subsequently, the increased production of endogenous NO, as shown by the increased plasma/serum, urine, and ascitic fluid nitrate levels, was demonstrated in patients with liver cirrhosis.[51,71–74] These results suggest that long-lasting increased local production of NO may contribute to maintenance of splachnic vasodilatation and, thus, worsen the hyperkinetic state in these patients. These reports also found that the degree of liver damage severity correlates with high serum nitrate levels. In addition, significantly higher peripheral and hepatic vein NO levels are observed in cirrhotic patients by the measurement of NO-hemoglobin complexes using electron paramagnetic resonance spectroscopy.[75] It has also been demonstrated that NO output is increased in the air exhaled by patients with cirrhosis, especially by patients with decompensated cirrhosis.[76] The NO might be produced by the vascular and bronchial tissues in the lungs of these patients instead of directly from the liver. Nevertheless, increases in the production of endogenous NO correspond to the progress of liver cirrhosis. At the cellular level, PMNs and monocytes isolated from cirrhotic patients have greater iNOS activity compared to that of healthy subjects,[77] and NOS activity in peripheral neutrophils of cirrhotic patients increases with increasing severity of liver dysfunction.[78] To provide further evidence that NO does play a role in the pathologic events of cirrhosis, Ryan *et al.*[79] and Campillo *et al.* demonstrated that infusion of NOS inhibitor, L-NMMA, improved the responsiveness to noradrenaline in forearm arteries of cirrhotic patients.[79]

According to the hypothesis of Vallance and Moncada,[58] as well as the results demonstrated by various researchers, one would expect that iNOS would be the NOS isoform responsible for the sustained production of large amounts of NO during liver cirrhosis. However, there is contradictory evidence. Thus, the exact role of iNOS in liver cirrhosis and portal hypertension remains in doubt. For example, Fernandez *et al.*[81] could not show any significant increase in iNOS activity in BDL rats since dexamethasone failed to modify the hyperdynamic circulation in the cirrhotic animal model. In addition, it seems that an increase of eNOS activity could be responsible for NO overproduction in animal models of portal hypertension.[82] Campillo *et al.*[83] reported that NO might be a consequence, rather than a cause, of hemodynamic abnormalities and that anemia might play a key role in NO production of patients with cirrhosis. Renal impairment and diminished urinary nitrate clearance might be another cause of increased NO observed since the major metabolic pathway of serum nitrate is through urinary excretion.[84]

Generally, all these observations still support a detrimental role of NO in the pathogenesis of liver cirrhosis, where increased NO leads to the vascular hyperdynamic state, and reduction of NO is beneficial. However, other findings do not support the Vallance and Moncada hypothesis.[58] Studies have shown that NO production was reduced in the cirrhotic rat liver.[85] In humans, Sarela *et al.*[86] demonstrated that the activity of constitutive NOS is substantially lower in the liver of cirrhotic patients as compared to the histologically normal liver from metastatic colorectal adenocarcinoma patients who underwent liver resection. The activity of iNOS was unaltered. Reduced hepatic cNOS activity, with a resultant decrease in NO release, is likely to facilitate the counteracting influence of the potent vasoconstricting agent, endothelin 1, especially in the cirrhotic liver, and thereby bring about an increase in the sinusoidal resistance. The cause of diminished cNOS may reflect a response to hepatocellular damage. Portal hypertension is a major complication of liver cirrhosis. An increase in the intrahepatic resistance to portal venous flow is an important factor in the development and maintenance of portal hypertension. A decrease of NO may, therefore, promote portal hypertension through an increase in the hepatic sinusoidal resistance. Overall, alternation of NO synthesis at various times and sites is detrimental for the host and aggravates the pathogenesis of liver cirrhosis.

HEPATOCELLULAR CARCINOMA (HCC)

HCC is one of the most common solid tumor malignancies affecting humans. It is responsible for about one million deaths per year worldwide. A major epidemiologic association between either HBV or HCV infections and HCC is evident.[87] However, the underlying mechanisms that lead to the development of HCC are still not clear. Although integration of virus DNA sequences into liver cell genome could activate cellular protooncogenes, this integration is very infrequent and does not explain the majority of virus-induced HCCs. Increased NO formation by liver cells has been shown to occur in many hepatic diseases, such as that caused by the parasites *Opisthorchis viverrini* (liver fluke),[88] by hepatitis viruses,[52,89] and in cirrhosis.[76,76] Increased NO could be the common underlying mechanism for the increased risk of liver cancer associated with these various forms of chronic liver diseases, where liver tissues that are exposed to high concentrations of NO over long periods of time could accumulate mutations.

In a woodchuck hepatitis virus model, chronic infection with hepatitis virus seems to enhance NO production and formation of carcinogenic nitrosamines.[45–47] Viral hepatitis might increase the risk of liver cancer through a mechanism of increased NO production. In man, NO levels are elevated in patients with chronic hepatitis and this has been linked with the predisposition to develop liver cancer.[90] Other data showed that plasma nitrite/nitrate concentrations in patients with HCC are correlated with tumor volume as well as tumor surface area.[50] Recently, it has been suggested that plasma nitrite/nitrate concentrations could be used as a tumor marker for HCC in conjunction with serum α-fetoprotein (AFP, a well established biochemical parameter for HCC).[91] In agreement with this study, Moussa *et al.*[51] also found that plasma nitrite/nitrate levels in patients with HCC are elevated. A possible explanation for increased plasma NO levels in HCC is that NO is reactively induced by the hepatic tissue surrounding HCC by three independent mechanisms: (1) tumor cells directly stimulate macrophages and Kupffer cells to produce NO, (2) HCC produces a variety of cytokines that may stimulate hepatocytes to produce NO, and

(3) a marked deterioration of liver function in HCC patients may be associated with increased portosystemic shunting and further development of hyperdynamic circulation, leading to an increase in NO production. In addition, preliminary work in our laboratory has demonstrated increased iNOS and eNOS mRNA expression by RT-PCR in the tumor tissue compared to the tissue surrounding the tumor of HCC patients (unpublished data). These may lead to the observed increased NO level.

The role of NO in cancer formation is controversial. It has mutagenic effects as well as known antitumor effects.[92] It is well known that high levels of NO can cause nitrosative deamination or oxidation of DNA bases, leading to DNA damage, as well as mutation in human cells.[56,93,94] In addition, as tumor growth progresses, NO may mediate capillary leakiness, support angiogenesis, and limit leukocyte infiltration. On the other hand, NO possesses antitumor properties because of its long known cytotoxic and cytostatic[95] effects toward tumor cells, as well as its anti-apoptotic role.[27]

Several specific types of mutations would be expected from DNA deamination, including GC to AT, GC to TA, and AT to GC base pair substitutions.[96] Sequence data on precore and core gene mutations in hepatitis B virus isolated from chronic HBV-infected individuals show GC to AT mutations.[97] Similar patterns of mutations of the p53 gene have been found in HCC.[98,99] These suggest that chronic inflammation and NO produced by NOS may mutate these genes through deamination of DNA bases. The p53 tumor suppressor is a 393-amino acid nuclear transcription factor that plays a central role in cell cycle regulation and apoptosis.[100] Inactivation of p53 function has been reported in approximately 50% of all human cancers, including malignancies of the liver. It has also been shown that NO is able to induce conformational changes of p53, decrease its specific DNA binding activity, and thus, increasing the risk of malignant cell transformation.[101,102]

Recently, Morbidelli *et al.*[103] and Ziche *et al.*[104] reported that NO plays a central role in the angiogenic cascade by demonstrating that vascular endothelial growth factor (VEGF), released by tumor cells, requires a functioning NO/cGMP pathway within the endothelial compartment to promote neovascular growth. Other research groups have found that NO induces VEGF expression in human Hep G2 hepatocarcinoma cells.[105] VEGF plays an important role in tumor biology in at least two ways: as a vascular permeability factor and/or endothelial growth factor. Elevated expression of VEGF has also been reported for tumors of the gastrointestinal tract,[106] kidney,[107] and breast[108] in humans. However, there are few reports on the expression of VEGF in HCC, which is well known for its hypervascularity.[109,110] The promotion of tumor growth by NO may involve the induction of angiogenic factors such as VEGF.

In contrast, there is also increasing evidence indicating that NO might also play an important role in antitumor mechanisms. It was demonstrated that Kupffer cell-derived NO suppresses proliferation and induces apoptosis of hepatoma cells.[111] These results indicate that in the event of HCC, Kupffer cells surrounding the tumor can induce apoptosis of the tumor cells via the production of NO.[112,113] In conjunction with the well known cytotoxic effects of NO, through its effect of nitrosating important proteins in tumor cells, the hepatoprotective role of NO in HCC cannot be ignored.

The effects of NO in various stages of cancer are widespread and often self contradictory. Its effects can be better understood based on timing, location, and concentration. Here, we hypothesize that NO plays an important role in HCC formation and its progression. During the pre-HCC period, viral infection (hepatitis B and C) or other unforeseen circumstances may lead to uncontrolled, prolonged, and/or massive production of NO by iNOS in the liver. Genotoxic properties of NO may lead to the mutation of certain oncogenes or tumor suppressor genes, such as p53, that allow the cells to escape from the cell cycle growth arrest mechanism and thus contribute to hepatic carcinogenesis. At the same time, NO regulates angiogenesis. NO released by the tumor cells might enhance angiogenesis, which can lead to accelerated growth of the primary tumor, as well as facilitate the process of metastasis. Angiogenic effects of NO may be due to an upregulation of proangiogenic factors and/or downregulation of their natural inhibitors (antiangiogenic factors). On the other hand, other functional cells surrounding the tumor might possess certain tumoristatic and tumoricidal effects on the tumor cells through the action of NO and, hence, provide hepatoprotective action.

LIVER TRANSPLANTATION AND HEPATIC ISCHEMIA-REPERFUSION INJURY

Orthotopic liver transplantation is now widely used for the treatment of end-stage liver disease. Cellular rejection is one of the common complications following liver transplantation and remains a significant cause of graft loss. The basis of cellular rejection is an alloantigen activation and proliferation of inflammatory cells that infiltrate the donor organ and mediate graft damage. Experimental work suggests that NO as a cytotoxic molecule may play an important role in the *in vivo* response to allogenic tissue.

Increased endogenous NO production, represented by the rise in systemic nitrate levels, have been described during clinical hepatic allograft rejection.[114–116] It was found that serum[114] and urinary[115] nitrate levels correlate well with the grade of allograft rejection in human liver transplantation. These results agree with the data reported during allograft rejection in animal models.[117,118] Hence, it was suggested that plasma and urinary nitrate levels could offer a sensitive predictor of acute allograft rejection in human liver transplantation, as well as its severity, and the evaluation of its resolution. Moreover, Sugioka *et al.*[119] found that changes in graft tissue NO preceded any other indicators of acute rejection, thus providing more concrete proof that NO is a mediator of acute rejection, and not a consequence. Corticosteroid (a well known iNOS inhibitor) treatment during the rejection episode leads to significant decline of the serum nitrate levels, hence, further confirms the involvement of NO in allograft rejection. Direct evidence was demonstrated by Romero *et al.*[120] using an immunohistochemical method. They found marked expression of iNOS in hepatocytes from liver biopsies of patients with acute rejection when compared with patients without rejection. Positive hepatocytes presented strong cytoplasmic staining, whereas no detectable iNOS reactivity was detected in any other cell types. After treatment with intravenous corticosteroids, iNOS reactivity decreased significantly.

This provides evidence that NO plays a role in clinical hepatic allograft rejection. However, it remains controversial whether NO generated during acute rejection is beneficial or harmful to the transplanted recipients. A beneficial role of NO was supported by the following findings during allograft rejection or other forms of liver injuries: (1) NO inhibits lymphocyte proliferation and acquisition of cytolytic T lymphocytes during alloimmune response;[121] (2) NO scavenges free oxygen radicals during hepatocyte injury;[122,123] and (3) blocking of iNOS activity impairs hepatic microvascular blood and increases tissue damage.[15,124] Hence, NO may be an important mediator of the graft-versus-host reaction and may serve as a marker of this process. On the other hand, a sudden burst of NO locally may produce damage to cells during rejection.[125,126] Nitrosylation of key enzymes involved in the respiratory cycle and the synthesis of DNA in target cells could mediate the cytotoxicty of NO. The differences in the effects of NO in rejection may be related to the amount of NO produced. Additional studies are warranted for an understanding of the exact function of NO in hepatic allograft rejection.

Liver ischemia-reperfusion is an important syndrome encountered in a number of clinical scenarios, such as trauma, hemorrhagic shock, liver resection, and liver transplantation. In particular, for liver transplantation, ischemia-reperfusion injury is an important cause of primary nonfunction of the liver leading to urgent retransplantation. The generation of reactive oxygen intermediates and inflammatory cytokines, as well as microcirculatory disturbances during the reperfusion phase, underlies the pathophysiology of this syndrome. In view of the capability of sinusoidal endothelial cells in producing a basal level of NO from eNOS in response to flow, and the increased concentration of oxygen free radicals and cytokines in inducing the expression of iNOS in other cell types of the liver,[127,128] the role of NO in hepatic ischemia-reperfusion has aroused tremendous research interest in the field.

NO maintains hepatic perfusion under physiologic conditions in normal rat livers, inhibition of NO synthesis causes a marked increase in perfusion pressure.[7] Administration of both non-selective and selective inhibitors of NOS into ischemic rat models results in marked aggravation of postischemic liver injury.[124,129,130] By increasing NO availability, using either an NO precursor or NO donors, reperfusion injury in animal models of hepatic ischemia-reperfusion was markedly reduced.[130–134] Hence, the liver dysfunction associated with reperfusion injury was suggested to be linked to a decrease in the release of NO. The beneficial role of NO during reperfusion is believed to be related to its ability to counteract the effects of endothelin,[135] reduced inflammatory cell activity, and expression of cytokines and adhesion molecules.[136,137]

Notwithstanding, some researchers have found that NO producing toxic injuries during reperfusion[138.139] may act as a lethal messenger in cell-mediated cytotoxicity. It has been shown that hypoxia also favors the transcription of iNOS through the engagement of hypoxia response elements present in the iNOS promoter.[140] Therefore, it is possible that ischemic episodes might contribute to iNOS expression and in this way account for the obvious sustained NO synthesis posttransplantation.

In conclusion, it appears that multiple factors, such as NO-superoxide radical ratios, hepatic stores of reduced glutathione, and length of ischemia, all determine whether NO will act as a cytoprotective or cytotoxic agent. The constitutive form of NOS predominates in acute conditions of ischemia releasing a small amount of NO that is important in the preservation of homeostasis, regulating microvascular permeability and neutrophil–endothelial cell interaction, and maintaining local perfusion. In severe chronic ischemia, the high levels of NO produced by iNOS, may be cytotoxic. NO may combine with the high levels of superoxides to form peroxynitrite, which becomes extremely toxic to cells.

In liver transplant recipients, L-arginine seems to be extremely deficient after reperfusion since high plasma arginase (an enzyme that hydrolyzes arginine to urea and ornithine) levels were observed following liver perfusion.[141,142] In view of the protective role of NO in ischemic-reperfusion, endogenous NO augmentation may be more effective in attenuating hepatic ischemia reperfusion than any other approach. Furthermore, therapeutic strategies that increase endogenous NO may be useful in improving the outcome of liver transplantation involving the use of suboptimal grafts, which are known to be more susceptible to ischemic-reperfusion injury and poorer outcomes. However, we need to bear in mind that NO also exerts a strong hypotensive effect and has the ability to produce toxic peroxynitrite radicals after reacting with superoxide radicals. Thus, treatment with high doses of NO can be hazardous. In the event of ischemia reperfusion, a combination of NO donors and endothelin blockers might provide an alternative intervention strategy since inhibition of enthothelin in animal ischemic models seems to improve microcirculation and circumvented reperfusion injury of the liver.[143,144] In view of the availability of the selective hepatic NO donor—V-PYRRO/NO (O^2-vinyl 1-[pyrrolodin-1-yl]diazen-1-ium-1,2-diolate), which selectively delivers NO directly to the hepatocyte by virtue of cytochrome P450 metabolism,[23] we are getting closer to the use of this molecule in a clinical setting.

LIVER REGENERATION

In adult vertebrates, the capacity for regeneration is limited to a few tissues, one of which is the liver. In the event of surgical resection of diseased livers, the ability of the remnant liver to initiate the regeneration process is a very important part of the recovery process for the patients. The mechanisms that permit the tissues to regenerate are not well understood. Initiation of liver regeneration requires injury-related cytokines, such as TNF-α[145] and IL-6,[146] and it involves the activation of cytokine-regulated transcription factors, such as NF-κβ[147] and STAT3.[148] During regeneration, these cytokines promote hepatocyte viability, as well as proliferation.[149] These observations suggest that the cytokines induce hepatoprotective factors in the regenerating liver. iNOS is a well known cytokine-inducible enzyme; thus, the product, NO might be involved in one way or the other for the action of cytokines accumulated during tissue damage.

Among the multiple changes in the immediate hours following partial hepatectomy, an induction of iNOS and the release of NO have been reported in rats.[150–152] The role of NO in liver regeneration was further tested by estimating the amount of nitrite accumulated during 24 hours in the culture media of hepatocytes, Kupffer cells, and sinusoidal endothelial cells isolated at various times following partial hepatectomy.[153,154] The time course of NO production was compared with the course of the proliferating activity of the same cells. During the time interval when liver cells passed through their first cell cycles, hepatocytes were the main producers of NO in the liver. The time-dependent changes of their NO production corresponded to those obtained with the whole liver and were inversely correlated with DNA-synthesizing activity. NO production by Kupffer and endothelial cells followed that by hepatocytes in this order; the time displacement between them corresponded to the schedule of their proliferating activity. Díaz-Guerra *et al.*[155] further confirmed the role of NO in partially hepatectomized mice and they postulated that NF-κβ was required for the transcriptional control of iNOS in the regenerating liver.

Subsequently, Wang and Lautt[156] hypothesized that the hemodynamic consequence of partial hepatectomy triggered the cascade of events that leads to liver regeneration. After partial hepatectomy, all the portal flow must go through the remaining vascular bed, thus producing increased shear stress and release of NO, which then initiates the next stages of the regeneration process. Their results in rat models demonstrated that the vascular shear stress-induced release of NO following partial hepatectomy serves as a primary trigger to initiate the regeneration process.

Recently, more significant results were demonstrated by Rai *et al.*[157] in iNOS knockout mice. These iNOS knockout animals exhibit a significant increase in hepatocyte apoptosis 24 hours after partial hepatectomy, indicating that NO is probably involved in preventing apoptosis and serves as an important hepatoprotective factor in the regenerating liver. With further advances in research, someday, NO or its analogues may be delivered directly to the liver to limit hepatic injury and facilitate liver cell regeneration.

CONCLUSIONS

The significance of NO in various diseases, whether it is a cause or merely a consequence, has not been fully determined. Depending on the experimental conditions, NO can attenuate or enhance certain pathophysiologic responses that reflect the complex behavior of this small molecule in the body. Most detrimental effects of NO are based on the concept that NO serves as a precursor, yielding potent toxic factors, such as peroxynitrite and hydroxyl radical, as a result of the interaction with the superoxide anion. The beneficial role of NO is normally based on its ability to maintain the integrity of microvascular function, its ability to inhibit platelet aggregation and neutrophil infiltration, as well as its unique ability to prevent apoptosis in hepatocytes and, hence, prevent further liver injury and promote liver regeneration.

Conflicting results may be due to concentration-dependent effects of NO, the variety of *in vivo* and *in vitro* experimental models and differences in cell sensitivity to NO. The effects of NO may also vary depending on acute or chronic inflammatory states and the animal species used in experiments. Posttranslational modifications may also modulate the amount of NO released. It should be borne in mind that studies evaluating mRNA or protein levels alone are insufficient for determination of true NOS-dependent NO production. This might also explain the contradictory results from various research groups, as they look at various regulation levels under different settings and at different times. A study that involves the direct measurement of NOS activity in the organ involved will give a clearer picture of what NO does in a particular pathologic event than those studies that only demonstrate the effects of NO donors or NOS inhibitors in the entire animal. Because NO has diverse physiologic functions, the overall effects of the donors and inhibitors might not represent the specific action of NO in the particular organ involved. Drugs that can modulate NO levels have been used as therapeutic agents for a long time. Nitrovasodilators, such as nitroprusside, that act by donating NO spontaneously, or glyceryl trinitrite and isosorbide dinitrate, that release NO after metabolic conversions are able to activate guanylate cyclase and elevate cGMP in the vasculature, thus providing a favorable adjustment of the vasculature environment. In view of the existence of more specific NOS inhibitors, as well as cell-directed NO donors, the use of NO donors or NOS inhibitors in the treatment of liver diseases may soon become reality.

REFERENCES

1. Moncada, S., R.M. Palmer & E.A. Higgs. 1991. Nitric oxide: physiology, pathophysiology, and pharmacology. Pharmacol. Rev. **43:** 109–142.
2. Schmidt, H.H., S.M. Lohmann & U. Walter. 1993. The nitric oxide and cGMP signal transduction system: regulation and mechanism of action. Biochim. Biophys. Acta **1178:** 153–175.
3. Nathan, C. 1992. Nitric oxide as a secretory product of mammalian cells. FASEB J. **6:** 3051–3064.
4. Bartosz, G. 1996. Peroxynitrite: mediator of the toxic action of nitric oxide. Acta Biochim. Pol. **43:** 645–659.
5. Davies, M.G., G.J. Fulton & P.O. Hagen. 1995. Clinical biology of nitric oxide. Br. J. Surg. **82:** 1598–1610.
6. Moncada, S. & A. Higgs. 1993. The L-arginine-nitric oxide pathway. N. Engl. J. Med. **329:** 2002–2012.
7. Mittal, M.K., T.K. Gupta, F.Y. Lee, *et al.* 1994. Nitric oxide modulates hepatic vascular tone in normal rat liver. Am. J. Physiol. **267:** G416–G422.
8. Geller, D.A., S.M. Di, A.K. Nussler, *et al.* 1993. Nitric oxide synthase expression is induced in hepatocytes *in vivo* during hepatic inflammation. J. Surg. Res. **55:** 427–432.
9. Geller, D.A., P.D. Freeswick, D. Nguyen, *et al.* 1994. Differential induction of nitric oxide synthase in hepatocytes during endotoxemia and the acute-phase response. Arch. Surg. **129:** 165–171.
10. Harbrecht, B.G., E.A. McClure, R.L. Simmons, *et al.* 1995. Prostanoids inhibit Kupffer cell nitric oxide synthesis. J. Surg. Res. **58:** 625–629.
11. Rockey, D.C. & J.J. Chung. 1996. Regulation of inducible nitric oxide synthase in hepatic sinusoidal endothelial cells. Am. J. Physiol. **271:** G260–G267.
12. Helyar, L., D.S. Bundschuh, J.D. Laskin, *et al.* 1994. Induction of hepatic Ito cell nitric oxide production after acute endotoxemia. Hepatology **20:** 1509–1515.
13. Kitade, H., K. Sakitani, K. Inoue, *et al.* 1996. Interleukin 1β markedly stimulates nitric oxide formation in the absence of other cytokines or lipopolysaccharide in primary cultured rat hepatocytes but not in Kupffer cells. Hepatology **23:** 797–802.
14. Kaplowitz, N. 2000. Cell death at the millennium. Implications for liver diseases. Clin. Liver Dis. **4:** 1–23.
15. Harbrecht, B.G., T.R. Billiar, J. Stadler, *et al.* 1992. Nitric oxide synthesis serves to reduce hepatic damage during acute murine endotoxemia. Crit. Care Med. **20:** 1568–1574.
16. Smets, D., H. Spapen, M. Diltoer, *et al.* 1999. Liver perfusion and hepatocellular inflammatory response in sepsis. Acta Clin. Belg. **54:** 201–206.
17. Vos, T.A., A.S. Gouw, P.A. Klok, *et al.* 1997. Differential effects of nitric oxide synthase inhibitors on endotoxin-induced liver damage in rats. Gastroenterology **113:** 1323–1333.

18. HARBRECHT, B.G., T.R. BILLIAR, J. STADLER, *et al.* 1992. Inhibition of nitric oxide synthesis during endotoxemia promotes intrahepatic thrombosis and an oxygen radical-mediated hepatic injury. J. Leukocyte Biol. **52:** 390–394.
19. HARBRECHT, B.G., B. WU, S.C. WATKINS, *et al.* 1995. Inhibition of nitric oxide synthase during hemorrhagic shock increases hepatic injury. Shock **4:** 332–337.
20. TAKEMURA, S., Y. MINAMIYAMA, M. INOUE, *et al.* 2000. Nitric oxide synthase inhibitor increases hepatic injury with formation of oxidative DNA damage and microcirculatory disturbance in endotoxemic rats. Hepatogastroenterology **47:** 1364–1370.
21. ZHU, W. & P.C. FUNG. 2000. The roles played by crucial free radicals like lipid free radicals, nitric oxide, and enzymes NOS and NADPH in CCl_4-induced acute liver injury of mice. Free Radical Biol. Med. **29:** 870–880.
22. KIM, Y.M., M.E. DE VERA, S.C. WATKINS, *et al.* 1997. Nitric oxide protects cultured rat hepatocytes from tumor necrosis factor-alpha-induced apoptosis by inducing heat shock protein 70 expression. J. Biol. Chem. **272:** 1402–1411.
23. SAAVEDRA, J.E., T.R. BILLIAR, D.L. WILLIAMS, *et al.* 1997. Targeting nitric oxide (NO) delivery in vivo. Design of a liver-selective NO donor prodrug that blocks tumor necrosis factor-alpha-induced apoptosis and toxicity in the liver. J. Med. Chem. **40:** 1947–1954.
24. ALBINA, J.E., S. CUI, R.B. MATEO, *et al.* 1993. Nitric oxide-mediated apoptosis in murine peritoneal macrophages. J. Immunol. **150:** 5080–5085.
25. FEHSEL, K., K.D. KRONCKE, K.L. MEYER, *et al.* 1995. Nitric oxide induces apoptosis in mouse thymocytes. J. Immunol. **155:** 2858–2865.
26. MCDANIEL, M.L., J.A. CORBETT, G. KWON, *et al.* 1997. A role for nitric oxide and other inflammatory mediators in cytokine-induced pancreatic beta-cell dysfunction and destruction. Adv. Exp. Med. Biol. **426:** 313–319.
27. LI, J. & T.R. BILLIAR. 1999. The anti-apoptotic actions of nitric oxide in hepatocytes. Cell Death Differ. **6:** 952–955.
28. BOHLINGER, I., M. LEIST, J. BARSIG, *et al.* 1995. Interleukin-1 and nitric oxide protect against tumor necrosis factor alpha-induced liver injury through distinct pathways. Hepatology **22:** 829–1837.
29. OU, J., T.M. CARLOS, S.C. WATKINS, *et al.* 1997. Differential effects of nonselective nitric oxide synthase (NOS) and selective inducible NOS inhibition on hepatic necrosis, apoptosis, ICAM-1 expression, and neutrophil accumulation during endotoxemia. Nitric Oxide **1:** 404–416.
30. KIM, Y.M., R.V. TALANIAN & T.R. BILLIAR. 1997. Nitric oxide inhibits apoptosis by preventing increases in caspase-3-like activity via two distinct mechanisms. J. Biol. Chem. **272:** 31138–31148.
31. FARINELLI, S.E., D.S. PARK & L.A. GREENE. 1996. Nitric oxide delays the death of trophic factor-deprived PC12 cells and sympathetic neurons by a cGMP-mediated mechanism. J. Neurosci. **16:** 2325–2334.
32. MOJENA, M., S. HORTELANO, A. CASTRILLO, *et al.* 2001. Protection by nitric oxide against liver inflammatory injury in animals carrying a nitric oxide synthase-2 transgene. FASEB J. **15:** 583–585.
33. MORIO, L.A., H. CHIU, K.A. SPROWLES, *et al.* 2001. Distinct roles of tumor necrosis factor-alpha and nitric oxide in acute liver injury induced by carbon tetrachloride in mice. Toxicol. Appl. Pharmacol. **172:** 44–51.
34. AOKI, K., M. OHMORI, M. TAKIMOTO, *et al.* 1997. Cocaine-induced liver injury in mice is mediated by nitric oxide and reactive oxygen species. Eur. J. Pharmacol. **336:** 43–49.
35. DIEZ-FERNANDEZ, C., N. SANZ, A.M. ALVAREZ, *et al.* 1998. Influence of aminoguanidine on parameters of liver injury and regeneration induced in rats by a necrogenic dose of thioacetamide. Br. J. Pharmacol. **125:** 102–108.
36. GARDNER, C.R., D.E. HECK, C.S. YANG, *et al.* 1998. Role of nitric oxide in acetaminophen-induced hepatotoxicity in the rat. Hepatology **27:** 748–754.
37. SASS, G., K. KOERBER, R. BANG, *et al.* 2001. Inducible nitric oxide synthase is critical for immune-mediated liver injury in mice. J. Clin. Invest. **107:** 439–447.
38. MANNICK, J.B. 1995. The antiviral role of nitric oxide. Res. Immunol. **146:** 693–697.
39. POWELL, K.L. & S.A. BAYLIS. 1995. The antiviral effects of nitric oxide. Trends Microbiol. **3:** 81–82.

40. CROEN, K.D. 1993. Evidence for antiviral effect of nitric oxide. Inhibition of herpes simplex virus type 1 replication. J. Clin. Invest. **91:** 2446–2452.
41. LOWENSTEIN, C.J., S.L. HILL, A. LAFOND-WALKER, *et al.* 1996. Nitric oxide inhibits viral replication in murine myocarditis. J. Clin. Invest. **97:** 1837–1843.
42. LIN, Y.L., Y.L. HUANG, S.H. MA, *et al.* 1997. Inhibition of Japanese encephalitis virus infection by nitric oxide: antiviral effect of nitric oxide on RNA virus replication. J. Virol. **71:** 5227–5235.
43. SCULLARD, G.H., C.I. SMITH, T.C. MERIGAN, *et al.* 1981. Effects of immunosuppressive therapy on viral markers in chronic active hepatitis B. Gastroenterology **81:** 987–991.
44. GUIDOTTI, L.G., H. MCCLARY, J.M. LOUDIS, *et al.* 2000. Nitric oxide inhibits hepatitis B virus replication in the livers of transgenic mice. J. Exp. Med. **191:** 1247–1252.
45. LIU, R.H., B. BALDWIN, B.C. TENNANT, *et al.* 1991. Elevated formation of nitrate and *N*-nitrosodimethylamine in woodchucks (*Marmota monax*) associated with chronic woodchuck hepatitis virus infection. Cancer Res. **51:** 3925–3929.
46. LIU, R.H., J.R. JACOB, B.C. TENNANT, *et al.* 1992. Nitrite and nitrosamine synthesis by hepatocytes isolated from normal woodchucks (*Marmota monax*) and woodchucks chronically infected with woodchuck hepatitis virus. Cancer Res. **52:** 4139–4143.
47. LIU, R.H., J.R. JACOB, J.H. HOTCHKISS, *et al.* 1994. Woodchuck hepatitis virus surface antigen induces nitric oxide synthesis in hepatocytes: possible role in hepatocarcinogenesis. Carcinogenesis **15:** 2875–2877.
48. MIHM, S., A. FAYYAZI & G. RAMADORI. 1997. Hepatic expression of inducible nitric oxide synthase transcripts in chronic hepatitis C virus infection: relation to hepatic viral load and liver injury. Hepatology **26:** 451–458.
49. AMARO, M.J., J. BARTOLOME, M. PARDO, *et al.* 1997. Decreased nitric oxide production in chronic viral hepatitis B and C. J. Med. Virol. **51:** 326–331.
50. MORIYAMA, A., A. MASUMOTO, H. NANRI, *et al.* 1997. High plasma concentrations of nitrite/nitrate in patients with hepatocellular carcinoma. Am. J. Gastroenterol. **92:** 1520–1523.
51. MOUSSA, Y.I., J.N. PLEVRIS & P.C. HAYES. 2000. Plasma nitrites/nitrates in HCV infection and hepatocellular carcinoma. Eur. J. Gastroenterol. Hepatol. **12:** 159–163.
52. MAJANO, P.L., C. GARCIA-MONZON, M. LOPEZ-CABRERA, *et al.* 1998. Inducible nitric oxide synthase expression in chronic viral hepatitis. Evidence for a virus-induced gene upregulation. J. Clin. Invest. **101:** 1343–1352.
53. SCHWEYER, S., S. MIHM, H.J. RADZUN, *et al.* 2000. Liver infiltrating T lymphocytes express interferon gamma and inducible nitric oxide synthase in chronic hepatitis C virus infection. Gut **46:** 255–259.
54. GARCIA-MONZON, C., P.L. MAJANO, I. ZUBIA, *et al.* 2000. Intrahepatic accumulation of nitrotyrosine in chronic viral hepatitis is associated with histological severity of liver disease. J. Hepatol. **32:** 331–338.
55. AMARO, M.J., J. BARTOLOME & V. CARRENO. 1999. Hepatitis B virus X protein transactivates the inducible nitric oxide synthase promoter. Hepatology **29:** 915–923.
56. BURNEY, S., J.L. CAULFIELD, J.C. NILES, *et al.* 1999. The chemistry of DNA damage from nitric oxide and peroxynitrite. Mutat. Res. **424:** 37–49.
57. PAN, J., L.X. DUAN, B.S. SUN, *et al.* 2001. Hepatitis B virus X protein protects against anti-Fas-mediated apoptosis in human liver cells by inducing NF-κ B. J. Gen. Virol. **82:** 171–182.
58. VALLANCE, P. & S. MONCADA. 1991. Hyperdynamic circulation in cirrhosis: a role for nitric oxide? Lancet **337:** 776–778.
59. BUSSE, R. & A. MULSCH. 1990. Induction of nitric oxide synthase by cytokines in vascular smooth muscle cells. FEBS Lett. **275:** 87–90.
60. REES, D.D., S. CELLEK, R.M. PALMER, *et al.* 1990. Dexamethasone prevents the induction by endotoxin of a nitric oxide synthase and the associated effects on vascular tone: an insight into endotoxin shock. Biochem. Biophys. Res. Commun. **173:** 541–547.
61. FONG, Y.M., M.A. MARANO, L.L. MOLDAWER, *et al.* 1990. The acute splanchnic and peripheral tissue metabolic response to endotoxin in humans. J. Clin. Invest. **85:** 1896–1904.

62. SUFFREDINI, A.F., R.E. FROMM, M.M. PARKER, *et al.* 1989. The cardiovascular response of normal humans to the administration of endotoxin. N. Engl. J. Med. **321:** 280–287.
63. PETROS, A., G. LAMB, A. LEONE, *et al.* 1994. Effects of a nitric oxide synthase inhibitor in humans with septic shock. Cardiovasc. Res. **28:** 34–39.
64. LUMSDEN, A.B., J.M. HENDERSON & M.H. KUTNER. 1988. Endotoxin levels measured by a chromogenic assay in portal, hepatic and peripheral venous blood in patients with cirrhosis. Hepatology **8:** 232–236.
65. YOKOTA, M., J. KAMBAYASHI, T. TANAKA, *et al.* 1989. A simple turbidimetric time assay of the endotoxin in plasma. J. Biochem. Biophys. Meth. **18:** 97–104.
66. SIEBER, C.C., J.C. LOPEZ-TALAVERA & R.J. GROSZMANN. 1993. Role of nitric oxide in the in vitro splanchnic vascular hyporeactivity in ascitic cirrhotic rats. Gastroenterology **104:** 1750–1754.
67. PIZCUETA, P., J.M. PIQUE, M. FERNANDEZ, *et al.* 1992. Modulation of the hyperdynamic circulation of cirrhotic rats by nitric oxide inhibition. Gastroenterology **103:** 1909–1915.
68. CLARIA, J., W. JIMENEZ, J. ROS, *et al.* 1994. Increased nitric oxide-dependent vasorelaxation in aortic rings of cirrhotic rats with ascites. Hepatology **20:** 1615–1621.
69. MIYÁSE, S., S. FUJIYAMA, H. CHIKAZAWA, *et al.* 1990. Atrial natriuretic peptide in liver cirrhosis with mild ascites. Gastroenterol. Jpn. **25:** 356–362.
70. MIDGLEY, S., I.S. GRANT, W.G. HAYNES, *et al.* 1991. Nitric oxide in liver failure. Lancet **338:** 1590.
71. GUARNER, C., G. SORIANO, A. TOMAS, *et al.* 1993. Increased serum nitrite and nitrate levels in patients with cirrhosis: relationship to endotoxemia. Hepatology **18:** 1139–1143.
72. HORI, N., T. OKANOUE, T. MORI, *et al.* 1996. Endogenous nitric oxide production is augmented as the severity advances in patients with liver cirrhosis. Clin. Exp. Pharmacol. Physiol. **23:** 30–35.
73. BORIES, P.N., B. CAMPILLO, L. AZAOU, *et al.* 1997. Long-lasting NO overproduction in cirrhotic patients with spontaneous bacterial peritonitis. Hepatology **25:** 1328–1333.
74. COSKUN, U., S. OZENIRLER, B. SANCAK, *et al.* 2001. Serum and ascitic fluid nitrate levels in patients with cirrhosis. Clin. Chim. Acta **306:** 127–132.
75. BATTISTA, S., F. BAR, G. MENGOZZI, *et al.* 1997. Hyperdynamic circulation in patients with cirrhosis: direct measurement of nitric oxide levels in hepatic and portal veins. J. Hepatol. **26:** 75–80.
76. MATSUMOTO, A., K. OGURA, Y. HIRATA, *et al.* 1995. Increased nitric oxide in the exhaled air of patients with decompensated liver cirrhosis. Ann. Intern. Med. **123:** 110–113.
77. LAFFI, G., M. FOSCHI, E. MASINI, *et al.* 1995. Increased production of nitric oxide by neutrophils and monocytes from cirrhotic patients with ascites and hyperdynamic circulation. Hepatology **22:** 1666–1673.
78. GALLEY, H.F., D. COOMANSINGH, N.R. WEBSTER, *et al.* 1998. Nitric oxide synthase activity is increased in relation to the severity of liver dysfunction. Clin. Sci. **95:** 355–359.
79. RYAN, J., G. JENNINGS, F. DUDLEY, *et al.* 1996. Smooth muscle-derived nitric oxide is elevated in isolated forearm veins in human alcoholic cirrhosis. Clin. Sci. **91:** 23–28.
80. CAMPILLO, B., P.E. CHABRIER, G. PELLE, *et al.* 1995. Inhibition of nitric oxide synthesis in the forearm arterial bed of patients with advanced cirrhosis. Hepatology **22:** 1423–1429.
81. FERNANDEZ, M., J.C. GARCIA-PAGAN, M. CASADEVALL, *et al.* 1995. Evidence against a role for inducible nitric oxide synthase in the hyperdynamic circulation of portal-hypertensive rats. Gastroenterology **108:** 1487–1495.
82. GADANO, A.C., P. SOGNI, J. HELLER, *et al.* 1999. Vascular nitric oxide production during the development of two experimental models of portal hypertension. J. Hepatol. **30:** 896–903.
83. CAMPILLO, B., P.N. BORIES, C. BENVENUTI, *et al.* 1996. Serum and urinary nitrate levels in liver cirrhosis: endotoxemia, renal function and hyperdynamic circulation. J. Hepatol. **25:** 707–714.

84. WENNMALM, A., G. BENTHIN, A. EDLUND, *et al.* 1993. Metabolism and excretion of nitric oxide in humans. An experimental and clinical study. Circ. Res. **73:** 1121–1127.
85. ROCKEY, D.C. & J.J. CHUNG. 1998. Reduced nitric oxide production by endothelial cells in cirrhotic rat liver: endothelial dysfunction in portal hypertension. Gastroenterology **114:** 344–351.
86. SARELA, A.I., F.M. MIHAIMEED, J.J. BATTEN, *et al.* 1999. Hepatic and splanchnic nitric oxide activity in patients with cirrhosis. Gut **44:** 749–753.
87. SCHAFER, D.F. & M.F. SORRELL. 1999. Hepatocellular carcinoma. Lancet **353:** 1253–1257.
88. OHSHIMA, H., T.Y. BANDALETOVA, I. BROUET, *et al.* 1994. Increased nitrosamine and nitrate biosynthesis mediated by nitric oxide synthase induced in hamsters infected with liver fluke (*Opisthorchis viverrini*). Carcinogenesis **15:** 271–275.
89. LANE, J.M., L.L. SHEARS, C. HIERHOLZER, *et al.* 1997. Chronic hepatitis C virus infection in humans: induction of hepatic nitric oxide synthase and proposed mechanisms for carcinogenesis. J. Surg. Res. **69:** 321–324.
90. OHSHIMA, H. & H. BARTSCH. 1994. Chronic infections and inflammatory processes as cancer risk factors: possible role of nitric oxide in carcinogenesis. Mutat. Res. **305:** 253–264.
91. MORIYAMA, A., A. TABARU, H. UNOKI, *et al.* 2000. Plasma nitrite/nitrate concentrations as a tumor marker for hepatocellular carcinoma. Clin. Chim. Acta **296:** 181–191.
92. WINK, D.A., Y. VODOVOTZ, J. LAVAL, *et al.* 1998. The multifaceted roles of nitric oxide in cancer. Carcinogenesis **19:** 711–721.
93. NGUYEN, T., D. BRUNSON, C.L. CRESPI, *et al.* 1992. DNA damage and mutation in human cells exposed to nitric oxide *in vitro*. Proc. Natl. Acad. Sci. U.S.A. **89:** 3030–3034.
94. WINK, D.A., K.S. KASPRZAK, C.M. MARAGOS, *et al.* 1991. DNA deaminating ability and genotoxicity of nitric oxide and its progenitors. Science **254:** 1001–1003.
95. HIBBS, J.B., JR., R.R. TAINTOR, Z. VAVRIN, *et al.* 1988. Nitric oxide: a cytotoxic activated macrophage effector molecule. Biochem. Biophys. Res. Commun. **157:** 87–94.
96. FELLEY-BOSCO, E. 1998. Role of nitric oxide in genotoxicity: implication for carcinogenesis. Cancer Metastasis Rev. **17:** 25–37.
97. CHUANG, W.L., M. OMATA, T. EHATA, *et al.* 1993. Precore mutations and core clustering mutations in chronic hepatitis B virus infection. Gastroenterology **104:** 263–271.
98. HOLLSTEIN, M., D. SIDRANSKY, B. VOGELSTEIN, *et al.* 1991. p53 mutations in human cancers. Science **253:** 49–53.
99. SHI, C.Y., T.W. PHANG, A. WEE, *et al.* 1995. Mutations of the tumour suppressor gene p53 in colorectal and hepatocellular carcinomas. Ann. Acad. Med. Singapore **24:** 204–210.
100. LEVINE, A.J., J. MOMAND & C.A. FINLAY. 1991. The p53 tumour suppressor gene. Nature **351:** 453–456.
101. CALMELS, S., P. HAINAUT & H. OHSHIMA. 1997. Nitric oxide induces conformational and functional modifications of wild-type p53 tumor suppressor protein. Cancer Res. **57:** 3365–3369.
102. FORRESTER, K., S. AMBS, S.E. LUPOLD, *et al.* 1996. Nitric oxide-induced p53 accumulation and regulation of inducible nitric oxide synthase expression by wild-type p53. Proc. Natl. Acad. Sci. U.S.A. **93:** 2442–2447.
103. MORBIDELLI, L., C.H. CHANG, J.G. DOUGLAS, *et al.* 1996. Nitric oxide mediates mitogenic effect of VEGF on coronary venular endothelium. Am. J. Physiol. **270:** H411–H415.
104. ZICHE, M., L. MORBIDELLI, R. CHOUDHURI, *et al.* 1997. Nitric oxide synthase lies downstream from vascular endothelial growth factor-induced but not basic fibroblast growth factor-induced angiogenesis. J. Clin. Invest. **99:** 2625–2634.
105. CHIN, K., Y. KURASHIMA, T. OGURA, *et al.* 1997. Induction of vascular endothelial growth factor by nitric oxide in human glioblastoma and hepatocellular carcinoma cells. Oncogene **15:** 437–442.

106. BROWN, L.F., B. BERSE, R.W. JACKMAN, *et al.* 1993. Expression of vascular permeability factor (vascular endothelial growth factor) and its receptors in adenocarcinomas of the gastrointestinal tract. Cancer Res. **53:** 4727–4735.
107. BROWN, L.F., B. BERSE, R.W. JACKMAN, *et al.* 1993. Increased expression of vascular permeability factor (vascular endothelial growth factor) and its receptors in kidney and bladder carcinomas. Am. J. Pathol. **143:** 1255–1262.
108. BROWN, L.F., B. BERSE, R.W. JACKMAN, *et al.* 1995. Expression of vascular permeability factor (vascular endothelial growth factor) and its receptors in breast cancer. Hum. Pathol. **26:** 86–91.
109. MISE, M., S. ARII, H. HIGASHITUJI, *et al.* 1996. Clinical significance of vascular endothelial growth factor and basic fibroblast growth factor gene expression in liver tumor. Hepatology **23:** 455–464.
110. SUZUKI, K., N. HAYASHI, Y. MIYAMOTO, *et al.* 1996. Expression of vascular permeability factor/vascular endothelial growth factor in human hepatocellular carcinoma. Cancer Res. **56:** 3004–3009.
111. KUROSE, I., H. HIGUCHI, Y. YONEI, *et al.* 1996. Rat Kupffer cell-derived nitric oxide suppresses proliferation and induces apoptosis of syngeneic hepatoma cells. Gastroenterology **111:** 1058–1070.
112. SAITO, H., I. KUROSE, H. EBINUMA, *et al.* 1996. Kupffer cell-mediated cytotoxicity against hepatoma cells occurs through production of nitric oxide and adhesion via ICAM-1/CD18. Int. Immunol. **8:** 1165-1172.
113. FUKUMURA, D., Y. YONEI, I. KUROSE, *et al.* 1996. Role in nitric oxide in Kupffer cell-mediated hepatoma cell cytotoxicity *in vitro* and *ex vivo*. Hepatology **24:** 141–149.
114. DEVLIN, J., R.M. PALMER, C.E. GONDE, *et al.* 1994. Nitric oxide generation. A predictive parameter of acute allograft rejection. Transplantation **58:** 592–595.
115. UCHIYAMA, H., K. YANAGA, K. KISHIHARA, *et al.* 2000. Increased urinary nitrate excretion associated with hepatic allograft rejection in experimental rat models and clinical cases. Transpl. Int. **13:** 260–265.
116. FABREGA, E., F. CASAFONT, J. DE LA PENA, *et al.* 1997. Nitric oxide production in hepatic cell rejection of liver transplant patients. Transplant. Proc. **29:** 505–506.
117. OHDAN, H., Y. FUKUDA, S. SUZUKI, *et al.* 1995. Simultaneous evaluation of nitric oxide synthesis and tissue oxygenation in rat liver allograft rejection using near-infrared spectroscopy. Transplantation **60:** 530–535.
118. KUO, P.C., E.J. ALFREY, K.Y. ABE, *et al.* 1996. Cellular localization and effect of nitric oxide synthesis in a rat model of orthotopic liver transplantation. Transplantation **61:** 305–312.
119. SUGIOKA, A., M. MORITA, T. ESAKI, *et al.* 1999. Kinetics and significance of graft tissue nitric oxide after rat and mouse liver transplantation. Transplant. Proc. **31:** 860–862.
120. ROMERO, M., C. GARCIA-MONZON, G. CLEMENTE, *et al.* 2001. Intrahepatic expression of inducible nitric oxide synthase in acute liver allograft rejection: evidence of modulation by corticosteroids. Liver Transpl. **7:** 16–21.
121. LANGREHR, J.M., R.A. HOFFMAN, T.R. BILLIAR, *et al.* 1991. Nitric oxide synthesis in the *in vivo* allograft response: a possible regulatory mechanism. Surgery **110:** 335–342.
122. KUO, P.C. & A. SLIVKA. 1994. Nitric oxide decreases oxidant-mediated hepatocyte injury. J. Surg. Res. **56:** 594–600.
123. KOKEN, T. & M. INAL. 1999. The effect of nitric oxide on ischemia-reperfusion injury in rat liver. Clin. Chim. Acta **288:** 55–62.
124. KOEPPEL, T.A., J.C. THIES, P. SCHEMMER, *et al.* 1997. Inhibition of nitric oxide synthesis in ischemia/reperfusion of the rat liver is followed by impairment of hepatic microvascular blood flow. J. Hepatol. **27:** 163–169.
125. AKIZUKI, E., T. AKAIKE, S. OKAMOTO, *et al.* 2000. Role of nitric oxide and superoxide in acute cardiac allograft rejection in rats. Proc. Soc. Exp. Biol. Med. **225:** 151–159.
126. ROZA, A.M., M. COOPER, G. PIEPER, *et al.* 2000. NOX 100, a nitric oxide scavenger, enhances cardiac allograft survival and promotes long-term graft acceptance. Transplantation **69:** 227–231.

127. KUO, P.C., K. ABE & R.A. SCHROEDER. 2000. Superoxide enhances interleukin 1β-mediated transcription of the hepatocyte-inducible nitric oxide synthase gene. Gastroenterology **118:** 608–618.

128. SHAH, V., F.G. HADDAD, G. GARCIA-CARDENA, *et al.* 1997. Liver sinusoidal endothelial cells are responsible for nitric oxide modulation of resistance in the hepatic sinusoids. J. Clin. Invest. **100:** 2923–2930.

129. WANG, Y., W.R. MATHEWS, D.M. GUIDO, *et al.* 1995. Inhibition of nitric oxide synthesis aggravates reperfusion injury after hepatic ischemia and endotoxemia. Shock **4:** 282–288.

130. COTTART, C.H., L. DO, M.C. BLANC, *et al.* 1999. Hepatoprotective effect of endogenous nitric oxide during ischemia-reperfusion in the rat. Hepatology **29:** 809–813.

131. KOBAYASHI, H., T. NONAMI, T. KUROKAWA, *et al.* 1995. Role of endogenous nitric oxide in ischemia-reperfusion injury in rat liver. J. Surg. Res. **59:** 772–779.

132. OHMORI, H., D.K. DHAR, Y. NAKASHIMA, *et al.* 1998. Beneficial effects of FK409, a novel nitric oxide donor, on reperfusion injury of rat liver. Transplantation **66:** 579–585.

133. SHIMAMURA, T., Y. ZHU, S. ZHANG, *et al.* 1999. Protective role of nitric oxide in ischemia and reperfusion injury of the liver. J. Am. Coll. Surg. **188:** 43–52.

134. RICCIARDI, R., D.P. FOLEY, S.H. QUARFORDT, *et al.* 2001. V-PYRRO/NO: an hepatoselective nitric oxide donor improves porcine liver hemodynamics and function after ischemia reperfusion. Transplantation **71:** 193–198.

135. PANNEN, B.H., F. AL ADILI, M. BAUER, *et al.* 1998. Role of endothelins and nitric oxide in hepatic reperfusion injury in the rat. Hepatology **27:** 755–764.

136. LIU, P., B. XU, C.E. HOCK, *et al.* 1998. NO modulates P-selectin and ICAM-1 mRNA expression and hemodynamic alterations in hepatic I/R. Am. J. Physiol. **275:** H2191–H2198.

137. DE CATERINA, R., P. LIBBY, H.B. PENG, *et al.* 1995. Nitric oxide decreases cytokine-induced endothelial activation. Nitric oxide selectively reduces endothelial expression of adhesion molecules and proinflammatory cytokines. J. Clin. Invest. **96:** 60–68.

138. MA, T.T., H. ISCHIROPOULOS & C.A. BRASS. 1995. Endotoxin-stimulated nitric oxide production increases injury and reduces rat liver chemiluminescence during reperfusion. Gastroenterology **108:** 463–469.

139. ISOBE, M., T. KATSURAMAKI, K. HIRATA, *et al.* 1999. Beneficial effects of inducible nitric oxide synthase inhibitor on reperfusion injury in the pig liver. Transplantation **68:** 803–813.

140. MELILLO, G., L.S. TAYLOR, A. BROOKS, *et al.* 1997. Functional requirement of the hypoxia-responsive element in the activation of the inducible nitric oxide synthase promoter by the iron chelator desferrioxamine. J. Biol. Chem. **272:** 12236–12243.

141. IKEMOTO, M., S. TSUNEKAWA, K. TANAKA, *et al.* 1998. Liver-type arginase in serum during and after liver transplantation: a novel index in monitoring conditions of the liver graft and its clinical significance. Clin. Chim. Acta **271:** 11–23.

142. LANGLE, F., E. ROTH, R. STEININGER, *et al.* 1995. Arginase release following liver reperfusion. Evidence of hemodynamic action of arginase infusions. Transplantation **59:** 1542–1549.

143. GOTO, M., Y. TAKEI, S. KAWANO, *et al.* 1994. Endothelin-1 is involved in the pathogenesis of ischemia/reperfusion liver injury by hepatic microcirculatory disturbances. Hepatology **19:** 675–681.

144. DHAR, D.K., A. YAMANOI, H. OHMORI, *et al.* 1998. Modulation of endothelin and nitric oxide: a rational approach to improve canine hepatic microcirculation. Hepatology **28:** 782–788.

145. YAMADA, Y., I. KIRILLOVA, J.J. PESCHON, *et al.* 1997. Initiation of liver growth by tumor necrosis factor: deficient liver regeneration in mice lacking type I tumor necrosis factor receptor. Proc. Natl. Acad. Sci. U.S.A. **94:** 1441–1446.

146. CRESSMAN, D.E., L.E. GREENBAUM, R.A. DEANGELIS, *et al.* 1996. Liver failure and defective hepatocyte regeneration in interleukin-6-deficient mice. Science **274:** 1379–1383.

147. CRESSMAN, D.E., L.E. GREENBAUM, B.A. HABER, *et al.* 1994. Rapid activation of post-hepatectomy factor/nuclear factor kappa B in hepatocytes, a primary response in the regenerating liver. J. Biol. Chem. **269:** 30429–30435.
148. CRESSMAN, D.E., R.H. DIAMOND & R. TAUB. 1995. Rapid activation of the Stat3 transcription complex in liver regeneration. Hepatology **21:** 1443–1449.
149. FAUSTO, N. 2000. Liver regeneration. J. Hepatol. **32:** 19–31.
150. OBOLENSKAYA, M.Y., A.F. VANIN, P.I. MORDVINTCEV, *et al.* 1994. Epr evidence of nitric oxide production by the regenerating rat liver. Biochem. Biophys. Res. Commun. **202:** 571–576.
151. HORTELANO, S., B. DEWEZ, A.M. GENARO, *et al.* 1995. Nitric oxide is released in regenerating liver after partial hepatectomy. Hepatology **21:** 776–786.
152. CARNOVALE, C.E., C. SCAPINI, M.L. ALVAREZ, *et al.* 2000. Nitric oxide release and enhancement of lipid peroxidation in regenerating rat liver. J. Hepatol. **32:** 798–804.
153. OBOLENSKAYA, M., A. SCHULZE-SPECKING, B. PLAUMANN, *et al.* 1994. Nitric oxide production by cells isolated from regenerating rat liver. Biochem. Biophys. Res. Commun. **204:** 1305–1311.
154. DECKER, K.F. & M.Y. OBOLENSKAYA. 1995. Cytokines, nitric oxide synthesis and liver regeneration. J. Gastroenterol. Hepatol. **10**(Suppl. 1): S12–S17.
155. DIAZ-GUERRA, M.J., M. VELASCO, P. MARTIN-SANZ, *et al.* 1997. Nuclear factor κB is required for the transcriptional control of type II NO synthase in regenerating liver. Biochem. J. **326**(3): 791–797.
156. WANG, H.H. & W.W. LAUTT. 1998. Evidence of nitric oxide, a flow-dependent factor, being a trigger of liver regeneration in rats. Can. J. Physiol. Pharmacol. **76:** 1072–1079.
157. RAI, R.M., F.Y. LEE, A. ROSEN, *et al.* 1998. Impaired liver regeneration in inducible nitric oxide synthase deficient mice. Proc. Natl. Acad. Sci. U.S.A. **95:** 13829–13834.

Nitric Oxide (NO) and Convulsions Induced by Pentylenetetrazol

ANKICA JELENKOVIĆ,[a] MARINA JOVANOVIĆ,[b] MILICA NINKOVIĆ,[b] M. MAKSIMOVIĆ,[b] D. BOKONJIĆ,[b] AND B. BOŠKOVIĆ[b]

[a]*Institute for Biological Research, Belgrade, FR Yugoslavia*

[b]*Military Medical Academy, Belgrade, FR Yugoslavia*

ABSTRACT: Data about the role of nitric oxide (NO) in epileptogenesis are contradictory. It is found to exert both proconvulsant and anticonvulsant effects. In an attempt to elucidate the role of NO in seizures, male Wistar rats were treated intraperitoneally by pentylenetetrazol (PTZ) (60, 80, and 100 mg/kg) and by a nitric oxide synthase antagonist, *N*-omega-nitro-L-arginine-methylester (L-NAME) (10, 40, and 70 mg/kg), applied before PTZ. The time to onset and incidence of forelimb dystonia (FLD), generalized clonic convulsions (GCC), clonic-tonic convulsions (CTC), and mortality were recorded. The most successful convulsive response and mortality prevention were found in PTZ (80 mg/kg)-treated groups, where L-NAME (70 mg/kg) decreased the incidence by 29, 50, 67 ($p = 0.052$), and 50%, respectively, and significantly prolonged the time to onset, except that for mortality. Unexpectedly, L-NAME (40 mg/kg) increased incidence of GCC and mortality by 16%, similar to L-NAME (10 mg/kg) in PTZ (60 mg/kg)-treated groups, where GCC, CTC, and mortality increased by 14, 14, and 28%, respectively. Convulsive latency was prolonged in some PTZ (100 mg/kg) + L-NAME (40 and 70 mg/kg)-treated groups. In the experimental model and protocol used, it is concluded that (1) the effects of NO are L-NAME- and PTZ-dose dependent; (2) clonic-tonic convulsions are more strongly influenced by NO than limbic, probably because of PTZ limbic structure overstimulation; (3) L-NAME decreases the incidence of CTC and prolongs FLD, GCC, and CTC times to onset, indicating that NO acts as a proconvulsant; and (3) increased GCC, CTC, and mortality that suggests an anticonvulsant effect of NO needs further investigation.

KEYWORDS: convulsions; seizures; nitric oxide; nitric oxide synthase antagonists; pentylenetetrazol

ABBREVIATIONS:

PTZ	pentylenetetrazol.
FLD	forelimb dystonia.
GCC	generalized clonic convulsions.
CTC	clonic-tonic convulsions.
NO	nitric oxide.

Address for correspondence: A. Jelenković, Department for Medical Research, Military Medical Academy, Crnotravska 17, Belgrade, Serbia, FR Yugoslavia.
aka.950@yahoo.com

INTRODUCTION

Underlying epilepsies, which involve about 50 syndromes, is a complex biochemical and structural heterogenicity. This offers a remarkable scientific, clinical, and pharmacologic challenge in understanding of their pathophysiology. Therefore, monotherapy, the preferred approach, is often unsuccessful in seizure prevention. However, even with polytherapy, about 20–30% of patients with epilepsy cannot be well protected. Thus, epilepsies offer a wide field for investigation. Better understanding of the cellular and molecular basis of epileptogenesis will enable development of more effective antiepileptic and antiseizure drugs and, thus, improve epilepsy treatment.

During the past decade scientific interest in understanding the physiologic role of endogenous nitric oxide (NO) has greatly increased. It is indicated that NO is involved in a cascade of various regulatory biological events and processes.[1–3] It has also become clear that changes in NO production take place in a number of pathophysiologic disturbances.[4–6]

The hypothesis concerning NO involvement in pathogenesis of epilepsy has been tested many times, with conflicting results. Highlighting and elucidating the role of NO in epileptogenesis potentially opens a new horizon for therapeutic approaches to epilepsies.

By using xenobiotics that release NO or modify its endogenous synthesis, the question of whether NO displays proconvulsant[7–9] or anticonvulsant[10,11] effects remains unresolved. As well as unequivocal results about the role of NO in various animal convulsive models, similar data have been obtained about the place of NO in pentylenetetrazol (PTZ)-evoked convulsions. It was found that NO could be without effect in PTZ-induced clonic[12,13] and tonic[12] seizures, and to be anticonvulsant,[14,15] or to intensify some convulsive patterns.[16] Such conflicting results are sometimes the consequence of inconsistent definitions of convulsive response and variations in animal species used in experiment.[17] Furthermore, the dose[18] and the route of PTZ and NO synthase antagonist administration, as well as its type, exert a strong influence on the outcome of PTZ-evoked behavior changes.[19] In view of this, a comparison between convulsive responses based on various animal species, various experimental models (*in vitro*, *ex vivo*, or *in vivo*), and protocols is not suggested. Great caution is required in order to escape the conclusive trap.

In an attempt to elucidate the role of NO in seizures, we examined its effects on chemically induced convulsions in rats and consequent mortality.

MATERIAL AND METHODS

Seizures were induced in 11–12-week-old male Wistar rats housed in Plexiglas cages (five rats/cage) at ambient temperature 23 ± 2°C, dark/light cycle 13 : 11 hours, with free access to food and water. Experiments were carried out from 9 A.M. to 2 P.M. Each group included seven rats, chosen randomly.

A chemoconvulsant, pentylenetetrazol (PTZ, Sigma), dissolved in 0.9% saline, was injected at doses of 60, 80, and 100 mg/kg body weight (BW) 30 minutes after saline or *N*ω-nitro-L-arginine methyl ester (L-NAME). Both were applied intraperitoneally (ip). Immediately after PTZ application, each rat was placed into an individual Plexiglas cage and observed for a period of 30 minutes.

A nitric oxide synthase (NOS) inhibitor, L-NAME (Sigma), dissolved in 0.9% saline was given at doses of 10, 40, and 70 mg/kg BW 30 minutes before PTZ. To see whether L-NAME, *per se*, exerted any influence on the convulsive behavior, another three groups were also treated with 10, 40, and 70 mg/kg BW of L-NAME, and with 0.9% saline (ip) instead of PTZ.

Drugs were applied at a volume of 1 ml/kg BW. The route and the time of L-NAME application were selected according to its pharmacokinetic characteristics.

To quantify PTZ convulsive response and to evaluate L-NAME effects, the time to onset and the intensity of behavioral alterations were recorded with respect to the appearance of forelimb dystonia (FLD), generalized clonic (GCC), and clonic-tonic convulsions (CTC). Mortality rate and time of death were also recorded.

The number of convulsive responses (incidences) observed was expressed as percentage of the number of animals in each group. A Student's t-test of proportion as well as Mann-Whitney U-test were used to analyze statistical differences between groups.

Median effective doses (ED_{50}) were calculated according to the well known Litchfield and Wilcoxon method.

RESULTS

In cases where a PTZ convulsive response was observed, the order was almost regular: the forelimb dystonia appeared and preceded generalized clonic convulsions. Generalized clonic convulsions were followed by clonic-tonic convulsions and death (see TABLE 1).

A low dose of PTZ (60 mg/kg BW) induced forelimb dystonia in all treated rats. Convulsive response and mortality were dose dependent and statistically significant in the case of CTC and mortality rate between PTZ 60- and 100 mg/kg-treated groups ($p < 0.05$). The latency agreed with the incidence of convulsive response. Latency was decreased (see TABLE 2), especially for GCC and CTC ($p < 0.01$ in PTZ 80 mg/kg-treated group), but did not reach a statistically significant level for mortality ($p = 0.06$). Although PTZ at a dose of 60 mg/kg was high enough to induce FLD

TABLE 1. Pentylenetetrazol-evoked convulsions and mortality in rats

PTZ (mg/kg)	Convulsive response and mortality (percent observed)			
	FLD	GCC	CTC	Mortality
60	100	57	43	43
80	100	86	86	86
100	100	100	100*	100*

NOTES: The incidence of convulsive response is expressed as a percentage seizured and dead rats of the total number in the group. *$p < 0.05$ versus PTZ (60 mg/kg) treated group. FLD, forelimb dystonia; GCC, generalized clonic convulsions; CTC, clonic-tonic convulsions.

TABLE 2. Effects of L-NAME on time to onset of pentylenetetrazol-evoked convulsions and mortality in rats

PTZ (mg/kg)	L-NAME (mg/kg)	Time to onset of convulsive response and mortality (seconds ± SD)			
		FLD	GCC	CTC	Mortality
60	0	178.0 ± 283.3	828.5 ± 908.8	1082.6 ± 894.8	1,312.9 ± 619.5
60	10	62.4 ± 29.8	663.1 ± 802.5	897.1 ± 861.5	1,119.3 ± 501.1
60	40	822.1 ± 915.1	857.9 ± 883.3	1080.0 ± 901	1,291.4 ± 642.5
60	70	817.1 ± 919.6	854.6 ± 886.9	1075.7 ± 903	1,245.7 ± 706.0
80	0	55.4 ± 8.3[a]	365.0 ± 638.2	383.6 ± 628.8	752.1 ± 532.6
80	10	57.6 ± 5.9	333.3 ± 647	343.1 ± 642	802.9 ± 467.5
80	40	57.3 ± 11.1	133.9 ± 89.9	367.1 ± 632.1	774.3 ± 333.3
80	70	585.7 ± 830.9[b]	1,113.6 ± 858.4[b]	1,339.2 ± 786.9[b]	1,318.6 ± 622.4
100	0	47.4 ± 7.9[a]	67.1 ± 24.8[a]	78.1 ± 25.0[a]	668.6 ± 247.0
100	10	48.1.6 ± 6.5	95.3 ± 55.4	99.1 ± 55.1	666.4 ± 313.0
100	40	59.0 ± 12.3	123.3 ± 40.5[c]	152.9 ± 105.3[c]	696.4 ± 81.1
100	70	553.6 ± 851	566.8 ± 842.6	605.0 ± 816.9[c]	971.4 ± 537.7

NOTES: In the case without convulsive response, latency was defined as 1,800 seconds. FLD, forelimb dystonia; GCC, generalized clonic convulsions; CTC, clonic-tonic convulsions.

[a] Significance to corresponding PTZ (60 mg/kg) values, $p < 0.05$ (0.01).

[b] Significance to corresponding PTZ (80 mg/kg) values, $p < 0.05$ (0.01).

[c] Significance to corresponding PTZ (100 mg/kg) values, $p < 0.05$ (0.01).

in all animals, the increased doses of PTZ (80 and 100 mg/kg) shortened the time to its onset ($p < 0.05$ and $p < 0.01$, respectively).

In the absence of noxious response, we observed the expected decrease of basal NO level induced by L-NAME application but not convulsive response. Motor behavior was influenced only if PTZ was applied. The efficacy of L-NAME strongly depended on the PTZ dose administered. The response to PTZ 60 mg/kg was quite different to that obtained with the PTZ 80- and 100 mg/kg-treated groups.

In PTZ (60 mg/kg)-treated groups L-NAME was completely ineffective, either in prevention of convulsive response (see FIGURE 1), or in increasing the time to onset of convulsive response and death (TABLE 2), even though L-NAME (40 and 70 mg/kg) prevented limbic motor response in 43% of animals ($p > 0.05$). However, pretreatment with L-NAME (10 mg/kg) was followed by a new, unexpected, phenomenon; GCC, CTC, and mortality became more severe. Their incidence increased by 14, 14, and 29%, respectively ($p > 0.05$), and the time to onset was shortened ($p > 0.05$).

Pretreatment with L-NAME was more effective in preventing convulsive response to PTZ (80 mg/kg). The highest dose of L-NAME decreased the incidence of monitored behavior alterations (FLD, GCC, and CTC), and mortality was reduced by 29%, 50%, 57%, and 50%, respectively (see FIGURE 2) ($p > 0.05$). We point out that

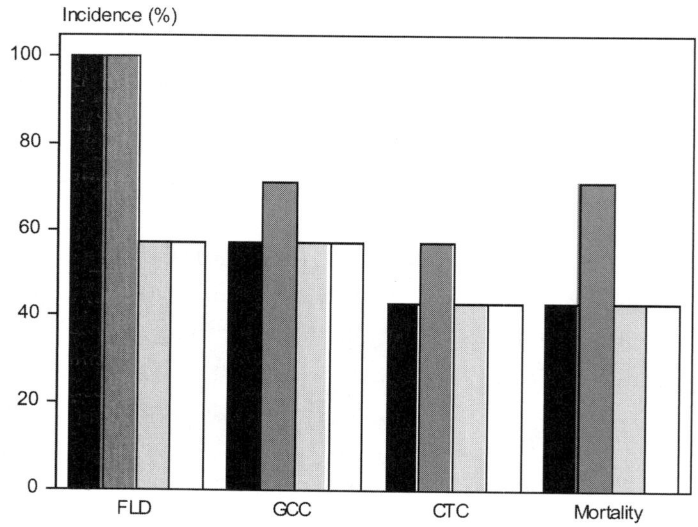

FIGURE 1. Effects of L-NAME on the incidence of pentylenetetrazol (60 mg/kg)-evoked convulsive response and mortality in rats. ■ PTZ; ▩ L-NAME 10 mg/kg; ▢ L-NAME 40 mg/kg; □ L-NAME 70 mg/kg. FLD, forelimb dystonia; GCC, generalized clonic convulsions; CTC, clonic-tonic convulsions.

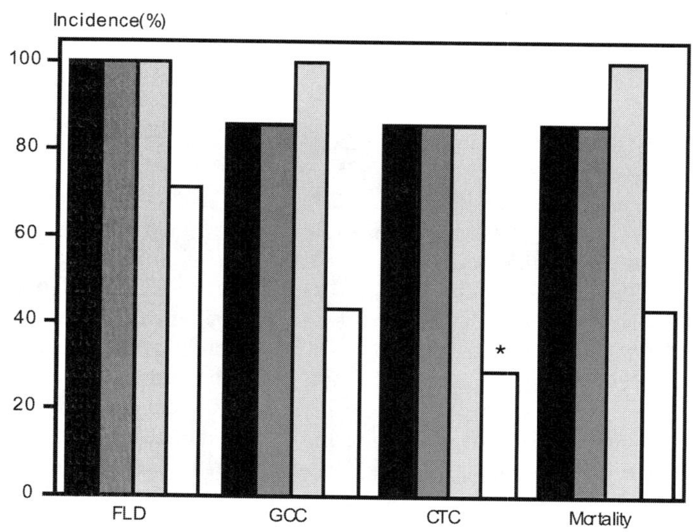

FIGURE 2. Effects of L-NAME on the incidence of pentylenetetrazol (80 mg/kg)-evoked convulsive response and mortality in rats. *Significance to corresponding PTZ (80 mg/kg) values, $p = 0.052$. ■ PTZ; ▩ L-NAME 10 mg/kg; ▢ L-NAME 40 mg/kg; □ L-NAME 70 mg/kg. FLD, forelimb dystonia; GCC, generalized clonic convulsions; CTC, clonic-tonic convulsions.

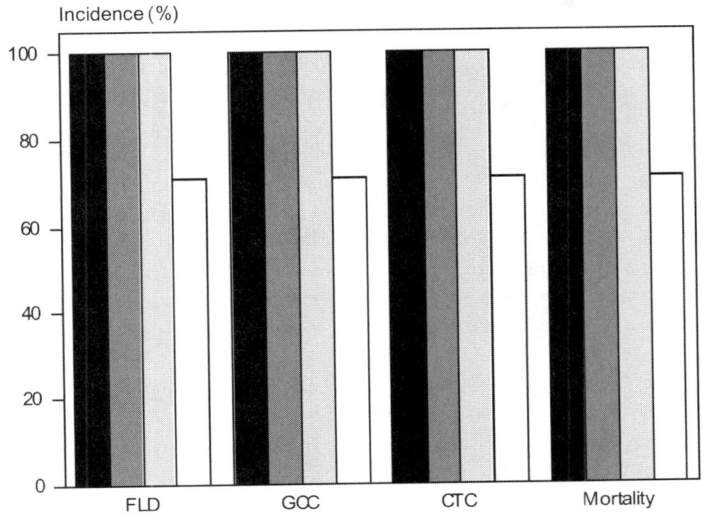

FIGURE 3. Effects of L-NAME on the incidence of pentylenetetrazol (100 mg/kg)-evoked convulsive response and mortality in rats. ■ PTZ; ■ L-NAME 10 mg/kg; □ L-NAME 40 mg/kg; □ L-NAME 70 mg/kg. FLD, forelimb dystonia; GCC, generalized clonic convulsions; CTC, clonic-tonic convulsions.

the reduction in CTC appearance almost reached a statistically significant level of ($p = 0.052$). The time to onset of convulsive response was prolonged with L-NAME, but only at a dose of 70 mg/kg, and corresponded to convulsive responses (TABLE 2). The latent period of FPD was prolonged ($p < 0.01$), as was GCC and CTC ($p < 0.05$, for both).

TABLE 3. Effects of L-NAME on convulsive and mortality pentylenetetrazol-ED_{50} in rats

	Convulsive response and mortality		
Treatment	GCC	CTC	Mortality
PTZ	58.46 (44.95–76.02)	62.98 (51–77.2)	62.98 (51–77.2)
PTZ + L-NAME 10 mg/kg	50.43 (34.53–73.64)	58.46 (44.95–76.02)	50.43 (34.53–73.64)
PTZ + L-NAME 40 mg/kg	—	62.98 (51.4–77.17)	—
PTZ + L-NAME 70 mg/kg	58.86 (23.7–146.18)	81.9 (52.06–128.85)	74.51 (48.35–114.83)

NOTES: Table data are ED_{50} values (mg/kg) with 95% confidence limits in parenthesis. GCC, generalized clonic convulsions; CTC, clonic-tonic convulsions.

In PTZ (100 mg/kg)-treated groups, FLD, GCC, CTC, and mortality incidence were decreased only at the highest dose of L-NAME, by 29%, in each case (see FIGURE 3). The latent period was also affected. The time to GCC onset was prolonged (L-NAME 40 mg/kg, $p < 0.01$), as well as that of CTC (L-NAME 40 and 70 mg/kg, $p < 0.05$ and $p < 0.01$, respectively). The FLD latency period was not significantly affected, although this was borderline ($p = 0.06$).

Even though the latency of all three convulsive responses was prolonged by L-NAME, the time to death was not influenced by L-NAME in any of the treated groups. However, there was a tendency to delay mortality.

According to ED_{50}, which, due to lack of dose-dependent effects, could not be calculated for all groups of animals, L-NAME did not influence convulsive noxious action or mortality (see TABLE 3).

DISCUSSION

In the applied experimental model and protocol, L-NAME influenced the PTZ convulsive response. It prevented CTC and prolonged the latency to the convulsive response.

According to our results, there is no doubt that NO plays an important role in PTZ-evoked convulsions. However, in the absence of convulsive noxious response, neither small, nor large reduction in NO synthesis, was *per se* sufficient to induce a convulsive response.

By using various doses of PTZ and L-NAME we found that the role of NO in the pathogenesis of convulsions is not equivocal. We registered the proconvulsant action of NO, but also resistance to it. Whether NO could act as a proconvulsant or to be without action was strongly related to the applied dose of PTZ. Worsened convulsive response and death observed in some groups could suggest an anticonvulsant action of NO, but they did not reach statistically significant levels. Further investigations are required to determine if this finding is in conflict with our other results.

From data available from many authors, summarized by Willoughby,[20] it is thought that primary generalized clonic-tonic (motor) convulsions are the result of brainstem stimulation (mesencephalic or pontine structures). This convulsive pattern represents an animal model of human grand mal epilepsy, found in our experiment to be the most sensitive among the convulsive responses evoked by PTZ (80 mg/kg) to the reduction of NO synthesis. In those groups, the appearance of convulsions was not changed with two lower doses of L-NAME, but a dose of 70 mg/kg prevented them in 57%, and prolonged their latent time. This finding suggests a proconvulsant role for NO. Osnoe and colleagues[15] also registered anticonvulsant effects of L-NAME against tonic convulsions in PTZ (80 mg/mg) in rats, but L-NAME was administered in much higher doses (500 mg/kg, ip).

Anticonvulsant effects of L-NAME could be achieved by decreasing NO levels and consequent changes in guanylate cyclase activity, as well as presynaptic glutamate release. A decreased level of NO is reflected in cyclic guanosine monophosphate synthesis, which is involved in the pathogenesis of PTZ-evoked convulsions.[21]

The potential contribution of blood pressure changes to convulsive response also needs to be considered. Namely, PTZ application typically produces an initial and transient decrease in aortic pressure and induces dilatation of brain blood vessels.[22] This PTZ effect is mediated through the NO-cGMP pathway, although there is evidence of cGMP-independent vasodilatation induced by NO.[23] In our experiments, PTZ-induced vasodilatation is limited by L-NAME administration since generalized vasoconstriction produced by NOS antagonists decreases cerebral blood flow despite a significant increase in blood pressure.[24] Furthermore, such changes may slow PTZ distribution in the brain and postpone its convulsant effects that are clinically expressed in prolonged latent periods.

PTZ ED_{50} did not support our results indicating proconvulsant NO effects. This is similar to the findings of Czuczwar and coworkers[25,26] in mice, showing that tonic-clonic PTZ ED_{50} is not changed by NOS antagonist *N*ω-nitro-L-arginine (NNA) given before PTZ.

The limbic region is highly vulnerable and susceptible to PTZ, compared to other brain regions, because of its specific neurochemical organization. In our experiments, the lowest dose of PTZ induced limbic convulsions in all animals. Inhibition of NO synthesis prevented their incidence only in 43% of the cases, probably because of the limbic structure overstimulation by PTZ. Limbic seizures consist of sniffing, forelimb dystonia, rearing, salivation, and other symptoms. They are evoked by limbic structure stimulation. In kainate-induced limbic convulsions, it seems that the hippocampal CA3 region operates as a pacemaker with output projection to the cortex, medial septum, substantia nigra, amygdala, ventral striatum, and other parts of the brain.[27] Limbic convulsions have been proposed to offer an animal model of complex partial seizures in humans.

Activation of extralimbic structures is followed by severe limbic convulsions where forelimb dystonia is their clinical pattern. Application of L-NAME influenced these convulsions by decreasing NO synthesis and inhibition of cholinergic transmission. Afferent neurons from basal forebrain transmit convulsive information from hippocampus to other brain regions, including some parts of the thalamus. Since alkyl esters of arginine (L-NAME) act as muscarinic receptor antagonists,[28] and muscarinic receptors are of great importance in limbic seizure initiation and spreading,[29] the time to convulsive onset is expected to be prolonged by L-NAME.

Our results for limbic convulsions are not in agreement with those obtained by Rondoulin *et al.*,[30] even though we registered a small increase in FLD incidence. They found dramatic worsening of kainate-evoked limbic convulsions by its local application into the amygdala and did not register any suppressive effect of L-NAME. The chemoconvulsants used, the affected brain regions, the doses of L-NAME, and the duration of the treatment could explain these differences. For example, we applied L-NAME only once, whereas they treated rats chronically (four days, twice a day).

There are dissimilarities in clonic convulsive response between our results and those obtained by Hara *et al.*,[13] who found that the NOS antagonist *N*ω-nitro-L-arginine (L-NA) did not influence clonic response to PTZ (100 mg/kg) in mice; neither their incidence, nor their latency. This is contrary to our finding for the latency. One of the reasons for these differences is likely to lie in the different definition of convulsions.

There is not much information available concerning NO involvement in convulsive mortality. By evaluating ED_{50} in mice, Czuczwar *et al.*[31] did not find a statistically significant influence of NNA on PTZ mortality. This is similar to our finding, even though mortality rate was decreased by 50%. Taking into account LD_{50}, incidence, and latency, it indicates that NO is without significant influence on PTZ mortality. It seems that NOS antagonists can influence only the early stage of PTZ-evoked disturbances. It is not clear if the applied doses of L-NAME were high enough, or even more, if they could influence a wide variety of changes in the brain that followed PTZ application.

CONCLUSION

Administration of L-NAME to rats led to decrease in basal NO levels without any convulsive response in itself, but the effects of NO in PTZ induced convulsions were L-NAME- and PTZ-dose dependent. In this respect, clonic-tonic convulsions were under stronger NO influence than limbic, probably because of PTZ limbic structure overstimulation. Inhibition of NO synthesis decreased incidence of CTC and prolonged the time to onset of FLD, GCC, and CTC, indicating that NO acts as a proconvulsant. Increased incidence of GCC, CTC and mortality, which suggests an anticonvulsant effect of NO, needs further investigation.

REFERENCES

1. CHIUEH, C. 1999. Neuroprotective properties of nitric oxide. Ann. N.Y. Acad. Sci. **890:** 301–311.
2. IKARASHI, Y., M. YUZURIHARA, A. TAKAHASHI, *et al.* 1998. Direct regulation of acetylcholine release by *N*-methyl-D-aspartic receptors in rat striatum. Brain Res. **795:** 215–220.
3. YAMADA, K., Y. NODA, S. NAKAYAMA, *et al.* 1995. Role of nitric oxide in learning and in monoamine metabolism in the rat brain. Br. J. Pharmacol. **115:** 852–858.
4. DAWSON, V., T. DAWSON, D. BARTLEY, *et al.* 1993. Mechanism of nitric oxide-mediated neurotoxicity in primary brain cultures. J. Neurosci. **13:** 2651.
5. SMITH, K., R. KAPOOR & P. FELTS. 1999. Demyelination: the role of reactive oxygen and nitrogen species. Brain Pathol. **9:** 69–92.
6. MARGAILL, I., M. ALLIX, R. BOULU & M. Plotkine. 1997. Dose- and time-dependence of L-NAME neuroprotection in transient focal cerebral ischaemia in rats. Br. J. Pharmacol. **120:** 160–163.
7. GROSS, P., D. WEAVER, R. BOWERS, *et al.* 1994. Neurotoxicity in conscious rats following intraventricular SNAP, a nitric oxide donor. Neuropharmacology **7:** 915–927.
8. JAYAKUMAR, A., R. SUJATHA, V. PAUL, *et al.* 1999. Involvement of nitric oxide and nitric oxide synthase activity in anticonvulsive action. Brain Res. Bull. **4:** 387–394.
9. MULSCH, A., R. BUSSE, P. MORDVINCEV, *et al.* 1994. Nitric oxide promotes seizure activity in kainate-treated rats. Neuroreport **5:** 2325–2328.
10. BUISSON, A., N. LAKHMECHE, C. VERRECCHIA, *et al.* 1993. Nitric oxide: an endogenous anticonvulsant substance. Neuroreport **4:** 444–446.
11. LALLEMENT, G., T. SHIN, I. PERNOT-MARINO, *et al.* 1996. The role of nitric oxide in soman-induced seizures, neuropathology, and lethality. Pharmacol. Bioch. Behav. **54**(4): 731–737.
12. PRZEGALINSKI, E., L. BARAN & J. SIANOWICZ. 1996. The role of nitric oxide in chemically- and electrically-induced seizures in mice. Neurosci. Lett. **217:** 145–148.
13. HARA, S., F. KURIIWA, N. IWATA,*et al.* 1996. Distinct effects of *N*-ω-nitro-L-arginine on seizures induced by several drugs in mice. Pharmacol. Biochem. Behav. **53:** 673–677.

14. NIDHI, G., S. BALAKRISHNAN & P. PANDHI. 1999. Role of nitric oxide in electroshock and pentylenetetrazole seizure threshold in rats. Meth. Find. Exp. Clin. Pharmacol. **21:** 609–612.
15. OSONOE, K., N. MORI, K. SUZUKI & M. OSONOE. 1994. Antiepileptic effects of inhibitors of nitric oxide synthase examined in pentylenetetrazol-induced seizures in rats. Brain Res. **663:** 338–340.
16. DEL-BEL, E., P. OLIVEIRA, J. OLIVEIRA, *et al.* 1997. Anticonvulsant and proconvulsant roles of nitric oxide in experimental epilepsy models. Braz. J. Med. Biol. Res. **30:** 971–979.
17. LOSCHER, W. 1999. Animal model of epilepsy and epileptic seizures. *In* Handbook of Experimental Pharmacology. M. Eadie & F. Vajda, Eds.: 19–62. Springer-Verlag, Berlin, Heildeberg.
18. RUNDFELDT, C., R. KOCH, A. RICHTER, *et al.* 1995. Dose-dependent anticonvulsant and proconvulsant effects of nitric oxide synthase inhibitors on seizure threshold in a cortical stimulation model in rats. Eur. J. Pharmacol. **274:** 73–81.
19. ALEXANDER, C., T. ELLMORE, T. KOKATE & R. KIRKBY. 1998. Further studies on anti- and proconvulsant effects of inhibitors of nitric oxide synthase in rodents. Eur. J. Pharmacol. **344:** 15–25.
20. WILLOUGHBY, J. 1999. Epileptogenesis: electrophysiology. *In* Handbook of Experimental Pharmacology. M. Eadie & F. Vajda, Eds.: 63–85. Springer-Verlag, Berlin, Heildeberg.
21. WOOD, P., M. EMMETT, T. RAO, *et al.* 1990. Inhibition of nitric oxide synthase blocks *N*-methyl-D-aspartate-, quisqualate-, kainate-, harmaline-, and pentylenetetrazole-dependent increases in cerebellar cyclic GMP *in vivo*. J. Neurochem. **55:** 346–348.
22. FARACI, F., K. BREESE & D. HEISTAD. 1993. Nitric oxide contributes to dilatation of cerebral arterioles during seizures. Am. J. Physiol. **265:** H2209–2212.
23. GOUD, C., A. DIPIERO, W. LOCKETTE, *et al.* 1999. Cyclic GMP-independent mechanisms of nitric oxide-induced vasodilatation. Gen. Pharmacol. **32**(1): 51–5.
24. REES, D., R. PALMER, R. SCHULZ, *et al.* 1990. Characterization of three inhibitors of endothelial nitric oxide synthase *in vitro* and *in vivo*. Br. J. Pharmacol. **101:** 746–752.
25. TUTKA, P., P. KLONOWSKI, J. DZIECIUCH, *et al.* 1996. N^G-nitro-L-arginine differentially affects glutamate- or kainate-induced seizures. Neuroreport **7:** 1605–1608.
26. URBANSKA, E., E. DRELEWSKA, K. BOROWICZ, *et al.* 1996. N^G-nitro-L-arginine, a nitric oxide synthase inhibitor, and a seizure susceptibility in four seizure models in mice. J. Neural. Transm. **103:** 1145–1152.
27. LOTHMAN, E. & R. COLLINS. 1981. Kainic acid induced limbic seizures: metabolic, behavioural, electroencephalographic and neuropathological correlates. Brain Res. **218:** 299–318.
28. BUXTON, I., D. CHEEK, D. ECKMAN, *et al.* 1993. N^G-nitro-L-arginine methyl ester and other alkyl esters of arginine are muscarinic receptor antagonists. Circ. Res. **72:** 387–391.
29. MRAOVITCH, S. & Y. CALANDO. 1999. Interactions between limbic, thalamo-striatal-cortical, and central anatomic pathways during epileptic seizure progression. J. Compar. Neurol. **411:** 145–161.
30. RONDOULIN, G., J. BOCKAERT & M. LERNER-NATOLI. 1993. L-Nitroarginine, an inhibitor of NO synthase, dramatically worsens limbic epilepsy in rats. NeuroReport **4:** 1187–1190.
31. CZUCZWAR, S., P. TUTKA, P. KLONOWSKI & Z. KLEINROK. 1999. N^G-nitro-L-arginine impairs the anticonvulsive action of ethosuximide against pentylenetetrazol. Eur. J. Pharmacol. **366:** 137–142.

Serum-Dependence of LPS-Induced Neurotoxicity in Rat Cortical Neurons

CYNTHIA L. COOPER,[a,b] GWANG-HO JEOHN,[b]
PETER TOBIAS,[c] AND JAU-SHYONG HONG[b]

[a]*Division of Science, Truman State University, Kirksville, Missouri, U.S.A.*

[b]*Neuropharmacology Section, Laboratory of Pharmacology and Chemistry, National Institute of Environmental Health Sciences, Research Triangle Park, North Carolina, U.S.A.*

[c]*Department of Immunology, The Scripps Research Institute, La Jolla, California, U.S.A.*

ABSTRACT: Previous studies have shown that the bacterial endotoxin, lipopolysaccharide (LPS), is neurotoxic both *in vitro* and *in vivo*. The rate of binding of LPS to a target cell is greatly enhanced by serum in general and by LPS binding protein (LBP) in particular. The purpose of the study described in this paper was to determine if microglia activation and LPS-induced neurotoxicity is serum or LBP dependent. A murine microglial cell line, BV2, was used to assess the serum dependence of nitric oxide production and tumor necrosis factor α release in microglia. Embryonic rat cortical neuron/glia mixed cultures were used to determine the serum dependence of LPS-induced neurotoxicity. Our results from both cell culture systems show that LPS-induced inflammatory responses are serum dependent at lower doses of LPS and progressively become serum independent above 10 ng/ml. Purified human recombinant LBP reconstitutes the lost LPS-induced inflammatory responses in primary and immortalized cell cultures treated with heat-denatured serum and appears to account for the serum dependence. These data suggest that the cell surface signaling receptor for LPS at the low and high concentrations are likely to differ, consistent with the existence of a variety of LPS receptors.

KEYWORDS: serum dependence; LPS-induced neurotoxicity; rat cortical neurons

ABBREVIATIONS:

DMEM	Dulbecco's modified Eagle's medium.
IL	interleukin.
MEM	minimal essential medium.
FBS	fetal bovine serum.
LPS	lipopolysaccharide.
PBS	phosphate-buffered saline.
TLR	toll-like receptor.
TNFα	tumor necrosis factor-α.
NO	nitric oxide.
iNOS	inducible nitric oxide synthase.
IFNγ	interferon γ.

Address for correspondence: Dr. Cynthia Cooper, Division of Science, Truman State University, Kirksville, MO 63501, U.S.A. Voice: 660-785-4082; fax: 660-785-4045.
clcooper@truman.edu

INTRODUCTION

Lipopolysaccharide (LPS), a major outer membrane glycolipid of gram-negative bacteria, is a potent pyrogen that triggers a large number of biological responses, including fever, disseminated intravascular coagulation, septic shock, and even death.[1–3] LPS is one of the most potent inducers of proinflammatory cytokines, such as tumor necrosis factor α (TNFα), interleukin-1β (IL-1β), interleukin-6 (IL-6), and arachidonic acid metabolites that recruit and activate other immune cells to help fight bacterial infections. In addition to the cytokines, LPS also elicits a characteristic oxidative burst that involves nitric oxide (NO) and superoxide anion and their highly reactive derivatives.[4] A description of NO events in the brain and their clinical implications are reviewed by Licinio *et al.*[5] The effects of LPS in the peripheral immune system are mediated primarily through neutrophils, monocytes, and macrophages, however, the acute inflammatory response to LPS in the central nervous system differs from that in the periphery.[6,7] Resident microglia and astrocytes appear to manage inflammatory events in the brain to a large extent.

Maximal stimulation of neutrophils by LPS occurs in the presence of a serum factor, LPS binding protein (LBP).[8–10] LBP appears to catalytically transfer LPS to CD14 on the cell surface without forming a ternary complex.[11–13] This step is quickly followed by activation of the toll-like receptor 4 (TLR4) and a transmembrane signal transduction event.[14–16] TLR4 receptor activation appears to function through MD-2 or MyD88 to activate NFκB nuclear translocation and phosphorylation of one or more mitogen activated protein kinases.[15,17,18] LPS and its analogues are also internalized through CD14 and β2 integrin adhesion receptors.[12,19] Perera *et al.* recently proposed that coordination among CD14, CD11b/CD18, and TLR4 receptors is necessary for maximal induction of inflammation related genes by LPS.[20]

In brain glia, LPS functions through a CD14 receptor[21] and TLR4[22] to increase the expression of inducible nitric oxide synthase (iNOS) and release of nitric oxide into the growth medium.[21] Extracellular signal-regulated kinase and p38 subgroups of mitogen-activated protein kinases regulate inducible nitric oxide synthase and TNFα gene expression in endotoxin-stimulated primary glial cultures.[23] LPS also activates nuclear factor-κB[24] and increases expression of IL-1α,[25] IL-1β,[26] IL-6,[25,27] and various factors involved in the inflammatory response. LPS-activated microglia have been shown to be neurotoxic *in vitro* and *in vivo,*[28,29] however, LPS is not directly toxic to neurons in the absence of microglia.[30] Furthermore, the regional distribution of resident microglia in brain tissue correlates with LPS-induced neurotoxicity.[28]

Given the severe consequences of infection-mediated inflammation in the central nervous system, a better understanding of LPS-induced neurotoxicity is needed. The purpose of the study described here was to determine the extent to which the inflammatory response to LPS in brain tissue is serum dependent. A murine microglial cell line, BV2, was used to assess the serum dependence of NO production and TNFα release in microglia. Embryonic rat cortical neuron/glia mixed cultures were used to determine the serum dependence of LPS-induced neurotoxicity, as well as NO and TNFα production. The results from both cell culture systems show that LPS-induced inflammatory responses were serum dependent at the lower doses of LPS and less so at higher doses. Furthermore, purified human recombinant LPS binding protein (LBP) restored the lost LPS-induced inflammatory responses in primary and immortalized cell cultures treated with heat-denatured serum. The data suggest that the cell surface signaling receptors for LPS at low and high concentrations are likely to differ, consistent with the existence of a variety of LPS receptors.

METHODS

Materials

LPS from *Escherichia coli* 0111:B4 was purchased from Calbiochem (La Jolla, CA). Dulbecco's modified Eagle's medium (DMEM) was purchased from the American Type Culture Collection (Manassas, VA). Minimal essential medium (MEM), fetal bovine serum (FBS), heat-inactivated horse serum, sodium pyruvate, nonessential amino acids, penicillin, and streptomycin were from Life Technologies (Rockville, MD). All culture flasks and multiwell plates were Costar® brand (Corning, Inc., Corning, NY). Human recombinant LBP (hrLBP) was cloned and purified as described by Theofan *et al.*[31] Rat interferon-γ (IFNγ) was obtained from R&D Systems Inc. (Minneapolis, MN).

Cell Culture and LPS Treatments

The mouse microglial cell line BV2[32,33] was kindly provided by Dr. M. McKinny of the Mayo Clinic (Jacksonville, FL). Cells were cultured in 75-cm^2 flasks as adherent cells in DMEM medium supplemented with 50 U/ml penicillin G, 50 μg/ml streptomycin sulfate and 10% heat-inactivated FBS (56°C for 30 min). Cultures were maintained in a humidified atmosphere of 5% CO_2/95% air at 37°C. Cells were seeded into 24-well tissue culture plates at 1.5×10^5 cells per well in one ml medium two days before each experiment. Before changing to treatment medium (as above except using 2% heat-inactivated FBS or 2% heat-denatured FBS), cells were quickly rinsed twice in ambient phosphate buffered saline (PBS). To assess the serum dependence of LPS responsiveness an aliquot of the heat-inactivated FBS was heat denatured at 72°C for 20 min. The LPS responses were compared in assays with control serum (FBS heat inactivated at 56°C) and heat-denatured FBS. NO and TNFα accumulations in media with control serum were proportional to cell number, between 1×10^5 and 3×10^5 per well.

Pregnant rats with E15/16 embryos were euthanized with CO_2 and embryonic cortical brain tissue was used to prepare cortical neuron/glia mixed cultures similar to those described by Kim *et al.*[28] Brains were aseptically removed and blood vessels and meninges were discarded. Tissues were gently triturated by 10-ml pipette fitted with P1000, then P200 micropipette tips in ambient culture medium (MEM supplemented with 10% heat-inactivated FBS, 10% heat-inactivated horse serum, 1 g/l D-glucose, 1 mM Na-pyruvate, 2 mM L-glutamine, 50 U/ml penicillin, and 50 μg/ml streptomycin sulfate). Neuron/glia mixtures were seeded into 24-well tissue culture plates at 5×10^5 cells/well and the medium was refreshed after three days. Treatments were initiated on day 6 by quickly rinsing each well twice with ambient PBS and adding MEM supplemented with 2% FBS (control or heat denatured), Na-pyruvate, L-glutamine, penicillin, and streptomycin.

Nitric Oxide and TNFα Assays

Nitric oxide production was assessed as nitrite accumulation in samples of culture medium using a colorimetric reaction with the Griess reagent.[34] TNFα was measured in samples of culture media using the enzyme linked immunosorbent assay kit for mouse TNFα from Genzyme (Cambridge, MA) as described by the supplier. Optical densities were measured with a UV MAX kinetic microplate reader (Molecular Devices, Menlo Park, CA).

Immunocytochemistry for Cortical Neurons

After treatment the cultured cells were rapidly rinsed twice in PBS, fixed in 4% formaldehyde solution and immunostained with antineuronal nuclear protein antibody (anti-NeuN; Chemicon International, Inc., Temecula, CA), as described by Liu *et al.*[35] Neurons were counted in eight areas (0.1 cm^2 total) of each well at 100× magnification.

Statistical Analysis

Data were expressed as mean ± standard error of the mean. Each experiment was performed using 2, 3, or 4 replicates and each experiment was repeated 2–4 times. A value of $p < 0.05$ was considered significant.

RESULTS

Serum Dependence of LPS-Induced Cytokine Accumulation

The ability of LPS to activate the production and/or release of cytokines depends on the presence of serum factors. This has been reported for neutrophils and macrophages,[8,36] but the serum dependence of LPS stimulation has not been characterized for brain tissue. FIGURE 1 shows that production of NO and release of TNFα in the microglial cell line was completely serum dependent at the low doses, 1 and 3 ng/ml LPS. Higher doses of LPS, 100 and 1000 ng/ml LPS, induced maximal NO accumulation, 67% and 75%, respectively, and were serum independent. A similar pattern was found for TNFα production with complete serum dependence at and below 10 ng/ml LPS, and 90–100% of the TNFα release was serum independent at 100 and 1000 ng/ml LPS.

Cortical neuron/glia mixed cultures responded to LPS in a similar manner to that of the BV2 culture system (see FIGURE 2). The response to the lowest level of LPS (1 ng/ml) was 100% serum dependent and largely serum independent at and above 10 ng/ml. Interestingly, the LPS response at 1,000 ng/ml was often greater in the heat-denatured treatments than in the control serum. The major difference in the two

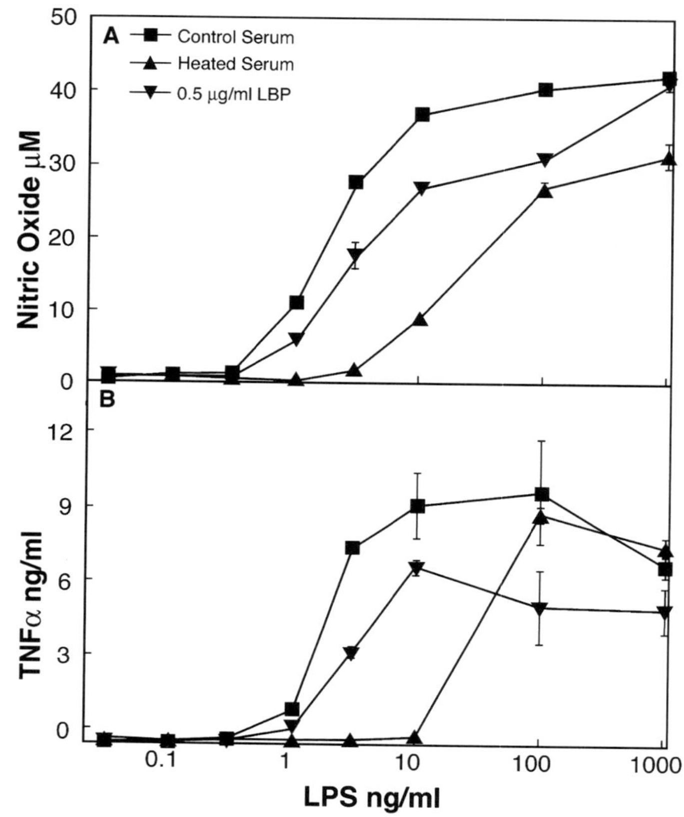

FIGURE 1. Serum-dependence of LPS-induced NO and TNFα production is restored with hrLBP in BV2. Cells (1.5×10^5) in duplicate wells were treated in DMEM with 2% control or heat-denatured serum, with or without 0.5 µg/ml hrLBP, for 20 min before LPS was added and media samples were analyzed for NO (24 h) and TNFα (6 h) as described in METHODS. Representative data from one of four experiments are shown.

culture systems was in the relative quantity of TNFα produced. Although the number of cells initially seeded was lower in the BV2 system (150,000 vs. 500,000) the actual number of microglia during treatment was not determined. Even so, the ratio of TNFα to NO was much lower in the neuron/glia mixed cultures than in the pure microglial cell line. This result may reflect the tempering effect of neurons on microglia reported by Chang *et al.*[37]

Recovery of Serum Dependence with LBP

The ability of human recombinant LBP (hrLBP) to recover the LPS-induced cytokine release in the presence of heat-denatured serum is shown in FIGURES 1

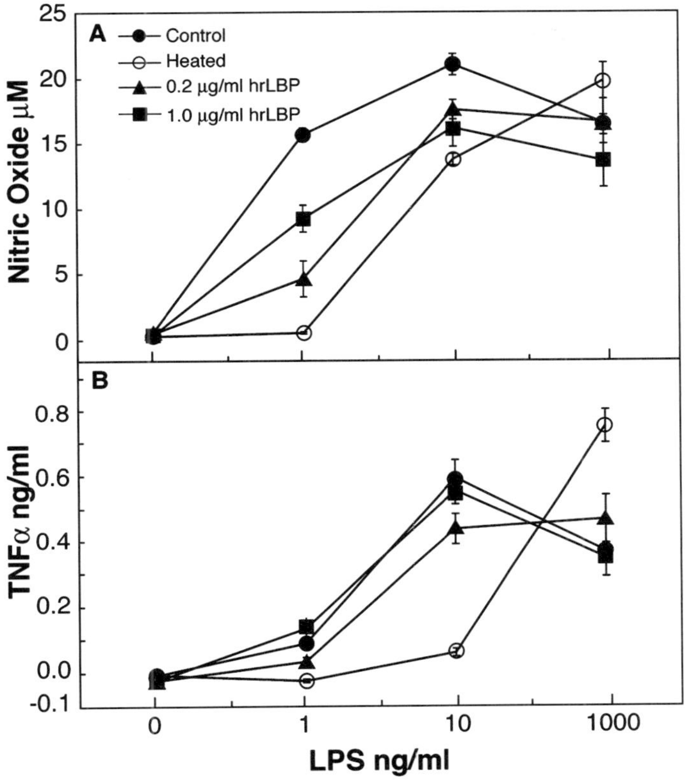

FIGURE 2. Serum dependence of LPS-induced NO and TNFα production is restored with hrLBP in primary cortical neuron/glia. Mixed cultures were treated in MEM with 2% control or heat-denatured FBS. hrLBP in a PBS vehicle was added first, followed by IFNγ (50 U/ml; 20 ng/ml), then LPS, and treatment continued for 72 h. Media samples, collected at 6 h for TNFα and 48 h for NO, were analyzed as described in METHODS. Data shown were combined from two experiments, with three or four replicates of each.

and 2. hrLBP (0.5 μg/ml) recovered 63% of the NO response at 3 ng/ml LPS and 43% of the TNFα production in BV2 cells. At 10 ng/ml the LPS effects were 75% serum dependent for NO and 100% serum dependent for TNFα. At this dose of LPS, the hrLBP recovered 73% of both the NO and TNFα accumulations. The ability of LBP to recover NO accumulation was generally better than that for TNFα in BV2 cells. In contrast, TNFα in the neuron/glia mixed cultures was somewhat more responsive to hrLBP than was NO.

The inflammation enhancement by hrLBP was concentration dependent in the BV2 cell system from 0.1 to 1.0 μg/ml (see FIGURE 3). A 100% recovery of NO or TNFα production was not realized with 1 μg/ml LBP in BV2 cells. The use of higher concentrations did not seem warranted because of cross-species differences in LBP,

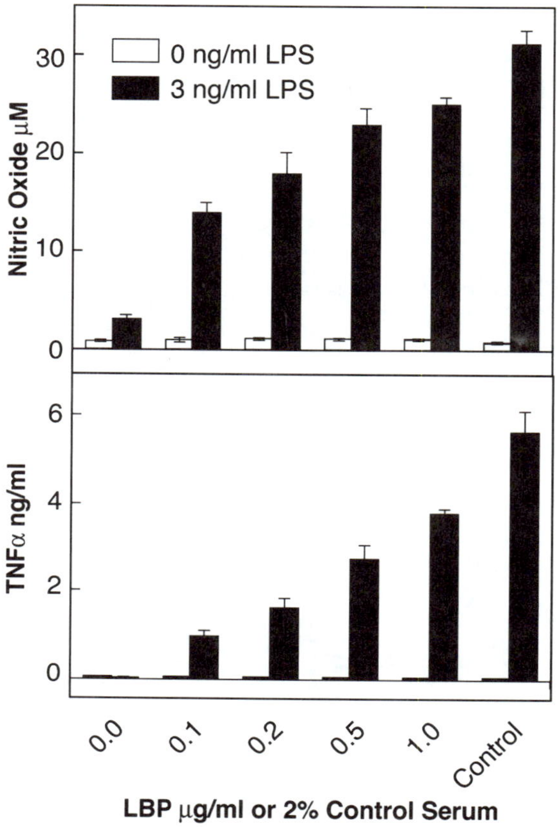

FIGURE 3. hrLBP reconstitution of inflammatory response is concentration dependent. BV2 cells (1.5×10^5) were treated in DMEM with 2% control FBS or heat-denatured FBS and the indicated amount of hrLBP was added 20 min before LPS (3 ng/ml). Media samples for NO (24 h) or TNFα (6 h) were analyzed as described in METHODS. The data shown were combined from four experiments, with two or three replicates of each.

and the 1.0 µg/ml was much higher than that normally found in human or bovine serum (see DISCUSSION). In contrast, 1 µg/ml hrLBP in the neuron/glia mixed cultures was 100% effective at reconstituting the neurotoxicity of LPS in medium with heat-denatured serum (see FIGURE 4C). Although that amount of hrLBP reconstituted only 60–70% of the NO accumulation, it recovered 150–200% of the TNFα release (FIGS. 2 and 4). Thus, the low ratio of TNFα to NO in the primary culture system did not impair the inflammation-mediated toxicity to neurons.

Taken together, we conclude from the data that LPS-induced neurodegeneration is serum dependent at low doses of LPS equal to or less than 1 ng/ml yet serum-independent at or above 10 ng/ml LPS.

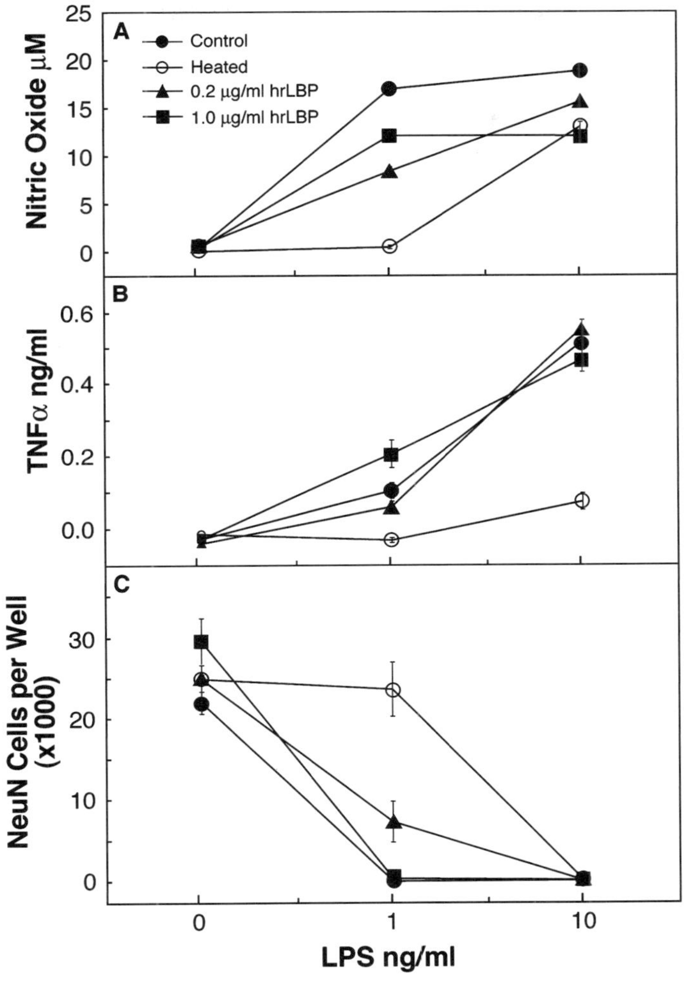

FIGURE 4. hrLBP reconstitutes heat-denatured serum factor to evoke LPS-induced neurotoxicity. Mixed cultures were treated in MEM with 2% control or heat-denatured FBS. hrLBP in PBS vehicle was added first, followed by IFNγ (50 U/ml, 20 ng/ml), then LPS, and treatment was continued for 72 h. Media samples, collected at 6 h for TNFα and 48 h for NO, were analyzed as described in METHODS. Neurons were fixed after 72 h and analyzed for NeuN protein as described in METHODS. Data shown are from one of three experiments ($n = 3$ for control and heated FBS; $n = 4$ for LBP groups).

DISCUSSION

Bacterial endotoxin is a potent inflammatory agent and a major cause of life threatening septic shock.[38] A toxic situation develops when the lipid A region of LPS activates immune cells in healthy organs, and the toxic compounds TNFα, NO, and $O_2^{-\bullet}$, cause collateral damage to liver, kidney, or brain. In the case of brain tissue, previous studies from our laboratory show that activated microglia kill neurons.[29–30] In the healthy mammal with an intact blood brain barrier, both LPS and serum proteins are largely excluded from the cerebral vasculature. However, once breached by acute or chronic inflammation at the meninges, both peripheral serum proteins and infectious agents (fragmented or intact) may penetrate the central nervous system.[39,40]

The data presented here show that LPS-induced inflammatory responses in murine microglia and primary rat cortical neuron/glia mixtures are serum dependent at low doses of LPS (0–1 ng/ml). Above 10 ng/ml the cytotoxic factors released in response to LPS are less serum dependent. The differential serum dependence of LPS-induced inflammatory responses *in vitro* most likely reflects the range of LPS-receptors on microglia since LPS is not directly toxic to neurons.[28,30] Microglia express CD14, TLR2, TLR4, and CD11b/CD18 receptors, all of which have been associated with LPS-mediated activation.[21,22,41–44] At low doses of LPS, the CD14-activated TLR4 receptor is most likely the primary site of transmembrane signal transduction. Above 10 ng/ml LPS, the serum-independent LPS receptor, CD11b/18 is most likely the primary transmembrane signal transducer, as well as phagocytosis receptor.[45,46] Although the role of TLR2 in LPS-induced signaling is not clear, studies have shown that TLR2 mRNA in murine brain is upregulated, whereas that for TLR4 is downregulated in response to LPS treatment.[22,44] Preliminary studies in this laboratory show a similar response to LPS in rat microglia (unpublished). Thus, microglia potentially respond to LPS over a range of concentrations using a variety of receptors.

Wilde *et al.* have shown that hrLBP replaces the serum requirement for LPS binding to and activation of human monocytes.[47] According to Tobias *et al.*,[48] the amount of LBP in serum is about 2.7 nM and 2% serum gives about 54 pM LBP or 3.2 ng/ml. Thus, the use of 1 μg/ml hrLBP in the current study is about 300 times more than the amount of LBP normally seen in control FBS serum. Since LBP has been shown to be effective in binding LPS and activating cells at a ratio of 1 : 1, the loss of serum dependence at high doses of LPS is most likely due to LBP independent activation events.

Prinz *et al.*[49] showed that activation responses to 100 ng/ml LPS in murine primary microglia are partially serum dependent and exogenously added murine recombinant LBP partially restored the serum dependence. Our data show a complete dependence on LBP for LPS-induced cytokine and NO production in the mouse microglial cell line, BV2, and primary rat cortical neuron/glia mixed cultures occurred at lower doses of LPS (1–3 ng/ml). However, above 10 ng/ml LPS, LPS-induced cytokine release was less serum dependent. The implication of these results for brain infections and microglia-mediated neurodegeneration is that bacterial endotoxin is inflammatory, with or without the presence of serum factors. This conclusion is supported by results from LBP knockout mice, which show that *in vitro*, LBP deficient serum is ineffective at promoting LPS-induced inflammatory responses, however, *in vivo,* LPS responsiveness was the same as in genetic control animals.[50] Thus, *in vivo,* the serum factor requirement for LPS may not be essential.

Although largely excluded from access to the brain, very low levels of LPS injected into the periphery of healthy animals can be detected in the brain.[7] Sporadic and low levels of LPS in the brain could be the source of both chronic and acute neuroinflammation. A thorough understanding of the role of LBP and endotoxin receptor mechanisms in the central nervous system will aid the development of new strategies to combat neurologic inflammation.

REFERENCES

1. HOLST, O., A.J. ULMER, H. BRADE, *et al.* 1996. Biochemistry and cell biology of bacterial endotoxins. FEMS Immunol. Med. Microbiol. **16:** 83–104.
2. RIETSCHEL, E.T., T. KIRIKAE, F.U. SCHADE, *et al.* 1993. The chemical structure of bacterial endotoxin in relation to bioactivity. Immunobiology **187:** 169–190.
3. RIETSCHEL, E.T., H. BRADE, O. HOLST, *et al.* 1996. Bacterial endotoxin: chemical constitution, biological recognition, host response, and immunological detoxification. Curr. Top. Microbiol. Immunol. **216:** 39–81.
4. DAWSON, V.L., T.M. DAWSON, D.A. BARTLEY, *et al.* 1993. Mechanisms of nitric oxide-mediated neurotoxicity in primary brain cultures. J. Neurosci. **13:** 2651–2661.
5. LICINIO, J., P. PROLO, S.M. MCCANN, *et al.* 1999. Brain iNOS: current understanding and clinical implications. Mol. Med. Today **5:** 225–232.
6. PERRY, V.H., P.B. ANDERSSON & S. GORDON. 1993. Macrophages and inflammation in the central nervous system. Trends Neurosci. **16:** 268–273.
7. ANDERSSON, P.B., V.H. PERRY & S. GORDON. 1992. The acute inflammatory response to lipopolysaccharide in CNS parenchyma differs from that in other body tissues. Neuroscience **48:** 169–186.
8. SCHUMANN, R.R., S.R. LEONG, G.W. FLAGGS, *et al.* 1990. Structure and function of lipopolysaccharide binding protein. Science **249:** 1429–1431.
9. SCHUMANN, R.R. 1992. Function of lipopolysaccharide (LPS)-binding protein (LBP) and CD14, the receptor for LPS/LBP complexes: a short review. Res. Immunol. **143:** 11–15.
10. ULEVITCH, R.J. & P.S. TOBIAS. 1995. Receptor-dependent mechanisms of cell stimulation by bacterial endotoxin. Annu. Rev. Immunol. **13:** 437–457.
11. HAILMAN, E., H.S. LICHENSTEIN, M.M. WURFEL, *et al.* 1994. Lipopolysaccharide (LPS)-binding protein accelerates the binding of LPS to CD14. J. Exp. Med. **179:** 269–277.
12. INGALLS, R.R., H. HEINE, E. LIEN, *et al.* 1999. Lipopolysaccharide recognition, CD14, and lipopolysaccharide receptors. Infect. Dis. Clin. North Am. **13:** 341–353, vii.
13. TOBIAS, P.S., K. SOLDAU, J.A. GEGNER, *et al.* 1995. Lipopolysaccharide binding protein-mediated complexation of lipopolysaccharide with soluble CD14. J. Biol. Chem. **270:** 10482–10488.
14. LIEN, E., T.K. MEANS, H. HEINE, *et al.* 2000. Toll-like receptor 4 imparts ligand-specific recognition of bacterial lipopolysaccharide. J. Clin. Invest. **105:** 497–504.
15. MEANS, T.K., D.T. GOLENBOCK & M.J. FENTON. 2000. The biology of Toll-like receptors. Cytokine Growth Factor Rev. **11:** 219–232.
16. ROCK, F.L., G. HARDIMAN, J.C. TIMANS, *et al.* 1998. A family of human receptors structurally related to Drosophila Toll. Proc. Natl. Acad. Sci. U.S.A. **95:** 588–593.
17. GOLENBOCK, D.T. & M.J. FENTON. 2001. Extolling the diversity of bacterial endotoxins. Nat. Immunol. **2:** 286–288.
18. YANG, H., D.W. YOUNG, F. GUSOVSKY, *et al.* 2000. Cellular events mediated by lipopolysaccharide-stimulated toll-like receptor 4. MD-2 is required for activation of mitogen-activated protein kinases and Elk-1. J. Biol. Chem. **275:** 20861–20866.
19. INGALLS, R.R., B.G. MONKS, R. SAVEDRA, JR., *et al.* 1998. CD11/CD18 and CD14 share a common lipid A signaling pathway. J. Immunol. **161:** 5413–5420.

20. PERERA, P.Y., T.N. MAYADAS, O. TAKEUCHI, *et al.* 2001. CD11b/CD18 acts in concert with CD14 and Toll-like receptor (TLR) 4 to elicit full lipopolysaccharide and taxol-inducible gene expression. J. Immunol. **166:** 574–581.
21. GALEA, E., D.J. REIS, E.S. FOX, *et al.* 1996. CD14 mediate endotoxin induction of nitric oxide synthase in cultured brain glial cells. J. Neuroimmunol. **64:** 19–28.
22. LAFLAMME, N. & S. RIVEST. 2001. Toll-like receptor 4: the missing link of the cerebral innate immune response triggered by circulating gram-negative bacterial cell wall components. FASEB J. **15:** 155–163.
23. BHAT, N.R., P. ZHANG, J.C. LEE, *et al.* 1998. Extracellular signal-regulated kinase and p38 subgroups of mitogen- activated protein kinases regulate inducible nitric oxide synthase and tumor necrosis factor-alpha gene expression in endotoxin-stimulated primary glial cultures. J. Neurosci. **18:** 1633–1641.
24. COLASANTI, M., T. PERSICHINI, M. MENEGAZZI, *et al.* 1995. Induction of nitric oxide synthase mRNA expression. Suppression by exogenous nitric oxide. J. Biol. Chem. **270:** 26731–26733.
25. KONG, L.Y., C. LAI, B.C. WILSON, *et al.* 1997. Protein tyrosine kinase inhibitors decrease lipopolysaccharide-induced proinflammatory cytokine production in mixed glia, microglia-enriched or astrocyte-enriched cultures. Neurochem. Int. **30:** 491–497.
26. DAS, K.P., M.K. MCMILLIAN, G. BING, *et al.* 1995. Modulatory effects of [Met5]-enkephalin on interleukin-1 beta secretion from microglia in mixed brain cell cultures. J. Neuroimmunol. **62:** 9–17.
27. KONG, L.Y., M.K. MCMILLIAN, P.M. HUDSON, *et al.* 1997. Inhibition of lipopolysaccharide-induced nitric oxide and cytokine production by ultralow concentrations of dynorphins in mixed glia cultures. J. Pharmacol. Exp. Ther. **280:** 61–66.
28. KIM, W.G., R.P. MOHNEY, B. WILSON, *et al.* 2000. Regional difference in susceptibility to lipopolysaccharide-induced neurotoxicity in the rat brain: role of microglia. J. Neurosci. **20:** 630–6316.
29. MCMILLIAN, M., L.Y. KONG, S.M. SAWIN, *et al.* 1995. Selective killing of cholinergic neurons by microglial activation in basal forebrain mixed neuronal/glial cultures. Biochem. Biophys. Res. Commun. **215:** 572–577.
30. BRONSTEIN, D.M., I. PEREZ-OTANO, V. SUN, *et al.* 1995. Glia-dependent neurotoxicity and neuroprotection in mesencephalic cultures. Brain Res. **704:** 112–116.
31. THEOFAN, G., A.H. HORWITZ, R.E. WILLIAMS, *et al.* 1994. An amino-terminal fragment of human lipopolysaccharide-binding protein retains lipid A binding but not CD14-stimulatory activity. J. Immunol. **152:** 3623–3629.
32. BLASI, E., R. BARLUZZI, V. BOCCHINI, *et al.* 1990. Immortalization of murine microglial cells by a v-raf/v-myc carrying retrovirus. J. Neuroimmunol. **27:** 229–237.
33. BOCCHINI, V., R. MAZZOLLA, R. BARLUZZI, *et al.* 1992. An immortalized cell line expresses properties of activated microglial cells. J. Neurosci. Res. **31:** 616–621.
34. GREEN, L.C., D.A. WAGNER, J. GLOGOWSKI, *et al.* 1982. Analysis of nitrate, nitrite, and [15N]nitrate in biological fluids. Anal. Biochem. **126:** 131–138.
35. LIU, B., L. DU & J.S. HONG. 2000. Naloxone protects rat dopaminergic neurons against inflammatory damage through inhibition of microglia activation and superoxide generation. J. Pharmacol. Exp. Ther. **293:** 607–617.
36. RAETZ, C.R., R.J. ULEVITCH, S.D. WRIGHT, *et al.* 1991. Gram-negative endotoxin: an extraordinary lipid with profound effects on eukaryotic signal transduction. FASEB J. **5:** 2652–2660.
37. CHANG, R.C., W. CHEN, P. HUDSON, *et al.* 2001. Neurons reduce glial responses to lipopolysaccharide (LPS) and prevent injury of microglial cells from over-activation by LPS. J. Neurochem. **76:** 1042–1049.
38. GLAUSER, M.P., G. ZANETTI, J.D. BAUMGARTNER, *et al.* 1991. Septic shock: pathogenesis. Lancet **338:** 732–736.
39. BOJE, K.M. 1996. Inhibition of nitric oxide synthase attenuates blood-brain barrier disruption during experimental meningitis. Brain Res. **720:** 75–83.
40. TUOMANEN, E. 1993. Breaching the blood-brain barrier. Sci. Am. **268:** 80–84.
41. BECHER, B. & J.P. ANTEL. 1996. Comparison of phenotypic and functional properties of immediately *ex vivo* and cultured human adult microglia. Glia **18:** 1–10.

42. FORD, A.L., A.L. GOODSALL, W.F. HICKEY, *et al.* 1995. Normal adult ramified microglia separated from other central nervous system macrophages by flow cytometric sorting. Phenotypic differences defined and direct *ex vivo* antigen presentation to myelin basic protein- reactive CD4+ T cells compared. J. Immunol. **154:** 4309–4321.
43. KATO, H., K. KOGURE, X.H. LIU, *et al.* 1996. Progressive expression of immunomolecules on activated microglia and invading leukocytes following focal cerebral ischemia in the rat. Brain Res. **734:** 203–212.
44. MATSUMURA, T., A. ITO, T. TAKII, *et al.* 2000. Endotoxin and cytokine regulation of toll-like receptor (TLR) 2 and TLR4 gene expression in murine liver and hepatocytes. J. Interferon Cytokine Res. **20:** 915–921.
45. INGALLS, R.R., M.A. ARNAOUT, R.L. DELUDE, *et al.* 1998. The CD11/CD18 integrins: characterization of three novel LPS signaling receptors. Prog. Clin. Biol. Res. **397:** 107–117.
46. FLAHERTY, S.F., D.T. GOLENBOCK, F.H. MILHAM, *et al.* 1997. CD11/CD18 leukocyte integrins: new signaling receptors for bacterial endotoxin. J. Surg. Res. **73:** 85–89.
47. WILDE, C.G., J.J. SEILHAMER, M. MCGROGAN, *et al.* 1994. Bactericidal/permeability-increasing protein and lipopolysaccharide (LPS)-binding protein. LPS binding properties and effects on LPS- mediated cell activation. J. Biol. Chem. **269:** 17411–17416.
48. TOBIAS, P.S., K. SOLDAU & R.J. ULEVITCH. 1989. Identification of a lipid A binding site in the acute phase reactant lipopolysaccharide binding protein. J. Biol. Chem. **264:** 10867–10871.
49. PRINZ, M., O. KANN, H.J. DRAHEIM, *et al.* 1999. Microglial activation by components of gram-positive and -negative bacteria: distinct and common routes to the induction of ion channels and cytokines. J. Neuropathol. Exp. Neurol. **58:** 1078–1089.
50. WURFEL, M.M., B.G. MONKS, R.R. INGALLS, *et al.* 1997. Targeted deletion of the lipopolysaccharide (LPS)-binding protein gene leads to profound suppression of LPS responses *ex vivo,* whereas *in vivo* responses remain intact. J. Exp. Med. **186:** 2051–2056.

Role of Nitric Oxide in Inflammation-Mediated Neurodegeneration

BIN LIU, HUI-MING GAO, JIZ-YUH WANG, GWANG-HO JEOHN, CYNTHIA L. COOPER, AND JAU-SHYONG HONG

Neuropharmacology Section, Laboratory of Pharmacology and Chemistry, National Institute of Environmental Health Sciences/National Institutes of Health, Research Triangle Park, North Carolina, U.S.A

ABSTRACT: Increasing evidence has suggested that inflammation in the brain is closely associated with the pathogenesis of several degenerative neurologic disorders, including Parkinson's disease, Alzheimer's diseases, multiple sclerosis, amyotrophic lateral sclerosis, and AIDS dementia. The hallmark of brain inflammation is the activation of glial cells, especially that of microglia that produce a variety of proinflammatory and neurotoxic factors, including cytokines, fatty acid metabolites, free radicals—such as nitric oxide (NO) and superoxide. Excessive production of NO, as a consequence of nitric oxide synthase induction in activated glia, has been attributed to participate in neurodegeneration. Using primary mixed neuron-glia cultures and glia-enriched cultures prepared from embryonic rodent brain tissues, we have systemically studied the relationship between the production of NO and neurodegeneration in response to stimulation by the inflammagen lipopolysaccharide. This review summarizes our recent findings on the kinetics of NO generation, the relative contribution of microglia and astrocytes to NO accumulation, the relationship between NO production and neurodegeneration, and points of intervention along the pathways associated with NO generation to achieve neuroprotection. We also describe our results relating to the effect of several opioid-related agents on microglial activation and neuroprotection. Among these agents, the opioid receptor antagonist naloxone, especially its non-opioid enantiomer (+)-naloxone, promises to be of potential therapeutic value for the treatment of inflammation-related diseases.

KEYWORDS: inflammation; lipopolysaccharide; naloxone; neurodegeneration; neuroprotection; nitric oxide; opioids; superoxide

INTRODUCTION

Nitric oxide (NO) is produced by the action of nitric oxide synthase (NOS), which catalyzes the reaction of arginine with molecular oxygen to form citrulline and NO.[1] To date, three types of NOS have been discovered: the constitutively expressed endothelial NOS, neuronal NOS, and the inducible NOS (iNOS). Research in the past decade demonstrates that NO possesses a diverse array of physiologic functions, such as muscle relaxation, neuronal activity, and immune modulations.[2] As a free

Address for correspondence: Bin Liu, M.D., Ph.D., F1-01, NIEHS, P.O. Box 12233, Research Triangle Park, NC 27710, U.S.A. Voice: 919-541-4024; fax: 919-541-0841. liu3@niehs.nih.gov radical, NO is one of the major contributors to the formation of reactive nitrogen species. On one hand, NO serves as an antioxidant by scavenging oxygen free radicals so as to protect cells from reactive oxidative insults. On the other hand, NO directly reacts with proteins, especially heme-containing enzymes, such as guanylate cyclase, cytochrome P450, hemoglobin, and cyclooxygenase, to modulate their functions. In addition, NO can react with membrane lipids to induce lipid peroxidation. Indirectly, the combination of NO and superoxide can form highly reactive intermediates, such as peroxynitrite, that can induce DNA strand breaks, lipid peroxidation, and protein nitration.[3]

Inflammation in the brain primarily involves the participation of the two types of glial cells, microglia and astrocytes.[4,5] Under physiologic conditions, microglia, the resident immune cells in the brain, serve a role of immune surveillance. Astrocytes, on the other hand, act to maintain ionic homeostasis, buffer the action of neurotransmitters, and secrete nerve growth factors. However, glia, especially microglia, readily become activated in response to immunologic challenge and injury.[4] Activation of glia, a process termed reactive gliosis, has been observed during the pathogenesis of Parkinson's disease, Alzheimer's disease, multiple sclerosis, and AIDS dementia complex, as well as postneuronal death in cerebral stroke and traumatic brain injury.[6–12] Activated astrocytes secrete trophic factors in an attempt to enhance neuronal survival. Activated microglia produce a variety of proinflammatory and neurotoxic factors, including cytokines, such as tumor necrosis factor-α (TNFα) and interleukin-1β (IL-1β); fatty acid metabolites, such as eicosanoids; and free radicals, such as NO and superoxide.[13,14] NO and IL-1β are also produced by activated astrocytes. The production and release of these factors constitute a portion of the innate immunity that enables the host to destroy invading pathogens. However, excessive production and accumulation of these factors are deleterious to neurons, although the precise mechanism(s) and relative contribution of individual factors to neurodegeneration remain poorly understood.

Using primary mixed neuron-glia cultures, we have demonstrated that stimulation of the cultures with the bacterial endotoxin lipopolysaccharide (LPS) activates glia, especially microglia, to produce TNFα, IL-1β, superoxide, and NO, leading to eventual death of neurons. This *in vitro* model of inflammation-mediated neurodegeneration offers a useful system in which to study the contribution of NO to neurodegeneration.

RELATIONSHIP BETWEEN LPS-INDUCED PRODUCTION OF PROINFLAMMATORY AND CYTOTOXIC FACTORS AND NEURODEGENERATION

Reactive gliosis has traditionally been considered to occur as a consequence of neuronal injury associated with brain ischemia, trauma, and neurodegenerative diseases. However, increasing evidence has suggested that activation of glia, especially microglia, may play a much more active role than previously thought in the pathogenesis of neurodegeneration. For example, microinjection of LPS or LPS plus interferon γ (IFNγ) into the substantia nigra or hippocampus of rat brain induces the activation of microglia followed by degeneration of neurons in that region.[15–18] Thus, at least in these *in vivo* models, activation of microglia and the resulting production of proinflammatory and cytotoxic factors is sufficient to induce neurodegeneration.

Several lines of evidence support the notion of inflammation-mediated neurodegeneration. First, LPS itself is ineffective in killing neurons without the participation of glial cells. Treatment of primary neuron-enriched cultures from embryonic midbrain with 1–10 μg/ml LPS for three days did not exhibit a significant effect on dopaminergic neurons, as judged by uptake capacity for [^{3}H]dopamine (see FIGURE 1), in agreement with cell-counting results obtained with neuron-enriched cultures treated with 50 μg/ml LPS for two days.[19] In contrast, in mixed neuron-glia cultures, significant neurodegeneration was observed after treatment with 10–100 ng/ml LPS for two days.[20,21] Second, analysis of the temporal relationship between the generation of cytotoxic factors by activated glia and neurodegeneration demonstrates that glial activation precedes neurodegeneration.[17,20,22] In the rat brain, significant microglial activation was observed one day after a single intranigral injection of LPS whereas no apparent loss of DA neurons was observed until three days after the LPS injection.[17] In mesencephalic neuron-glia cultures, production of cytokines, such as TNFα and IL-1β, and free radicals, such as superoxide and NO, reached a maximum at 6–12 h after stimulation with LPS, whereas neuronal death was not observed until 48–72 h later (see FIGURE 2).[20–24,90] Third, proinflammatory and cytotoxic factors produced by glia, especially microglia, have been shown to induce neurodegeneration. However, it is increasingly more appreciated that these factors may work synergistically to damage neurons.[25–27] Fourth, inhibition of the

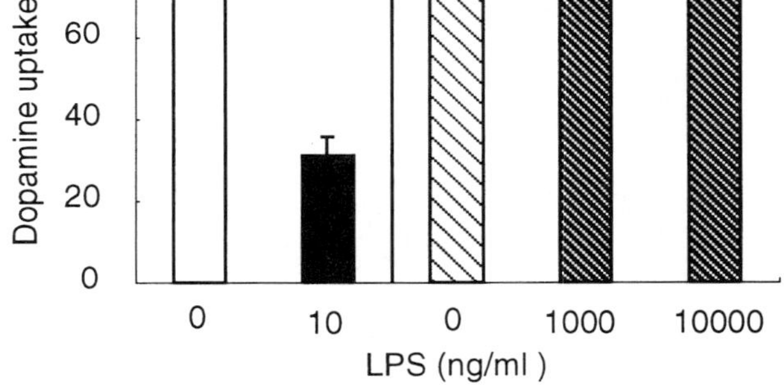

FIGURE 1. Differential sensitivity of neuron-glia and neuron-enriched cultures to LPS neurotoxicity. Primary rat mesencephalic neuron-glia or neuron-enriched cultures were prepared from embryonic day-14 Fisher 344 rat brain tissues as described[20,91] and treated for 72 h with the indicated concentrations of LPS. Degeneration of dopaminergic neurons was assessed by [^{3}H]dopamine uptake assays.[20,91]

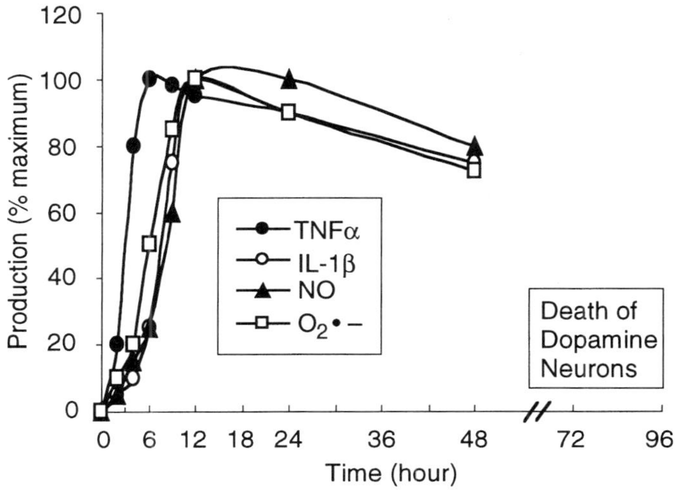

FIGURE 2. Temporal relationship between the generation of cytokines and free radicals and degeneration of neurons, based on results obtained with rat primary mesencephalic neuron-glia cultures treated with 10 ng/ml of LPS.[20,21,48]

production or neutralization of the reactivity of glia-produced factors affords neuroprotection. Because of the involvement of multiple glia-originated factors in neurodegeneration, inhibition of the production or neutralization of the activity of a single factor often affords only partial protection.[17,21,27–33,75,91]

RELATIVE CONTRIBUTION OF MICROGLIA AND ASTROCYTES TO LPS-INDUCED NO PRODUCTION

Although both astrocytes and microglia are involved in inflammation-mediated neurodegeneration, the repertoire and kinetics of proinflammatory and cytotoxic factors they produce in response to various immune stimuli may differ dramatically. In response to LPS stimulation, microglia appear to be significantly more responsive than astrocytes in the production of NO. Using the Greiss reagent to measure the levels of accumulated nitrite and nitrate, as an indicator of NO production,[35] primary rat microglia-enriched cultures produced significant quantities of NO as early as 6–9 h following treatment with concentrations of LPS as low as 0.1–1 ng/ml (see FIGURE 3A). In contrast, in rat astrocyte-enriched cultures, significant production of NO (above the assay detection limit of approximately 0.5 μM) was not detected except for the cultures treated for 48 h with 10 μg/ml LPS (FIG. 3B). These results are in agreement with those reported by others. McNaught and Jenner reported a 30–50% increase in NO production by primary rat astrocytes treated with 10–50 μg/ml LPS for up to three days,[36] although the possibility of microglia contamination was not excluded.[37] The inability of astrocytes to produce NO at the magnitude

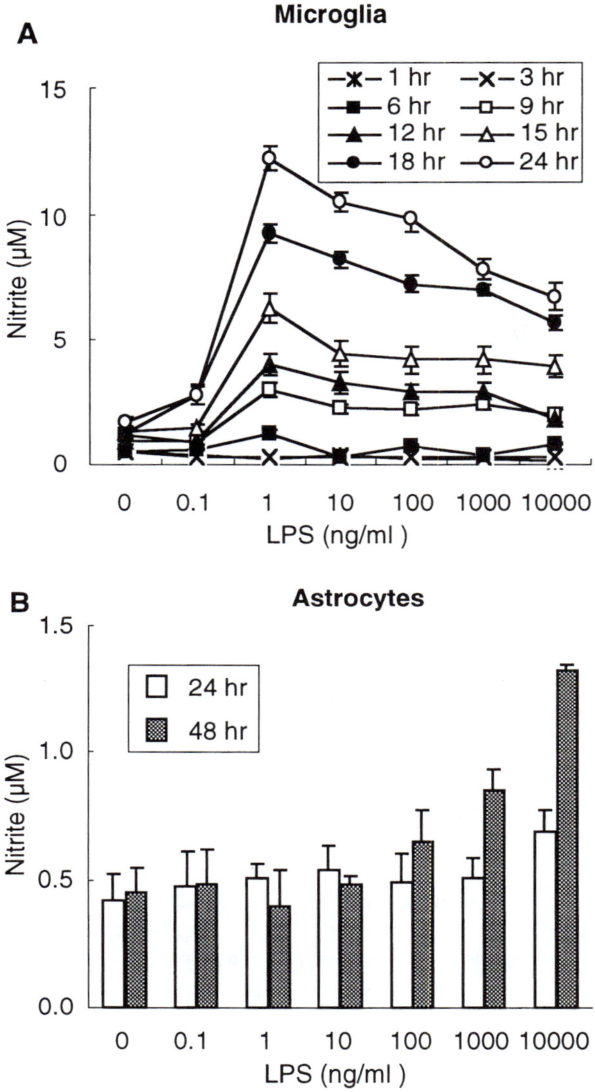

FIGURE 3. Comparison of LPS-stimulated nitric oxide production in microglia and astrocytes. Microglia- or astrocyte-enriched cultures were prepared from one-day-old pups of Fisher 344 rats as described.[91] Cultures were treated with the indicated concentrations of LPS and culture supernatants were removed at the indicated times for measurement of accumulated nitrite with the Griess reagent.[20]

observed for microglia may be due, at least in part, to the absence of iNOS upregulation following LPS stimulation.[38] Stronger stimulation of iNOS expression is achieved with the combination of LPS and interferon-γ.[39–41] Furthermore, the temporal relationship is equally important in understanding the relative contribution of iNOS activation in microglia and astrocytes to inflammation in the brain. Garcion and associates reported that intrahippocampal injection of LPS first induced the expression of iNOS in microglia at 15–24 h, followed by iNOS expression in astrocytes at 72 h.[42] It is, therefore, reasonable to speculate that microglia may be the primary source of NO in the initial stage of brain inflammation and that astrocytes join force with microglia in contributing to NO production at a later stage (FIG. 3). Understanding the temporal and spatial relationship of iNOS activation is important in gaining further insights into its contribution to brain inflammation and neurodegeneration.

CONTRIBUTION OF NO TO INFLAMMATION-RELATED NEURODEGENERATION

It is generally accepted that excessive production and accumulation of NO by activated glial cells contribute to neurodegeneration. However, it was not clear if NO alone was sufficient to cause the death of neurons. Studies using NO donors indicated that significant neuronal death was achieved often with high μM to low mM concentrations of NO donors,[43–65] compared to low μM concentrations of NO produced by cultures treated with agents such as LPS and the HIV goat protein gp120. On the other hand, inhibition of NO production by NOS inhibitors afforded significant neuroprotection.[20,21,26,47–49,50] These results suggest that NO, at pathophysiologic concentrations, is necessary but not sufficient to induce neurodegeneration. One explanation is that NO works in concert with other proinflammatory and cytotoxic factors to induce neurodegeneration because blockade of individual glia-originated factors only provides partial neuroprotection.[20,26,27,34,50,51] Another possibility is that NO reacts with oxygen free radicals, such as superoxide, to form more toxic intermediates, such as peroxynitrite.[52–55] Using an experimental ischemic model, Eliasson and colleagues reasoned that the conversion to peroxynitrite, but not NO itself, was associated with neuronal death.[56] Goureau and coworkers reported that partial inhibition of NO-mediated neurotoxicity was obtained with peroxynitrite scavengers and superoxide dismutase, again suggesting the involvement of oxygen free radicals in NO toxicity.[57]

INTERVENTION OF GLIAL NO PRODUCTION AND RELEVANCE TO NEUROPROTECTION

The fact that the glial iNOS-mediated NO generation involves transcription and translation of iNOS gene and enzymatic synthesis of NO offers multiple points for intervention (see FIGURE 4). Inhibitors of NOS such as *N*-nitro-L-arginine methyl ester (L-NAME) have been used to effectively inhibit NO production.[58–59] Using mixed murine cortical neuron-glia cultures treated with LPS/IFNγ, Jeohn and colleagues demonstrated that the rate of NO generation is linearly correlated with the

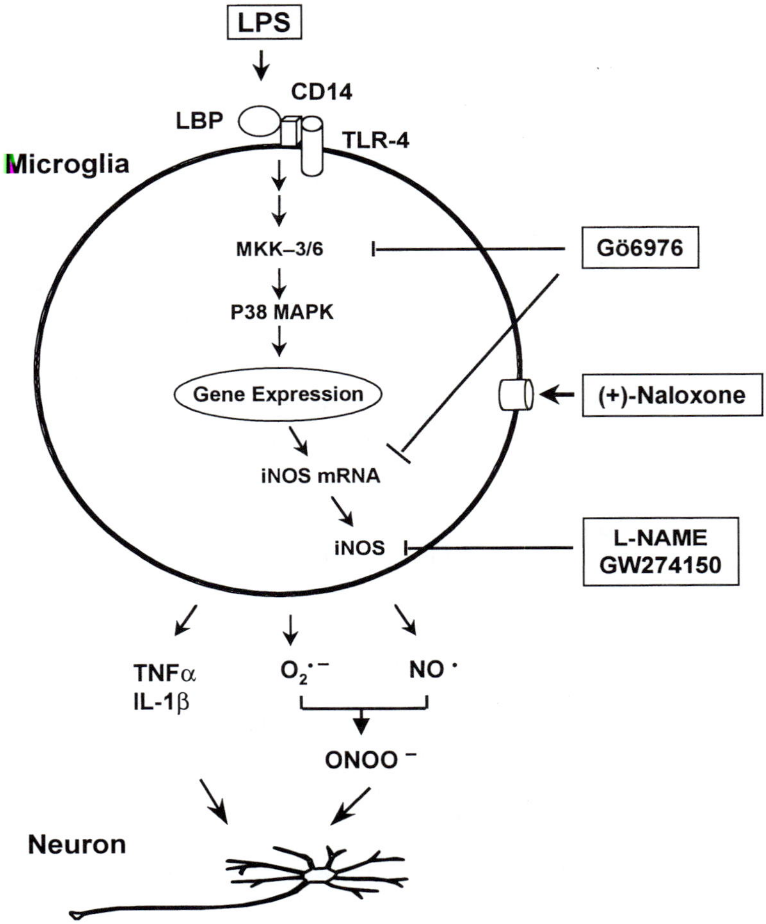

FIGURE 4. Points of intervention for inhibition of LPS-induced production of nitric oxide in microglia. LBP, LPS binding protein; TLR-4, Toll like receptor-4; MKK, MAP kinase kinase.

degree of neurodegeneration.[23] The initial stage of NO production following LPS/IFNγ treatment was the critical window for achieving neuroprotection since L-NAME was effective only if added during the first 12 hours. A recently identified selective inhibitor for iNOS, GW274150, was significantly more potent than L-NAME. The EC_{50} values for inhibition of LPS-induced NO production and neuroprotection by GW274150 were between 5–50 μM, compared to 1–2 mM for L-NAME.[49] An additional benefit of using GW274150 is the possibility to attribute the involvement of LPS neurotoxicity specifically to iNOS, not NOS in general.

In studying the transcriptional and translational regulation of iNOS gene, Jehon and coworkers found, by surprise, that an indolocarbazol compound, Gö6976, which had previously been considered a selective protein kinase C inhibitor, posttranscriptionally downregulated the expression of iNOS through destabilization of its mRNA.[48] Further analysis revealed that Gö6976 was a very effective neuroprotective agent in that it nearly completely prevented LPS/IFNγ from damaging neurons in a cortical neuron-glia culture system.[47] Inhibition of NO production appeared to be only partially responsible for its neuroprotective efficacy since NO production was never completely abolished by Gö6976,[47] suggesting the existence of additional targets. Indeed, subsequent studies discovered that Gö6976 prevented the LPS-induced activation of p38 mitogen activated protein kinase (MAPK), a member of the stress activated kinase superfamily.[60] Since p38 MAPK is involved in the induction of genes for both TNFα and iNOS,[61,62] inhibition of p38 MAPK activation by Gö6976, therefore, results in reduced production of both TNFα and NO.[63] Mechanistically, Gö6976 most likely inhibits p38 MAPK phosphorylation by interfering with the activity of a critical upstream kinase yet to be identified (FIG. 4).

OPIOIDS, BRAIN INFLAMMATION, AND NEURODEGENERATION

Enzyme inhibitors, such as those of iNOS, are certainly valuable tools for studying the mechanism of action for inflammation-mediated neurotoxicity. However, side effects and toxicity often dampen the enthusiasm for their eventual development into agents of therapeutic value. Analysis of the modulation of the inflammatory process by physiologically relevant agents, in contrast, may prove to be a critical step in the search for clinically useful remedies. One example is the study of opioids and related compounds in the regulation of brain inflammatory process and neuroprotection. Endogenous opioid peptides represent a family of peptides that include dynorphins, enkephalins, and β-endorphins. They are widely distributed throughout the central nervous system, as well as in peripheral tissues, such as cardiac myocytes and lymphatic cells.[64–66] Through interaction with distinct G-protein coupled transmembrane opioid receptors (delta, kappa, and mu), these peptides regulate a wide range of biological activities, including respiration, immune responses, and ion channel activity.[67–69] In 1995, Das and colleagues discovered that enkephalins were capable of reducing the LPS-induced IL-1β secretion from microglia.[70] Subsequently, another group of endogenous opioid peptides, dynorphins, were found to inhibit LPS-induced NO production and to afford partial protection against LPS-induced neurotoxicity.[32,71,91] A rather unexpected finding from these studies was the discovery that naloxone, a synthetic and nonselective opioid receptor antagonist, was capable of inhibiting LPS-induced microglial activation and production of nitric oxide and TNFα.[20,50,70,71] Most surprising of all was that, (+)-naloxone, which is ineffective in blocking the stereospecific binding of opioids to their receptors,[72,73] was as effective as the opioid receptor antagonist (−)-naloxone in inhibiting microglial activation and affording neuroprotection.[17,20,50] This result suggests that the effect of naloxone isomers on microglial activation does not directly involve the opioid receptor mechanism. Analysis of the mechanism of action for naloxone isomers demonstrated that inhibition of microglia superoxide generation best correlated with their effectiveness in neuroprotection.[20,74] Moreover, naloxone may have prevented the formation of peroxynitrite by inhibiting the production of both superoxide and NO. In a rodent model of inflammation-mediated Parkinsonism,[15,17,18] damage to dopaminergic neurons in the substantia nigra induced by intracerebrally injected LPS was reduced by systemically infused (–)-naloxone[17,18] or (+)-naloxone.[17] Naloxone, specifically the (+)-naloxone isomer, which does not bind conventional opioid receptors, may be a promising prototype for drugs with potential therapeutic value in the prevention and treatment of inflammation-mediated neurodegenerative disease such as Parkinson's disease.

During the past twenty years, naloxone has been tested for beneficial effects in a variety of experimental disease models and patients of septic shock,[76–78] injuries to spinal cord or brain,[79,80] myocardiac and cerebral stroke,[81–85] and Alzheimer's disease.[86] Although the use of naloxone is more encouraging in some disease conditions, such as ischemia,[81–85] conflicting results have been obtained in others. For example, interest in the beneficial effect during the treatment of patients with Alzheimer's disease was high in the early 1980s.[86–89] A significant portion of the negative results appear to have come from trials with patients in an advanced stage of the disease.[92–93] In younger subjects, the use of naloxone, especially at higher doses, seemed to have a beneficial effect on their cognitive ability.[94] The increasing realization of the involvement of inflammation in the pathogenesis of neurodegenerative diseases, including Alzheimer's disease, and the recent findings on the inflammation-inhibitory effect of naloxone may revitalize our interest in examining the therapeutic value of naloxone. The bridging of inflammation and neurodegeneration may pave a new path in the search for clinically useful agents for these diseases. In fact, common to at least some of these pathologic conditions, for which naloxone has been evaluated, is the involvement of the inflammatory process at certain stage(s). For example, activation of peripheral immune cells and production of cytotoxic factors, such as TNFα and superoxide free radical, is certainly part of the manifestation of septic shock[95] and antiinflammatory therapy has been proposed as part of the treatment regimen.[96,97] Since naloxone is known to inhibit the production of superoxide by activated neutrophils and macrophages,[98,99] a large component of the beneficial effect of naloxone in the experimental treatment of septic shock may be its inhibitory effect on immune cell activation. Similarly, since inflammation contributes to secondary cellular damage in ischemic stroke, the suppressive effect of naloxone on inflammation may at least partially explain the protective effect of naloxone.[100,101]

Since (+)-naloxone does not bind to opioid receptors, there may exist a cell surface receptor for (+)-naloxone to which (–)-naloxone is also capable of binding. The association of naloxone to its non-opioid receptor may negatively affect the activation of microglia and, hence, inhibits the production of proinflammatory and cytotoxic factors, such as the superoxide free radical (FIG. 4). Further characterization of the putative non-opioid receptor for (+)-naloxone should advance our quest for the precise mechanism of action for this potentially promising compound.

In summary, excessive production and accumulation of nitric oxide is deleterious to neurons in the inflammation-mediated neurodegenerative process. Agents that are capable of blocking the upregulation of iNOS and production of NO afford neuroprotection. Because of its established low toxicity, naloxone, especially the

(+)-enantiomer, warrants further study as a potential therapeutic agent for the treatment of neurodegenerative and other inflammation-mediated diseases.

REFERENCES

1. BREDT, D.S. 1999. Endogenous nitric oxide synthesis: biological functions and pathophysiology. Free Radic. Res. **31:** 577–596.
2. GRISHAM, M.B., D. JOURD'HEUIL & D.A. WINK. 1999. Nitric oxide. I. Physiological chemistry of nitric oxide and its metabolites:implications in inflammation. Am. J. Physiol. **276:** G315–321.
3. BECKMAN, J.S. 1996. Oxidative damage and tyrosine nitration from peroxynitrite. Chem. Res. Toxicol. **9:** 836–844.
4. KREUTZBERG, G.W. 1996. Microglia: a sensor for pathological events in the CNS. Trend. Neurosci. **19:** 312–318.
5. ALOISI, F. 1999. The role of microglia and astrocytes in CNS immune surveillance and immunopathology. Adv. Exp. Med. Biol. **468:** 123–133.
6. MCGEER, P.L., S. ITAGAKI, B.E. BOYES & E.G. MCGEER. 1988. Reactive microglia are positive for HLA-DR in the substantia nigra of Parkinson's and Alzheimer's disease brains. Neurology **38:** 1285–1291.
7. DICKSON, D.W., S.C. LEE, L.A. MATTIACE, *et al.* 1993. Microglia and cytokines in neurological disease, with special reference to AIDS and Alzheimer's disease. Glia **7:** 75–83.
8. RAINE, C.S. 1994. Multiple sclerosis: immune system molecule expression in the central nervous system. J. Neuropathol. Exp. Neurol. **53:** 328–337.
9. O'BANION, M.K. & C.E. FINCH. 1996. Inflammatory mechanisms and anti-inflammatory therapy in Alzheimer's disease. Neurobiol. Aging. **17:** 669–671.
10. HAUSS-WEGRZYNIAK, B., P. DOBRZANSKI, J.D. STOEHR & G.L. WENK. 1998. Chronic neuroinflammation in rats reproduces components of the neurobiology of Alzheimer's disease. Brain Res. **780:** 294–303.
11. GONZALEZ-SCARANO, F. & G. BALTUCH. 1999. Microglia as mediators of inflammatory and degenerative diseases. Annu. Rev. Neurosci. **22:** 219–240.
12. GIULIAN, D. 1999. Microglia and the immune pathology of Alzheimer disease. Am. J. Hum. Genet. **65:** 13–18.
13. BANATI, R.B., J. GEHRMANN, P. SCHUBERT & G.W. KREUTZBERG. 1993. Cytotoxicity of microglia. Glia **7:** 111–118.
14. MINGHETTI, L. & G. LEVI. 1998. Microglia as effector cells in brain damage and repair: focus on prostanoids and nitric oxide. Prog. Neurobiol. **54:** 99–125.
15. CASTANO, A., A.J. HERRERA, J. CANO & A. MACHADO. 1998. Lipopolysaccharide intranigral injection induces inflammatory reaction and damage in nigrastriatal dopaminergic system. J. Neurochem. **70:** 1584–1592.
16. MATSUOKA, Y., Y. KITAMURA, H. TAKAHASHI, *et al.* 1999. Interferon-gamma plus lipopolysaccharide induction of delayed neuronal apoptosis in rat hippocampus. Neurochem. Int. **34:** 91–99.
17. LIU, B., J. W. JIANG, B. C. WILSON, L. *et al.* 2000. Systemic infusion of naloxone reduces degeneration of rat substantia nigral dopaminergic neurons induced by intranigral injection of lipopolysaccharide. J. Pharmacol. Exp. Ther. **295:** 125–132.
18. LU, X., G. BING & T. HAGG. 2000. Naloxone prevents microglia-induced degeneration of dopaminergic substantia nigra neurons in adult rats. Neuroscience **97:** 285–291.
19. BRONSTEIN, D.M., I. PEREZ-OTANO, V. SUN, *et al.* 1995. Glia-dependent neurotoxicity and neuroprotection in mesencephalic cultures. Brain Res. **704:** 112–116.
20. LIU, B., L. DU & J.-S. HONG. 2000. Naloxone protects rat dopaminergic neurons against inflammatory damage through inhibition of microglia activation and superoxide generation. J. Pharmacol. Exp. Ther. **293:** 607–617.
21. KIM, W.G., R.P. MOHNEY, B. WILSON, *et al.* 2000. Regional difference in susceptibility to lipopolysaccharide-induced neurotoxicity in the rat brain: role of microglia. J. Neurosci. **20:** 6309–6316.

22. SZCZEPANIK, A.M., R.J. FISHKIN, D.K. RUSH & C.A. WILMOT. 1996. Effects of chronic intrahippocampal infusion of lipopolysaccharide in the rat. Neuroscience **70:** 57–65.
23. JEOHN, G.-H., W.G. KIM & J.S. HONG. 2000. Time dependency of the action of nitric oxide in lipopolysaccharide-interferon-gamma-induced neuronal cell death in murine primary neuron-glia co-cultures. Brain Res. **880:** 173–177.
24. CHANG, R.C., W. CHEN, P. HUDSON, *et al.* 2001. Neurons reduce glial responses to lipopolysaccharide (LPS) and prevent injury of microglial cells from over-activation by LPS. J. Neurochem. **76:** 1042–1049.
25. SKAPER, S.D., L. FACCI, S. ROMANELLO & A. LEON. 1996. Mast cell activation causes delayed neurodegeneration in mixed hippocampal cultures via the nitric oxide pathway. J. Neurochem. **66:** 1157–1166.
26. JEOHN, G.-H., L.-Y. KONG, B. WILSON, *et al.* 1998. Synergistic neurotoxic effects of combined treatments with cytokines in murine primary mixed neuron/glia cultures. J. Neuroimmunol. **85:** 1–10.
27. CHAO, C.C., S. HU, L. EHRLICH & P.K. PETERSON. 1995. Interleukin-1 and tumor necrosis factor-alpha synergistically mediate neurotoxicity: involvement of nitric oxide and of *N*-methyl-D-aspartate receptors. Brain Behav. Immun. **9:** 355–365.
28. KLOCKER, N., A. CELLERINO & M. BAHR. 1998. Free radical scavenging and inhibition of nitric oxide synthase potentiates the neurotrophic effects of brain-derived neurotrophic factor on axotomized retinal ganglion cells *in vivo.* J. Neurosci. **18:** 1038–1046.
29. KONG, L.Y., J.L. MADERDRUT, G.H. JEOHN & J.S. HONG. 1999. Reduction of lipopolysaccharide-induced neurotoxicity in mixed cortical neuron/glia cultures by femtomolar concentrations of pituitary adenylate cyclase-activating polypeptide. Neuroscience **91:** 493–500.
30. ALLAN, S.M., L.C., PARKER, B. COLLINS, *et al.* 2000. Cortical cell death induced by IL-1 is mediated via actions in the hypothalamus of the rat. Proc. Natl. Acad. Sci. U.S.A. **97:** 5580–5585.
31. ESTEVEZ, A.G., J.B. SAMPSON, Y.X. ZHUANG, *et al.* 2000. Liposome-delivered superoxide dismutase prevents nitric oxide-dependent motor neuron death induced by trophic factor withdrawal. Free Radic. Biol. Med. **28:** 437–446.
32. KONG, L.-Y., G.H. JEOHN, P.M. HUDSON, *et al.* 2000. Reduction of lipopolysaccharide-induced neurotoxicity in mouse mixed cortical neuron/glia cultures by ultralow concentrations of dynorphins. J Biomed. Sci. **7:** 241–247.
33. TERRADO, J., D. MONNIER, D. PERRELET, *et al.* 2000. Soluble TNF receptors partially protect injured motoneurons in the postnatal CNS. Eur. J. Neurosci. **12:** 3443–3447.
34. TEZEL, G. & M.B. WAX. 2000. Increased production of tumor necrosis factor-alpha by glial cells exposed to simulated ischemia or elevated hydrostatic pressure induces apoptosis in cocultured retinal ganglion cells. J. Neurosci. **20:** 8693–8700.
35. GREEN, L.C., D.A. WAGNER, J. GLOGOWSKI, *et al.* 1982. Analysis of nitrate, nitrite, and $[^{15}N]$nitrate in biological fluids. Anal. Biochem. **126:** 131–138.
36. McNaught, K. S. & P. Jenner. 2000. Extracellular accumulation of nitric oxide, hydrogen peroxide, and glutamate in astrocytic cultures following glutathione depletion, complex I inhibition, and/or lipopolysaccharide-induced activation. Biochem. Pharmacol. **60:** 979–988.
37. CHAO, C.C., S. HU & P.K. PETERSON. 1996. Glia: the not so innocent bystanders. J. Neurovirol. **2:** 234–239.
38. POSSEL, H., H. NOACK, J. PUTZKE, *et al.* 2000. Selective upregulation of inducible nitric oxide synthase (iNOS) by lipopolysaccharide (LPS) and cytokines in microglia: *in vitro* and *in vivo* studies. Glia **32:** 51–59.
39. HENEKA, M.T., L. DUMITRESCU, P.A. LOSCHMANN, *et al.* 2000. Temporal, regional, and cell-specific changes of iNOS expression after intrastriatal microinjection of interferon gamma and bacterial lipopolysaccharide. Chem. Neuroanat. **18:** 167–179.
40. CALABRESE, V., A. COPANI, D. TESTA, *et al.* 2000. Nitric oxide synthase induction in astroglial cell cultures: effect on heat shock protein 70 synthesis and oxidant/antioxidant balance. J. Neurosci. Res. **60:** 613–622.
41. NOMURA, Y. 2001. NF-kappaB activation and IkappaB alpha dynamism involved in iNOS and chemokine induction in astroglial cells. Life Sci. **68:** 1695–1701.

42. GARCION, E., L. SINDJI, C. MONTERO-MENEI, *et al.* 1998. Expression of inducible nitric oxide synthase during rat brain inflammation: regulation by 1,25-dihydroxyvitamin D3. Glia **22:** 282–294.
43. BONFOCO, E., B. ZHIVOTOVSKY, A.D. ROSSI, *et al.* 1996. BCL-2 delay apoptosis and PARP cleavage induced by NO donors in GT1-7 cells. Neuroreport **8:** 273–276.
44. GHATAN, S., S. LARNER, Y. KINOSHITA, *et al.* 2000. p38 MAP kinase mediates bax translocation in nitric oxide-induced apoptosis in neurons. J. Cell Biol. **150:** 335–347.
45. CANALS, S., M.J. CASAREJOS, E. RODRIGUEZ-MARTIN, *et al.* 2001. Neurotrophic and neurotoxic effects of nitric oxide on fetal midbrain cultures. J. Neurochem. **76:** 56–68.
46. WEI, T., C. CHEN, J. HOU, *et al.* 2000. Nitric oxide induces oxidative stress and apoptosis in neuronal cells. Biochim. Biophys. Acta **1498:** 72–79.
47. JEOHN, G.-H., B. WILSON, W.C. WETSEL & J.S. HONG. 2000. The indolocarbazole Go6976 protects neurons from lipopolysaccharide/interferon-gamma-induced cytotoxicity in murine neuron/glia co-cultures. Brain Res. Mol. Brain Res. **79:** 32–44.
48. JEOHN, G.H., R.C. CHANG, W.G. KIM, B. *et al.* 2000. Post-transcriptional inhibition of lipopolysaccharide-induced expression of inducible nitric oxide synthase by Go6976 in murine microglia. Brain Res. Mol. Brain Res. **79:** 18–31.
49. JEOHN, G.H., C.L. COOPER, B. WILSON, *et al.* 2002. p38 MAP kinase is involved in lipopolysaccharide-induced dopaminergic neuronal cell death in rat mesencephalic neuron-glia cultures. Ann. N.Y. Acad. Sci. **962:** this volume.
50. LIU, B., L. DU, L.Y. KONG, *et al.* 2000. Reduction by naloxone of lipopolysaccharide-induced neurotoxicity in mouse cortical neuron-glia co-cultures. Neuroscience **97:** 749–756.
51. HU, S., P.K. PETERSON & C.C. CHAO. 1997. Cytokine-mediated neuronal apoptosis. Neurochem. Int. **30:** 427–431.
52. LIPTON, S.A., Y.B. CHOI, Z.H. PAN, *et al.* 1993. A redox-based mechanism for the neuroprotective and neurodestructive effects of nitric oxide and related nitroso-compounds. Nature **364:** 626–632.
53. DAWSON, V.L. 1995. Nitric oxide: role in neurotoxicity. Clin. Exp. Pharmacol. Physiol. **22:** 305–308.
54. ESTEVEZ, A.G., N. SPEAR, S.M. MANUEL, *et al.* 1998. Nitric oxide and superoxide contribute to motor neuron apoptosis induced by trophic factor deprivation. J. Neurosci. **18:** 923–931.
55. GOW, A.J., Q. CHEN, M. GOLE, *et al.* 2000. Two distinct mechanisms of nitric oxide-mediated neuronal cell death show thiol dependency. Am. J. Physiol. Cell. Physiol. **278:** C1099–1107.
56. ELIASSON, M.J., Z. HUANG, R.J. FERRANTE, *et al.* 1999. Neuronal nitric oxide synthase activation and peroxynitrite formation in ischemic stroke linked to neural damage. J. Neurosci. **19:** 5910–5918.
57. GOUREAU, O., F. REGNIER-RICARD & Y. COURTOIS. 1999. Requirement for nitric oxide in retinal neuronal cell death induced by activated Muller glial cells. J. Neurochem. **72:** 2506–2515.
58. DAWSON, V.L., T.M. DAWSON, E.D. LONDON, *et al.* 1991. Nitric oxide mediates glutamate neurotoxicity in primary cortical cultures. Proc. Natl. Acad. Sci. U.S.A. **88:** 6368–6371.
59. KUBES, P., M. SUZUKI & D.N. GRANGER. 1991. Nitric oxide: an endogenous modulator of leukocyte adhesion. Proc. Natl. Acad. Sci. U.S.A. **88:** 4651–4655.
60. SCHAEFFER, H.J. & M.J. WEBER. 1999. Mitogen-activated protein kinases: specific messages from ubiquitous messengers. Mol. Cell Biol. **19:** 2435–2444.
61. DA SILVA, J., B. PIERRAT, J.L. MARY & W. LESSLAUER. 1997. Blockade of p38 mitogen-activated protein kinase pathway inhibits inducible nitric-oxide synthase expression in mouse astrocytes. J. Biol. Chem. **272:** 28373–28380.
62. BHAT, N.R., P. ZHANG, J.C. LEE & E.L. HOGAN. 1998. Extracellular signal-regulated kinase and p38 subgroups of mitogen- activated protein kinases regulate inducible nitric oxide synthase and tumor necrosis factor-alpha gene expression in endotoxin-stimulated primary glial cultures. J. Neurosci. **18:** 1633–1641.

63. JEOHN, G.H., C.L. COOPER, K.-J. JANG, *et al.* 2002. Gö6976 inhibits LPS-induced TNFa release by suppressing p38 MAP kinase activation. Neoroscience. In press.
64. SMITH, A.P. & N.M. LEE. 1988. Pharmacology of dynorphin. Annu. Rev. Pharmacol. Toxicol. **28:** 123–140.
65. HERZ, A. 1993. Opioids, Vol. 1. Springer-Verlag, Berlin.
66. BARRON, B.A. 2000. Cardiac Opioids. Proc. Soc. Exp. Biol. Med. **224:** 1–7.
67. MINAMI, M. & M. SATOH. 1995. Molecular biology of the opioid receptors: structures, functions and distributions. Neurosci. Res. **23:** 121–145.
68. ROY, S. & H. LOH. 1996. Effects of opioids on the immune system. Neurochem. Res. **21:** 1375–1386.
69. JORDAN, B.A., S. CVEJIC & L.A. DEVI. 2000. Opioids and their complicated receptor complexes. Neuropsychopharmacol. **23:** S5–S18.
70. DAS, K.P., M.K. MCMILLIAN, G. BING & J.S. HONG. 1995. Modulatory effects of [Met^5]-enkephalin on interleukin-1 beta secretion from microglia in mixed brain cell cultures. J. Neuroimmunol. **62:** 9–17.
71. KONG, L.-Y., M.K. MCMILLIAN, P.M. HUDSON, *et al.* 1997. Inhibition of lipopolysaccharide-induced nitric oxide and cytokine production by ultralow concentrations of dynorphins in mixed glia cultures. J. Pharmacol. Exp. Ther. **280:** 61–66.
72. IJIMA, I., J.-I. MINAMIKAWA, A.E. JACOBSEN, *et al.* 1978. Studies in the (+)-morphinin series-5. Synthesis and biological properties of (+)-naloxone. J. Med. Chem. **21:** 398–400.
73. MARCOLI, M., G. RICEVUTI, A. MAZZONE, *et al.* 1989. A stereoselective blockade by naloxone of opioid and non-opioid-induced granulocyte activation. Int. J. Immunopharmacol. **11:** 57–61.
74. CHANG, R.C., C. ROTA, R.E. GLOVER, *et al.* 2000. A novel effect of an opioid receptor antagonist, naloxone, on the production of reactive oxygen species by microglia: a study by electron paramagnetic resonance spectroscopy. Brain Res. **854:** 224–229.
75. NEW, D.R., S.B. MAGGIRWAR, L.G. EPSTEIN, *et al.* 1998. HIV-1 Tat induces neuronal death via tumor necrosis factor-alpha and activation of non-*N*-methyl-D-aspartate receptors by a NFkappaB-independent mechanism. J. Biol. Chem. **273:** 17852–17858.
76. HOLADAY, J.W. & A.I. FADEN. 1978. Naloxone reversal of endotoxin hypotension suggests role of endorphins in shock. Nature **275:** 450–451.
77. HOLADAY, J.W. & D.S. MACOLM. 1986. Endogenous opioids and their antagonist in endotoxic shock. *In* 9th International Congress of Infectious and Parasitic Diseases: Bacterial Infection, Antibacterial Chemotherapy. W. Marget, W. Lang & E. Gabler-Sanberger, Eds.: 425–435. MMV Medzin, Munich.
78. NAPOLITANO, L.M. 2000. Naloxone therapy in shock: the controversy continues. Crit. Care Med. **28:** 887–888.
79. SEIDL, E.C. 2000. Promising pharmacological agents in the management of acute spinal cord injury. Pharm. Pract. Manag. Q. **20:** 21–27.
80. WHITELAW, A. 2000. Systematic review of therapy after hypoxic-ischaemic brain injury in the perinatal period. Semin. Neonatol. **5:** 33–40.
81. HOSOBUCHI, T., D.S. BASKIN & S.K. WOO. 1982. Reversal of induced ischemic neurologic deficit in gerbil by the opiate antagonist naloxone. Science **215:** 69–71.
82. FALLIS, R.J., M. FISHER & R.A. LOBO. 1983. A double blind trial of naloxone in the treatment of stroke. Stroke **15:** 627–629.
83. FADEN, I.A. & S. SALZMAN. 1992. Pharmacological strategies in CNS trauma. Trend. Pharmacol. Sci. **13:** 29–35.
84. KAN, M.N., Y.T. CHEN & A.Y.S. LEE. 1992. Naloxone reversal of ischemic arrhythmia is stereospecific and suggesting role of endogenous opioid peptides in ischemic heart disease. Proc. Soc. Expt. Biol. Med. **200:** 518–521.
85. CHEN, C.J., F.C. CHENG, S.L. LIAO, *et al.* 2000. Effects of naloxone on lactate, pyruvate metabolism and antioxidant enzyme activity in rat cerebral ischemia/reperfusion. Neurosci. Lett. **287:** 113–116.
86. TARIOT, P.N., A. UPADHYAYA, T. SUNDERLAND, *et al.* 1999. Physiologic and neuroendocrine responses to intravenous naloxone in subjects with Alzheimer's disease and age-matchedcontrols. Biol. Psychiatry **46:** 412–419.

87. REISBERG, B., S.H. FERRIS, R. ANAND, *et al.* 1983. Effects of naloxone in senile dementia: a double-blind trial. N. Engl. J. Med. **308:** 721–722.
88. RATHMANN, K.L. & C.S. CONNER. 1984. Alzheimer's disease: clinical features, pathogenesis, and treatment. Drug Intell. Clin. Pharm. **18:** 684–691.
89. TARIOT, P.N., T. SUNDERLAND, H. WEINGARTNER, *et al.* 1985. Low- and high-dose naloxone in dementia of the Alzheimer type. Psychopharmacol. Bull. **21:** 680–682.
90. LIU, B., K. WANG, H.M. GAO, *et al.* 2001. Molecular consequences of activated microglia in the brain: overactivation induces apoptosis. J. Neurochem. **77:** 182–189.
91. LIU, B., L. QIN, S.-N. YANG, *et al.* 2001. Femtomolar concentrations of dynorphins protect rat mesencephalic dopaminergic neurons against inflammatory damage. J. Pharmacol. Exp. Ther. **298:** 1133–1141.
92. SERBY, M., R. RESNICK, B. JORDAN, *et al.* 1986. Naltrexone and Alzheimer's disease. Prog. Neuropsychopharmacol. Biol. Psychiatry. **10:** 587–590.
93. HENDERSON, V.W., E. ROBERTS, C. WIMER, *et al.* 1989. Multicenter trial of naloxone in Alzheimer's disease. Ann. Neurol. **25:** 404–406.
94. TARIOT, P.N., M. GROSS, T. SUNDERLAND, *et al.* 1988. High-dose naloxone in older normal subjects: implications for Alzheimer's disease. J. Am. Geriatr. Soc. **36:** 681–686.
95. BREEN, G. & A.R. TUNKEL. 1999. Adjunctive therapies for sepsis and septic shock. Curr. Infect. Dis. Rep. **1:** 224–229.
96. FREEMAN, B.D. & C. NATANSON. 2000. Anti-inflammatory therapies in sepsis and septic shock. Expert Opin. Investig. Drugs. **9:** 1651–1663.
97. HOWE, L,M. 2000. Novel agents in the therapy of endotoxic shock. Expert Opin. Investig. Drugs. **9:** 1363–1372.
98. SIMPKINS, C.O., N. IVES, E, TATE & M. JOHNSON. 1985. Naloxone inhibits superoxide release from human neutrophils. Life Sci. **37:** 1381–1386.
99. SIMPKINS, C.O., S.T. ALAILIMA & E.A. TATE. 1986. Inhibition by naloxone of neutrophil superoxide release: a potentially useful antiinflammatory effect. Circ. Shock. **20:** 181–191.
100. LYSLE, D.T. & T. HOW. 1999. Endogenous opioids regulate the expression of inducible nitric oxide synthase by splenocytes. J. Pharmacol. Exp. Ther. **288:** 502.
101. SCHULTZ, J.E. & G.J. GROSS. 2001. Opioids and cardioprotection. Pharmacol. Ther. **89:** 123–237.

p38 MAP Kinase Is Involved in Lipopolysaccharide-Induced Dopaminergic Neuronal Cell Death in Rat Mesencephalic Neuron-Glia Cultures

GWANG-HO JEOHN,[a] CYNTHIA L. COOPER,[b] BELINDA WILSON,[a] RAYMOND C.C. CHANG,[a] KYUNG-JIN JANG,[c] HYOUNG-CHUN KIM,[b] BIN LIU[a] AND JAU-SHYONG HONG[a]

[a]*Neuropharmacology Section, Laboratory of Pharmacology and Chemistry, National Institute of Environmental Health Sciences, National Institute of Health, Research Triangle Park, North Carolina, U.S.A.*

[b]*Division of Science, Truman State University, Kirksville, Missouri, U.S.A.*

[c]*Department of Pharmacy, Kangwon National University, Chuncheon, 200-701, Korea*

ABSTRACT: Immune stimulants, such as the bacterial endotoxin, lipopolysaccharide (LPS), the human immunodeficiency virus-1 coat protein gp120, or β-amyloid peptides, lead to glial activation and production of various immune mediators, such as nitric oxide (NO) and proinflammatory cytokines in the brain. These mediators appear to contribute to neuronal cell death in neurodegenerative diseases. However, the signaling pathways, which mediate the neurotoxic effect by the endotoxin, are not understood. The purpose of this study was to determine the role of mitogen-activated protein kinase (MAPK) in LPS-induced neurodegeneration using mesencephalic dopaminergic neuron/glia cultures. We have found that the p38 MAPK is important in LPS-induced death of mesencephalic neurons in rat neuron-glia mixed cultures. Upon treatment with 10 ng/ml LPS, the number of dopaminergic neurons decreased by 30% within 48 h, preceded by a significant production of NO by glia. Neuroprotection by selective inhibition of p38 MAPK activity paralleled a decrease in LPS-induced inducible nitric oxide synthase (iNOS) expression. These events were significantly reduced by the selective p38 MAPK inhibitor, SB202190, but not by the inactive analogue SB202474. Inhibition of iNOS activity and NO production by treatment with GW274150 was also neuroprotective. Although the p38 MAPK inhibitor afforded significant neuroprotection from LPS toxicity in the neuron-glia mixed culture, it failed to protect dopaminergic neurons from 6-hydroxy-dopamine-induced toxicity, which acts directly on dopaminergic neurons by inducing hydroxyl radical formation from the mitochondria. The results suggest that p38 MAPK in glia plays a significant role in the LPS-induced death of mesencephalic neurons through induction of nitric oxide synthase and resulting NO production.

KEYWORDS: neurotoxicity; glia; lipopolysaccharide; endotoxin; nitric oxide; p38 MAP kinase; SB202190; GW274150; iNOS

Address for correspondence: J.S. Hong, Neuropharmacology Section, LPC, National Institute of Environmental Health Sciences, National Institutes of Health, P.O. Box 12233, Research Triangle Park, NC 27709, U.S.A. Voice: 919-541-2358; fax: 919-541-0841.
hong3@niehs.nih.gov

ABBREVIATIONS:

ABC	avidin-biotin-peroxidase complex.
DAB	3,3′-diaminobenzidine.
GW274150	(*S*)-2-amino-(1-iminoethylamino)-5-thioheptanoic acid.
iNOS	inducible nitric oxide synthase.
LPS	lipopolysaccharide.
MAPK	mitogen-activated protein kinase.
NO	nitric oxide.
6-OHDA	6-hydroxy-dopamine.
PBS	phosphate buffered saline.
PCR	polymerase chain reaction.

INTRODUCTION

Microglia and astrocytes play important roles in host defense against immunologic challenges or environmental stress in the brain. Glia respond to toxic agents such as bacterial endotoxin (lipopolysaccharide, LPS), HIV-1 coat protein gp120, and β-amyloid-related peptides, by releasing nitric oxide (NO), reactive oxygen intermediates, tumor necrosis factor-α, interleukin-1, arachidonic acid metabolites, and quinolinic acid.[1–10] Glia activation is thought to lead directly to neurodegenerative processes through the release of proinflammatory and/or cytotoxic factors.[11–18] Glia activation and production of inflammatory factors appear to be involved in several types of neurologic disorders, including Parkinson's disease, Alzheimer's disease, multiple sclerosis, and AIDS dementia complex.[13,19–24]

Understanding the mechanisms underlying glia-mediated neuron death is a prerequisite to developing therapeutic interventions for immune-mediated neurodegenerative diseases. LPS has been widely used as a tool to trigger production of inflammatory factors and, consequently, neuronal cell death. LPS binds to a CD14 receptor in brain glia,[25] and appears to stimulate a Toll-like receptor 4.[26,27] Several types of protein kinases, including protein kinase C and protein tyrosine kinases, are then activated.[28–31] Many protein kinases, such as protein kinase C, protein tyrosine kinase, and mitogen-activated protein kinases (MAPKs, ERK, and p38 MAPK) have been implicated in the release of immune-related cytotoxic factors, such as NO and proinflammatory cytokines in brain glia.[6,32–37] However, it is not clear what types of protein kinases are important in glia-mediated or inflammation-related neuronal cell death. p38 MAPK may play a role in inflammation-mediated neurotoxicity, since it is activated after forebrain ischemia and also in the brains of Alzheimer's disease patients.[38,39] p38 MAPK is also activated by endotoxin in the central nervous system.[27] Furthermore, inhibition of p38 MAPK reduced production of inflammatory factors after exposure to LPS or combined cytokines in glia[32,40] and monocytes.[41,42] However, it is still not clear if p38 MAPK is involved in inflammation-mediated neuronal cell death.

We previously reported that the treatment of mesencephalic neuron-glia cocultures with LPS resulted in glia-mediated death of dopaminergic neurons.[43,44] Although 10 ng/ml LPS was sufficient to elicit neurotoxicity in mesencephalic neurons, 10 μg/ml of LPS failed to produce similar effects in hippocampal or cortical neuron-glia cultures. In the study we report here, we asked if the activation of p38 MAPK by LPS contributes to inflammation-mediated neuronal cell death in primary mesencephalic neuron-glia cultures. We found that the selective inhibition of p38 MAPK affords marked protection of neurons from LPS-elicited cytotoxicity by suppressing the endotoxin-induced release of NO in primary mesencephalic neuron-glia cultures.

MATERIALS AND METHODS

Materials

LPS (*E. coli* O111:B4) was purchased from List Biological Laboratories (Campbell, CA); poly-D-lysine and the TRI-reagent™ were bought from Sigma-Aldrich (St. Louis, MO). Tissue culture plates (24-well) were obtained from Costar (Cambridge, MA). Cell culture components and the Superscript™ preamplification system were obtained from Life Technologies (Gaithersburg, MD). SB202190 and SB202474 were purchased from Calbiochem-Novabiochem International (La Jolla, CA), and (*S*)-2-amino-(1-iminoethylamino)-5-thioheptanoic acid (GW274150) from Alexis Biochemicals (San Diego, CA). The Gene Amp polymerase chain reaction (PCR) kit with AmpliTaq DNA polymerase came from Perkin Elmer (Foster City, CA). Primers for inducible nitric oxide synthase (iNOS) and β-actin and the competitive DNA fragments for iNOS (MIMIC™) were from Clontech Laboratories (Palo Alto, CA). The polyclonal antityrosine hydroxylase (TH) antiserum was a gift from GlaxoSmithKline (Research Triangle Park, NC). The phospho-p38 and p38 rabbit polyclonal antibodies, HRP-conjugated antirabbit secondary antibody, Phototope-HRP Western detection kit, immobilized phospho-p38 MAPK monoclonal antibody, ATF-2 fusion protein, and the phospho-ATF-2 antibody were from New England Biolabs (Beverly, MA). Biotinylated secondary antibodies, the avidin-biotin-peroxidase complex (ABC) kit, and 3,3′-diaminobenzidine (DAB) came from Vector Laboratories (Burlingame, CA). All other reagents were of the purest grade available.

Culture and Stimulation of Primary Microglia and Neuronal/Glial Mixed Cultures

Primary microglia enriched cultures were prepared from brains of postnatal day-1 rats as described elsewhere by Liu *et al.*[45] Primary mixed neuron-glia cultures were prepared from brains of embryonic day 14/15 rats (Fischer 344) from Charles River Laboratories (Raleigh, NC) by modifying the culture methods we describe elsewhere.[43] For mixed neuron-glia, the ventral parts of midbrain were dissected, pooled and mechanically dissociated by mild trituration in ice-cold calcium and magnesium free W3 buffer (15 mM HEPES, pH 7.4, 145 mM NaCl, 5.4 mM KCl, 1 mM NaH_2PO_4, and 11 mM glucose). Each poly-D-lysine-coated well of the 24-well culture plates were seeded with $5{\times}10^5$ cells in 0.5 ml of culture medium, and cultures were maintained at 37°C in a humidified atmosphere of 5% CO_2 and 95% air. The culture medium consisted of Minimal Essential Medium supplemented with 10% heat-inactivated horse serum, 10% heat-inactivated fetal bovine serum, 2 mM L-glutamine, 1 mM sodium pyruvate, 100 μM non-essential amino acids, 15 mM KCl, 50 U/ml penicillin, 50 μg/ml streptomycin, and 1 g/L D-glucose. The cultures were replenished by adding 0.5 ml fresh medium three days after plating. Seven days after plating, the cells were treated in culture medium for which both serum supplements were reduced to 2% and no additional glucose or non-essential amino acids were included. The inhibitors SB202190, SB202474, or GW274150 were added one hour before treatment with LPS or 6-hydroxy dopamine (6-OHDA). When antibody against tyrosine hydroxylase was used to detect the number of dopaminergic neurons the mesencephalic neuron-glia culture showed about 13 tyrosine hydroxylase-positive (TH^+) neurons/mm^2.

Microglial Cell Line BV2 Cultures

The immortalized murine microglial BV2 cell line[46,47] was generously provided by Dr. M. McKinny of the Mayo Clinic (Jacksonville, FL). BV2 cells were seeded at a density of 1×10^6 cells into Costar 75 cm^2 flasks using Dulbecco's minimal essential medium (ATCC, Manassas, VA) with 10% heat-inactivated fetal bovine serum, 50 U/ml penicillin, and 50 μg/ml streptomycin and cultured in a humidified atmosphere of 5% CO_2 and 95% air at 37°C until reaching confluence, which typically occurred within six days. Cells were split and seeded onto 24-well tissue culture plates (0.5×10^5 cells/0.5 ml/well) or six-well plates (2.0×10^5 cells/2.0 ml/well). After three days in culture, cells were treated with specialized reagents as indicated in the figure legends.

Immunoblot Analysis of p38 and Phospho-p38 MAPKs

Cells in six-well plates were lysed by adding 300 μl sample buffer (62.5 mM Tris-Cl, pH 6.8, 2% sodium dodecyl sulfate, 10% glycerol, 50 mM dithiothreitol, 0.1% bromophenol blue) to each well and the cell lysates were immediately transferred to microfuge tubes and placed on ice. The extracts were sonicated for two seconds to shear the DNA and reduce sample viscosity. The samples were then heated to 95–100°C for five minutes and placed on ice. After microcentrifuging for five minutes at room temperature, 20 μl of each sample was used for gel electrophoresis. Proteins were electroblotted onto nitrocellulose membranes (Osmonics Inc.) in a transfer buffer (25 mM Tris base, 0.2 M glycine, 20% methanol) at 100 V for 1.5 h. After transfer, the membranes were prepared for immunoblotting as described in the product literature for *p38 MAP kinase assay kit* (New England BioLabs, Beverly, MA). Briefly, the membranes were washed and blocked and then incubated overnight in 10 ml primary antibody (1 : 1,000) against either phospho-p38 MAPK or p38 MAPK. The membranes were then washed and incubated with HRP-conjugated antirabbit secondary antibody (1:2,000). The phospho-p38 and p38 MAPKs were visualized with LumiGlo chemiluminescent reagent and X-ray film.

Measurement of p38 MAPK Activity

After cell cultures were treated with LPS and/or drugs in 6-well plates, media were removed and cells were harvested and processed for lysis, immunoprecipitation, kinase assay, immunoblotting, and detection as detailed in the product literature for *p38 MAP kinase assay kit*.

Immunocytochemistry

Cultured cells were fixed in 3.7% formaldehyde solution (Fisher Scientific, Fair Lawn, NJ) for 10 min at room temperature, followed by washing in phosphate-buffered saline (PBS). Dopaminergic neurons were stained with a polyclonal anti-TH-antibody. Briefly, cultures were blocked for 20 min with PBS containing 0.4% Triton X-100, 1% bovine serum albumin, and 4% normal goat serum, followed by an overnight incubation with the primary antibody, anti-TH (1:20,000 dilution). The cells in wells were washed (three changes of PBS) and treated for 10 minutes with 1% H_2O_2, followed by washing and incubation with biotinylated secondary antibody (goat antirabbit), and TH^+ neurons were visualized using the ABC method and DAB as the chromagen according to the manufacturer's instructions. Images of immunostained cells were recorded using a CCD camera (DAGE-MTI, Michigan City, IN) and captured by Metamorph Image System software (Universal Image and Co., West Chester, PA).

Competitive Reverse-Transcription-PCR Analysis

Following treatment of primary neuron-glia cultures, the total RNA was prepared using the TRI-reagent™ according to the method of Chomczynski and Sacchi.[48] One microgram of the total RNA from each treatment was used for the reverse transcription reaction in 20 µl with the Superscript™ preamplification system. The level of target mRNA was determined by a modification of a reverse-transcription-PCR for iNOS mRNA described elsewhere.[49] One and one-half microliters of the reverse transcription reaction (corresponding to 0.075 µg RNA of template) were used for each PCR analysis using the Gene Amp PCR kit with AmpliTaq DNA polymerase. The standard PCR reactions for iNOS were run with the upper primer 5′-CCCTTC-CGAAGTTTCTGGCAGCAGC-3′ and a lower primer 5′-GGCTGTCAGAGC-CTCGTGGCTTTGG-3′ that gave a product of 496 bp; whereas the reactions for β-actin used as upper primer 5′-GTGGGCCGCTCTAGGCACCAA-3′ and lower primer 5′-CTCTTTGATGTCACGCACGATTTC-3′ that gave a product of 540 bp. The competitive PCR reactions were run with 0.125 attomoles of 272 bp iNOS MIMIC™. All primers were added at a concentration of 1 µM each. The temperatures for denaturing, annealing and extension reactions during PCR were as follows: the first cycle was set for three minutes at 95°C, followed by one minute at 62°C (iNOS) or 56°C (β-actin), and then 1.5 min at 72°C; the second cycle was for one minute at 95°C, followed by one minute at 60°C (iNOS) or 56°C (β-actin), and then 1.5 minutes at 72°C; all subsequent cycles were for 30 sec at 94°C, followed by 30 sec at 60°C (iNOS) or 56°C (β-actin), and then 45 sec at 72°C. One and one-half microliters of the reverse transcription reaction were used for each PCR analysis, employing the Gene Amp PCR kit with AmpliTaq DNA polymerase, and the template was amplified for 25 cycles (iNOS) or 21 cycles (β-actin). The amplified DNA fragments were quantified by densitometry using the MULTI-ANALYST program of the Model GS-700 Imaging Densitometer (Bio-Rad, Hercules, CA).

Measurement of NO Production

The production of NO was assessed as the accumulation of nitrite in medium using a colorimetric reaction with Griess reagent.[50] Briefly, the culture supernatants were collected and mixed with an equal volume of Griess reagent (0.1% *N*-[1-naphthyl] ethylenediamine dihydrochloride, 1% sulfanilamide, and 2.5% H_3PO_4). The absorbance at 540 nm was measured with a UV MAX kinetic microplate reader (Molecular Devices, Menlo Park, CA).

Statistical Analysis

The data are expressed as the mean ± standard error of the mean. Statistical significance was assessed by an analysis of variance followed by Bonferroni's *t*-test using the StatView program (Abacus Concepts, Inc., Berkeley, CA). A value of $p < 0.05$ was considered statistically significant.

RESULTS

A Selective p38 Inhibitor Protects Neurons from LPS-Induced Cytotoxicity in Mesencephalic Neuron-glia Cultures

The bacterial endotoxin LPS elicited dopaminergic neuronal cell death in the ventral midbrain primary neuron-glia cultures. Treatment of the cultures with 10 ng/ml LPS destroyed more than 80% of the dopaminergic neurons by 48 h. Treatment of the cultures with a selective p38 MAPK inhibitor (SB202190) before exposure to LPS noticeably protected neurons in a concentration dependent manner from LPS-induced cytotoxicity (see FIGURE 1 A and B). In contrast, significant neuroprotection was not observed by pretreatment with the inactive analogue, SB202474 (FIG. 1 A and B).

To show that SB202190 was an effective p38 MAPK inhibitor in the neuron-glia primary culture system, we looked at the induction of phosphorylation of p38 MAPK (phospho-p38) by LPS. Although the level of non-phosphorylated form (p38) was relatively constant over the time intervals examined, an induction of p38 phosphorylation was clearly observed 30–90 min after LPS-treatment (FIG. 1 C). Parallel analysis of enzyme activity indicated a marked increase in p38 MAPK activity in both primary microglia and the immortalized microglial cell line BV2, and the p38 activity that was induced by LPS was significantly inhibited by SB202190 (FIG. 1 D).

The Selective p38 MAPK Inhibitor does not Protect Neurons from 6-Hydroxy Dopamine-Induced Cytotoxicity

Glia respond to LPS by producing immune-modulators or inflammatory factors that are thought to cause neuronal cell degeneration, a process known as glia-mediated or immune-mediated neurotoxicity. In contrast, 6-OHDA is taken up through the dopamine transporter and directly kills dopaminergic neurons by producing free hydroxyl radicals in the neuron. Thus, glia do not mediate the neurotoxic action of 6-OHDA.[51]

Treatment of the neuron-glia culture with 40 μM 6-OHDA killed about 80% of the dopaminergic neurons after 48 hours of exposure (see FIGURE 2). Pretreatment with the p38 MAPK inhibitor SB202190 did not show a significant protection of TH^+ neurons from 6-OHDA-induced cytotoxicity, unlike the marked protection of

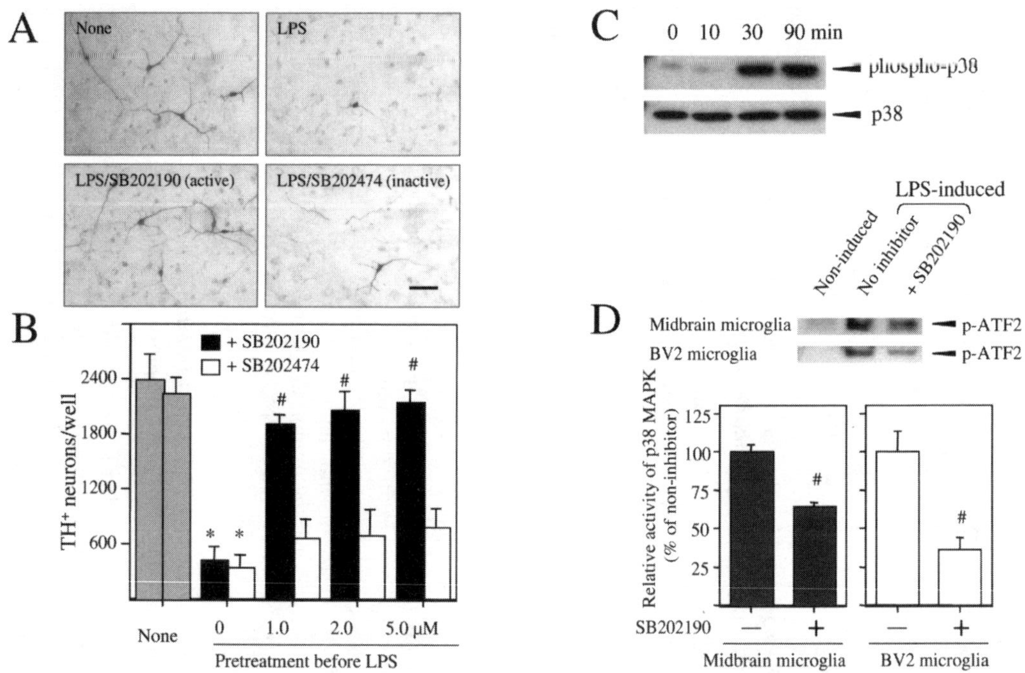

FIGURE 1. p38 MAPK inhibitor SB202190 protects neurons from LPS-induced toxicity in rat mesencephalic neuron/glia mixed cultures. **(A)** Primary neuron/glia cultures were treated with 2 μM SB202190, SB202474, or vehicle for one hour before LPS (10 ng/ml) for 48 h. Neurons were stained with anti-TH-antibody as described in MATERIALS AND METHODS (*bar* = 100 μm). **(B)** Concentration dependent effect of SB202190 on LPS-induced neurotoxicity. **(C)** Time dependent induction of p38 MAPK phosphorylation by LPS. Neuron-glia cultures were treated with LPS for the indicated time points. Immunoblot analyses for phospho-p38 and p38 were performed as described in the MATERIALS AND METHODS. **(D)** Inhibition of p38 MAPK activity. After 90 min (microglia) or 30 min (BV2) exposure to LPS, cell extracts were prepared and the activity of p38 MAPK was measured as described in MATERIALS AND METHODS.

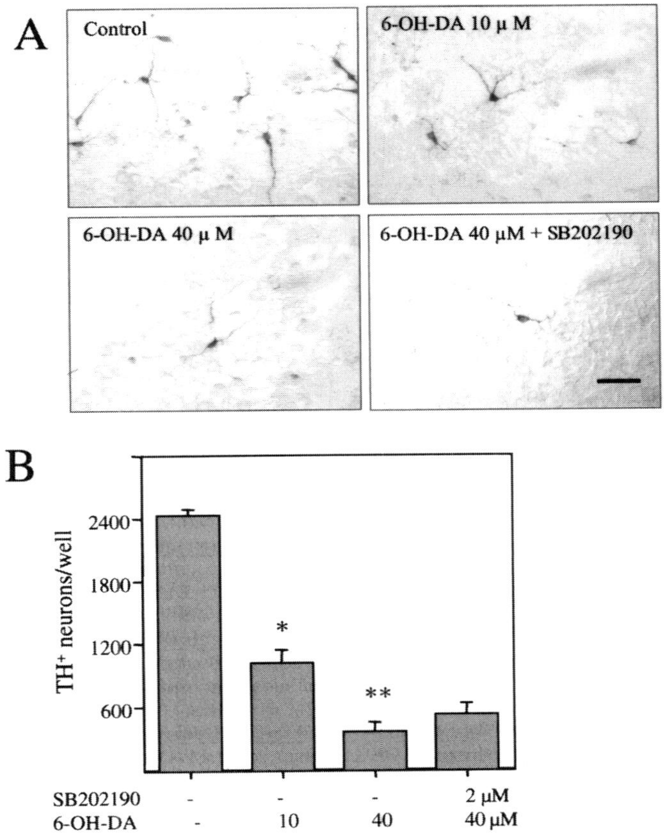

FIGURE 2. p38 MAPK inhibitor SB202190 does not protect neurons from 6-OHDA-induced neuronal cell death in neuron/glia mixed cultures. The neuron/glia cultures were exposed to 6-OHDA for 48 h in the presence or absence of SB202190. Neurons were identified with anti-TH antibody. $*p < 0.05$ and $**p < 0.0001$ compared to untreated control group. *Bar* = 100 μm.

neurons from LPS-induced cytotoxicity. This indicates that neural p38 MAPK played no role in the 6-OHDA-induced neuronal cell death, and that non-neuronal p38 MAPK contributed to the inflammation-mediated neuronal cell death shown in FIGURE 1.

A Selective iNOS Inhibitor Protects Neurons from LPS-Induced Dopaminergic Neuronal Cell Death

GW274150 has been reported to be a potent and highly selective inhibitor for iNOS (IC_{50} = 1.4 μM for iNOS, 466 μM for eNOS, 145 μM for nNOS).[52] In primary neuron/glia cultures, the inducible subtype of NOS is expressed mainly in microglia and also in astrocytes upon exposure to LPS.[4,6,9,53] Treatment of the neuron/glia culture with the selective iNOS inhibitor GW274150 conferred marked protection of neurons from LPS-induced cytotoxicity (see FIGURE 3). Neuroprotection by GW274150 paralleled the inhibition of the LPS-induced increase in NO production. LPS (10 ng/ml) increased NO production to 8.5 μM 12 hours after treatment and was suppressed by the selective iNOS inhibitor GW274150 in a concentration dependent manner. GW274150 (100 μM) suppressed more than 90% of the LPS-induced increase in NO production (FIG. 3). This suggests that the major source of NO production was from glial cells, in which LPS induced the expression of iNOS. The data also indicate that NO was important in LPS-elicited neurotoxicity.

Inhibition of iNOS Expression and NO Production by the p38 MAPK Inhibitor SB202190

Because of the neuroprotective effects of the selective p38 MAPK inhibitor SB202190 and the selective iNOS inhibitor GW274150, we examined the inhibitory effect of SB202190 on the LPS-induced increase in NO production. The LPS

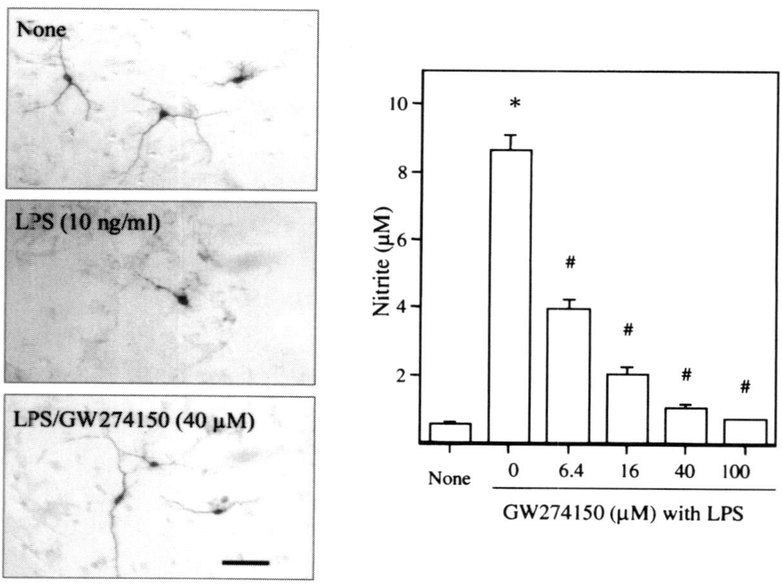

FIGURE 3. Inhibition of iNOS activity reduces LPS-induced neuronal cell death in neuron/glia cultures. The neuron/glia cultures were exposed to 10 ng/ml LPS for 48 h in the presence or absence of 40 μM GW274150. Dopaminergic neurons were immunostained with anti-TH antibody. Nitrite was assessed in culture media from neuron/glia treated with GW274150 for one hour before treatment with LPS (10 ng/ml). The control group received no LPS or GW274150. $^{*}p < 0.0001$ compared to untreated control group; $^{\#}p < 0.0001$ compared to LPS-treatment. *Bar* = 100 μm.

induced increase in NO production was significantly inhibited by SB202190, in a concentration dependent manner (see FIGURE 4). The marked inhibitory effect of SB202190 on the LPS-induced increase in NO production at concentrations as low as 320 nM was in marked contrast to the inability of high concentrations of SB202474 (2 μM) to suppress NO production. PD98059, a potent inhibitor of ERK, did not produce a significant inhibition of the LPS-induced NO production within the same concentration ranges as that used for SB202190. The results indicate that p38 MAPK rather than ERK regulates NO production in the rat ventral mesencephalic neuron/glia cultures.

LPS induced iNOS mRNA transcription in a time dependent manner (FIGURE 5A). In the neuron/glia cultures, LPS induced iNOS mRNA to a detectable level three hours after exposure and the iNOS mRNA level reached a maximum at six hours. The LPS-induced increase in iNOS mRNA transcription was suppressed by SB202190 at all time points examined. Following the notable increase in iNOS mRNA, NO production increased markedly 6–12 h after exposure to LPS (FIG. 5B). The LPS-induced increase in NO production was suppressed by SB202190 (FIG. 5B). The results indicate that the p38 MAPK inhibitor reduces nitric oxide production by reducing iNOS expression at the level of mRNA synthesis.

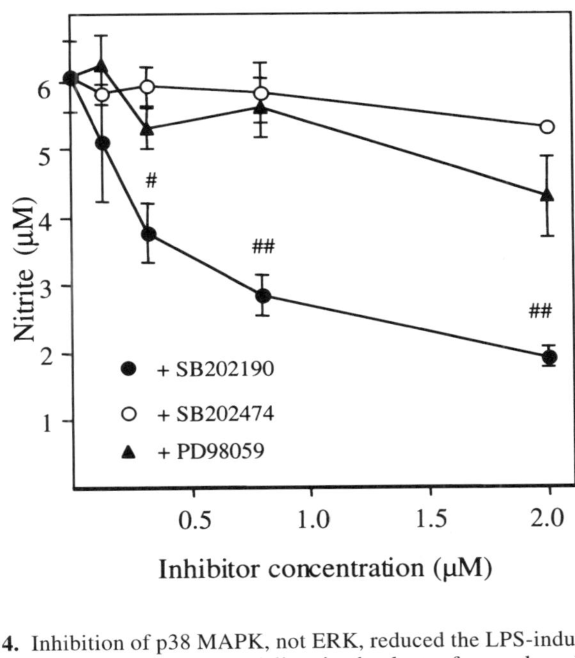

FIGURE 4. Inhibition of p38 MAPK, not ERK, reduced the LPS-induced NO production. Each inhibitor was added to neuron/glia mixed cultures for one hour before exposure to 10 ng/ml LPS for 12 h. Aliquots of culture media were removed and analyzed for nitric oxide as described in MATERIALS AND METHODS. $^{\#}p < 0.05$ and $^{\#\#}p < 0.0001$ compared to LPS-treated group.

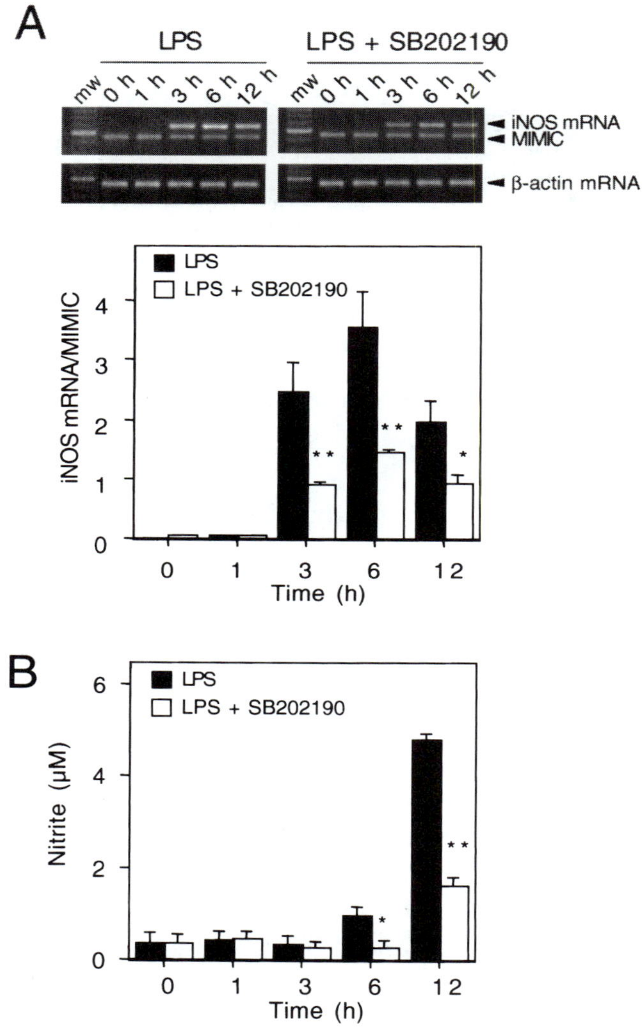

FIGURE 5. Inhibition of p38 MAPK blocked iNOS gene expression and NO production induced by LPS. (**A**) The neuron/glia cultures were incubated with 2 μM SB202190 for one hour before exposure to 10 ng/ml LPS. Total RNA was isolated at each time point and a competitive reverse transcription-polymerase chain reaction for iNOS was performed as described in MATERIALS AND METHODS. $*p < 0.05$ compared to corresponding LPS treatments. (**B**) Same experiment as in **A** for analysis of NO accumulation in the culture medium.

DISCUSSION

Our results demonstrate that an inhibitor for p38 MAPK affords significant protection against LPS-elicited neurotoxicity. Although p38 MAPK is assumed to be involved in the apoptosis of neurons, our results show that a potent and selective inhibitor of p38 MAPK did not provide neuroprotection against 6-OHDA, which directly generates oxidative stress in dopaminergic neurons. The data suggest that p38 MAPK is important in glial production of inflammatory factors such as NO. Indeed, inhibition of p38 MAPK significantly reduced iNOS expression and subsequent NO production. Furthermore, inhibition of iNOS by a specific iNOS inhibitor, GW274150, afforded protection against LPS-elicited neurotoxicity. All these findings lend support to the idea that production of NO is important in inflammation-mediated neurotoxicity in the mesencephalon. Once iNOS synthesis was blocked and NO production reduced by SB202190, neuroprotection is achieved. The results also point to the importance of p38 MAPK in iNOS gene expression as well as in NO production.

p38 MAPK is expressed in both neurons and glia. One might question whether this kinase plays a role in LPS-induced neurotoxicity by triggering apoptosis in neurons.[54,55] We previously reported that 6-OHDA directly kills dopaminergic neurons without affecting glia, whereas LPS requires stimulation of glial cells and production of inflammatory factors to exert its neurotoxicity.[51] In the present study, a p38 MAPK inhibitor SB202190 did not alter the neurotoxicity of 6-OHDA to dopaminergic neurons. Yet, SB202190 did suppress glia-mediated and NO-mediated neurotoxicity. These results provide evidence that the neuroprotective effect of SB202190 is not achieved through direct effects on the neurons, rather, SB202190 reduces the production of NO by glia.

Under physiologic conditions, NO is involved in intercellular and intracellular signaling.[3,56] However, excessive levels of NO induce neuronal injury. A number of studies have shown that addition of NO synthase inhibitors to neuron/glia cultures inhibits the LPS-induced accumulation of nitrite and the associated loss of neurons.[1,12,51] The present study also indicates that NO is important in LPS-elicited neurotoxicity in the neuron/glia culture, perhaps by forming the more toxic peroxynitrite from NO and superoxide radical.[57] Indeed, our results demonstrate that the selective inhibitor of iNOS described by Young,[52] GW274150, afforded protection from LPS-elicited neurotoxicity by reducing NO production (FIG. 3). The results reinforce the notion that NO is important in endotoxin-elicited neurotoxicity. The p38 MAPK inhibitor exerts its neuroprotective effect against LPS by inhibiting iNOS expression and subsequent NO production from microglia.

The use of a selective inhibitor of p38 MAPK significantly reduced iNOS gene expression, suggesting that p38 MAPK is an essential component in the signaling pathway for induction of iNOS. It has been reported that inhibitors of p38 MAPK and ERK (SB203580 and PD98059) synergistically enhance the inhibitory effects on LPS-induced release of NO and tumor necrosis factor α.[32,58] However, in the ventral mesencephalic cultures, significant inhibition of LPS-stimulated NO production was observed only by the selective inhibition of p38 MAPK. In contrast, the ERK inhibitor PD98059 did not reduce NO production (FIG. 4). These results suggest that the p38 MAPK plays a significant role in regulating the expression of iNOS and NO production in mesencephalic cultures, whereas ERK does not.

Taken together, the results of the present study indicate that inhibition of p38 MAPK prevents neurotoxicity from LPS by suppressing iNOS gene expression and subsequent NO production. Further understanding of the control mechanism(s) underlying this neuroprotection by inhibition of p38 MAPK should provide new therapeutic interventions for inflammation-mediated neurodegenerative disorders, such as Parkinson's disease.

REFERENCES

1. BOJE, K.M. & P.K. ARORA. 1992. Microglial-produced nitric oxide and reactive nitrogen oxides mediate neuronal cell death. Brain Res. **587:** 250–256.
2. CHAO, C.C., S. HU & P.K. PETERSON. 1995. Glia, cytokines, and neurotoxicity. Crit. Rev. Neurobiol. **9:** 189–205.
3. DAWSON, T.M. & S.H. SNYDER. 1994. Gases as biological messengers: nitric oxide and carbon monoxide in the brain. J. Neurosci. **14:** 5147–5159.
4. GALEA, E., D.L. FEINSTEIN & D.J. REIS. 1992. Induction of calcium-independent nitric oxide synthase activity in primary rat glial cultures. Proc. Natl. Acad. Sci. U.S.A. **89:** 10945–10949.
5. II, M., M. SUNAMOTO, K. OHNISHI, *et al.* 1996. beta-Amyloid protein-dependent nitric oxide production from microglial cells and neurotoxicity. Brain Res. **720:** 93–100.
6. KONG, L.Y., M.K. MCMILLIAN, R. MARONPOT, *et al.* 1996. Protein tyrosine kinase inhibitors suppress the production of nitric oxide in mixed glia, microglia-enriched or astrocyte-enriched cultures. Brain Res. **729:** 102–109.
7. LIPTON, S.A. 1994. Neuronal injury associated with HIV-1 and potential treatment with calcium-channel and NMDA antagonists. Dev. Neurosci. **16:** 145–151.
8. MEDA, L., M.A. CASSATELLA, G.I. SZENDREI, *et al.* 1995. Activation of microglial cells by beta-amyloid protein and interferon- gamma. Nature **374:** 647–650.
9. SIMMONS, M.L. & S. MURPHY. 1992. Induction of nitric oxide synthase in glial cells. J. Neurochem. **59:** 897–905.
10. WILLARD, L.B., B. HAUSS-WEGRZYNIAK & G.L WENK. 1999. Pathological and biochemical consequences of acute and chronic neuroinflammation within the basal forebrain cholinergic system of rats. Neuroscience **88:** 193–200.
11. BROSNAN, C.F., L. BATTISTINI, C.S. RAINE, *et al.* 1994. Reactive nitrogen intermediates in human neuropathology: an overview. Dev. Neurosci. **16:** 152–161.
12. CHAO, C.C., S. HU, T.W. MOLITOR, *et al.* 1992. Activated microglia mediate neuronal cell injury via a nitric oxide mechanism. J. Immunol. **149:** 2736–2741.
13. DICKSON, D.W., S.C. LEE, L.A. MATTIACE, *et al.* 1993. Microglia and cytokines in neurological disease, with special reference to AIDS and Alzheimer's disease. Glia **7:** 75–83.
14. ESPEY, M.G., O.N. CHERNYSHEV, J.F. REINHARD, *et al.* 1997. Activated human microglia produce the excitotoxin quinolinic acid. Neuroreport **8:** 431–434.
15. LEE, S.C., W. LIU, D.W. DICKSON, *et al.* 1993. Cytokine production by human fetal microglia and astrocytes. Differential induction by lipopolysaccharide and IL-1β. J. Immunol. **150:** 2659–2667.
16. MATSUO, M., Y. HAMASAKI, F. FUJIYAMA, *et al.* 1995. Eicosanoids are produced by microglia, not by astrocytes, in rat glial cell cultures. Brain Res. **685:** 201–204.
17. MERRILL, J.E., L.J. IGNARRO, M.P. SHERMAN, *et al.* 1993. Microglial cell cytotoxicity of oligodendrocytes is mediated through nitric oxide. J. Immunol. **151:** 2132–2141.
18. MINGHETTI, L. & G. LEVI. 1995. Induction of prostanoid biosynthesis by bacterial lipopolysaccharide and isoproterenol in rat microglial cultures. J. Neurochem. **65:** 2690–2698.
19. MATSUMOTO, Y., K. OHMORI & M. FUJIWARA. 1992. Microglial and astroglial reactions to inflammatory lesions of experimental autoimmune encephalomyelitis in the rat central nervous system. J. Neuroimmunol. **37:** 23–33.

20. McGeer, P.L., S. Itagaki, B.E. Boyes, *et al.* 1988. Reactive microglia are positive for HLA-DR in the substantia nigra of Parkinson's and Alzheimer's disease brains. Neurology **38:** 1285–1291.
21. Raine, C.S. 1994. Multiple sclerosis: immune system molecule expression in the central nervous system. J. Neuropathol. Exp. Neurol. **53:** 328–337.
22. Rogers, J., J. Luber-Narod, S.D. Styren, *et al.* 1988. Expression of immune system-associated antigens by cells of the human central nervous system: relationship to the pathology of Alzheimer's disease. Neurobiol. Aging **9:** 339–349.
23. Sherman, M.P., J.M. Griscavage & L.J. Ignarro. 1992. Nitric oxide-mediated neuronal injury in multiple sclerosis. Med. Hypoth. **39:** 143–146.
24. Youdim, M.B., L. Lavie & P. Riederer. 1994. Oxygen free radicals and neurodegeneration in Parkinson's disease: a role for nitric oxide. Ann. N.Y. Acad. Sci. **738:** 64–68.
25. Galea, E., D.J. Reis, E.S. Fox, *et al.* 1996. CD14 mediate endotoxin induction of nitric oxide synthase in cultured brain glial cells. J. Neuroimmunol. **64:** 19–28.
26. Laflamme, N. & S. Rivest. 2001. Toll-like receptor 4: the missing link of the cerebral innate immune response triggered by circulating gram-negative bacterial cell wall components. FASEB J. **15:** 155–163.
27. Yang, H., D.W. Young, F. Gusovsky, *et al.* 2000. Cellular events mediated by lipopolysaccharide-stimulated toll-like receptor 4. MD-2 is required for activation of mitogen-activated protein kinases and Elk-1. J. Biol. Chem. **275:** 20861–20866.
28. Boulet, I., S. Ralph, E. Stanley, *et al.* 1992. Lipopolysaccharide- and interferon-gamma-induced expression of hck and lyn tyrosine kinases in murine bone marrow-derived macrophages. Oncogene **7:** 703–710.
29. Shapira, L., S. Takashiba, C. Champagne, *et al.* 1994. Involvement of protein kinase C and protein tyrosine kinase in lipopolysaccharide-induced TNF-α and IL-1β production by human monocytes. J. Immunol. **153:** 1818–1824.
30. Weinstein, S.L., M.R. Gold & A.L. DeFranco. 1991. Bacterial lipopolysaccharide stimulates protein tyrosine phosphorylation in macrophages. Proc. Natl. Acad. Sci. U.S.A. **88:** 4148–4152.
31. Ziegler, S.F., C.B. Wilson & R.M. Perlmutter. 1988. Augmented expression of a myeloid-specific protein tyrosine kinase gene (hck) after macrophage activation. J. Exp. Med. **168:** 1801–1810.
32. Bhat, N.R., P. Zhang, J.C. Lee, *et al.* 1998. Extracellular signal-regulated kinase and p38 subgroups of mitogen- activated protein kinases regulate inducible nitric oxide synthase and tumor necrosis factor-alpha gene expression in endotoxin-stimulated primary glial cultures. J. Neurosci. **18:** 1633–1641.
33. Chen, C.C., J.K. Wang, W.C. Chen, *et al.* 1998. Protein kinase C eta mediates lipopolysaccharide-induced nitric-oxide synthase expression in primary astrocytes. J. Biol. Chem. **273:** 19424–19430.
34. Feinstein, D.L., E. Galea, J. Cermak, *et al.* 1994. Nitric oxide synthase expression in glial cells: suppression by tyrosine kinase inhibitors. J. Neurochem. **62:** 811–814.
35. Kitamura, Y., R. Imaizumi, Y. Kitayama, *et al.* 1996. Possible involvement of tyrosine kinase activation in lipopolysaccharide-induced expression of Ca^{2+}-insensitive but calmodulin-coupling nitric oxide synthase in rat glial cells. J. Neurosci. Res. **43:** 235–245.
36. Schumann, R.R., D. Pfeil, D. Freyer, *et al.* 1998. Lipopolysaccharide and pneumococcal cell wall components activate the mitogen activated protein kinases (MAPK) erk-1, erk-2, and p38 in astrocytes. Glia **22:** 295–305.
37. Simmons, M.L. & S. Murphy. 1994. Roles for protein kinases in the induction of nitric oxide synthase in astrocytes. Glia **11:** 227–234.
38. Hensley, K., R.A. Floyd, N.Y. Zheng, *et al.* 1999. p38 kinase is activated in the Alzheimer's disease brain. J. Neurochem. **72:** 2053–2058.
39. Walton, K.M., R. DiRocco, B.A. Bartlett, *et al.* 1998. Activation of p38MAPK in microglia after ischemia. J. Neurochem. **70:** 1764–1767.
40. Da Silva, J., B. Pierrat, J.L. Mary, *et al.* 1997. Blockade of p38 mitogen-activated protein kinase pathway inhibits inducible nitric-oxide synthase expression in mouse astrocytes. J. Biol. Chem. **272:** 28373–28380.

41. BALDASSARE, J.J., Y. BI & C.J. BELLONE. 1999. The role of p38 mitogen-activated protein kinase in IL-1 beta transcription. J. Immunol. **162:** 5367–5373.
42. LEE, J.C., J.T. LAYDON, P.C. MCDONNELL, *et al.* 1994. A protein kinase involved in the regulation of inflammatory cytokine biosynthesis. Nature **372:** 739–746.
43. KIM, W.G., R.P. MOHNEY, B. WILSON, *et al.* 2000. Regional difference in susceptibility to lipopolysaccharide-induced neurotoxicity in the rat brain: role of microglia. J. Neurosci. **20:** 6309–6316.
44. LIU, B., L. DU & J.S. HONG. 2000. Naloxone protects rat dopaminergic neurons against inflammatory damage through inhibition of microglia activation and superoxide generation. J. Pharmacol. Exp. Ther. **293:** 607–617.
45. LIU, B., K. WANG, H.M. GAO, *et al.* 2001. Molecular consequences of activated microglia in the brain: overactivation induces apoptosis. J. Neurochem. **77:** 182–189.
46. BLASI, E., R. BARLUZZI, V. BOCCHINI, *et al.* 1990. Immortalization of murine microglial cells by a v-raf/v-myc carrying retrovirus. J. Neuroimmunol. **27:** 229–237.
47. BOCCHINI, V., R. MAZZOLLA, R. BARLUZZI, *et al.* 1992. An immortalized cell line expresses properties of activated microglial cells. J. Neurosci. Res. **31:** 616–621.
48. CHOMCZYNSKI, P. & N. SACCHI. 1987. Single-step method of RNA isolation by acid guanidinium thiocyanate- phenol-chloroform extraction. Ann. Biochem. **162:** 156–159.
49. JEOHN, G.H., R.C. CHANG, W.G. KIM, *et al.* 2000. Post-transcriptional inhibition of lipopolysaccharide-induced expression of inducible nitric oxide synthase by Go6976 in murine microglia. Brain Res. Mol. Brain Res. **79:** 18–31.
50. GREEN, L.C., D.A. WAGNER, J. GLOGOWSKI, *et al.* 1982. Analysis of nitrate, nitrite, and [15N]nitrate in biological fluids. Ann. Biochem. **126:** 131–138.
51. BRONSTEIN, D.M., I. PEREZ-OTANO, V. SUN, *et al.* 1995. Glia-dependent neurotoxicity and neuroprotection in mesencephalic cultures. Brain Res. **704:** 112–116.
52. YOUNG, R.J., R.M. BEAMS, K. CARTER, *et al.* 2000. Inhibition of inducible nitric oxide synthase by acetamidine derivatives of hetero-substituted lysine and homolysine. Bioorg. Med. Chem. Lett. **10:** 597–600.
53. KOPROWSKI, H., Y.M. ZHENG, E. HEBER-KATZ, *et al.* 1993. In vivo expression of inducible nitric oxide synthase in experimentally induced neurologic diseases. Proc. Natl. Acad. Sci. U.S.A. **90:** 3024–3027.
54. KUMMER, J.L., P.K. RAO & K.A. HEIDENREICH. 1997. Apoptosis induced by withdrawal of trophic factors is mediated by p38 mitogen-activated protein kinase. J. Biol. Chem. **272:** 20490–20494.
55. XIA, Z., M. DICKENS, J. RAINGEAUD, *et al.* 1995. Opposing effects of ERK and JNK-p38 MAP kinases on apoptosis. Science **270:** 1326–1331.
56. GARTHWAITE, J. 1991. Glutamate, nitric oxide and cell-cell signalling in the nervous system. Trend. Neurosci. **14:** 60–67.
57. BECKMAN, J.S., T.W. BECKMAN, J. CHEN, *et al.* 1990. Apparent hydroxyl radical production by peroxynitrite: implications for endothelial injury from nitric oxide and superoxide. Proc. Natl. Acad. Sci. U.S.A. **87:** 1620–1624.
58. PYO, H., I. JOU, S. JUNG, *et al.* 1998. Mitogen-activated protein kinases activated by lipopolysaccharide and beta-amyloid in cultured rat microglia. Neuroreport **9:** 871–874.

Gö6976 Protects Mesencephalic Neurons from Lipopolysaccharide-Elicited Death by Inhibiting p38 MAP Kinase Phosphorylation

GWANG-HO JEOHN,[a] CYNTHIA L. COOPER,[b] KYUNG-JIN JANG,[c] HYOUNG-CHUN KIM,[c] AND JAU-SHYONG HONG[a]

[a]*Neuropharmacology Section, Laboratory of Pharmacology and Chemistry, National Institute of Environmental Health Sciences, National Institutes of Health, Research Triangle Park, North Carolina, U.S.A.*

[b]*Division of Science, Truman State University, Kirksville, Missouri, U.S.A.*

[c]*Department of Pharmacy, Kangwon National University, Chuncheon, 200-701, Korea*

ABSTRACT: Glial activation is associated with inflammation-related neuron degeneration in the brain. A variety of protein kinases are assumed to contribute to the expression of inflammation-related products, such as nitric oxide (NO) and proinflammatory cytokines, however, the mechanisms of glial activation and glia-mediated neurotoxicity remain unclear. We found that the indolocarbazole, Gö6976, originally known as a selective protein kinase C (PKC) inhibitor, protects neurons from glia-mediated damage and suppresses lipopolysaccharide (LPS)-induced microglial production of inflammatory factors. The purpose of the study we report here was to determine the mechanism underlying the neuroprotective effect of Gö6976 in mesencephalic neuron/glia cultures. Gö6976 suppressed LPS-induced neurotoxicity in mesencephalic neuron/glia cultures and the protective effect of Gö6976 paralleled the suppression of p38 mitogen activated protein kinase (MAPK) activation and inhibition of NO production. Gö6976 did not directly inhibit the activity of p38 MAPK; rather, the inhibitor suppressed the phosphorylation of p38 MAPK, suggesting that the target of Gö6976 is a signaling event upstream of p38 MAPK. Although Gö6976 was originally known to be a selective PKC inhibitor, the neuroprotection was not mediated through its reputed effects on PKC activity. This paper demonstrates that the neuroprotective effect of Gö6976 against LPS-induced damage is mediated through the inhibition of proinflammatory factors, such as NO from microglia, by inhibiting the phosphorylation of p38 MAPK.

KEYWORDS: BV2; microglia; tumor necrosis factor α; endotoxin; nitric oxide; inflammation; neurodegeneration

ABBREVIATIONS:

LPS	lipopolysaccharide.
MAP2	microtubule associated protein 2.
MAPK	mitogen-activated protein kinase.

Address for correspondence: J.-S. Hong, Ph.D., Neuropharmacology Section, National Institute of Environmental Health Sciences, National Institutes of Health P.O. Box 12233, Research Triangle Park, NC 27709, U.S.A. Voice: 919-541-2358; fax: 919-541-0841.
hong3@niehs.nih.gov

NO	nitric oxide.
PKC	protein kinase C.
TH	tyrosine hydroxylase.
TNFα	tumor necrosis factor α.

INTRODUCTION

Glial cells readily respond to environmental stress or immunologic challenges to release cytotoxic factors that are associated with inflammation-related neurodegenerative processes.[1–8] Glia respond to toxins, such as lipopolysaccharide (LPS), the HIV-1 coat protein gp120, and β-amyloid-related peptides; and release nitric oxide (NO), reactive oxygen intermediates, tumor necrosis factor-α, (TNFα), interleukin-1, arachidonic acid metabolites, and quinolinic acid.[9–17] Recent reports suggest the immune-elicited inflammatory processes may underlie the pathologies of several neurological disorders, including Parkinson's disease, Alzheimer's disease, multiple sclerosis, and the AIDS dementia complex.[3,18–23]

To develop potential interventions for immune-mediated neurodegenerative diseases it is essential to understand the mechanism that underlies glia-mediated neuronal cell death. The bacterial endotoxin LPS has been widely used as a tool to induce inflammation-mediated neuronal cell degeneration. LPS binds to a CD14-like receptor in brain glia,[24] and activates a Toll-like receptor 4.[25–27] Several types of protein kinases, including protein kinase C (PKC) and other protein tyrosine kinases, are activated by LPS.[28–31] Protein kinases, such as PKC, protein tyrosine kinases, and mitogen-activated protein kinases (MAPKs), including extracellular signal-related kinases and p38 MAPK, have been implicated in the release of immune-related cytotoxic factors, such as NO and proinflammatory cytokines in brain glia.[14,32–37] Under neurotoxic conditions, such as exposure of brain tissue to endotoxins, the roles of protein kinases in neurotoxicity are not known.

Recently, we reported that Gö6976, previously known as a selective PKC inhibitor, suppressed the LPS-induced increase in microglial inflammatory function.[38,39] The indolocarbazole was shown to protect neurons from immune-mediated cytotoxicity induced by exposure to endotoxin or a combination of cytokines. Because the neuroprotective effect of Gö6976 was not mediated through its inhibition of PKC activity, it was suggested that a kinase other than PKC contributes to neuronal cell degeneration. Here, we report that the protective action of Gö6976 on LPS-induced neurotoxicity in rat mesencephalic glia-neuron cultures is mediated through the inhibition of phosphorylation of p38 MAPK.

MATERIALS AND METHODS

Materials

LPS (*E. coli* O111:B4) was from List Biological Laboratories (Campbell, CA), and poly-D-lysine was from Sigma Chemical Co. (St. Louis, MO). Tissue culture plates (24-well) were Costar® from Corning, Inc. (Corning, NY). Minimal essential medium (MEM), heat-inactivated horse serum, fetal bovine serum, and the Superscript Preamplification System were from Life Technologies (Gaithersburg, MD).

Gö6976, SB202190, Ro-31-8220, and H89 were from Calbiochem-Novabiochem International (La Jolla, CA). All other reagents were of the purest grade available.

Neuron/Glia Cell Culture and Treatment

Primary cultures of mixed neuron/glia were prepared from embryonic day 14/15 rats (Fischer 344) from Charles River Laboratories (Raleigh, NC), modifying the culture method of Liu *et al.*,[40] by dissecting the ventral region of each midbrain. To each poly-D-lysine-coated well (1.88 cm^2) were seeded 5×10^5 cells in 0.5 ml of culture medium and cultures were maintained at 37°C in an atmosphere of 5% CO_2 and 95% air. MEM was supplemented, as described by Liu *et al.,*[40] with the addition of 15 mM KCl. Cultures were replenished with fresh medium three days after plating. At seven days after plating, the cells were treated in a modified culture medium, where the sera were reduced to 2% each for horse and fetal bovine (both heat inactivated). The inhibitors added one hour before treatment with 10 ng/ml LPS were 0.5 μM Gö6976, 2 μM SB202190, 2 μM Ro-31-8220, or 2 μM H89.

BV2 Microglial Cell Cultures

The immortalized murine microglial cell line, BV2,[41,42] was generously provided by Dr. M. McKinny of the Mayo Clinic (Jacksonville, FL). BV2 cultures were seeded at 1×10^6 cells into 75 cm^2 Costar® flasks using Dulbecco's minimal essential medium (ATCC, Manassas, VA) with 10% heat-inactivated fetal bovine serum, 50 U/ml penicillin, and 50 μg/ml streptomycin and grown in a humidified atmosphere of 5% CO_2/95% air at 37°C. Cells were seeded into 24-well tissue culture plates (0.5×10^5 cells/0.5 ml/well) or six-well plates (2.0×10^5 cells/2.0 ml/well). After three days in culture, the cells were treated with 10 ng/ml LPS, with or without a pretreatment with 0.5 μM Gö6976, 2 μM SB202190, 2 μM Ro-31-8220, or 2 μM H89.

Nitric Oxide and TNFα Assays

Nitric oxide production was assessed as nitrite accumulation in samples of culture medium using a colorimetric reaction with the Griess reagent.[43] TNFα was measured in samples of culture media using the enzyme linked immunosorbent assay kit for mouse TNFα from Genzyme (Cambridge, MA) as described by the supplier. Optical densities were measured with an UV MAX kinetic microplate reader (Molecular Devices, Menlo Park, CA).

Immunoblot Analysis of p38 and Phospho-p38 MAPKs

Cells in six-well plates were lysed by adding 300 μl sample buffer (62.5 mM Tris-Cl, pH 6.8, 2% sodium dodecyl sulfate, 10% glycerol, 50 mM dithiothreitol, and 0.1% bromophenol blue) to each well and cell lysates were immediately transferred to microfuge tubes on ice. The extracts were sonicated for two seconds to shear DNA and reduce sample viscosity. Samples were then heated to 95–100°C for five minutes and placed on ice. After microcentrifuging for five minutes at room temperature, 20 μl of each sample was used for gel electrophoresis. Proteins were electroblotted onto nitrocellulose membranes (Osmonics Inc.) in transfer buffer (25 mM Tris base, 0.2 M glycine and 20% methanol) at 100 V for 1.5 h. After transfer, membranes were prepared for immunoblotting as described in the product literature for *p38 MAP kinase assay kit* (New England BioLabs, Beverly, MA). Briefly, membranes were washed and blocked and then incubated in 10 ml primary antibody (1 : 1000) against either phospho-p38 MAPK or p38 MAPK overnight. The membranes were then washed and incubated with HRP-conjugated antirabbit secondary antibody (1 : 2000). The phospho-p38 and p38 MAPKs were visualized with LumiGlo chemiluminescent reagent and X-ray film.

Measurement of p38 MAPK Activity

After cell cultures were treated with LPS and/or drugs in six-well plates, media were removed and cells were harvested and processed for lysis, immunoprecipitation, kinase assay, immunoblotting, and detection as detailed in the product literature for *p38 MAP kinase assay kit* (New England BioLabs, Beverly, MA).

Immunocytochemistry

Neuron/glia cells were fixed in 3.7% formaldehyde solution and immunostained with neuronal cell markers: polyclonal antibody against tyrosine hydroxylase (TH) provided courtesy of GlaxoSmithKline (Research Triangle Park, NC) or monoclonal antibody against microtubule associated protein 2 (MAP2) from Boehringer Mannheim (Indianapolis, IN) as described by Liu *et al.*[40]

Statistical Analysis

The data were expressed as mean ± SEM. Statistical significance was assessed by an analysis of variance followed by Bonferroni's *t*-test using the StatView program (Abacus Concepts, Inc., Berkeley, CA). A value of $p < 0.05$ was considered statistically significant.

RESULTS

Suppression of LPS-Induced Neurotoxicity by Gö6976

LPS was used to induce inflammation-related neurotoxicity in mesencephalic neuron/glia cultures. LPS at 10 ng/ml caused a significant decrease in both TH-positive and MAP2-positive cells. When Gö6976 was added one hour before LPS treatment, a marked protection of both types of neurons was observed (see FIGURE 1). In the mesencephalic neuron/glia culture, LPS induced a significant increase in NO production at six hours and the level of NO metabolite, nitrite (NO_2^-), increased up to 24 hours after treatment. Gö6976 suppressed LPS-induced NO production by more than 90% throughout the observation period. Secretion of TNFα, a representative immune-related product was significant at three hours after LPS exposure, and reached a maximum at six hours. Gö6976 suppressed TNFα secretion to a lesser extent than NO production (see FIGURE 2).

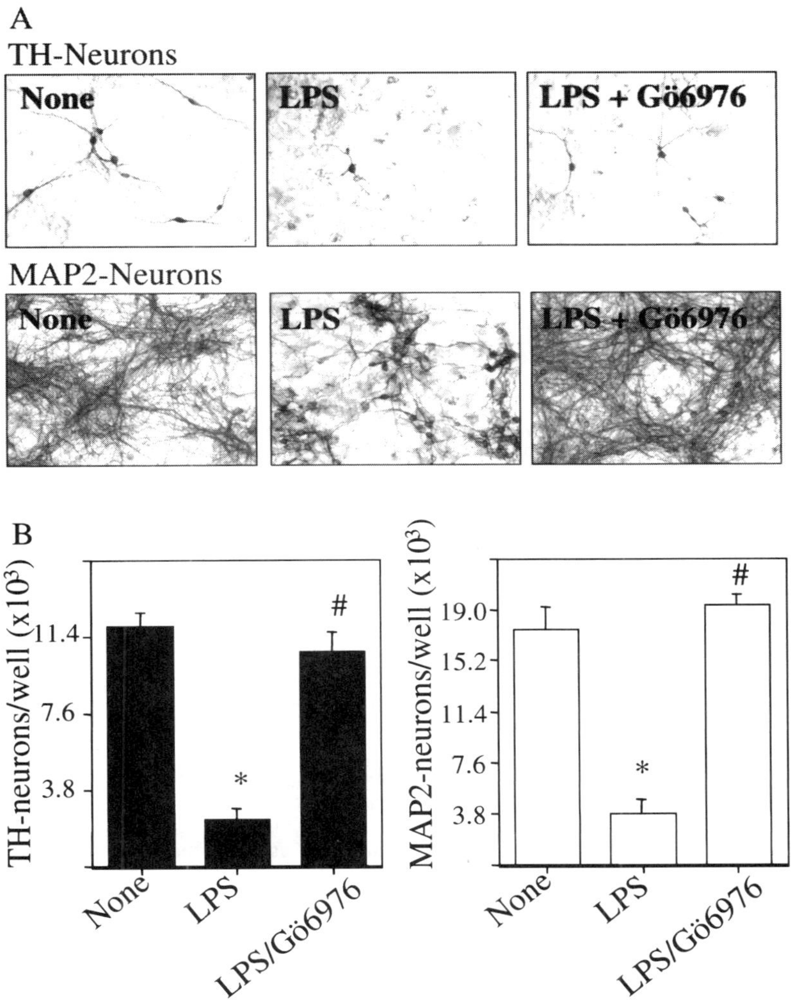

FIGURE 1. Neuroprotective effect of Gö6976 on LPS-induced neurotoxicity. Mesencephalic neuron/glia cultures were pretreated with Gö6976 for one hour before exposure to LPS (10 ng/ml) for 48 h and immunostaining with anti-TH and anti-MAP2 antibodies. **A. Upper panel**, TH-positive cells; **Lower panel**, MAP2-positive cells. **B.** Quantification of TH- and MAP2-positive cells. Experiments were repeated at least five times. $^{*}p < 0.0001$ compared to non-treatment; $^{\#}p < 0.0001$ compared to LPS-treatment.

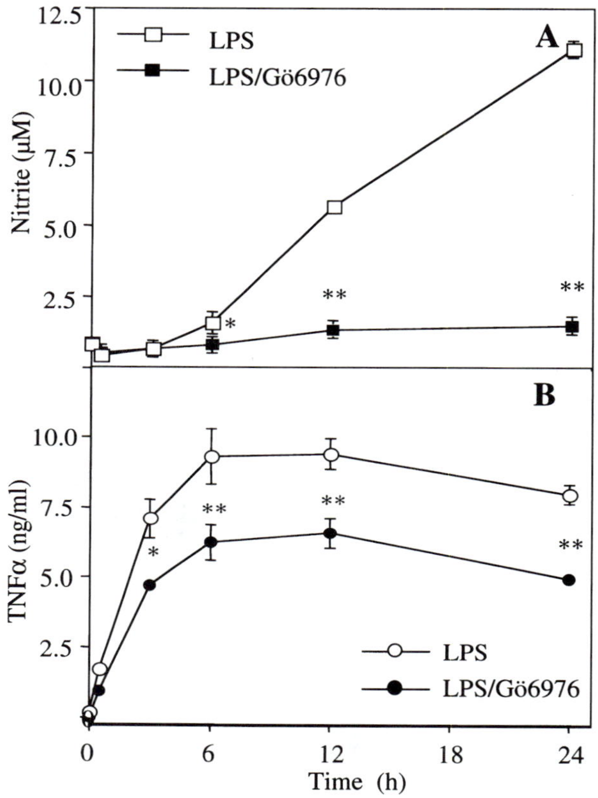

FIGURE 2. Inhibition by Gö6976 of LPS-induced increase in the production of NO and the secretion of TNFα. Mesencephalic neuron/glia cultures were pretreated with Gö6976 (0.5 μM) one hour before exposure to LPS (10 ng/ml) for the indicated time intervals. NO and TNFα in culture media were analyzed as described in *Methods* (n = 3). $^{\#}p < 0.05$ $^{\#\#}p < 0.0001$ compared to each corresponding LPS-treatment.

Gö6976 Suppressed LPS-Induced Phosphorylation of p38 MAPK

To determine if the action of Gö6976 correlated with the regulation of p38 MAPK, the effects of Gö6976 on the activation of p38 MAPK was examined. Although the non-phosphorylated form of p38 MAPK was relatively constant over 270 min, phosphorylation of p38 MAPK markedly increased after 30–90 min exposure to LPS in mesencephalic neuron/glia cultures (see FIGURE 3). Gö6976 inhibited the LPS-induced increase in the phosphorylation of p38 MAPK in a concentration dependent manner. At 0.5 μM, the inhibitor decreased the LPS-induced increase in phospho-p38 by more than 30% and 1.5 μM Gö6976 decreased the LPS effect by more than 55% (data not shown). The inhibitor Gö6976 decreased the LPS-induced increase in phospho-p38 at all time points (FIG. 3), indicating that Gö6976 affects a process that leads to the phosphorylation of p38 MAPK.

Effects of Other Protein Kinase Inhibitors on LPS-Induced Increase in p38 MAPK Activation

Treatment of ventral midbrain neuron/glia with a selective p38 MAPK inhibitor, SB202190, before LPS-exposure did not reduce the LPS-induced increase in phospho-p38 (see FIGURE 4). SB202190 also failed to suppress the LPS-induced increase in p38 MAPK activity, as measured by ATF-2 phosphorylation. Instead, SB202190 stimulated the LPS-induced increase in total p38 MAPK activity (see FIGURE 5). On the other hand, pretreatment with Gö6976 not only decreased the amount of phospho-p38 (FIG. 4), but also suppressed the LPS-induced increase in p38 MAPK activity (FIG. 5). Since Gö6976 has been used as an inhibitor for several PKC subtypes, including

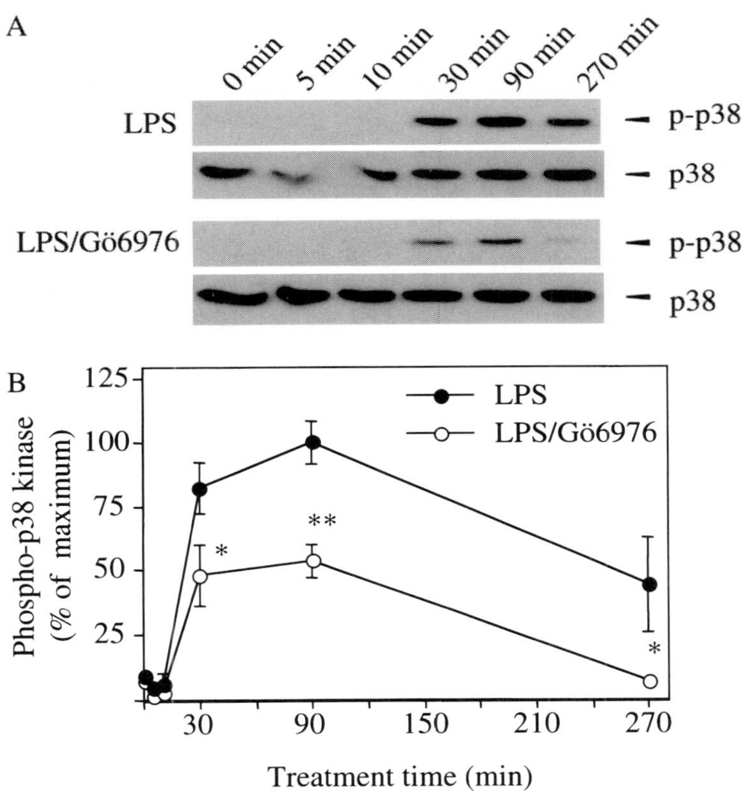

FIGURE 3. Effect of Gö6976 on LPS-induced increase in phosphorylation of p38 MAPK. Mesencephalic neuron/glia cultures were pretreated with 0.5 μM Gö6976 for one hour before exposure to LPS (10 ng/ml). Cell lysates were prepared and phospho-p38 and -p38 were analyzed by immunoblot as described in METHODS. **A.** LPS-induced phosphorylation of p38 MAPK in the absence (**upper panel**) and presence (**lower panel**) of Gö6976. **B.** Densitometric quantification of phospho-p38. Experiments were repeated at least three times. $^{*}p < 0.05$ and $^{**}p < 0.01$ compared to each corresponding LPS-treatment.

Ca^{2+}-dependent PKCs, PKCη, and PKCμ,[33,44–47] we studied the suppressive effect of Gö6976 on LPS-induced phosphorylation of p38 MAPK to see if this was through the inhibition of PKC activity (FIGS. 4 and 5). Ro-31-8220, which inhibits a broad range of PKC subtypes, including Ca^{2+}-dependent PKC subtypes and PKCη,[46–48]

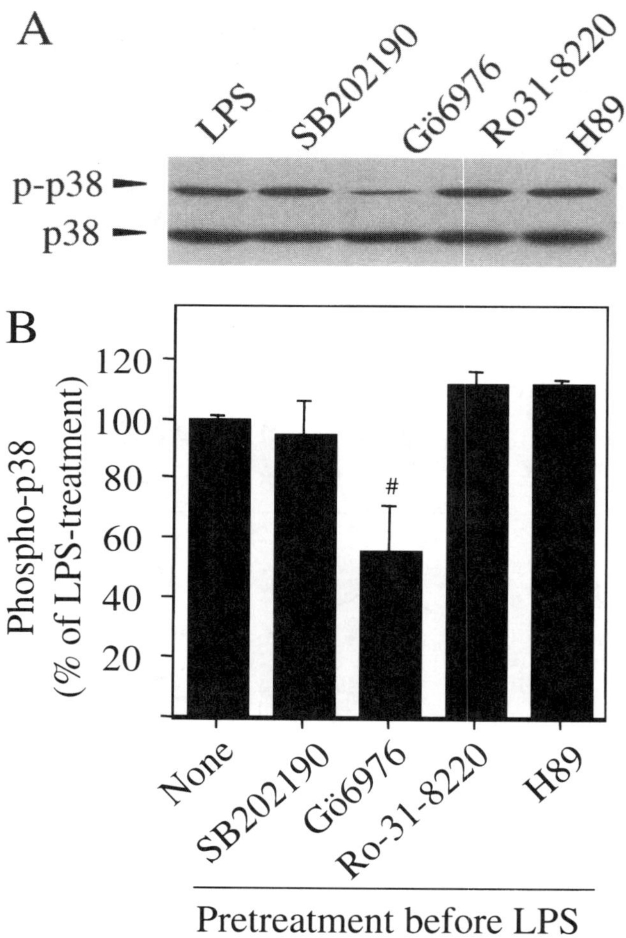

FIGURE 4. Comparison of Gö6976 and other inhibitors on the LPS-induced phosphorylation of p38 MAPK. **A.** Each inhibitor; 92 μM SB202190, 0.5 μM Gö6976, 2 μM Ro31-8220, or 2 μM H89) was added before exposure of neuron/glia to LPS (10 ng/ml for 90 min). Cells were then lysed with sample buffer and the lysate was analyzed for p38 MAPK and phospho-p38 MAPK as described in METHODS. **B.** Densitometric quantification of phospho-p38 MAPK. Experiments were repeated at least three times, and the mean is shown. $^{\#}p < 0.05$ compared to LPS-treatment.

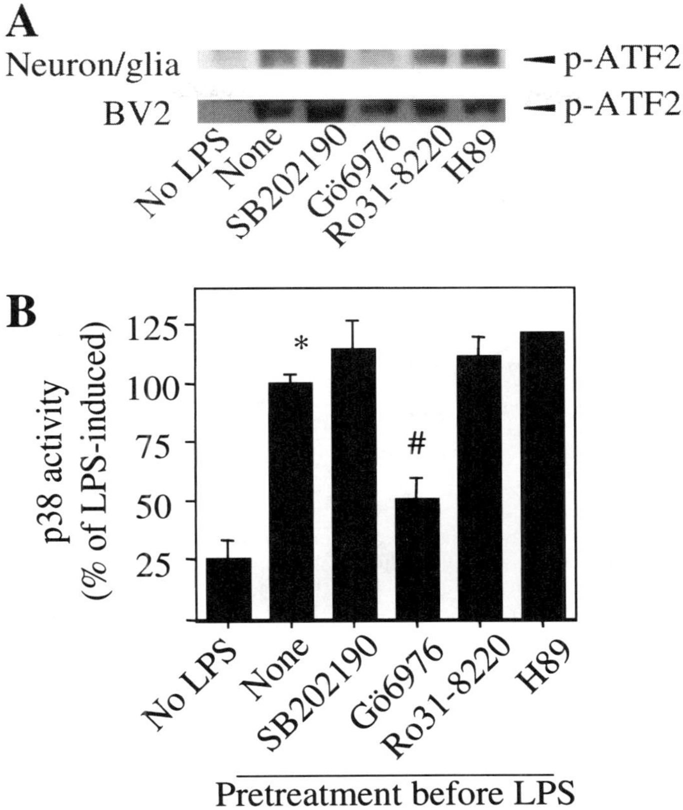

FIGURE 5. Effects of Gö6976 on LPS-induced increase in p38 MAPK activity. **A.** Inhibitor (2 μM SB202190, 0.5 μM Gö6976, 2 μM Ro31-8220, or 5 μM H89) was added before LPS treatment (10 ng/ml for 90 min). Cell extracts were then prepared and p38 MAPK activity was measured, as described in METHODS. **B.** Densitometric quantification of p-ATF2 in BV2 cells. Experiments were repeated at least three times; $^{*}p < 0.001$ compared to non-treatment; and $^{\#}p < 0.01$ compared to LPS-treatment.

suppressed neither the LPS-induced increase in phospho-p38 nor the activity of p38 MAPK after exposure to LPS. H89, a potent inhibitor of PKCμ as well as protein kinase A,[45] also suppressed neither the LPS-induced increase in phospho-p38 nor the activity of p38 MAPK after exposure to LPS (FIGS. 4 and 5).

The results indicate that Gö6976 plays a previously unknown role in the suppression of synthesis of inflammation-related products, a role that is distinct from the inhibition of PKC subtypes. Gö6976 inhibited the phosphorylation of p38 MAPK in microglia and, consequently, interrupted inflammation-mediated neurotoxicity (see FIGURE 6).

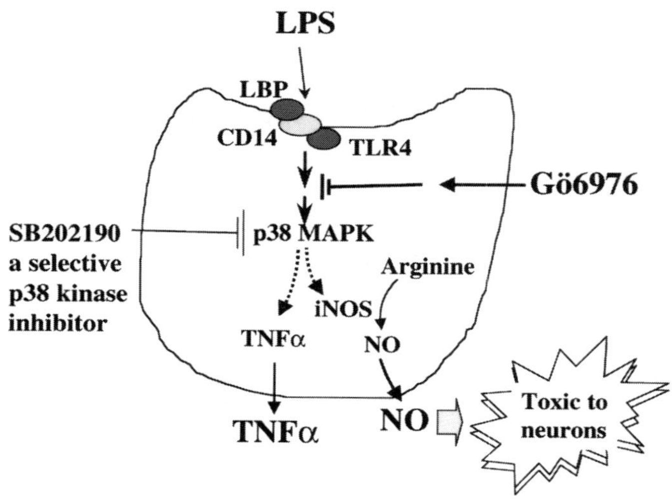

FIGURE 6. Summary of the neuroprotection by Gö6976 in mesencephalic neuron/glia culture. Gö6976 suppresses the LPS-induced increase in NO production by inhibiting the activation of p38 MAPK in glial cells, which, in turn, leads to the protection of neurons from the LPS-elicited neurotoxicity. TLR4, toll-like receptor 4; LBP, LPS binding protein; iNOS, inducible nitric oxide synthase.

DISCUSSION

Activation of p38 MAPK has been observed in brain tissues of patients who suffered from cerebral ischemia or Alzheimer's disease.[49,50] Inhibition of p38 MAPK reduces LPS- or cytokine-induced production of inflammatory factors from glial cells.[32,51] In a previous study, we found that the inhibition of p38 MAPK by the selective inhibitor SB202190 led to a significant protection from LPS-elicited neurotoxicity by suppressing NO production.[52] It was also shown that a selective inhibitor of the inducible form of nitric oxide synthase, GW274150, showed neuroprotection against LPS-elicited neurotoxicity. The suppression of the p38 MAPK phosphorylation by Gö6976 paralleled its inhibition of NO production. These results led us to conclude that the protection of mesencephalic neurons by Gö6976 was mediated through suppression of p38 MAPK.

The present study indicates that the target site for the action of Gö6976 is upstream of p38 MAPK. Gö6976 decreased the amount of phospho-p38 induced by LPS, but this inhibitor did not suppress the p38 MAPK activity assayed *in vitro*.[53] This suggests that Gö6976 interrupts a signaling event upstream of p38 MAPK to decrease the activation of p38 MAPK and suppress the production of NO.

We also provided evidence that PKC is not a target for Gö6976 in protecting neurons against LPS-induced damage. Gö6976 has been used as a selective inhibitor to several PKC subtypes, including Ca^{2+}-dependent PKCs and PKCη.[46–48] However, Ro-31-8220 (an inhibitor to a wide range of PKC subtypes including Ca^{2+}-

dependent PKCs and PKCη) or H89 (an inhibitor to PKCμ, which is also inhibited by Gö6976) failed to inhibit the phosphorylation of p38 MAPK and showed no neuroprotective effects. Among the various PKC inhibitors studied, only Gö6976 exhibited a significant inhibition of the phosphorylation of p38 MAPK, a decrease in cellular p38 MAPK activity, and significant neuroprotection against LPS-induced damage. This study also suggests that inhibition of microglial p38 MAPK phosphorylation is the major target of Gö6976-mediated neuroprotection, rather than astrocytes or neuronal p38 MAPK. We previously reported that microglia are responsible for the initial increase in iNOS expression and NO production in mixed glia cultures.[38] Furthermore, analysis of the effect of Gö6976 on the phosphorylation of p38 MAPK using a pure microglial cell line (BV2) indicated that a similar manner of inhibition occurs in both culture systems. Thus, we conclude that inhibition of p38 MAPK activation in microglia is responsible for the neuroprotective effect of Gö6976 against LPS.

In conclusion, our results indicate that Gö6976 suppresses the phosphorylation of p38 MAPK and the consequent p38 MAPK activation induced by LPS in mesencephalic neuron/glia cultures. Inhibition of p38 MAPK activation led to the inhibition of LPS-induced increase in NO production and consequently, interrupted the LPS-induced neurotoxicity. The inhibitory function of Gö6976 on the phosphorylation of p38 MAPK was not mediated through the inhibition of PKC. Rather, one of the signaling events upstream of p38 MAPK was the target of Gö6976. Further studies on the functional target of Gö6976 in the p38 MAPK-related cascade are needed to provide insights into the mechanism of inflammation-related neurotoxicity and potential targets for intervention.

REFERENCES

1. Brosnan, C.F., L. Battistini, C.S. Raine, *et al.* 1994. Reactive nitrogen intermediates in human neuropathology: an overview. Dev. Neurosci. **16:** 152–161.
2. Chao, C.C., S. Hu, T.W. Molitor, *et al.* 1992. Activated microglia mediate neuronal cell injury via a nitric oxide mechanism. J. Immunol. **149:** 2736–2741.
3. Dickson, D.W., S.C. Lee, L.A. Mattiace, *et al.* 1993. Microglia and cytokines in neurological disease, with special reference to AIDS and Alzheimer's disease. Glia **7:** 75–83.
4. Espey, M.G., O.N. Chernyshev, J.F. Reinhard, *et al.* 1997. Activated human microglia produce the excitotoxin quinolinic acid. Neuroreport **8:** 431–434.
5. Lee, S.C., W. Liu, D.W. Dickson, *et al.* 1993. Cytokine production by human fetal microglia and astrocytes. Differential induction by lipopolysaccharide and IL-1 beta. J. Immunol. **150:** 2659–2667.
6. Matsuo, M., Y. Hamasaki, F. Fujiyama, *et al.* 1995. Eicosanoids are produced by microglia, not by astrocytes, in rat glial cell cultures. Brain Res. **685:** 201–204.
7. Merrill, J.E., L.J. Ignarro, M.P. Sherman, *et al.* 1993. Microglial cell cytotoxicity of oligodendrocytes is mediated through nitric oxide. J. Immunol. **151:** 2132–2141.
8. Minghetti, L. & G. Levi. 1995. Induction of prostanoid biosynthesis by bacterial lipopolysaccharide and isoproterenol in rat microglial cultures. J. Neurochem. **65:** 2690–2698.
9. Boje, K.M. & P.K. Arora. 1992. Microglial-produced nitric oxide and reactive nitrogen oxides mediate neuronal cell death. Brain Res. **587:** 250–256.
10. Chao, C.C., S. Hu & P.K. Peterson. 1995. Glia, cytokines, and neurotoxicity. Crit. Rev. Neurobiol. **9:** 189–205.

11. DAWSON, V.L., H.P. BRAHMBHATT, J.A. MONG, *et al.* 1994. Expression of inducible nitric oxide synthase causes delayed neurotoxicity in primary mixed neuronal-glial cortical cultures. Neuropharmacology **33:** 1425–1430.
12. GALEA, E., D.L. FEINSTEIN & D.J. REIS. 1992. Induction of calcium-independent nitric oxide synthase activity in primary rat glial cultures. Proc. Natl. Acad. Sci. U.S.A. **89:** 10945–10949.
13. II, M., M. SUNAMOTO, K. OHNISHI, *et al.* 1996. β-Amyloid protein-dependent nitric oxide production from microglial cells and neurotoxicity. Brain Res. **720:** 93–100.
14. KONG, L.Y., M.K. MCMILLIAN, R. MARONPOT, *et al.* 1996. Protein tyrosine kinase inhibitors suppress the production of nitric oxide in mixed glia, microglia-enriched or astrocyte-enriched cultures. Brain Res. **729:** 102–109.
15. LIPTON, S.A. 1994. Neuronal injury associated with HIV-1 and potential treatment with calcium-channel and NMDA antagonists. Dev. Neurosci. **16:** 145–151.
16. MEDA, L., M.A. CASSATELLA, G.I. SZENDREI, *et al.* 1995. Activation of microglial cells by β-amyloid protein and interferon-γ. Nature **374:** 647–650.
17. SIMMONS, M.L. & S. MURPHY. 1992. Induction of nitric oxide synthase in glial cells. J. Neurochem. **59:** 897–905.
18. MATSUMOTO, Y., K. OHMORI & M. FUJIWARA. 1992. Microglial and astroglial reactions to inflammatory lesions of experimental autoimmune encephalomyelitis in the rat central nervous system. J. Neuroimmunol. **37:** 23–33.
19. MCGEER, P.L., S. ITAGAKI, B.E. BOYES, *et al.* 1988. Reactive microglia are positive for HLA-DR in the substantia nigra of Parkinson's and Alzheimer's disease brains. Neurology **38:** 1285–1291.
20. RAINE, C.S. 1994. Multiple sclerosis: immune system molecule expression in the central nervous system. J. Neuropathol. Exp. Neurol. **53:** 328–337.
21. ROGERS, J., J. LUBER-NAROD, S.D. STYREN, *et al.* 1988. Expression of immune system-associated antigens by cells of the human central nervous system: relationship to the pathology of Alzheimer's disease. Neurobiol. Aging. **9:** 339–349.
22. SHERMAN, M.P., J.M. GRISCAVAGE & L.J, IGNARRO. 1992. Nitric oxide-mediated neuronal injury in multiple sclerosis. Med. Hypotheses **39:** 143–146.
23. YOUDIM, M.B., L. LAVIE & P. RIEDERER. 1994. Oxygen free radicals and neurodegeneration in Parkinson's disease: a role for nitric oxide. Ann. N.Y. Acad. Sci. **738:** 64–68.
24. GALEA, E., D.J. REIS, E.S. FOX, *et al.* 1996. CD14 mediate endotoxin induction of nitric oxide synthase in cultured brain glial cells. J. Neuroimmunol. **64:** 19–28.
25. BEUTLER, B. 2000. Tlr4: central component of the sole mammalian LPS sensor. Curr. Opin. Immunol. **12:** 20–26.
26. LAFLAMME, N. & S. RIVEST. 2001. Toll-like receptor 4: the missing link of the cerebral innate immune response triggered by circulating gram-negative bacterial cell wall components. FASEB J. **15:** 155–163.
27. YANG, H., D.W. YOUNG, F. GUSOVSKY, *et al.* 2000. Cellular events mediated by lipopolysaccharide-stimulated toll-like receptor 4. MD-2 is required for activation of mitogen-activated protein kinases and Elk-1. J. Biol. Chem. **275:** 20861–20866.
28. BOULET, I., S. RALPH, E. STANLEY, *et al.* 1992. Lipopolysaccharide– and interferon–gamma-induced expression of hck and lyn tyrosine kinases in murine bone marrow-derived macrophages. Oncogene **7:** 703–710.
29. SHAPIRA, L., S. TAKASHIBA, C. CHAMPAGNE, *et al.* 1994. Involvement of protein kinase C and protein tyrosine kinase in lipopolysaccharide-induced TNF-α and IL-1β production by human monocytes. J. Immunol. **153:** 1818–1824.
30. WEINSTEIN, S.L., M.R. GOLD & A.L. DEFRANCO. 1991. Bacterial lipopolysaccharide stimulates protein tyrosine phosphorylation in macrophages. Proc. Natl. Acad. Sci. U.S.A. **88:** 4148–4152.
31. ZIEGLER, S.F., C.B. WILSON & R.M. PERLMUTTER. 1988. Augmented expression of a myeloid-specific protein tyrosine kinase gene (hck) after macrophage activation. J. Exp. Med. **168:** 1801–1810.
32. BHAT, N.R., P. ZHANG, J.C. LEE, *et al.* 1998. Extracellular signal-regulated kinase and p38 subgroups of mitogen- activated protein kinases regulate inducible nitric oxide synthase and tumor necrosis factor-alpha gene expression in endotoxin-stimulated primary glial cultures. J. Neurosci. **18:** 1633–1641.

33. CHEN, C.C., J.K. WANG, W.C. CHEN, *et al.* 1998. Protein kinase C eta mediates lipopolysaccharide-induced nitric-oxide synthase expression in primary astrocytes. J. Biol. Chem. **273:** 19424–19430.
34. FEINSTEIN, D.L., E. GALEA, J. CERMAK, *et al.* 1994. Nitric oxide synthase expression in glial cells: suppression by tyrosine kinase inhibitors. J. Neurochem. **62:** 811–814.
35. KITAMURA, Y., R. IMAIZUMI, Y. KITAYAMA, *et al.* 1996. Possible involvement of tyrosine kinase activation in lipopolysaccharide-induced expression of Ca^{2+}-insensitive but calmodulin-coupling nitric oxide synthase in rat glial cells. J. Neurosci. Res. **43:** 235–245.
36. SCHUMANN, R.R., D. PFEIL, D. FREYER, *et al.* 1998. Lipopolysaccharide and pneumococcal cell wall components activate the mitogen activated protein kinases (MAPK) erk-1, erk-2, and p38 in astrocytes. Glia **22:** 295–305.
37. SIMMONS, M.L. & S. MURPHY. 1994. Roles for protein kinases in the induction of nitric oxide synthase in astrocytes. Glia **11:** 227–234.
38. JEOHN, G.H., R.C. CHANG, W.G. KIM, *et al.* 2000. Post-transcriptional inhibition of lipopolysaccharide-induced expression of inducible nitric oxide synthase by Go6976 in murine microglia. Brain Res. Mol. Brain Res. **79:** 18–31.
39. JEOHN, G.H., B. WILSON, W.C. WETSEL, *et al.* 2000. The indolocarbazole Go6976 protects neurons from lipopolysaccharide/interferon-gamma-induced cytotoxicity in murine neuron/glia co-cultures. Brain Res. Mol. Brain Res. **79:** 32–44.
40. LIU, B., L. DU & J.S. HONG. 2000. Naloxone protects rat dopaminergic neurons against inflammatory damage through inhibition of microglia activation and superoxide generation. J. Pharmacol. Exp. Ther. **293:** 607–617.
41. BLASI, E., R. BARLUZZI, V. BOCCHINI, *et al.* 1990. Immortalization of murine microglial cells by a v-raf/v-myc carrying retrovirus. J. Neuroimmunol. **27:** 229–237.
42. BOCCHINI, V., R. MAZZOLLA, R. BARLUZZI, *et al.* 1992. An immortalized cell line expresses properties of activated microglial cells. J. Neurosci. Res. **31:** 616–621.
43. GREEN, L.C., D.A. WAGNER, J. GLOGOWSKI, *et al.* 1982. Analysis of nitrate, nitrite, and [15N]nitrate in biological fluids. Ann. Biochem. **126:** 131–138.
44. GSCHWENDT, M., S. DIETERICH, J. RENNECKE, *et al.* 1996. Inhibition of protein kinase C μ by various inhibitors. Differentiation from protein kinase c isoenzymes. FEBS Lett. **392:** 77–80.
45. JOHANNES, F.J., J. PRESTLE, S. DIETERICH, *et al.* 1995. Characterization of activators and inhibitors of protein kinase C μ. Eur. J. Biochem. **227:** 303–307.
46. MARTINY-BARON, G., M.G. KAZANIETZ, H. MISCHAK, *et al.* 1993. Selective inhibition of protein kinase C isozymes by the indolocarbazole Go 6976. J. Biol. Chem. **268:** 9194–9197.
47. QATSHA, K.A., C. RUDOLPH, D. MARME, *et al.* 1993. Go 6976, a selective inhibitor of protein kinase C, is a potent antagonist of human immunodeficiency virus 1 induction from latent/low-level-producing reservoir cells *in vitro*. Proc. Natl. Acad. Sci. U.S.A. **90:** 4674–4678.
48. WILKINSON, S.E., P.J. PARKER & J.S. NIXON. 1993. Isoenzyme specificity of bisindolylmaleimides, selective inhibitors of protein kinase C. Biochem. J. **294:** 335–337.
49. HENSLEY, K., R.A. FLOYD, N.Y. ZHENG, *et al.* 1999. p38 kinase is activated in the Alzheimer's disease brain. J. Neurochem. **72:** 2053–2058.
50. WALTON, K.M., R. DIROCCO, B.A. BARTLETT, *et al.* 1998. Activation of p38MAPK in microglia after ischemia. J. Neurochem. **70:** 1764–1767.
51. DA SILVA, J., B. PIERRAT, J.L. MARY, *et al.* 1997. Blockade of p38 mitogen-activated protein kinase pathway inhibits inducible nitric-oxide synthase expression in mouse astrocytes. J. Biol. Chem. **272:** 28373–28380.
52. JEOHN, G.-H., C.L. COOPER, B. WILSON, *et al.* 2002. p38 MAP kinase is involved in lipopolysaccharide induced dopaminergic neuronal cell death in rat mesencephalic neuron-glia cultures. Ann. N.Y. Acad. Sci. **962:** this volume.
53. JEOHN, G.H., C. COOPER, K.-J. JANG, *et al.* 2002. Gö6976 inhibits LPS-induced microglial TNFα release by suppressing p38 MAP kinase activation. Neuroscience. In press.

Nitric Oxide-Releasing Drugs

A Novel Class of Effective and Safe Therapeutic Agents

JEAN-LUC BURGAUD,[a] ENNIO ONGINI,[b] AND PIERO DEL SOLDATO[a]

[a]*NicOx SA, Espace Gaia II - Bâtiment I BP 313, Sophia Antipolis, France*

[b]*Nicox Research Institute, Bresso, Milan, Italy*

ABSTRACT: Nitric oxide (NO) deficiency has been implicated in many pathologic processes, thus providing a solid biological basis for the use of NO replacement therapy. Exogenous NO sources constitute a powerful way to supplement NO when the body cannot generate sufficient NO for normal biological functions. This theory has opened up the possibility of designing new drugs that are capable of delivering NO into tissues and the bloodstream in a sustained and controlled manner. This objective has been achieved by grafting an organic nitrate structure onto existing drugs through chemical spacers, such as aliphatic, aromatic, or a heterocyclic chain. The approach has led to the synthesis of several new chemical entities whose pharmacologic profile challenges the parent drug, not only on the basis of new properties, but also with respect to a better safety profile. In this article, a specific class of NO donors is reviewed, the nitric oxide-releasing non-steroidal antiinflammatory drugs, NO-NSAIDs. Recently discovered compounds, whose action depends on the combined properties of both the known drug and NO release, are illustrated. Two examples are described in detail: (1) nitric oxide-releasing aspirin, which has demonstrable innovative properties for treatment of vascular disorders and cancer; (2) nitro-derivatives of flurbiprofen that have shown encouraging results in models of Alzheimer's disease.

KEYWORDS: nitric oxide-releasing drugs; nitro-asprin; NCX 4016; nitro-(flurbiprofen); HCT 1026; NCX 2216; nitro-NSAIDs

NITRIC OXIDE DONOR DRUGS

Several chemical donors of nitric oxide (NO) are currently being used in various types of experimental study. Indeed, these drugs are the most convenient approach to study the role of NO in physiology and pathology. There are a number of pharmacologic tools available, ranging from *S*-nitro-acetylpenicillamine (SNAP) and nitroprusside to the most long-acting NO donors, such as NOC-18 and others.[1] By contrast, other experimental tools are available that make it possible to inhibit NO synthesis. The most useful strategy remains that of using the L-arginine analogues, such as *N*-nitro-L-arginine methyl ester (L-NAME).

In therapy, NO has received much attention.[2] It is now appreciated that NO can be inhaled at low concentration to produce clinically useful bronchodilation. Thus,

Address for correspondence: Dr. Piero Del Soldato, NicOx SA 2455, routes des Dolines, Espace Gaia II - Bâtiment I, BP 313 06906 - Sophia, Antipolis, France. Voice: +33 4 92 38 70 20; fax: +33 4 92 38 70 30.
delsol@tin.it

administration of NO through the inhaled route has proved effective in disorders, such as adult distress syndrome or pulmonary hypertension. Although the use of NO as a therapeutic gas is recent, NO donors have been used in patients for more than a century. Initially, amyl nitrite and, later, glyceryl trinitrate (see FIGURE 1) were widely used as antianginal drugs. Efforts to increase the action duration of glyceryl trinitrate have led to the synthesis of several related organic nitrates, such as isosorbide mononitrate or dinitrate.[2,3] Other, so called, nitrovasodilators also possess the property of generating NO, therefore, increasing cGMP formation intracellularly. One such compound is sodium nitroprusside (FIG. 1).

Recently, a wide range of novel NO donor classes have emerged. Some of these drug classes appear to be of interest for their potential therapeutic applications. Worth mentioning are the *S*-nitrosothiols, the diazeniumdiolates (formerly NONOates) and mesoionic oxatriazoles. There are excellent reviews available describing both the chemistry and the pharmacology of the new NO donor compounds.[2–4] In this paper we illustrate in detail a class of hybrid NO donors that combine a nitrate moiety with pharmacologically active structures. These drugs look promising for their advantages over conventional NO donors.[5,6]

NITRIC OXIDE-RELEASING NON-STEROIDAL ANTIINFLAMMATORY DRUGS

Non-steroidal antiinflammatory drugs (NSAIDs) are among the most widely prescribed medicines worldwide; they are the drugs of first choice for treatment of rheumatic disorders and other degenerative inflammatory joint diseases. However, their use is associated with gastrointestinal (GI) effects, which are among the most common serious adverse drug-related events.

A novel class of NSAID, the nitro-NSAIDs (NO-NSAIDs), have recently been developed by using an innovative technology.[5,6] that is, by grafting an organic nitrate

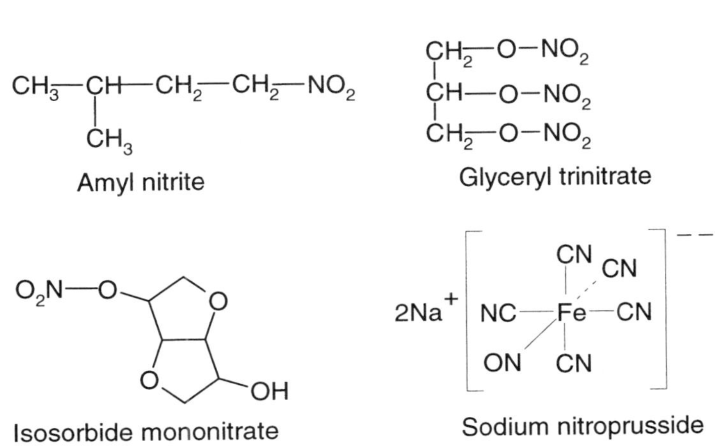

FIGURE 1. Structures of classical nitric oxide (NO) donors.

moiety onto a well established NSAID using various chemical spacers (see FIGURE 2). Generally, the NO-NSAIDs maintain the antiinflammatory properties of the parent compounds while showing enhanced tolerability and a wider range of pharmacologic activities. The organic nitrate moiety slowly releases NO, therefore, adding the beneficial properties of NO and improving the safety profile of the new drug. Specifically, NO-NSAIDs are characterized by a dramatic reduction in GI side effects[7,8] in comparison with the parent compounds, due to the protective effects of NO on gastric mucosa. There is evidence that NO increases blood flow in the gastric mucosa and inhibits the adherence of leukocytes to the endothelium within the GI microcirculation.[9–12] Other significant advantages derive from the chemical spacer used in each molecule. The use of aliphatic, aromatic, or heterocyclic spacers makes it possible to achieve various chemical-physical characteristics and various kinetics of NO release. Moreover, spacers with antioxidant properties, such as ferulic acid, have also been included in some drugs, thereby improving their overall pharmacologic properties.

Animal and human studies show that NO-releasing derivatives of NSAIDs spare gastrointestinal mucosa and that any dosage of nitrofenac, NO-naproxen, NO-flurbiprofen, NO-ketoprofen, or NO-aspirin causes significantly less gastric mucosal injury in rats than conventional NSAIDs.[6]

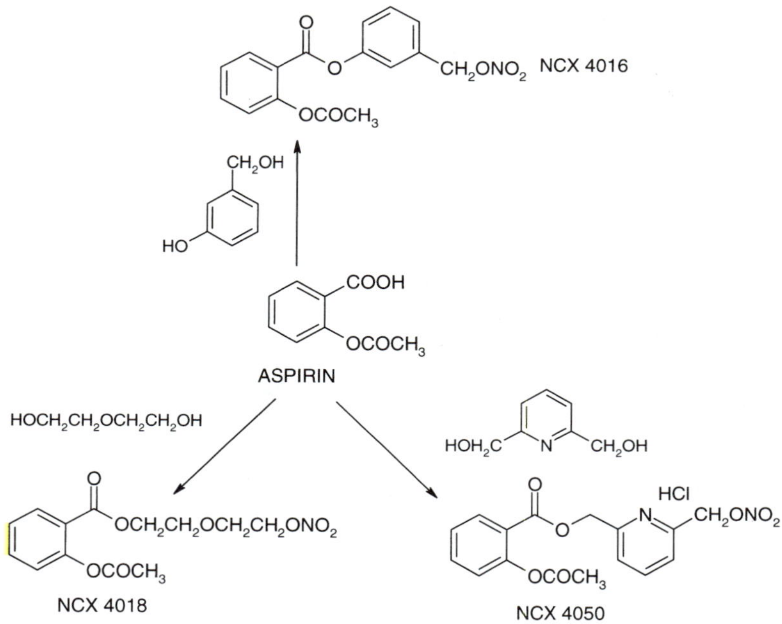

FIGURE 2. Structures of selected nitric oxide-releasing aspirins. Note that various linkers are used to combine the nitric ester moiety with acetylsalicylate.

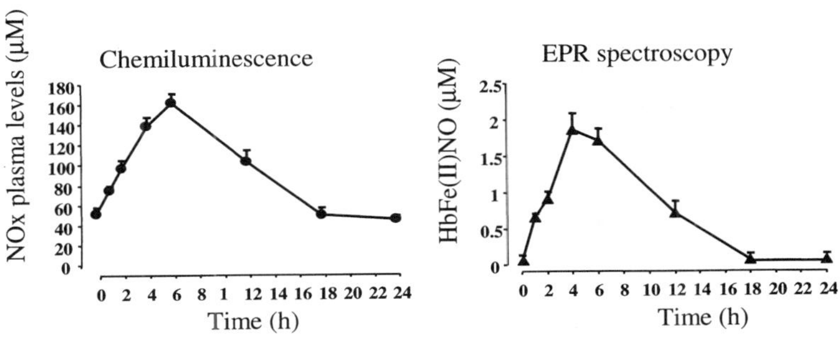

FIGURE 3. Kinetics of NO formation following oral administration of the nitric oxide-releasing aspirin, NCX 4016, in the rat. NO release was assessed by a chemiluminescence approach (**left**) and electron paramagnetic resonance (EPR) spectroscopy (**right**) (Figure modified from Ref. 29, with permission.)

Despite the NO effects on gastric microcirculation, it is noteworthy that these drugs do not change arterial blood pressure even when administered intravenously in large doses, whereas an equimolar dose of a conventional NO donor, sodium nitroprussiate, for example, causes marked fall of blood pressure.[13,14] Following NO-NSAID administration there is a significant increase in plasma levels of nitrate/nitrite, consistent with release of a nitric oxide in the systemic circulation, but the absence of hypotensive effects could be explained by the fact that *in vivo* hydrolysis of NO compounds occurs slowly (see FIGURE 3).[14,15]

NITRIC OXIDE-RELEASING ASPIRIN

The design of an aspirin derivative that is able to release NO, therefore counterbalancing the negative effect due to prostaglandin inhibition at the gastrointestinal level, has led to the synthesis of 2-(acetyloxy)benzoic acid 3-(nitrooxymethyl)phenyl ester NCX 4016 (FIG. 2).[5,6,13] This compound exhibits better pharmacologic activity than aspirin itself thanks to the safe GI profile and more effective activity at the vascular level.[15]

The biological activity of NCX 4016 was evaluated using a variety of experimental models. Clearly, NCX 4016 retains the antiinflammatory and analgesic activities of aspirin but, more interestingly, it exhibits better activity in the thrombotic process and cancer proliferation, both *in vitro* and *in vivo*.[6]

NCX 4016 is more potent than aspirin in inhibiting platelet activation induced by thrombin. The antithrombotic activity of NCX 4016 was studied *in vivo* in different models of platelet-dependent and platelet-independent pulmonary thromboembolism.[17] Whereas aspirin only protects mice from death induced by collagen plus epinephrin, NCX 4016 also protects mice from death induced by intravenous injection of U46619 (9,11-dideoxy-1,9-epoxymethano-prostaglandin F2). Moreover, NCX 4016 is more effective than the parent compound in decreasing the platelet count and the number of lung emboli. Unlike aspirin, NCX 4016 protects mice from mechanical pulmonary embolism induced by injection of hardened red blood cells. NCX 4016 exerts a significant inhibitory effect in reducing the accumulation of platelets in the rabbit pulmonary vasculature induced by both collagen and thrombin, whereas aspirin is ineffective on the thrombin-induced accumulation.[17]

The effects of NCX 4016 versus aspirin on leukocyte adherence to the vascular endothelium was studied in rats. Pretreatment with NCX 4016 does not affect basal leukocyte adherence, whereas pretreatment with aspirin shows a five-fold increase in the number of leukocytes adhering to the vessel wall. Moreover, NCX 4016, but not aspirin, is able to inhibit *N*-formyl-Met-Leu-Phe-induced leukocyte adherence to postcapillary mesenteris venules. Taken together, these data confirm that NCX 4016 exerts a more pronounced antithrombotic activity than aspirin, largely due to a marked inhibitory effect on platelets.

The use of aspirin as a means of reducing mortality from acute myocardial infarct is well documented, however, its effects are most likely due to prevention of reinfarction rather than any direct cardioprotective effect. Thus, should coronary occlusion occur, aspirin would not provide protection against either arrhythmias or myocardial injury. NO donors have been shown to reduce myocardial infarct size and to exert moderate antiarrhythmic activity. The effects of NO-aspirin were compared with those of native aspirin on ventricular arrhythmias and myocardial infarct size in anesthetized pigs.[18] This experiment showed that NCX 4016 reduces infarct size by a mechanism that is independent of either its antiplatelet effects or its effects on leukocyte adhesion.

NCX 4016 was also shown to improve postischemic ventricular dysfunction and to reduce myocardial infarct size in rabbit.[19,20] Cardioprotection conferred by NCX 4016 is evident in rats subjected to myocardial ischemia/reperfusion, as shown in both reduction of ventricular premature beats and incidence of ventricular tachycardia and fibrillation.[13] In these animals, infarct size is restricted proportionally to the dose of NCX 4016 associated with diminution of both plasma creatine phosphokinase and cardiac myeloperoxidase activities. Thus, NO release associated with aspirin confers protection against cardiac ischemia.

NCX 4016 is also effective on vascular smooth muscle cell proliferation, such as that observed following restenosis after balloon dilation of carotid arteries in mice. Using LDL-receptor deficient mice having hypercholesterolemia, NCX 4016 inhibits macrophage deposition at the site of injury more effectively than aspirin.[21] Overall, NCX 4016 significantly reduces neointimal hyperplasia and macrophage accumulation, therefore showing beneficial effects on development of restenosis.[21] These effects are also evident when the nitro-aspirin is administered one week prior to injury. The mechanism underlying vascular smooth cell proliferation was investigated *in vitro* using rat aorta. There, it was shown that the arginine-NO pathway has an inhibitory role and the action of NO donors occurs through cGMP-independent mechanisms.[22]

Aspirin has been shown to decrease the risk of colorectal cancer. To assess the potential interest of NCX 4016 in colon cancer chemoprevention, the effect of NO-aspirin on colonocyte kinetics was studied *in vitro*. NCX 4016 inhibits cell proliferation, induces apoptosis, and alters the cell cycle phase distribution, often producing a G_0/G_1 cell cycle block.[23] NCX 4016 is consistently more effective than aspirin in reducing the growth of colon and pancreatic cancer cells *in vitro*. These antiproliferative, proapoptotic, and cell cycle effects account for its action on cell growth, likely contributing to its putative chemopreventive and/or chemotherapeutic activity.

The chemopreventive effects of aspirin versus NCX 4016 were also tested *in vivo*. In a rat model of colon adenocarcinoma, NCX 4016 is significantly more effective than aspirin in preventing preneoplastic lesions in the colon.[24] Interestingly, the effects of NCX 4016 are not associated with inhibition of COX-1 or COX-2; instead they appear to be related to the release of NO.

Release of NO by NCX 4016

The pharmacologic properties of the NO-releasing aspirin are likely due to the combination of two major actions: the effects of aspirin *per se* and generation of NO in tissues. To elucidate this hypothesis, studies were performed using both *in vitro* and *in vivo* assays. In the isolated blood vessel preparation, the compound induces a concentration dependent relaxation of both intact and endothelium-denuded, adrenaline-precontracted arteries through an NO-dependent mechanism.[25] Using platelets, it was shown that NO is released following drug treatment, as measured by studying either stimulation of cGMP or inhibition of thrombin-induced aggregation.[26,27] In peripheral mononuclear blood cells (PMBC) exposed to lipopolysaccharide (LPS) stimulation, an *in vitro* model that is sensitive to NO modulation, NCX 4016 produces an increase of nitrate/nitrite in the supernatant.[28] Moreover, since NO inhibits ICE/caspase 1-like proteases and cytokine release in PMBC stimulated by LPS, the effects produced by NCX 4016 were compared with those of SNAP, an experimental compound that rapidly releases NO. Incubating PMBC with NCX 4016 induced a time dependent increase in nitrite/nitrate released in cell supernatants. However, in contrast to SNAP, which causes a rapid increase of nitrite/nitrate generation, production induced by NCX 4016 had a delayed onset and required about 3–6 hours to reach a steady state.[13,29] Measurements of intracellular signaling through cGMP-dependent pathways provide indirect evidence of NO formation, assessed as [cGMP]i, in PMBC challenged with SNAP, NCX 4016, or aspirin. Incubating human monocytes with SNAP or NCX 4016, but not aspirin, resulted in a time-dependent increase in [cGMP]i, to various extents. Exposure to NCX 4016 causes a 10-fold increase in [cGMP]i. Incubating cells with ODQ, to inhibit cyclic guanylyl cyclase, completely prevents the stimulatory effect exerted by SNAP and NCX 4016 on [cGMP]i. Thus, PMBC are able to release NO-derived compounds and generate cGMP through a guanylyl cyclase-dependent pathway in response to NCX 4016. The addition of SNAP or NCX 4016, but not aspirin, to the medium resulted in a time dependent increase in intracellular NO accumulation, as measured by DAF-2 generated fluorescence, indicating that NCX 4016 penetrates cell membranes and is metabolized to release free NO.

Hemoglobin (Hb) regulates the chemistry of NO and maintains it in a bioactive state as nitrosylhemoglobin, Fe(II)-HbNO. Since nitrosyl heme proteins are paramagnetic, electron paramagnetic resonance (EPR) was used to quantify the NO released from the NCX 4016.[28] For both *in vitro* and *in vivo* experiments the paramagnetic complex was determined in rat venous blood. In *in vivo* studies with NCX 4016, NO-Fe(II)-Hb was detectable one hour after dosing and its formation was maximal at 4–6 hours, at which time the antithrombotic activity peaks. Plasma nitrosothiol level peaks occurred later, at 12 hours. (FIG. 3).[29]

NITRIC OXIDE-RELEASING FLURBIPROFEN

Several epidemiologic observations show that prolonged use of NSAIDs can be beneficial in Alzheimer's disease (AD), the most common form of dementia among the elderly.[30] Neuropathologic examination of an AD brain reveals extensive brain atrophy due to neuronal loss, accumulation of neurofibrillary tangles (NFT) composed of τ protein, and amyloid deposits consisting of fibrils of variously sized β-amyloid peptides. In addition, an AD brain presents features typical of chronic neuroinflammation, such as astrocytosis, microglial activation, and altered levels of pro- and antiinflammatory mediators, including cytokines, oxygen free radicals, nitrogen species, and products of activation of the complement system.[31] The inflammatory reaction is localized mainly in the periplaque area and potentially contributes to neurodegeneration through imbalance of the levels of pro- and antiinflammatory mediators.

It has been reported that the risk of AD diminishes with increased duration of NSAID use,[31,32,33] and that the onset of dementia is delayed in NSAID users.[32] AD patients using NSAIDs exhibit a significantly slower progression of dementia over a one-year longitudinal study period.[33] Furthermore, NSAID use is associated with a reduced severity of cognitive symptoms, as shown by one double-blind placebo-controlled trial.[34] AD patients using indomethacin exhibit a significantly milder cognitive decline than the placebo group during a six-month of evaluation period. Postmortem evaluation of the effects of administered NSAIDs on AD pathology has been conducted with controversial results. The degree of amyloid plaques and NFT pathology is similar in AD brains from NSAID users versus nonusers, although NSAID use is associated with a significant reduction in the number of activated microglial cells surrounding plaques in one report,[35] but not in another.[36] Emerging evidence in animal studies indicates that one of many possible therapeutic actions of NSAID in the AD brain may be the suppression of microglial activation associated with β-amyloid deposits. It has been shown that indomethacin administration significantly reduces the number of activated microglial cells surrounding the intraparenchymal β-amyloid deposits developing after infusion in the lateral ventricle of rats.[37]

Furthermore, it has been demonstrated that ibuprofen has an antiinflammatory activity in aging HuAPPSw (Tg2576)[38] transgenic mice and that it significantly reduces AD pathology in the brains of these mice.[39] Chronic administration of ibuprofen in the diet significantly reduces the CNS levels of IL-1, GFAP, and activated microglia and it significantly decreases β-amyloid immunoreactive plaques and total β-amyloid burden in this mouse model.

The two nitro-flurbiprofen derivatives, HCT 1026 (2-fluoro α-methyl [1,1′-biphenyl] 4-acetic acid 4-[nitrooxy] butyl ester) and NCX 2216 (3-[4-{2-fluoro α-methyl 1,1′-biphenyl-4-acetyloxy}-3-methoxyphenyl]-2-propenoic acid 4-[nitrooxy] butyl ester) (see FIGURE 4), have been studied for their effects on AD pathology. NCX 2216, in particular, has the additional benefit of a spacer with antioxidant properties; that is, ferulic acid (4-hydroxy-3-methoxycinnamic acid), a phenolic compound

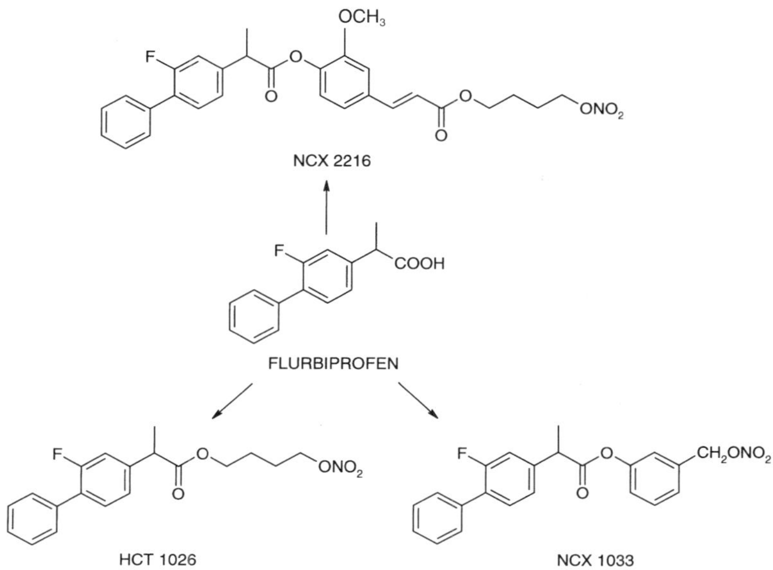

FIGURE 4. Structures of selected nitric oxide-releasing derivatives of flurbiprofen. Note that different linkers are used to combine the nitric ester moiety with flurbiprofen. The linker of NCX 2216 is ferulic acid, which possesses antioxidant properties.[41]

present in a variety of plants that has potent antioxidant and antiinflammatory activities.[40] A recent study reports that long-term administration of ferulic acid protects against β-amyloid peptide toxicity *in vivo*.[41] Administration of 0.006% ferulic acid in the drinking water for four weeks protects mice from the cognitive impairment induced by intracerebroventricular injection of Aβ42, suggesting that ferulic acid is potentially of benefit for AD.

Recently, promising results have been obtained by using HCT 1026 and NCX 2216 in animal models of neuroinflammation and AD. Thus, either HCT 1026 or NCX 2216 significantly decreases the inflammation and the loss of cholinergic neurons induced by LPS infusion in the basal forebrain of young rats.[42] Neuroinflammation was induced by continuous infusion of 0.25 μg/h of LPS for 37 days into the basal forebrain of three-month-old Fisher 344 male rats by using an Alzet osmotic minipump. HCT 1026 (15 mg/kg/day) or NCX 2216 (7.5 mg/kg/day) was administered by *sc* injection in LPS-treated rats. Chronic infusion of LPS impairs memory and learning performance in young rats and induces a neuroinflammatory reaction characterized by extensive microglial activation in the hippocampus, subiculum and enthorinal, and piriform cortices.[43] Furthermore, LPS-treated rats present significantly reduced choline acetyltransferase activity in the nucleus basalis and this is accompanied by increased activity of caspases 3, 8, and 9 in the ventral region of caudate/putamen. Coadministration of HCT 1026 or NCX 2216 greatly decreases the number of activated microglial cells entering the basal forebrain, attenuates the toxicity of LPS for cholinergic neurons, and reduces caspase activity to control level. In addition, HCT 1026 improves cognitive functions of LPS-infused young rats.[44]

Evidence of the beneficial effects of NCX 2216 was also reported in a mouse model of AD pathology; that is, APPsw/PS1mt doubly transgenic mice.[45] These mice obtained by breeding Tg2576 (APP695.K670N/M671L)[38]mice with FAD4 (PS1 M146L) mice, develop extensive amyloid pathology and cognitive impairment.[46] Chronic administration (five months *p.o.*) of NCX 2216 (375 ppm in the food; about 60 mg/Kg/day) reduces by about 45% the immunoreactivity for β-amyloid and the area occupied by Congo red-stained β-amyloid deposits (−35%). Surprisingly, the reduction in β-amyloid deposits by NCX 2216 is accompanied by a significant increase in the number of activated microglial cells in the periplaque region, as suggested by the increased immunoreactivity of several markers of microglial activation, MHC II, CR3, and CD11c. Short term administration of NCX 2216 (*sc* daily injection of 5 mg/kg for 12 days) does not alter β-amyloid deposits, but leads to a significant activation of microglial cells. However, administration of NCX 2216 to non-transgenic mice does not significantly activate microglia, suggesting that the presence of amyloid deposits is a prerequisite for microglial activation. The effect of NCX 2216 on microglial activation is currently a matter of debate and raises concerns since, until now, it was widely believed that activated microglia, as a component of the neuroinflammatory reaction in AD brain, could be harmful for neurons. However, recent findings, obtained by using β-amyloid vaccine, suggest that this assumption may not always be true. Vaccination using β-amyloid-42 peptide or anti-β-amyloid antiserum greatly reduces the β-amyloid plaque burden and it activates microglia in the brain of transgenic PDAPP mice.[47,48] There is evidence that microglia activation is indeed required to clear β-amyloid deposits through Fc-mediated phagocytosis. Administration of the vaccine results in a significant improvement in cognitive performance in memory tasks indicating the beneficial activity of the vaccine in the PDAPP[49] and doubly transgenic APPsw/PS1mt[50] mice. A recent report shows that microglial activation is induced by the cytokine TGF, suggesting that it is important for the clearance of β-amyloid plaques by mechanisms other than Fc-mediated phagocytosis.[51] It is still not known how NCX 2216 exerts its effects, on amyloid deposits although a potential role of microglia in β-amyloid clearance is likely. Further investigations are in hand to thoroughly explore the significance of the microglia activation observed with these compounds and their potential effect on cognitive performance.

CONCLUSION

The large body of evidence available to show that NO plays a critical role under a variety of physiologic and pathologic conditions has opened up the possibility of designing new drugs that are capable of delivering NO into tissues and the bloodstream in a sustained and controlled manner. Thus, nitro-derivatives of NSAIDs are a new class of compounds that combine the pharmacologic action of both NO and the native drugs. The release of NO accounts for both safe profile and ability to interfere with the pathologic process. Currently, some NO-releasing compounds are in clinical phase I–II testing and have been safely given both to healthy volunteers and patients. No serious and severe adverse events have been reported.

The lack of GI side effects, together with the potential for an increase therapeutic benefit due to their additional pharmacologic activities, thereby significantly improving the risk/benefit ratio, opens the possibility for a wide range of applications for these promising compounds. In the central nervous system, there are no encouraging data in animal models to support a potential for protypical NO-NSAID derivatives for their activity in reducing the damage associated with neurodegenerative disorders, such as Alzheimer's disease.

REFERENCES

1. MONCADA, S., A. HIGGS & R. FURCHGOTT. 1997. International Union of Pharmacology nomenclature in nitric oxide research. Pharmacol. Rev. **49:** 137–142.
2. HOROWITZ, J.D. 2000. Nitrovasodilators. *In* Contemporary Cardiology, Vol. 4. J. Loscalzo & J.A. Vita, Eds.: 383–409. Humana Press, Totowa.
3. MEGSON, I.L. 2000. Nitric oxide donor drugs. Drug. Future **25:** 701–715.
4. GASCO, A., *et al.* 1996. NO-donors: an emerging class of compounds in medicinal chemistry. Il Farmaco **51:** 617–635.
5. DEL SOLDATO, P., R. SORRENTINO & A. PINTO. 1999. NO-aspirins, a class of new anti-inflammatory and antithrombotic agents. Trend. Pharmacol. Sci. **20:** 319–323.
6. BURGAUD, J.L., J.P. RIFFAUD & P. DEL SOLDATO. 2002. Nitric-oxide releasing molecules: a new class of drugs in several major indications. Curr. Pharm. Design. **8:** 201–213.
7. WALLACE, J.L., *et al.* 1999. Nitric oxide-releasing NSAIDs: GI-safe antithrombotics. IDrugs **2:** 321–326.
8. FIORUCCI, S., *et al.* 1999. Gastrointestinal safety of nitric oxide-derived aspirin is related to inhibition of ICE-like cysteine proteases in rats. Gastroenterology **116:** 1089–1106.
9. TAKEUCHI, K., *et al.* 1998. Cyclooxygenase-2 selective and nitric oxide-releasing nonsteroidal anti-inflammatory drugs and gastric mucosal responses. J. Physiol. Pharmacol. **49:** 501–513.
10. FIORUCCI. S., *et al.* 1998. TNFα processing enzyme inhibitors prevent aspirin-induced TNFα release and protect against gastric mucosal injury in rats. Aliment. Pharmacol. Ther. **12:** 1139–1153.
11. UKAWA. H., *et al.* 1998. Effects on cyclooxygenase-2 selective and nitric oxide-releasing nonsteroidal anti-inflammatory drugs on mucosal ulcerogenic and healing responses of the stomach. Digest. Dis. Sci. **43:** 2003–2011.
12. UKAWA, H., *et al.* 1998. Effects of NO-releasing aspirin on functional and mucosal ulcerogenic responses in the rat stomach: comparison with aspirin. J. Pharmacol. Exp. Ther. **206:** 115–121.
13. ROSSONI, G., *et al.* 2001. The nitroderivative of aspirin, NCX 4016, reduces infarct size caused by myocardial ischemia/reperfusion in the anesthetized rat. J. Pharmacol. Exp. Ther. **297:** 380–387.
14. MUSCARÀ, M.N., *et al.* 2001. Vasorelaxant effects of a nitric oxide-releasing aspirin derivative in normotensive and hypertensive rats. Br. J. Pharmacol. **133:** 13314–1322.
15. WALLACE, J.L., *et al.* 1995. Anti-thrombotic effects of a nitric oxide-releasing, gastric-sparing aspirin derivative. J. Clin. Investig. **96:** 2711–2718.
16. LECHI, C., *et al.* 1996. In vitro study of the anti-aggregating activity of two nitroderivatives of acetylsalicylic acid. Blood Coag. Fibrinol. **7:** 206–209.
17. MOMI, S., *et al.* 2000. Prevention of pulmonary thromboembolism by NCX 4016, a nitric oxide-releasing aspirin. Eur. J. Pharmacol. **397:** 177–185.
18. WAINWRIGHT, C.L., A. MILLER & L. WORK. 2000. NO-aspirin, but not native aspirin, reduces myocardial infarct size in anaesthetised pigs subjected to ischaemia/reperfusion. Presented at the BPS Meeting, Bedford, U.K. September 5–7.

19. YAMAMOTO, T., *et al.* 2000. The effect of aspirin and two nitric oxide donors on the infarcted heart *in situ*. Life Sci. **67:** 839–846.
20. ROSSONI, G., *et al.* 2000. Myocardial protection by the nitroderivative of aspirin, NCX 4016: *in vitro* and *in vivo* experiments in the rabbit. Ital. Heart J. **1:** 146–155.
21. NAPOLI, C., *et al.* 2001. Effects of nitric oxide-releasing aspirin versus aspirin on restenosis in hypercholesterolemic mice. Proc. Natl. Acad. Sci. U.S.A. **98:** 2860–2864.
22. IGNARRO, L.J., *et al.* 2001. Role of the arginine-nitric oxide pathway in the regulation of vascular smooth muscle cell proliferation. Proc. Natl. Acad. Sci. U.S.A. **98:** 4202–4208.
23. WILLIAMS, J.L., *et al.* 2001. Nitric oxide-releasing nonsteroidal anti-inflammatory drugs (NSAIDs) alter the kinetics of human colon cancer lines more effectively than traditional NSAIDs: implications for colon cancer chemoprevention. Cancer Res. **61:** 3285–3289.
24. BAK, A.W., *et al.* 1998. Cyclooxygenase-independent chemoprevention with an aspirin derivative in a rat model of colonic adenocarcinoma. Life Sci. **63:** 367–373.
25. MINUZ, P., *et al.* 1998. NO-Aspirins: antithrombotic activity of derivatives of acetyl salicylic acid releasing nitric oxide. Cardiovasc. Drug Rev. **16:** 31–47.
26. FIORUCCI, S., *et al.* 2000. NO-aspirin protects from T cell-mediated liver injury by inhibiting caspase-dependent processing of Th1-like cytokines. Gastroenterology **118:** 404–421.
27. WALLACE, J.L., *et al.* 1999. *In vivo* antithrombotic effects of a NO-releasing aspirin derivative, NCX 4016. Thromb. Res. **93:** 43–50.
28. FIORUCCI, S., *et al.* 2000. IL-1β converting enzyme is a target for nitric oxide-releasing aspirin: new insights in the antiinflammatory mechanism of nitric oxide-releasing non steroidal antiinflammatory drugs. J. Immunol. **165:** 5245–5254.
29. CARINI, M., *et al.* 2001. Nitrosylhemoglobin, an unequivocal index of nitric oxide release from nitroaspirin: *in vitro* and *in vivo* studies in the rat by electron spin resonance (ESR) spectroscopy. J. Pharm. Biom. Anal. **26:** 503–518.
30. MCGEER, P.L., *et al.* 1996. Arthritis and anti-inflammatory agents as possible protective factors for Alzheimer's disease: a review of 17 epidemiologic studies. Neurology **47:** 425–432.
31. THE NEUROINFLAMMATION WORKING GROUP. 2000. Inflammation and Alzheimer's disease. Neurobiol. Aging **21:** 383–421.
32. BREITNER, J.C., *et al.* 1995. Delayed onset of Alzheimer's disease with nonsteroidal anti-inflammatory and histamine H2 blocking drugs. Neurobiol. Aging **16:** 523–530.
33. RICH, J.B., *et al.* 1995. Nonsteroidal anti-inflammatory drugs in Alzheimer's disease. Neurology **45:** 51–55.
34. ROGERS, J., *et al.* 1993. Clinical trial of indomethacin in Alzheimer's disease. Neurology **43:** 1609–1611.
35. MACKENZIE, I.R.A. & D. MUNOZ. 1998. Non-steroidal anti-inflammatory drug use and Alzheimer-type pathology in aging. Neurology **50:** 986–990.
36. Halliday, G.M. et al. 2000. Effect of anti-inflammatory medications on neuropathological findings in Alzheimer disease. Arch. Neurol. **57:** 831–836.
37. NETLAND, E.E., *et al.* 1998. Indomethacin reverses the microglial response to amyloid-beta-protein. Neurobiol. Aging **19:** 201–204.
38. HSIAO, K., *et al.* 1996. Correlative memory deficits, Abeta elevation, and amyloid plaques in transgenic mice. Science **274:** 99–102.
39. LIM, G.P., *et al.* 2000. Ibuprofen suppresses plaque pathology in a mouse model for Alzheimer's disease. J. Neurosci. **20:** 5709–5714.
40. GRAF, E. 1992. Antioxidant potential of ferulic acid. Free Rad. Biol. Med. **13:** 435–448.
41. YAN, J.-J., *et al.* 2001. Protection against amyloid peptide toxicity *in vivo* with long-term administration of ferulic acid. Br. J. Pharm. **133:** 89–96.
42. WENK, G.L., *et al.* 2000. Mechanisms to prevent the toxicity of chronic neuroinflammation on forebrain cholinergic neurons. Eur. J. Pharm. **402:** 77–85.
43. HAUSS-WEGRZYNIAK, B., *et al.* 1998. Chronic neuroinflammation in rats reproduces components of the neurobiology of Alzheimer's disease. Brain Res. **780:** 294–303.

44. HAUSS-WEGRZYNIAK, B., P. VRANIAK & G.L. WENK. 1999. The effects of a novel NSAIDS on chronic neuroinflammation are age dependent. Neurobiol. Aging **20:** 305–313.
45. JANTZEN, P.T., *et al.* 2000. Modification of microglial reactivity in MAPP/MPS1 transgenic mice using NCX 2216 (nitroflurbiprofen). Soc. Neurosci. Abs. **26:** 397.14.
46. HOLCOMB, L., *et al.* 1998. Accelerated Alzheimer-type phenotype in transgenic mice carrying both mutant amyloid precursor protein and presenilin 1 transgenes. Nat. Med. **4:** 97–100.
47. SCHENK, D., *et al.* 1999. Immunization with amyloid-β attenuates Alzheimer-disease-like pathology in the PDAPP mouse. Nature **400:** 173–177.
48. BARD, F., *et al.* 2000. Peripherally administered antibodies against amyloid β-peptide enter the central nervous system and reduce pathology in a mouse model of Alzheimer disease. Nat. Med. **6:** 916–919.
49. JANUS, C., *et al.* 2000. A peptide immunization reduces behavioural impairment and plaques in a model of Alzheimer's disease. Nature **408:** 979–982.
50. MORGAN, D., *et al.* 2000. A peptide vaccination prevents memory loss in an animal model of Alzheimer's disease. Nature **408:** 982–985.
51. WYSS-CORAY, T., *et al.* 2001.TGF-β 1 promotes microglial amyloid-β clearance and reduces plaque burden in transgenic mice. Nat. Med. **7:** 612–618.

Ginseng, Sex Behavior, and Nitric Oxide

LAURA L. MURPHY[a] AND TONY JER-FU LEE[b]

[a]*Department of Physiology, Southern Illinois University, School of Medicine, Carbondale, Illinois, U.S.A.*

[b]*Department of Pharmacology, Southern Illinois University, School of Medicine, Springfield, Illinois, U.S.A.*

ABSTRACT: **In Asia, ginseng is commonly included in herbals used for the treatment of sexual dysfunction. Recent studies in laboratory animals have shown that both Asian and American forms of ginseng enhance libido and copulatory performance. These effects of ginseng may not be due to changes in hormone secretion, but to direct effects of ginseng, or its ginsenoside components, on the central nervous system and gonadal tissues. Indeed, there is good evidence that ginsenosides can facilitate penile erection by directly inducing the vasodilatation and relaxation of penile corpus cavernosum. Moreover, the effects of ginseng on the corpus cavernosum appear to be mediated by the release and/or modification of release of nitric oxide from endothelial cells and perivascular nerves. Treatment with American ginseng also affects the central nervous system and has been shown to significantly alter the activity of hypothalamic catecholamines involved in the facilitation of copulatory behavior and hormone secretion. Recent findings that ginseng treatment decreased prolactin secretion also suggested a direct nitric oxide-mediated effect of ginseng at the level of the anterior pituitary. Thus, animal studies lend growing support for the use of ginseng in the treatment of sexual dysfunction and provide increasing evidence for a role of nitric oxide in the mechanism of ginsenoside action.**

KEYWORDS: **ginseng; sex behavior; nitric oxide**

INTRODUCTION

Ginseng has been an essential herbal in traditional Chinese medicine for many years. There is considerable literature describing the pharmacognosy of ginseng, including effects on the central nervous system, metabolism, immune function, and cardiovascular system. Two species of ginseng widely studied with respect to their medicinal benefits are *Panax ginseng* (Asian ginseng) and *Panax quinquefolium* (American ginseng). The principle active constituents in ginseng are ginsenosides, saponin glycosides unique to the *Panax* species. Over 20 different ginsenosides have been identified from extracts of the roots, leaves, and seeds of the ginseng plant. The different ginsenoside compositions of Asian and American ginsengs may be responsible for the unique medicinal properties they elicit. Crude ginsenoside extracts of ginseng, or the purified ginsenoside components themselves, have been found to

Address for correspondence: Dr. Laura L. Murphy, Department of Physiology, Southern Illinois University, School of Medicine, Carbondale, IL 62901, U.S.A. Voice: 618-453-8212. lmurphy@siumed.edu elicit a wide variety of actions similar to that of ginseng *in toto,* confirming the role of ginsenosides in the actions of ginseng. There are many reported mechanisms of ginsenoside action that suggest individual ginsenosides may exert specific tissue-dependent actions. This review focuses on the recently reported aphrodisiac effects of ginseng and discusses the role ginsenosides may play in the nitric oxide-mediated mechanisms of ginseng's actions on sex behavior and performance.

GINSENG EFFECTS ON SEX BEHAVIOR AND HORMONE SECRETION

Both Asian and American ginsengs have been reported to significantly enhance copulatory behavior parameters in male rats[1,2] and mice.[3] In male rats these behaviors include latency (in seconds) to mount a sexually receptive female rat (indicative of sexual arousal or libido), and latency to ejaculate (a measure of performance). For both parameters, ingestion of ginseng significantly reduced the latencies to mount and ejaculate.[1,2] Oral doses of powdered ginseng root sufficient to alter sexual behavior in male rats were as low as 10 mg/kg body weight and significant changes in libido were measured within one week of daily oral ginseng treatment.[2] Male mice exposed to prolonged individual housing fail to exhibit copulatory behavior when exposed to an estrous female mouse.[3] Daily treatment of these male mice with an extract of Asian ginseng (25–100 mg/kg, ip) or ginsenoside Rg1 (2.5–10 mg/kg, ip) produced a dose related increase in the number of mice displaying copulatory behaviors (mounting, intormission, and penis licking). Ginsenosides Rb1, Rb2, and Ro were without effect. These data present evidence that specific ginsenosides may be responsible for mediating the actions of ginseng on copulatory behavior.

Hormone analyses revealed no changes in blood testosterone or luteinizing hormone (LH) levels following acute or chronic exposure to American ginseng.[2] A previous study reported that long term administration of Asian ginseng produced a dose related increase in serum testosterone levels in male rats.[4] This suggests that Asian and American ginsengs exert different actions on hormone secretion that may, in part, depend on doses and duration of treatment. Interestingly, plasma levels of the hormone prolactin were significantly reduced by the doses of American ginseng that altered copulatory behavior in male rats.[2] Thus, American ginseng or its ginsenoside components may have direct actions on the anterior pituitary gland and/or on the hypothalamic dopaminergic mechanisms involved in the regulation of prolactin secretion. As well as producing a decrease in plasma prolactin levels, an increase in dopaminergic activity,[5] and/or noradrenergic activity,[6] ginsengs have also been shown to facilitate male rat copulatory behavior.

There is considerable evidence that ginseng or its ginsenoside constituents affect central nervous system function.[7–12] Behavioral studies have demonstrated that *Panax ginseng* and specific ginsenosides increase spontaneous motor activity,[11] inhibit food intake,[12] affect learning and memory,[7] and exert anxiolytic actions.[9] Moreover, ginseng and/or ginsenosides have been shown to enhance striatal dopaminergic activity,[11] and to modulate GABAergic[10] and cholinergic[8] neurotransmission. In a recent study, hypothalamic dopamine and norepinephrine turnover were measured in male rats treated with American ginseng (100 mg/kg, po) for four weeks, a treatment regimen that resulted in significantly enhanced copulatory behavior.[2]

Rats were injected with α-methyl-*p*-tyrosine (α-MPT; 250 mg/kg, ip), a tyrosine hydroxylase inhibitor, one hour before sacrifice so that catecholamine turnover could be determined using established procedures.[13] In ginseng-treated male rats, hypothalamic norepinephrine turnover was significantly increased in comparison with control rats (0.79 ± 0.08 versus 0.38 ± 0.02 ng/mg tissue/h in ginseng versus controls, respectively). Since increases in hypothalamic norepinephrine have been shown to facilitate male sexual arousal and copulatory performance in rats,[6] these results present a potential mechanism by which ginseng increases male copulatory behavior parameters. Unexpectedly, despite reduced plasma prolactin levels in these animals (4.2 ± 0.2 versus 7.1 ± 0.5 ng/ml in ginseng versus control, respectively; $p < 0.05$), hypothalamic dopaminergic activity was significantly reduced in ginseng treated animals when compared with controls (0.33 ± 0.4 versus 0.64 ± 0.09 ng/mg tissue/h, respectively; $p \leq 0.05$). These results suggest that the ability of ginseng to alter prolactin release is most likely through a direct action on the pituitary gland. A decrease in circulating prolactin levels would feedback to reduce the activity of hypothalamic dopamine neurons and the negative influence of dopamine on prolactin secretion.

Ginseng has been shown to exert direct effects on penile tissue that could also be responsible for its copulatory performance-enhancing actions. An extract of *Panax ginseng* directly caused endothelium-dependent vasodilatation and relaxation of rabbit penile corpus cavernosum *in vitro*.[14] Evidence described below addresses the purported role of nitric oxide in mediating both the penile and central nervous system responses to ginseng that may lead to increased libido and copulatory performance.

GINSENG, GINSENOSIDES, AND NITRIC OXIDE

Several lines of evidence support a role of nitric oxide in mediating the actions of ginseng and ginsenosides on copulatory parameters. Nitric oxide (NO) plays an important role in the mechanism of penile erection.[15] NO is released from the nonadrenergic, non-cholinergic (NANC) nerves, causing relaxation of the penile corpus cavernosum and penile erection.[16] Both endothelial cells and perivascular nerves in the corpus cavernosum contain NO synthase and may be stimulated by ginseng to release endogenous NO. Indeed, a ginsenoside-containing extract of *Panax ginseng* has been shown to produce a dose-related, endothelium-dependent, NO-mediated dilatation and relaxation of rabbit penile corpus cavernosum.[14] Treatment with *N*-nitro-L-arginine, a nitric oxide synthase (NOS) inhibitor blocked the ability of ginseng to induce relaxation. Ginsenosides also enhanced acetylcholine-induced and transmural nerve stimulaton-elicited relaxations in rabbit corpus cavernosum.[14] Both the neurogenic and endothelium-mediated relaxations were enhanced by superoxide dismutase (SOD), a scavenger of superoxide anion that is known to inactivate NO, and were blocked by *N*-nitro-L-arginine and oxy Hb, which inhibit NOS and trap NO, respectively.[14] These results suggest that the effects of ginseng extract on corpus cavernosum tissue are mediated by the release and/or the modification of release of NO from endothelial cells and perivascular nerves. The consequent vasodilatation of penile corpus cavernosum by ginsenosides may contribute to the copulatory performance-enhancing action of ginseng.

Nitric oxide has been implicated in many actions of ginseng,[17] including the ability of ginseng and ginsenosides to relax rabbit pulmonary[18] and monkey cerebral arteries,[19] decrease neuronal ischemia,[20] and inhibit stress-induced glucocorticoid secretion in mice.[21] Nitric oxide also plays an important role in neuroendocrine functions.[22] It has recently been demonstrated that NO, by a paracrine mechanism, inhibits prolactin release from rat anterior pituitaries *in vitro*.[23–25] Nitric oxide synthase is present in the folliculostellate cells of the anterior pituitary, which are in close proximity to the prolactin-secreting lactotrophs.[23] The ability of chronic ginseng treatment to reduce plasma prolactin levels, concomitant with decreased hypothalamic dopamine activity, suggests that ginsenosides directly inhibit prolactin secretion from the anterior pituitary gland. Because ginsenosides have been shown to stimulate the synthesis and release of NO in other systems,[14,18,20] in the absence of direct experimental evidence, it is tempting to speculate that ginsenosides induce NO release at the pituitary level, causing a decrease in prolactin secretion.

Increased hypothalamic noradrenergic receptor activation has been shown to facilitate copulatory behavior in males and females.[6,26] Although ginseng treatment produced a significant increase in hypothalamic norepinephrine activity, the role of NO as mediator of this ginseng effect is uncertain. Instead, evidence indicates that NO mediates the effects of norepinephrine on neuroendocrine function[27]and copulatory behavior.[28] Since ginseng has been shown to increase NO availability,[14,17,18] perhaps the copulatory-enhancing effects of norepinephrine are augmented by ginseng-induced NO synthesis/release. Unfortunately, the mechanisms of ginsenoside action in eliciting changes in neurotransmitter activity are unknown.

CONCLUSIONS

Ginseng treatment significantly enhances male libido and copulatory performance in rats. There is good evidence that ginseng actions at the level of both the central nervous system and the penis are responsible for its copulatory-enhancing effects. Moreover, NO appears to mediate the actions of ginseng and its ginsenosides on relaxation of penile corpus cavernosum, which may lead to enhanced copulatory performance. Much still needs to be learned regarding the mechanisms of central nervous system action of ginsenosides in their purported ability to facilitate libido in males.

In Asia, ginseng is commonly included in herbals that are used for treatment of sexual dysfunction in humans. Although formal clinical testing of the effects of ginseng on sexual behavior in humans has not been performed to date, animal studies lend growing support for the medicinal use of ginseng in treatment of sexual dysfunction.

REFERENCES

1. Kim, C., H. Choi,, C.C. Kim, *et al.* 1976. Influence of ginseng on mating behavior of male rats. Am. J. Chin. Med. **4:** 163–168.
2. Murphy, L.L., R.S. Cadena, D. Chávez, *et al.* 1998. Effect of American ginseng (*Panax quinquefolium*) on male copulatory behavior in the rat. Physiol. Behav. **64:** 445–450.

3. YOSHIMURA, H., N. KIMURA & K. SUGIURA. 1998. Preventive effects of various ginseng saponins on the development of copulatory disorder induced by prolonged individual housing in male mice. Meth. Find. Exp. Clin. Pharmacol. **20:** 59–64.
4. FAHIM, M.S., Z. FAHIM, J.M. HARMAN, *et al.* 1982. Effect of *Panax ginseng* on testosterone level and prostate in male rats. Arch. Androl. **8:** 261–263.
5. PLEIM, E.T., J.A. MATOCHIK, R.J. BARFIELD, *et al.* 1990. Correlation of dopamine release in the nucleus accumbens with masculine sexual behavior in rats. Brain Res. **524:** 160–163.
6. MALLICK, H., S.K. MANCHANDA & V.M. KUMAR. 1996. Beta-adrenergic modulation of male sexual behavior elicited from the medial preoptic area in rats. Behav. Brain Res. **74:** 181–187.
7. ABE, K., S.I. CHO, I. KITAGAWA, *et al.* 1994. Differential effects of ginsenoside Rb1 and malonylginsenoside Rb1 on long-term potentiation in the dentate gyrus of rats. Brain Res. **649:** 7–11.
8. BENISHIN, C.G. 1992. Actions of ginsenoside Rb1 on choline uptake in central cholinergic nerve endings. Neurochem. Int. **21:** 1–5.
9. BHATTACHURYA, S.K. & S.K. MITRA. 1991. Anxiolytic activity of *Panax ginseng* roots: an experimental study. J. Ethnopharmacol. **34:** 87–92.
10. KIMURA, T., P.A. SAUNDERS, H.S. KIM, *et al.* 1994. Interactions of ginsenosides with ligand-bindings of GABA(A) and GABA(B) receptors. Gen. Pharmacol. **25:** 193–199.
11. WATANABE, H., H. OHTA, L. IMAMURA, *et al.* 1991. Effect of *Panax ginseng* on age-related changes in the spontaneous motor activity and dopaminergic nervous system in the rat. Jpn. J. Pharmacol. **55:** 51–56.
12. YOSHIMATSU, H., T. SAKATA, H. MACHIDORI, *et al.* 1993. Ginsenoside Rg1 prevents histaminergic modulation of rat adaptive behavior from elevation of ambient temperature. Physiol. Behav. **53:** 1–4.
13. MURPHY, L.L., J. GHER, R.W. STEGER, *et al.* 1994. Effects of Δ^9-tetrahydrocannabinol on copulatory behavior and neuroendocrine responses of male rats to female conspecifics. Pharmacol. Biochem. Behav. **48:** 1011–1017.
14. CHEN, X. & T.J.-F. LEE. 1995. Ginsenosides-induced nitric oxide-mediated relaxation of the rabbit corpus cavernosum. Br. J. Pharmacol. **115:** 15–18.
15. ANDERSSON, K.E. 2001. Pharmacology of penile erection. Pharmacol. Rev. **53:** 417–450.
16. RAJFER, J., W.J. ARONSON, P.A. BUSH, *et al.* 1992. Nitric oxide as a mediator of relaxation of the corpus cavernosum in response to nonadrenergic, noncholinergic neurotransmission. New Engl. J. Med. **326:** 90–94.
17. GILLIS, C.N. 1997. *Panax ginseng* pharmacology: a nitric oxide link? Biochem. Pharmacol. **54:** 1–8.
18. CHEN, X., C.N. GILLIS & R. MAOLLI. 1984. Vascular effects of ginsenosides *in vitro.* Br. J. Pharmacol. **82:** 485–491.
19. TODA, N., K. AYAJIKI, H. FUJIOKA, *et al.* Ginsenoside potentiates NO-mediated neurogenic vasodilatation of monkey cerebral arteries. J. Ethnopharmacol. **76:** 109–113.
20. LIM, J.H., T.C. WEN, S. MATSUDA, *et al.* 1997. Protection of ischemic hippocampal neurons by ginsenoside Rb1, a main ingredient of ginseng root. Neurosci. Res. **28:** 191–200.
21. KIM, D.H., J.S. JUNG, H.W. SUH, *et al.* Inhibition of stress-induced plasma corticosterone levels by ginsenosides in mice: involvement of nitric oxide. Neuroreport **9:** 2261–2264.
22. BRANN, D.W., G.K. BHAT, C.A. LAMAR, *et al.* 1997. Gaseous transmitters and neuroendocrine regulation. Neuroendocrinol. **65:** 385–395.
23. DUVILANSKI, B.H., C. ZAMBRUNO, A. SEILICOVICH, *et al.* 1995. Role of nitric oxide in control of prolactin release by adenohypophysis. Proc. Natl. Acad. Sci. U.S.A. **92:** 170–174.
24. VANKELECOM, H., P. MATTHYS & C. DENEF. 1997. Involvement of nitric oxide in the interferon-gamma-induced inhibition of growth hormone and prolactin secretion in anterior pituitary cell cultures. Mol. Cell. Endocrinol. **129:** 157–167.
25. VELARDEZ, M.O., A. DE LAURENTIS, M. DEL CARMEN DÍAZ, *et al.* 2000. Role of phosphodiesterase and protein kinase G on nitric oxide-induced inhibition of prolactin release from the rat anterior pituitary. Eur. J. Endocrinol. **143:** 279–284.

26. ETGEN, A.M., M.A. ANSONOFF & A. QUESADA. 2001. Mechanisms of ovarian steroid regulation of norepinephrine receptor-mediated signal transduction in the hypothalamus: implications for female reproductive physiology. Horm. Behav. **40:** 169–177.
27. CANTEROS, G., V. RETTORI, A. GENARO, *et al.* 1996. Nitric oxide synthase content of hypothalamic explants: increase by norepinephrine and inactivated by NO and cGMP. Proc. Natl. Acad. Sci. U.S.A. **93:** 4246–4250.
28. CHU, H.P. & A.M. ETGEN. 1999. Ovarian hormone dependence of alpha(1)-adrenoreceptor activation of the nitric oxide-cGMP pathway: relevance for hormonal facilitation of lordosis behavior. J. Neurosci. **19:** 7191–7197.

The Clinical Safety of Viagra

PETER H.C. LIM,[a] P. MOORTHY,[a] AND KENNETH G.F. BENTON[b]

[a]*Department of Urology, Changi General Hospital, Singapore*

[b]*Department of Medicine, Changi General Hospital, Singapore*

ABSTRACT: Erectile dysfunction (ED) in men is amenable to correction with Viagra in a majority of patients. The accumulated experience of prescribing Viagra across the broad continuum of men suffering from ED is sufficient for a meaningful assessment of the safety of Viagra in clinical practice. The use of Viagra necessitates caution in cardiac failure and when used within six months of acute myocardial infarction and stroke. It is inadvisable in patients with unstable angina pectoris. The co-administration of Viagra with organic nitrates, for example, glyceryl trinitrate or isosorbide dinitrate, is unsafe. The relative contraindications to Viagra in cardiovascular disease are uncontrolled hypertension and impaired cardiac reserve. With respect to interactions with other drugs, the potential influence on the metabolism of Viagra by medications that affect the cytochrome-P-450 system does not translate into clinical effects. The vasodilatory properties of sildenafil citrate are largely responsible for unwanted effects. The most common side effects are headache, flushing (due to vasodilation), and dyspepsia (due to relaxation of the smooth muscle of the gastroesophageal sphincter with reflux). In the recommended single-dose range (25–100 mg), the use of Viagra for erectile dysfunction, in the absence of contraindications, is extremely safe provided the drug is taken under proper conditions.

KEYWORDS: erectile dysfunction; nitrates; nitric oxide; sildenafil citrate; Viagra

INTRODUCTION

The performance of sex for males is the *sine qua non* of sexual intercourse. A perception of sexual proficiency by men, commensurate with sexual expectations and social conditioning, is unsurprisingly related to masculine identity and male self esteem. Erectile dysfunction (ED), which is defined as *the inability to achieve and maintain an erection sufficient for satisfactory sexual activity*,[1] is a distressing problem. Afflicting a remarkable number of men, ED is the most common sexual difficulty encountered by pleasure-seeking males, estimated to affect at least 30 million men in the United States alone.[2] Furthermore, ED is age dependent.[3] The Massachusetts Male Aging Study (MMAS) reports that 5% of men at 40 years and 15% of men at 70 years are suffering from complete erectile dysfunction. The prevalence of ED in some degree is reported to be 52% of men for the age range 40–70 years, of which 25% are moderate and 10% completely dysfunctional.[3] In conjunction

[a]Address for correspondence: Peter H.C. Lim, Department of Urology, Changi General Hospital, 2 Simei Street 3, Singapore 529889.

with increasing longevity and rising public expectations, and as opportunities for effective therapeutic intervention in ED multiply, the implications of male erectile dysfunction in advancing societies are predicted to be immense.

THE NECESSITY OF VIAGRA

When erectile dysfunction is remedied by pharmacologic modalities, an adequate column of penile blood—that is, an erection, of sufficient rigidity and duration for satisfactory sexual intercourse—is achieved. Several therapeutic approaches to ED are presently attempted. They include: intracavernosal injections, medicated urethral suppositories, vacuum constriction devices (VCDs), and penile implants. The number and variety of options reflect that every method has its drawbacks.[5] These treatments are unsatisfactory because of patient reluctance, unpredictable efficacy, inconvenience, and nonavailability; inevitably they all lack spontaneity,[5] thereby, enhancing the need for Viagra. A thorough evaluation of drug safety and efficacy, together with a well defined clinical profile are prerequisites to the prescription of all medicines; Viagra (sildenafil citrate), a candidate *par excellence* for notoriety and public scrutiny, is certainly no exception. The emergence of Viagra as a therapeutic option for ED is widely recognized as a revolutionary advance; its position and importance in the treatment of ED are beyond refute. The millions of prescriptions already written for Viagra are testimony to its efficacy in practice and to fulfill genuine patient need.[6] They constitute an extensive basis of accumulated evidence upon which the safety of the drug may be evaluated. At present, Viagra is approved for prescription in more than 100 countries worldwide.

THE PHYSIOLOGY AND MECHANISM OF PENILE ERECTION

An erection of the penis is a hemodynamic event, effected by relaxation of smooth muscle cells in the corpora cavernosa and the arteries of the penis, coincident with restriction of penile venous outflow.[4,7] The arterial blood flow into the penis is, therefore, increased, by way of the cavernosal arteries and the helicine arteries that deliver blood directly to the cavernosal spaces.[4] There is consequently a sustained elevation of the intracavernosal arterial pressure with relaxation of smooth muscle trabeculæ and pooling of blood within the corpora cavernosa, leading to engorgement of cavernosal tissue and hence a penile erection.

The relaxation of smooth muscle cells in the corpora cavernosa is mediated by 3′5′-cyclic guanosine monophosphate (c-GMP), and this critical event is brought about by nitric oxide.[7,8] The response to a sexual stimulus is the release of nitric oxide at the non-adrenergic, non-cholinergic nerve endings innervating the penile arterioles resulting in the relaxation of smooth muscle cells.[8] Subsequently, cGMP is hydrolyzed by a cGMP specific phosphodiesterase-5 (PDE-5) isoenzyme, which is abundant in the cavernosal tissues of the penis.[4,9]

THE PHARMACOLOGIC ACTION OF SILDENAFIL CITRATE (VIAGRA)

Sildenafil citrate specifically and selectively inhibits cGMP-specific phosphodiesterase type 5, the isoenzyme of phosphodiesterase that is responsible for the degradation of cavernosal cGMP.[9] Since cGMP is the second messenger of NO it is the crucial mediator of vasodilatation and arteriole muscle cell relaxation in the penis.[8] By inhibiting the PDE-5 enzyme, sildenafil citrate protracts penile cGMP activity and, therefore, a sustained erection ensues. It is clear from the foregoing that sexual stimulation is nonetheless essential to initiate penile erection, and that sildenafil is capable solely of augmenting an erection that occurs in response to a natural sexual stimulus. In order for Viagra to be effective, therefore, sexual arousal is essential.

THE SAFETY OF VIAGRA IN CARDIOVASCULAR DISEASE

There is a well-recognized and self-evident relationship between erectile dysfunction and cardiovascular disease. In remarking on the adverse effects of Viagra in cardiovascular disease, it is imperative to emphasize that Viagra must *never* be administered to patients taking organic nitrates (like glyceryl trinitrate and isosorbide dinitrate) that are capable of donating nitric oxide, and vice versa. Furthermore, among the adverse effects attributed to Viagra, to date, cardiovascular mishaps have constituted the principal focus of clinical inquiries.[10–12] Sildenafil citrate has diminished selectivity for cAMP-specific phosphodiesterase isoform-3, which is involved in cardiac contractility. Moreover, potentially precipitous decreases in blood pressure, resulting from the vasodilating properties of sildenafil citrate, necessitate caution when prescribing Viagra to patients with cardiovascular disease. In this respect, previous myocardial infarction and angina pectoris, uncontrolled hypertension, and previous transient cerebral ischemic attacks and strokes are particularly important. Indeed, acute myocardial infarction, sudden cardiac death, symptomatic non-fatal ventricular arrhythmia, transient cerebral ischemic attacks *per se*, and hemorrhagic stroke are clearly documented in patients engaging in sexual exertion by means of Viagra.[13] However, the same effect has also been observed with placebo. The incidence of adverse cardiovascular events correlates directly with the severity of pre-existing cardiovascular disease and the dose of Viagra prescribed.[10,11,13] An assessment of individual cardiovascular risk is indispensable to optimizing patient-safety when contemplating the prescription of Viagra or any other treatment for ED.

DEATH AND SILDENAFIL

Death occurring in the act of sexual intercourse is extremely uncommon, and invariably is cardiovascular or cerebrovascular in origin. In the literature and reports concerning patients in the U.S.A. during the initial four-month period of post-marketing surveillance (Viagra launched in March 1998), 69 deaths with a temporal association to use of Viagra were recorded. Of these, 46 deaths were associated with probable cardiovascular events, and 12 deaths were attributable to possible interaction between Viagra and organic nitrates.[14] A high-degree of cardiovascular risk in these patients was identifiable, with hindsight. Death occurred during sexual intercourse itself (61%) or shortly after coitus. The majority of the deaths occurred within four to five hours of taking Viagra.[14] Any contribution of Viagra, *per se*, to such deaths cannot be differentiated from the undoubted contribution of sexual exertion. There is no evidence from standardized mortality ratios of greater incidence of fatal myocardial infarction, cerebrovascular accidents, and ischemic cardiac events amongst men taking sildenafil citrate.[15] Given that cardiovascular risk in patients is identifiable, cardiovascular mishaps in association with the usage of Viagra may perhaps be eliminated by careful preprescription screening.

CONTRAINDICATIONS TO SILDENAFIL USE

Absolute Contraindications

Therapeutic intervention with Viagra in erectile dysfunction is contraindicated in a number of concomitant medical conditions. In this respect, the conditions where Viagra needs to be used with caution are: cardiovascular instability within the preceding six months; that is, unstable angina pectoris, acute myocardial infarction, ventricular arrhythmia, cerebrovascular accidents, or cardiac failure.[13] A patient judged unsuitable for the physical exertion of sexual intercourse is clearly not a candidate for any treatment of ED, including Viagra. Data on the safety of Viagra are unavailable from patients with genetic disorders of the retinal phosphodiesterases, retinitis pigmentosa (potential risk of retinal bleeding), advancing hepatic impairments, and previous hepatic encephalopthy. For these conditions, therefore, caution is needed until information to the contrary becomes available. Stated previously, the co-administration of Viagra and organic nitrates is strictly contraindicated. A washout period of at least 24 hours is recommended for separation of these drugs.[12,13] Specific questioning for Viagra ingestion is necessary in every man presenting with chest pain, if inadvertent therapy with medicines containing organic nitrates. is to be avoided. On the other hand, the dietary sources of inorganic nitrates, nitrites, and the amino acid, L-arginine (NO substrate) present in nutritional supplements, for example, are by no means contraindicated; they make no contribution to nitric oxide in the circulation.[13]

Relative Contraindications

Relative contraindications on the other hand, are circumstances under which caution with Viagra is required. They include: resting hypertension exceeding 170/110 mmHg,[12] symptoms of hypotension on current blood pressure lowering medication,[13,16] and patients wishing to include Viagra with other treatments for ED. Special caution is required for patients with anatomical deformities of the penis, such as Peyronie's disease, cavernosal fibrosis, or developmental and acquired angulation of the penis. A consideration of the risk/benefit ratio, and determination of the minimum effective dose of any medical treatment for ED, including Viagra, is important in patients predisposed to prolonged erections and priapism under conditions such as myelomatosis, leukemia, sickle cell anemia, and paraproteinemia. Caution when prescribing Viagra is required in glaucoma, conditions of altered perfusion of the optic nerve head, and in patients who have ischemic problems of the retinæ. Since Viagra is associated with dyspepsia, and under certain circumstances it may change platelet aggregation, assessment of the risk/benefit ratio is prudent with active peptic ulceration unless information to the contrary becomes available.

Interaction of Viagra With Other Drugs

Evidence based intervention in cardiovascular disease and diabetes mellitus increasingly demands polypharmacy. Interactions between sildenafil citrate and other medications are theoretically interesting but clinically insignificant. Following ingestion, sildenafil citrate is metabolized by the hepatic microsomal cytochrome P450 enzyme system, which includes its isoforms.[13] The metabolic products of sildenafil degradation are excreted via the biliary system (80%) and via the kidney.[13] In elderly subjects, and in hepatic and severe renal impairment (i.e., creatinine clearance less than 30 ml/min), the plasma clearance of *n*-desmethyl, the active metabolite of sildenafil citrate, is delayed. Plasma concentrations of sildenafil are also increased by medications that inhibit cytochrome P450. Notable examples include (1) erythromycin, (2) cimetidine, (3) cyp-2C9 inhibitors, such as tolbutamide and phenytoin, and possibly (4) cyp-2D6 inhibitors, encompassing several drugs in common use. Cyp-2D6 inhibitors include selective serotonin reuptake inhibitors (SSRI) and tricyclic antidepressants, and antihypertensives, such as thiazides, calcium channel blocking agents, and angiotensin converting enzyme (ACE) inhibitors.[12,13,16] On the other hand, sildenafil is a weak inhibitor of several cytochrome P450 isoenzymes and, therefore, may occasionally itself interfere with the degradation of substrates cleared by that enzyme system. In clinical practice, sildenafil concentrations in the plasma rarely exceed 1 μmol/ml when Viagra is administered as a single dose in the recommended dose range, and significant interactions are unlikely to occur. It is noteworthy that the blood pressure reduction (not exceeding 10 mmHg) induced by sildenafil is added to the reduction brought about by antihypertensive drugs. The lowering of blood pressure attributable to the antihypertensive medication *per se* remains unchanged. With respect to circumstances of usage, Viagra is not reported to enhance the hypotensive action of alcohol, and the absorption of Viagra is not affected by antacid preparations. A delay in absorption of Viagra from the gut may be brought about by ingestion together with high fat food. The prolongation of bleeding time induced by antiplatelet agents, including aspirin, is not increased; significant interactions with oral hypoglycemic agents cleared by cytochrome P450, or disturbances of warfarin metabolism, do not occur.

THE UNWANTED EFFECTS OF VIAGRA

The adverse effects of sildenafil citrate are mainly symptoms of vasodilation. The cumulative experience of unwanted effects in several general populations confirms that headache (15.6%) and facial flushing (10.4%) are the most common untoward symptoms, followed by dyspepsia (4.6%), nasal stuffiness (1.1%) and conjunctival suffusion (see FIGURE 1 and TABLE 1).[17–35] In a general population, these unwanted effects are rarely troublesome in the context of sexual activity, and are short lived.[18] In patients with Parkinson's disease, headache is reported as the most common side

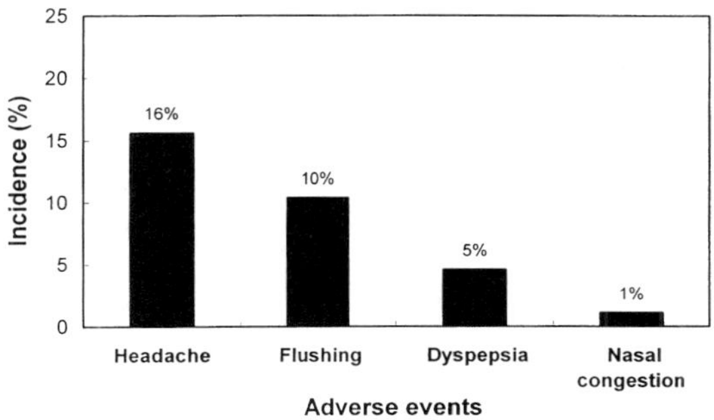

FIGURE 1. Adverse events profile of Viagra in clinical studies from various populations (pooled data from References 17–35).

TABLE 1. Profile of major adverse events in various clinical studies of Sildenafil citrate (table continues on next two pages)

No.	Major Side Effects	Percentage	Study Population	Reference	Nature of Study
1	headache	50 (combined)	general (English)	17	double-blind, placebo controlled
	dyspepsia				
	pelvic-musculo skeletal pain				
2	headache	30	general (American)	11	dose escalation
	flushing	30			
	dyspepsia	9			
	rhinitis	8			
	visual disturbances	4			
3	headache	28	general (American)	11	dose escalation open label
	flushing	22			
	dyspepsia	12			
	rhinitis	4			
	visual disturbances	9			
4	headache	32	general (American)	11	dose response*
	flushing	21			
	dyspepsia	17			
	rhinitis	12			
	visual disturbances	10			
5	headache	16	general (American)	18	double-blind, placebo controlled
	flushing	10			
	dyspepsia	7			

TABLE 1/continued-1.

No.	Major Side Effects	Percentage	Study Population	Reference	Nature of Study
6	muscular pains, headache, dyspepsia,	—	diabetic (English)	19	double-blind, placebo controlled
7	headache, flushing, dyspepsia, UTI, nasal congestion, abnormal vision, diarrhea, dizziness, and rash	> 2	general (American)	20	double-blind, placebo controlled
8	headache, flushing, dyspepsia	—	general (American)	21	double-blind, placebo controlled
9	adverse events	26	general (Spanish)	22	prospective
10	flushing	19	CA prostate (after radiotherapy) (American)	23	follow up
	abnormal color vision	14			
	headache	10			
11	flushing	37	CA prostate (after radiotherapy) (Swiss)	24	prospective plot
	headache	17			
	dyspepsia	9			
12	headache	11	diabetic (American)	25	double-blind, placebo controlled
	dyspepsia	9			
	respiratory tract disorder	6			
13	headache	25	ischæmic heart disease (American)	10	double-blind, placebo controlled
	flushing	14			
	dyspepsia	12			
	respiratory tract infection	8			
	rhinitis	5			
	myocardial infarction	3			
14	facial flushing	34	general (Australian)	26	prospective
	headache	23			
	nasal congestion	13			
	dyspepsia	10			
	dizziness	11			

TABLE 1/continued-2.

No.	Major Side Effects	Percentage	Study Population	Reference	Nature of Study
15	flushing	7	general (Norwegian)	27	open label, flexible dose
	headache	6			
	dyspepsia	5			
16	flushing	31	general (American)	28	open label, flexible dose
	headache	25			
	nasal congestion	19			
	heartburn	11			
17	headache	10 (combined)	spinal cord injured (Swedish)	29	prospective
	dizziness				
18	headache	3 (combined)	general (African)	30	prospective
	dyspepsia				
	flushing				
	nausea				
	vomiting				
19	headache	10	Parkinson's (American)	31	—
20	flushing	13	general (Hungarian)	32	non-placebo controlled
	headache	7			
	headache and flushing	1			
	nasal congestion and visual disturbances	1			
	prolonged erection**	1			
21	treatment-related events	27	general (European	33	double-blind, placebo controlled
22	headache	9	general (Asian)	34	double-blind, placebo controlled
	flushing	7			
	rhinitis	3			
	dizziness	1			
	dyspepsia	2			
	visual disturbances	3			
23	all causes	14	general (Asian)	35	retrospective

NOTES: Percentages were rounded off to the nearest integer. *At 100 mg Viagra. **Two hours maximum erection.

effect of Viagra (TABLE 1).[31] In patients undergoing radiotherapy for prostate cancer, flushing is the principal side effect of Viagra (TABLE 1).[11,14] Fair-skinned men in Sweden and Australia report flushing more commonly than dark-skinned men. Muscle aches and pains, more usually associated with a dosage *exceeding* the recommended dose range of Viagra, are more frequent complaints of patients with diabetes mellitus (TABLE 1).[19,25] There is a clinical impression that exceeding the recommended dosage of Viagra results in increased severity and incidence of unwanted effects. An enhanced perception of light and alterations of vision are well recognized; these include a blue color tinge, increased whiteness of light, and, occasionally, blurred vision that is invariably mild and transient. These visual side effects are mediated by the reaction of sildenafil with PDE-6 in the transduction pathway of retinal photoreceptors.

CONCLUSIONS

Erectile dysfunction (impotence) is exceptionally common in men older than 40 years. This is a disorder and inconvenience but it never amounts to an actual disease. The introduction of Viagra as an effective treatment for ED is undoubtedly responsible for the increased prevalence of ED reports, although this phenomenon is unlikely to reflect a real increase in the true frequency of the disorder. When patients are physically fit enough to withstand the exertion of sexual intercourse, Viagra (sildenafil citrate) has a wide safety margin. In normal volunteers, single 800-mg doses of Viagra are well tolerated.[13] In the recommended dose range (50–100 mg), the use of Viagra for genuine erectile dysfunction is safe under proper conditions, provided that contraindications to the drug are not present. Viagra is neither licensed for recreational purposes (for example, as a performance enhancer) nor is it effective when taken as an aphrodisiac. The misuse of Viagra (for example, coincident with use of amylnitrate as an orgasm mediator) is potentially harmful. As Viagra gains further acceptance in the medical community, its misuse is likely to diminish. As with all effective drugs, Viagra has side effects. The unwanted effects of sildenafil citrate, however, make no impact on clinical safety and, moreover, they are predictable, transient, and invariably insufficient to result in discontinuance of the drug. These effects may perhaps encourage the patient and physician to seek the minimum effective dose.

REFERENCES

1. NIH CONSENSUS CONFERENCE. 1993. Impotence: NIH Consensus Development Panel on Impotence. JAMA **270:** 83–90.
2. FURLOW, W.L. 1985. Prevalence of impotence in the United States. Med. Aspects Hum. Sex **19:** 1316.
3. FELDMAN, H.A., 1. GOLDSTEIN, D.G. HATZICHRISTOU, *et al.* 1994. Impotence and its medical and psycho-social correlates: results of the Massachusetts Male Aging Study. J. Urol. **151:** 5461.
4. ANDERSEN, K.E. & G. WAGNER. 1998. Physiology of penile erection. Physiol. Rev. **75:** 191–236.
5. ROSEN, J.C. 1996. Erectile dysfunction: the medicalization of male sexuality. Clin. Psychol. Rev. **16:** 497–519.

6. PERRIN, J. 1998. And then came Viagra. Pharmaceut. J. **260:** 707.
7. NAYLOR, A.M. 1998. Endogenous neurotransmitters mediating penile erection. Br. J. Urol. **81:** 424–431.
8. RAJFER, J., W.J. ARONSON, P.A. BUSH, *et al.* 1992. Nitric oxide as a mediator of relaxation of the corpus cavernosum in response to nonadrenergic, noncholinergic neurotransmission. N. Engl. J. Med. **326:** 90–94.
9. BEAVO, J.A. 1995. Cyclic nucleotide phosphodiesterases: functional implications of multiple isoforms. Physiol. Rev. **75:** 725–748.
10. CONTI, C.R., C.J. PEPINE & M. SWEENEY. 1999. Efficacy and safety of sildenafil citrate in the treatment of erectile dysfunction in patients with ischaemic heart disease. J. Am. Coll. Cardiol. **83:** 29C–34C.
11. GOLDSTEIN, I., T.F. LUE, H. PADMA-NATHAN, *et al.* 1998. Oral sildenafil in the treatment of erectile dysfunction. N. Engl. J. Med. **338:** 1397–1404.
12. ZUSMAN, R.M., A. MORALES, D.B. GLASSER & I.H. OSTERLOH. 1999. Overall cardiovascular profile of sildenafil citrate. J. Am. Coll. Cardiol. **83:** 35C–44C.
13. CHEITLIN, M.D., A.M. HUTTER, R.G. BRINDIS, *et al.* 1999. Use of sildenafil (Viagra) in patients with cardiovascular disease. J. Am. Coll. Cardiol. **33:** 273–282.
14. CENTRE FOR DRUG EVALUATION AND RESEARCH. 1998. Food and Drug Administration. Post marketing safety of sildenafil citrate (Viagra): summary of reports of death in Viagra users received from marketing (late March through June 1998).
15. SHAKIR, S.A.W., L.V. WILTON, A. BOSHIER, *et al.* 2001. Cardiovascular events in users of sildenafil: results from first phase of prescription event monitoring in England. Brit. Med. J. **322:** 651–652.
16. WEBB, D.J., S. FREESTONE, M.J. ALLEN & G.J. MUIRHEAD. 1999. Sildenafil citrate and blood-pressure lowering drugs: results of drug interaction studies with an organic nitrate and a calcium antagonist. Am. J. Cardiol. **83:** 21C–28C.
17. BOOLELL, M., S. GEPI-ATTEE, J.C. GINGELL, M.J. ALLEN. 1996. Sildenafil, a novel effective oral therapy for male erectile dysfunction. Br. J. Urol. **78:** 257–261.
18. MORALES, A., C. GINGELL, M. COLLINS, *et al.* 1998. Clinical safety of sildenafil citrate (Viagra) in the treatment of erectile dysfunction. Int. J. Impot. Res. **10:** 69–74.
19. PRICE, D.E., J.C. GINGELL, S. GEPI-ATTEE, *et al.* 1998. Sildenafil: study of a novel oral treatment for erectile dysfunction in diabetic men. Diabet. Med. **15:** 821–825.
20. GOLDENBERG, M.M. 1998. Safety and efficacy of sildenafil citrate in the treatment of male erectile dysfunction. Clin. Ther. **20:** 1033–1048.
21. PADMA-NATHAN, H., W.D. STEERS & P.A. WICKER. 1998. Efficacy and safety of oral sildenafil in the treatment of erectile dysfunction: a doubleblind, placebo controlled study of 329 patients. Sildenafil Study Group. Int. J. Clin. Pract. **52:** 375–379.
22. MAYZ, B.O., E. VILARASAU, A.P. MARTINEZ & J.M.P. MONSENY. 1999. Our clinical experience with the use of sildenafil citrate to treat erectile dysfunction. Acta Urol. Esp. **23:** 859–863.
23. KEDIA, S., C.D. ZIPPE, A. AGARWAL, *et al.* 1999. Treatment of erectile dysfunction with sildenafil citrate (Viagra) after radiation therapy for prostate cancer. Urology **54:** 308–312.
24. WEBER, D.C., S. BIERI, J.M. KURTZ & R. MIRALBELL. 1999. Prospective pilot study of sildenafil for treatment of post-radiotherapy erectile dysfunction in patients with prostate cancer. J. Clin. Oncol. **17:** 3444–3449.
25. RENDELL, M.S., J. RAJFER, P.A. WICKER & M.D. SMITH. 1999. Sildenafil for treatment of erectile dysfunction in men with diabetes: a randomized controlled trial. JAMA **281:** 421–465.
26. M.C. MAHON, C.G., R. SAMALI & H. JOHNSON. 2000. Efficacy, safety and patient acceptance of sildenafil citrate as treatment for erectile dysfunction. J. Urol. **164:** 1192–1196.
27. CHRISTANSEN, E., W.R. GUIRGUIS & I.H. OSTERLOH. 2000. Long-term efficacy and safety of oral Viagra (sildenafil citrate) in men with erectile dysfunction and the effect of randomized treatment withdrawal. Sildenafil Multicentre Study Group. Int. J. Impot. Res. **12:** 177–182.
28. MOREIRA, S.G., R.E. BRANNINGAN, A. SPITZ, F.J. *et al.* 2000. Side-effect profile of sildenafil citrate (Viagra) in clinical practice. Urology **56:** 474–476.

29. Schmid, D.M., B. Schurch & D. Hauri. Sildenafil in the treatment of sexual dysfunction in spinal-cord injured male patients. Eur. Urol. **38:** 184–193.
30. Magoha, G.A. 2000. Sildenafil (Viagra) in the treatment of male erectile dysfunction in Nairobi. East Afr. Med. J. **77:** 76–79.
31. Zesiewicz, T.A., M. Helal & R.A. Hauser. 2000. Sildenafil citrate (Viagra) for the treatment of erectile dysfunction in men with Parkinson's disease. Mov. Disord. **15:** 305–308.
32. Fekete, F & E. Fabian. 2000. Effectiveness and adverse effects of sildenafil in erectile dysfunction. Orv. Hetil. **141:** 115–118.
33. Meuleman, E., B. Cuzin, R.J. Opsomer, *et al.* 2001. A dose escalation study to assess the efficacy and safety of sildenafil citrate in men with erectile dysfunction. Br. J. Urol. **87:** 75–81.
34. Tan, H.M., C.M. Clarence Lai, J.B. Mendoza, *et al.* 2001. Asian sildenafil efficacy and safety (ASSESS-1): a double-blind, placebo controlled, flexible-dose study of oral sildenafil in Malaysian, Singaporean, and Filipino men with erectile dysfunction. Urology **56:** 635–640.
35. Lim, P.H.C., M.K. Li, F.C. Ng, *et al.* 2002. Clinical safety and efficacy of sildenafil citrate (Viagra) in a multi-racial population in Singapore: a retrospective study of 1520 patients. Int. J. Urol. **9:** in press.

Nitric Oxide

An Antioxidant and Neuroprotector

KOCHUPURACKAL P. MOHANAKUMAR,[a] BOBBY THOMAS,[a] SUDARSHANA M. SHARMA,[b] DHANASEKHARAN MURALIKRISHNAN,[a] RUKHSANA CHOWDHURY,[a] AND CHUANG C. CHIUEH[c]

[a]*Laboratory of Experimental and Clinical Neuroscience, Indian Institute of Chemical Biology, Calcutta, India*

[b]*Biophysics Division, Indian Institute of Chemical Biology, Calcutta, India*

[c]*Unit of Neurotoxicity and Neuroprotection, Laboratory Clinical Sciences, NIMH, National Institutes of Health, Bethesda, U.S.A.*

ABSTRACT: Indirect evidence, including neuroprotection against 1-methyl-4-phenyl-1,2,3,6-tetrahydropyridine (MPTP)-neurotoxicity by nitric oxide synthase (NOS) inhibitors and resistance of transgenic animals deficient in NOS, is controversial. We have reviewed evidence in favor of oxidative stress during the development of MPTP-neurotoxicity and the influence of antioxidants, including nitric oxide (NO) and NO donors, on MPTP-induced dopaminergic neurotoxicity. Systemic administration of MPTP causes dose-dependent generation of hydroxyl radicals (OH) *in vivo* in the striatum in mice; OH scavengers protect dopaminergic neurons from this insult. On the other hand the role of NO in MPTP-neurotoxicity is controversial. Hitherto, no direct evidence for the involvement of NO in MPTP neurotoxicity has been available. MPTP does not affect inducible-NOS mRNA level or its expression in SN or the striatum. Nitroglycerine, a NO donor, can attenuate MPTP-induced dopamine depletion in the striatum by virtue of its OH scavenging action. Several other NO donors have also been shown to scavenge the OH generated, following Fenton chemistry *in vitro*, and to protect against *in vivo* dopaminergic neurotoxicity by small mass iron complex formation. This evidence suggests that NO renders protection against MPTP-induced OH-mediated nigrostriatal lesions, acting as an antioxidant.

KEYWORDS: free radical scavenger; inducible nitric oxide synthase mRNA; nitroglycerine; MPTP; hydroxyl radicals; nitric oxide donors; ferrous citrate

INTRODUCTION

Oxidative stress has been implied in many neurodegenerative diseases, including Parkinson's disease (PD), Alzheimer's dementia, Huntington's disease, and amyotropic lateral sclerosis. The basis of this contention largely stems from findings in animal models, however several pieces of clinical evidences have recently become

Address for correspondence: Dr. K. P. Mohanakumar, Laboratory of Experimental and Clinical Neuroscience, Indian Institute of Chemical Biology, 4, Raja S.C. Mullick Road. Calcutta- 700 032, India. Voice: (91) (33) 473 6793/3491, ext. 104; fax: (91) (33) 473 0284/5197.
mohankumar@iicb.res.in

available. Nitric oxide (NO) and hydroxyl radicals (OH) are implicated in dopaminergic neurotoxicity in PD. Although OH is recognized as a potent neurotoxicant, NO is impressively linked to both normal and abnormal functions of the brain.[1] This gaseous molecule, with free radical chemical properties, has been recognized as a neurotransmitter since the discovery that NO is synthesized and released in cerebellar neurons.[2] It has been suggested that the major role of NO is signaling in cell-to-cell communication. This action of NO has been linked to its neurotoxic action, which is believed to be through the *N*-methyl-D-aspartate (NMDA) excitotoxicity in a calcium–calmodulin dependent fashion.[3]

PD is characterized by progressive degeneration of nigrostriatal dopaminergic neurons in the substantia nigra (SN) compacta region, resulting in severe loss of striatal dopamine (DA). The potent parkinsonian neurotoxin 1-methyl-4-phenyl-1,2,3,6-tetrahydropyridine (MPTP) causes lesions of the DA-ergic neurons of the A_9 SN neurons in humans, primates, and rodents. The mechanisms underlying this neurotoxicity are multifaceted, and involve selective uptake of the active metabolite, 1-methyl-4-phenyl pyridinium ion (MPP^+), through a DA uptake mechanism.[4] It has been suggested that the sequestered MPP^+ in monoaminergic neurons[5] causes an energy crisis[6] by inhibiting mitochondrial complex-I leading to ATP depletion.[7] Autoxidation of DA in the neuromelanin rich SN in presence of iron can also lead to oxidative stress and OH generation.[8] A mitochondrial energy crisis can result in excess free radical generation,[9] which can be directly linked to elevated calcium.[10]

Administration of the parkinsonian neurotoxin or its active metabolite in animals has been shown to produce nigrostriatal damage following OH generation in SN and striatum.[11,12] Interestingly, free radical scavengers, such as salicylic acid (SA) and some antiparkinsonian drugs, protected against this insult via their potent OH scavenging action.[9] On the other hand, the role of NO in this toxicity is unclear, but some reports show resistance of animals to MPTP neurotoxicity following pharmacologic manipulation of NO and in transgenic animals deficient in nitric oxide synthase (NOS).[13–15]

We report here that nitric oxide and NO donors like nitroglycerin can suppress OH generation both *in vivo* and *in vitro.* We also report that MPTP does not affect inducible NOS (iNOS) mRNA and its expression in SN or nucleus caudatus putamen (NCP) at a time when MPTP causes generation of OH and disturbances in endogenous antioxidant enzymes and GSH in these regions.

The results reviewed in this paper are the outcome of several investigations on rodents, both Sprague-Dawley rats and Balb/c mice. The Animal Ethics Committee of our Institute approved the experimental protocols. In these studies, systemic injections of MPTP (i.p.) in mice and intranigral infusions of MPP^+ or ferrous citrate in rats were studied. These treatments resulted in dose dependent depletion of DA in the striatum and typical behavioral abnormalities, including stereotypical turning in rats with unilateral lesions within 48 hours.

In vitro generation of OH was studied by employing a ferrous citrate model involving a cell free system containing 1mM sodium salicylate as described elsewhere.[16] Salicylate hydroxylation products, 2,3- and 2,5-dihydroxybenzoic acid (DHBA), were measured to provide an index of OH generation, using a sensitive HPLC electrochemical procedure.[11] Several NO donors (NO '24 hrs', spermine NONOate, *S*-nitroso-*N*-acetylpenicillamine, diethylamine NONOate, 3-morpholinosydnonimine, sodium nitroprusside [SNP], 4-hydroxy-2,2,6,6-tetramethylpiperidine-1-oxyl, methylamine NONOate, 53'A' 46'B' 63'A' 64'A' and nitroglycerine) and NO dissolved in PBS were tested for their effects on OH production in this *in vitro* system. Intranigrally administered ferrous citrate has been shown to cause significant decrease of tyrosine hydroxylase mRNA levels in the SN and DA concentrations in the striatum.[17]

In vivo generation of OH following systemic MPTP was monitored in adult Balb/c mice as described elsewhere.[12] SA was dissolved in saline and administered 30 min after MPTP. OH was measured in SN and in NCP by detecting the hydroxylation products of salicylate. OH generation following MPP^+ or ferrous citrate in living rats were also monitored using *in vivo* brain microdialysis.[8,16] Two injections of MPTP (30 mg/kg) at 16-h intervals were found to deplete striatal DA to about 50% by the third day. Several drugs, including bromocriptine, L- and D-deprenyl, salicylic acid, 7-nitroindazole, nitroglycerine, and melatonin were tested for their OH scavenging action and neuroprotective effects *in vitro* (ferrous citrate model) and *in vivo* (MPTP-model).

NO measurement, employing a sensitive fluorimetric procedure,[18] using micro plates failed to detect NO in the SN or striatum of mice treated with MPTP at 1–6 h, or after 24 or 48 h (Thomas and Mohanakumar; unpublished results). Therefore we tried to estimate the iNOS mRNA levels before and after MPTP in the SN and NCP of mice following RT-PCR. Total RNA was isolated from SN and NCP using a guanidium-isothiocyanate-phenol-chloroform single step method.[19] First strand DNA was synthesized at 42°C for one hour, using 1 μg of total RNA, random hexanucleotide primer, and Moloney murine leukemia virus reverse transcriptase. Gene specific primers were designed and used for amplification of inducible NOS gene, with hypoxanthine phosphate ribosyl transferase (HPRT) as an internal control. We used various dilutions of purified cDNA and Taq DNA polymerase, amplified by using a three-temperature PCR system. Primer annealing at 55°C for two minutes and extension at 72°C for one minute. With 1 μg of total RNA converted to cDNA, HPRT reached a plateau at 32 cycles. We examined for inducible NOS until 40 cycles were tested.

RESULTS AND DISCUSSION

Experimental Parkinsonism: Methodology

In this review, we discuss two animal models for PD—MPTP-induced Parkinsonism and ferrous citrate-induced nigral neuronal damage. These are well-accepted animal models and have direct relevance to idiopathic PD. The former neurotoxin is known to cause the disease in humans and primates, and in rodents it causes all the biochemical abnormalities akin to the disease. The latter model represents a reflection of the human disease, where accumulation of iron in the SN is reported. Both the models typically deplete striatal DA following inhibition of tyrosine hydroxylase activity in the nigral neurons.[8,11,17] They both cause oxidative stress in the brain as shown by changes in antioxidant enzymes and significant inhibition in the levels of reduced glutathione (GSH)[9,20] as well as increased OH production and lipid peroxidation.[16] These biochemical changes (see TABLES 1 and 2) are comparable to those reported in idiopathic PD.

Oxidative Stress, Nigrostriatal Degeneration, and PD

Oxidative stress could be one of the major contributory factors that leads to neurodegeneration of DA-ergic neurons because postmortem brains of PD patients showed increased levels of iron, decreased levels of GSH, impaired mitochondrial function in the SN, as well as wide spread lipid, protein and DNA oxidation products. Most clinical and experimental evidence that has been accumulated suggests a perpetual process of oxidative stress in the SN forms the etiopathology of DA-ergic

TABLE 1. Effects of MPTP on striatal dopamine, hydroxyl radicals and antioxidant molecules

Treatment ($n = 7$)	DA (pmol/mg)	GSH (ng/mg)	CAT (ΔAbs/min/ mg protein)	SOD (Unit/mg protein)	2,3- and 2,5-DHBA (pmol/mg)	Lipid Peroxidation (AFU)
Control	40.2 ± 1.4	165.4 ± 3.0	0.29 ± 0.05	1.8 ± 0.15	0.61 ± 0.17	0.1 ± 0.01
MPTP	20.2 ± 1.0*	83.5 ± 1.4*	0.79 ± 0.07*	2.7 ± 0.17*	1.48 ± 0.12*	0.5 ± 0.01*

NOTES: MPTP (30 mg/kg, i.p.) was administered twice, at an interval of 16 h, in Balb/C mice. Dopamine (DA) was analyzed following HPLC-electrochemistry. Reduced glutathione (GSH) levels were measured employing a fluorimetric procedure involving an *O*-pthaldialdehyde condensation reaction. Catalase (CAT) activity was measured by monitoring H_2O_2 breakdown. A pyragallol autoxidation procedure was used to measure superoxide dismutase (SOD) activity. MPTP caused an increase in SOD activities both in cytosolic and particulate fractions. Hydroxyl radicals (2,3- and 2,5-DHBA) were measured by using a salicylate hydroxylation procedure employing HPLC electrochemistry, where animals were treated with salicylic acid (SA), 100 mg/kg, 30 min after MPTP, and were sacrificed two hours after SA. Lipid peroxidation following intranigral MPP^+ was measured employing a sensitive fluorimetric procedure in which peroxidized products of polyunsaturated fatty acids were extracted into chloroform-methanol. Data given are mean ± SEM, $n = 5–7$, $^*p \leq 0.05$ as compared to control. AFU, arbitrary fluorescence unit.

TABLE 2. Effects intranigral infusion of ferrous citrate on striatal dopamine, nigral hydroxyl radical generation, and lipid peroxidation

Treatments ($n = 7$)	DA (pmol/mg)	Lipid Peroxidation (AFU)	2,3 and 2,5-DHBA (pmol/mg)
Control	100.5 ± 1.8	0.54 ± 0.04	1.55 ± 0.84
Fe2+-citrate	44.22 ± 1.08*	0.83 ± 0.09*	3.55 ± 1.09*

NOTES: Unilateral intranigral infusion of ferrous citrate (8.4 nmol) resulted in significant reduction of dopamine (DA) in the ipsilateral striatum by the seventh day. Oxidized products of polyunsaturated fatty acids in the SN, as detected by a sensitive fluorimetric assay procedure,[17] are significantly increased by four hours. *In vivo*, OH trapping by sodium salicylate employing an intranigral perfusion procedure following ferrous citrate, showed significantly higher levels of OH (2,3- and 2,5-DHBA) as compared to saline. Sodium salicylate (2 mM) was continuously infused (1 μl/min) through a microdialysis probe for 90 min following ferrous citrate (1 μl). The dialysis samples were collected over a period of 15 min into a tube containing ice-cold 0.1 N perchloric acid and 0.01% EDTA. Salicylate hydroxylation products, 2,3-, and 2,5-DHBA levels were measured employing HPLC-electrochemistry.[11,20] Data given are mean ± SEM, $n = 4$, $^*p \leq 0.05$ as compared to control. AFU, arbitrary fluorescence unit.

neurodegeneration in this area to cause parkinsonism. Recently, this aspect has been reviewed by many authors.[21,22] The moot question here is whether oxidative stress is the primary event that leads to the disease, or it is only a secondary effect. Data available from animal studies, including our *in vivo* brain microdialysis study, suggest that reactive oxygen species are generated prior to neuronal lesions and DA depletion (see FIGURE 1). In our brain microdialysis studies (either in the NCP or SN) we demonstrated increased production of OH (FIG. 1) as an initial event, followed by lipid peroxidation and altered antioxidant molecule status in the brain after treatment with MPTP (TABLE 1) or ferrous citrate (TABLE 2). Complex I inhibition or activation of glial cells, alone or in combination with GSH depletion, can lead to the extracellular accumulation of glutamate and the formation of NO and H_2O_2. These in turn may result in formation of highly toxic peroxynitrite and OH radicals leading to oxidative stress and excitotoxicity, contributing to the initiation or progression of neuronal death in SN in PD.[23] Thus, oxidative processes clearly contribute to the pathology of PD but are probably secondary to some other primary unidentified cause, presumably environmental or genetic.

Free radical scavengers, iron chelators, dopamine agonists, NOS inhibitors, and certain calcium channel antagonists have been shown to provide neuroprotection against MPTP-induced degeneration if given prior to the insult. We examined pre- and posttreatment of free radical scavengers for their ability to block MPTP-induced neurodegeneration in mice. Many of the drugs that are currently used in PD (selegiline, bromocriptine) and other diseases (melatonin, aspirin) have been shown

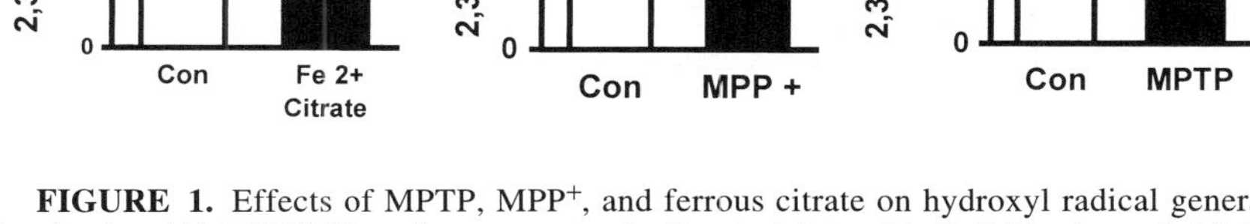

FIGURE 1. Effects of MPTP, MPP^+, and ferrous citrate on hydroxyl radical generation *in vivo*. **(A)**. MPTP 30 mg/kg was systemically administered in Balb/C mice and striatal OH measured by salicylate hydroxylation using an *ex vivo* procedure.[12] SA (100 mg/kg, i.p) given 30 min after MPTP, and the animals sacrificed two hours after MPTP, produced a significantly higher level of 2,3-DHBA, the most reliable indicator of OH adduct of salicylate. **(B)**. MPP^+ (6.72 nmols) by intrastriatal perfusion and **(C)** ferrous citrate (8.4 nmols) by intranigral perfusion with sodium salicylate significantly increased 2,3-DHBA levels. Sodium salicylate (2 mM) was continuously infused (1 μl/min) through a microdialysis probe 90 min after intrastriatal MPP^+ (4 μl) and intranigral ferrous citrate (1 μl). The dialysis samples were collected over a period of 15 min into a tube containing ice-cold 0.1 N perchloric acid and 0.01% EDTA. 2,3-DHBA levels were measured employing HPLC-electrochemistry.[20] Data shown are mean ± SEM, $n = 4–8$, $^*p \leq 0.05$ as compared to control.

to possess potent OH scavenging activity, and to protect against MPTP-induced neurotoxicity[9] (TABLE 2).

NO and Idiopathic Parkinson's Disease

The involvement of NO in neurodegenerative processes remains incompletely characterized. Although NO has been reported to be an important mediator in neuronal degeneration in various models of cell death that involve NMDA-receptor activation,[3] increasing evidence for protective mechanisms has been obtained. Indeed, in idiopathic PD itself, there is little evidence for the involvement of NO in nigral pathology. Except for a solitary report that showed an upregulation of NOS in microglia from parkinsonian, but not from neurologically normal SN and NCP, there are no other reports that directly or indirectly link idiopathic PD in humans with NO.[24] Analyses of the CSF samples from 108 PD patients showed no traces of citrulline, arginine, or glutamate.[25] Although the NOS inhibitor, 7-nitroindazole (7-NI), was successful in protecting against MPTP-induced neuronal lesions,[13–15] it has to be admitted that hitherto no study has directly demonstrated the status of NO or NOS

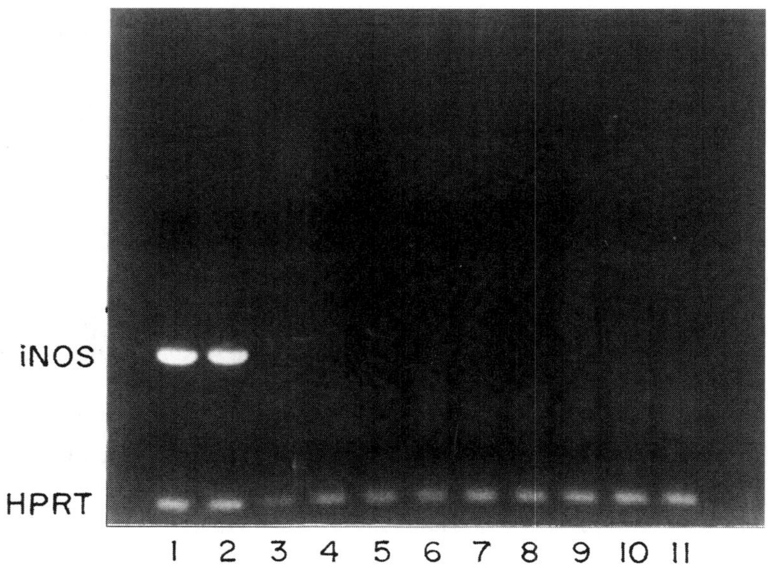

FIGURE 2. Effect of MPTP and/or melatonin on iNOS levels in SN (*Lanes 3–6*) and NCP (*Lanes 7–10*) at 48 h. Balb/C mice were grouped: Group I, received 5% ethanol (*Lanes 3* and *6*); Group II, received MPTP (30 mg/kg, i.p.) 16 h apart (*Lanes 4* and *8*); Group III, received melatonin (20 mg/kg, i.p.; *Lanes 5* and *9*) ; and Group IV, received pretreatment with melatonin (20 mg/kg) followed by MPTP after 30 min (*Lanes 6* and *10*). Animals were sacrificed at 48 h. RNA isolation and RT-PCR procedures[19] are explained in the text. Gene specific primers were designed and used for amplification of inducible nitric oxide synthase (iNOS) gene with hypoxanthine phosphate ribosyl transferase (HPRT) as an internal control (PCR - 40 cycles). This is a representative figure from four experiments. *Lanes 1* and *2* are infected macrophages (positive control) and *Lane 11* is a reload of *Lane 3*.

in SN or NCP, *per se*, following MPTP administration. We failed to detect iNOS mRNA by employing RT-PCR in the SN or NCP at any time (1, 2, 4, 24, or 48 hours; see FIGURE 2) following two doses of MPTP to cause significant depletion of the striatal DA (see TABLE 3). We do accept that failure to detect a response does not necessarily mean its absence. However, we did detect NOS proteins in the same assay in macrophages that had been infected with parasites (see FIG. 2; positive controls). Thus, our results indicate failure of MPTP to upregulate iNOS in SN or NCP following the neurotoxic insult. However, upregulation of iNOS was demonstrated (mRNA levels and by immunohistochemistry) in the ventral midbrain area of C57/BL mice following MPTP administration.[15]

Is NO Toxic to Substantia Nigra Neurons?

Let us examine the evidence that supports the action of NO as a neurotoxin. One of the earliest pieces of evidence came from the study with NOS inhibitor, 7-NI, which has been shown to render protection against MPTP-induced severe injury to dopaminergic neurons.[26] This was subsequently confirmed. It was further demonstrated that neuronal NOS deficient mutant mice are resistant to MPTP, suggesting that neuronally derived NO may be involved in the neurotoxic action of the neurotoxin.[14] Simultaneously, the ability of 7-NI to protect against MPTP-induced DA-ergic lesions and behavioral abnormalities was also demonstrated in higher mammals, such as Baboons.[13] This NOS inhibitor also completely prevented the loss of striatal ATP that occurs in mice during the initial hours after MPTP administration.[27] The best ever evidence for a direct involvement of NO came from the study that showed significant

TABLE 3. Effects of various drugs on striatal dopamine, reduced glutathione and hydroxyl radicals

	DA (pmol/mg)	2,3- + 2,5-DHBA (pmol/mg)	GSH (ng/mg)
Control	40.2 ± 1.42	0.61 ± 0.17	165.4 ± 2.96
MPTP	20.2 ± 0.98@	1.48 ± 0.12@	83.5 ± 1.42@
SA + MPTP	37.8 ± 2.3*	0.78 ± 0.14*	144.05 ± 2.32*
BRC + MPTP	35.38 ± 1.8*	0.91 ± 0.1*	134.2 ± 3.47*
L-DEP + MPTP	36.18 ± 2.2*	0.74 ± 0.09*	n.d.
D-DEP + MPTP	35.82 ± 1.9*	0.83 ± 0.12*	n.d.
MEL + MPTP	19.64 ± 2.85@	0.62 ± 0.06*	145.14 ± 1.49*
7-NI + MPTP	30.2 ± 2.8*	0.86 ± 0.10*	140.2 ± 0.32*

NOTES: Animals were administered salicylate (SA), bromocriptine (BRC), selegiline (L- and D-DEP), melatonin (MEL), or 7-nitroindazole (7-NI) 30 min prior to MPTP (30 mg/kg).[12,20] Data for OH (2,3- and 2,5-DHBA) are from animals treated with SA (100 mg/kg) 30 min prior to MPTP and sacrificed two hours after administering the neurotoxin. Data for reduced glutathione (GSH) and DA are for animals sacrificed on the seventh day. All the drugs effectively scavenged MPTP-induced OH production *in vivo* and blocked MPTP-induced striatal DA depletion (except melatonin) and GSH. Data reported are mean ± SEM; $n = 8$; $p \leq 0.05$, as compared to MPTP (*) or as compared to control (@); n.d., not done.

upregulation of iNOS in the midbrain that preceded or paralleled MPTP-induced dopaminergic neurodegeneration.[15] The toxic action of NO in neurodegeneration including culture systems has been reviewed recently.[1,28,29]

NO as an Antioxidant

In a biological system with physiologic pH, the presence of OH at the time of NO production can lead to HNO_2, thus inhibiting the damaging effects of both the radicals. It is possible that MPTP-induced OH formation is a consequence of NO production. This can happen by one-electron reduction of NO to produce nitroxyl anions, which in turn react with another molecule of NO to give N_2O and OH. However, we have two systems—(1) MPTP or MPP^+, which can probably generate both NO and OH, and (2) ferrous citrate, a system that probably generates OH alone

TABLE 4. *In vitro* effects of NO donors on Fe^{2+}-citrate–induced OH (2,3- and 2,5-DHBA)

Drug (2 nmol)	2 hours	24 hours
Fe^{2+}-citrate (4.2 nmol)	37.15 ± 0.72	486.08 ± 12.21
NO in PBS (5%)	16.86 ± 1.63*	49.21 ± 3.37*
Nitroglycerine (5 g)	10.44 ± 1.28*	257.99 ± 19.01*
NO '24 hrs'	26.40 ± 5.30*	104.78 ± 8.59*
SPERNO	17.19 ± 1.31*	118.68 ± 12.84*
SNAP	15.84 ± 1.33*	137.33 ± 20.48*
DEANO	13.57 ± 1.44*	151.67 ± 20.63*
SIN-I	41.91 ± 1.4*	564.21 ± 70.44
MEANO	14.52 ± 0.28*	142.22 ± 6.4*
53 'A'	14.12 ± 0.38*	123.3 ± 4.48*
46 'B'	12.71 ± 0.37*	84.28 ± 3.26*
63 'A'	12.66 ± 0.23*	91.21 ± 3.00*
64 'A'	12.14 ± 0.29*	91.88 ± 2.5*
TEMPOL	75.19 ± 9.98**	646.57 ± 47.35**
SNP	496.36 ± 18**	613.08 ± 16.41**

NOTES: Ferrous citrate complex was prepared by mixing 1:1 solutions of ferrous ammonium sulfate and citric acid in filtered Kreb's Ringer solution containing no glucose. The control was kept in the dark and in ice. The sample tubes were kept under incandescent light at room temperature. The 1-ml reaction mixture of contained 210 nmol of Fe^{2+}-citrate, 1 mM salicylic acid, and 100 nmols of the NO donors: NO '24 hrs', spermine NONOate (SPERNO), *S*-nitroso-*N*-acetylpenicillamine (SNAP), diethylamine NONOate (DEANO), 3-morpholinosydnonimine (SIN-I), sodium nitroprusside (SNP), 4-hydroxy-2,2,6,6-tetramethylpiperidine-1-oxyl (TEMPOL), methylamine NONOate (MEANO), 53'A' 46'B' 63'A', and 64'A'. NO in PBS (5%) and nitroglycerine (5 μg) were used. The reaction mixtures were incubated for 2 h and 24 h, and were stopped by acidifying the reaction mixture with ice-cold 0.1 M $HClO_4$ containing 0.01% EDTA. Twenty 1-μl samples were injected into the HPLC.[20] All the NO donors significantly reduced the generation of OH (*). However, TEMPOL and SNP caused a significant augmentation of the effect (**).

due to iron-citrate chelation.[16] Thus, we have two systems, one with the potency to generate both radicals and the other generating only OH. We tested various NO donors and NO in PBS on ferrous citrate-induced OH generation in an *in vitro* system and report here the antioxidant action of NO as revealed by reduced levels of OH (see TABLE 4). A typical example of a dose-dependent antioxidant action of nitroglycerine, an NO donor, in this *in vitro* system is depicted in FIGURE 3. We also asked if ferrous citrate-induced OH-mediated neurodegeneration could be blocked by NO in this model, where involvement of NO is limited in comparison with the classic MPTP model. *S*-nitroso-glutathione (GSNO), *S*-nitroso-*N*-acetylpenicillamine (SNAP), and NO in PBS have successfully blocked the oxidative stress induced nigrostriatal DA depletion in this model (see FIGURE 4). Nitroglycerine has been tested in the MPTP model and the neuroprotection was partial for the dose used (see FIGURE 5). Interestingly sodium nitroprusside (SNP), which contains iron, failed to protect against ferrous citrate-induced DA depletion (FIG. 2), but an iron chelator, such as deferoxamine, was successful in rendering protection and blocking the SNP-induced OH generation.[30] The foregoing account amply illustrates antioxidant action of NO.

NO as a Neuroprotector

Similarly, evidence for NO as a neuroprotector with reference to PD is also accumulating in the literature. One of us recently reviewed the neuroprotective properties of NO.[30] OH scavenging action, as well as its potent reactivity with peroxyl radicals, NO itself, heme protein, H_2O_2, and with amino acid radicals, provide this radical with the potency to act as a good neuroprotector. We have demonstrated the OH scavenging actions of NO donors[16,30,31] to render protection against ferrous citrate-induced neurotoxicity (FIG. 4). The present results, obtained by using nitroglycerine

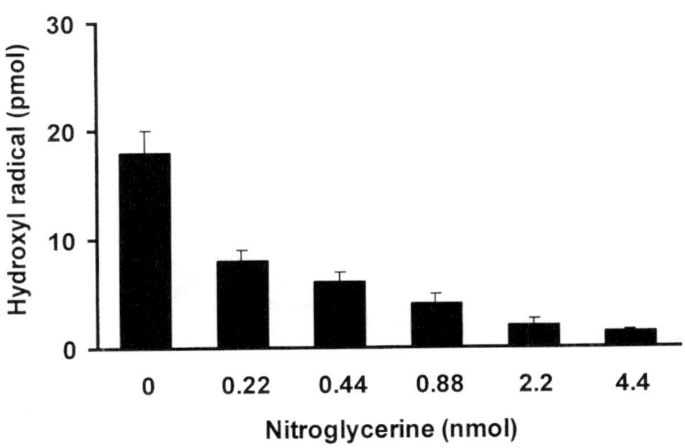

FIGURE 3. Effect of nitroglycerine (0–4.4 nmol) on the generation of OH caused by ferrous citrate complex (4.2 nmol). Ferrous iron chelated with citrate was incubated in plastic test tubes under incandescent light and the total OH was measured after a two-hour incubation period as adducts of salicylic acid, 2,3- and 2,5-DHBA.[17] Nitroglycerine scavenged ferrous citrate induced OH in a dose dependent manner. The results given are mean ± SEM, $n = 4$.

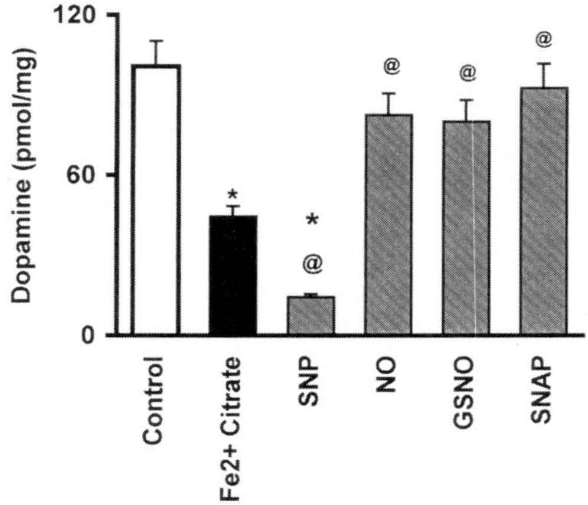

FIGURE 4. Effects of NO and NO donors on striatal dopamine levels on the seventh day after intranigral infusion of ferrous citrate in Sprague-Dawley rats. Rats were intranigrally infused with ferrous citrate alone or in combination with NO in PBS and NO donors, such as sodium nitroprusside (SNP), *S*-nitrosoglutathione (GSNO), and *S*-nitroso-*N*-acetylpenicillamine (SNAP) in 1 μl. The contralateral nigra received Ringer's solution (1 μl), which acted as control. Ferrous citrate (8.4 nmol) caused significant (*) depletion of striatal DA. SNP alone exhibited ferrous citrate like action (data not shown). SNP potentiated ferrous citrate-induced nigral injury. Other NO donors when administered alone did not exert any effect on striatal DA levels, however they significantly attenuated ferrous citrate-induced DA depletion. The results shown are mean ± SEM for six animals in each of the groups, $p \leq 0.05$. $^{@}$Significance compared to MPTP-treated animals; *Significance compared to control.

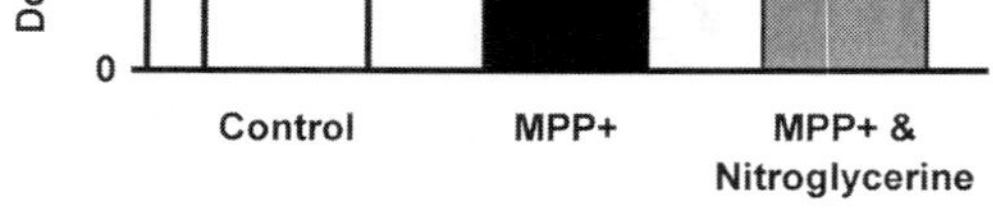

FIGURE 5. Effects of nigral infusion of nitroglycerine on striatal dopamine following intranigral infusion of MPP^+ on the seventh day in Sprague-Dawley rats. Intranigral infusion of MPP^+ (6.72 nmol) reduced dopamine levels by 50% in the striatum and was significantly reversed by intranigral infusion of nitroglycerine (6.72 nmol). Treatment with nitroglycerine alone had no effect on DA levels (data not shown). Results shown are mean ± SEM for six animals. Statistical significance ($p \leq 0.05$) compared with control group* or with the group treated with $MPP^{+@}$.

as a potent NO releaser, also demonstrate the ability of NO to attenuate the MPP^+-induced DA depletion in rats (FIG. 5). In an interesting study, it has been shown that erythropoietin, a cytokine-hormone, is neuroprotective in MPTP-induced nigral DA-ergic neuronal toxicity by inducing NO generation in SN and NCP in C57/BL mice.[32] MPP^+-induced damage in astrocytes has been shown to be protected by incubation with NO donor SNAP indicating neuroprotective action of NO.[33] These studies strongly suggest an antioxidant action of NO in the brain of rodents that effectively protects neurons against oxidative insults.

On the other hand, evidence in favor of NO as a neurotoxicant, as demonstrated by using pharmacological agents (e.g., NOS inhibitors) are probably influenced by their varied actions other than NOS inhibition. For example, *S*-methylthiocitrulline, 7-NI, and nitro-L-arginine (L-NAME), all potent NOS inhibitors, have been shown to possess potent MAO-B inhibitory action,[34] the capability to reverse ATP depletion caused by MPTP,[27] and an OH scavenging action,[35,36] by virtue of which they can protect against MPTP-neurotoxicity. Further support for the neuroprotective action of NO is obtained from a study in which neuronal cultures were exposed to NO producing astrocytes. No cytotoxicity was observed, despite aberration in the mitochondrial electron transport chain.[37]

There is evidence in the literature that throws light on hitherto unknown beneficial actions of NO. In iNOS deficient mice, MPTP caused microglial activation, but no increase in NOS expression and no difference between the levels of DA depletion following MPTP in transgenic and wildtype mice. This indicates mechanisms other than NO that contribute to the loss of DA-ergic terminals in MPTP neurotoxicity.[38] In another study, MPTP injections resulted in a similar decrease in striatal DA in iNOS deficient and wildtype mice, demonstrating that NO may not be directly responsible for DA neurotoxicity following MPTP.[39]

CONCLUDING REMARKS

We have briefly reviewed the role of NO as a neurotoxicant, as an antioxidant, and as a neuroprotector in idiopathic PD and two well accepted models of the disease. Fresh findings on the actions of an NO donor, nitroglycerine, on its OH scavenging and neuroprotective effects are described. The possibility that neurodegeneration in the nigrostriatal system following MPTP, MPP^+, or ferrous citrate results from OH generation, which is independent of NO, has been substantiated. We have also demonstrated the OH scavenging action of NO and its ability to reverse DA depletion caused by ferrous citrate or MPP^+, thus providing credence to the hypothesis that NO is not harmful to the nigrostriatal neurons, and that the actual foe of the nigral neurons is OH.

ACKNOWLEDGMENTS

The work was partially financed by DST, Government of India. BT and SMS received Sr. Research Fellowships from CSIR, Government of India. The work was the outcome of an Indo-US S&T program of USAID.

REFERENCES

1. MOHANAKUMAR, K.P. & H.W. STEINBUSCH. 1998. Hydroxyl radicals and nitric oxide in neurotoxicity and neuroprotection. J. Chem. Neuroanat. **14:** 125–127.
2. JAFFREY, S.R. & S.H. SNYDER. 1995. Nitric oxide: a neural messenger. Ann. Rev. Cell. Dev. Biol. **11:** 417–440.
3. DAWSON, V.L., *et al.* 1991. Nitric oxide mediates glutamate toxicity in primary cortical cultures. Proc. Natl. Acad. Sci. U.S.A. **88:** 6368–6371.
4. JAVITCH, J.A., *et al.* 1985. Parkinsonism-inducing neurotoxin *N*-methyl-4-phenyl-1,2,3,6-tetrahydro-pyridine: uptake of the metabolite *N*-methyl-4-phenylpyridinium by dopamine neurons explains selective toxicity. Proc. Natl. Acad. Sci. U.S.A. **82:** 2173–2177.
5. YELIN, R. & S. SCHULDINER. 1995. The pharmacological profile of the vesicular monoamine transporter resembles that of multidrug transporters. FEBS Lett. **377:** 201–217.
6. COSI, C. & M. MARIEN. 1999. Implication of poly (ADP-ribose) polymerase (PARP) in neurodegeneration and brain energy metabolism. Decreases in mouse brain NAD+ and ATP caused by MPTP are prevented by the PARP inhibitor benzamide. Ann. N.Y. Acad. Sci. **890:** 227–239.
7. GREENAMYRE, J.T., *et al.* 1999. Mitochondrial dysfunction in Parkinson's disease. Biochem. Soc. Symp. **66:** 85–97.
8. CHIUEH, C.C., D.L. GILBERT & C.A. COLTON. 1994. The neurobiology of NO and OH. Ann. N.Y. Acad. Sci. **738:** 1–471.
9. MOHANAKUMAR K.P & D. MURALIKRISHNAN. 2000. Neurochemical mechanisms underlying neuroprotective actions of bromocriptine, salicylate, D- and L-deprenyl in neurodegeneration caused by MPTP. *In* Neurotoxic Factors in Parkinson's Disease and Related Disorders. M.A. Collins & A. Storch, Eds.: 289–293. Kluwer Academic/Plenum, New York.
10. DYKENS, J.A. 1994. Isolated cerebral and cerebellar mitochondria produce free radicals when exposed to elevated Ca^{2+} and Na^{+}: implications for neurodegeneration. J. Neurochem. **63:** 584–591.
11. CHIUEH, C.C., H. MIYAKE & M.T. PENG. 1993. Role of dopamine autoxidation, hydroxyl radical generation, and calcium overload in underlying mechanisms involved in MPTP-induced parkinsonism. Adv. Neurol. **60:** 251–258.
12. THOMAS, B., D. MURALIKRISHNAN & K.P. MOHANAKUMAR. 2000. *In vivo* hydroxyl radical generation in the striatum following systemic administration of 1-methyl-4-phenyl-1,2,3,6-tetrahydropyridine in mice. Brain Res. **852:** 221–224.
13. HANTRAYE, P., *et al.* 1996. Inhibition of neuronal nitric oxide synthase prevents MPTP-induced parkinsonism in baboons. Nat. Med. **2:** 1017–1021.
14. PRZEDBORSKI, S., *et al.* 1996. Role of neuronal nitric oxide in 1-methyl-4-phenyl-1,2,3,6-tetrahydropyridine (MPTP)-induced dopaminergic neurotoxicity. Proc. Natl. Acad. Sci. U.S.A. **93:** 4565–4671.
15. LIBERATORE, G.T., *et al.* 1999. Inducible nitric oxide synthase stimulates dopaminergic neurodegeneration in the MPTP model of Parkinson disease. Nat. Med. **5:** 1403–1409.
16. MOHANAKUMAR, K.P., I. HANBAUER & C.C. CHIUEH. 1998. Neuroprotection by nitric oxide against hydroxyl radical-induced nigral neurotoxicity. J. Chem. Neuroanat. **14:** 195–205.
17. MOHANAKUMAR, K.P., *et al.* 1994. Ferrous-citrate complex and nigral degeneration: evidence for free-radical formation and lipid peroxidation. Ann. N.Y. Acad. Sci. **738:** 392–399.
18. MISKO, T.P., *et al.* 1993. A fluorometric assay for the measurement of nitrite in biological samples. Anal. Biochem. **214:** 11–16.
19. CHOMCZYNSKI, P. & N. SACCHI. 1987. Single-step method of RNA isolation by acid guanidinium thiocyanate-phenol-chloroform extraction. Anal. Biochem. **162:** 156–159.
20. MURALIKRISHNAN, D. & K.P. MOHANAKUMAR. 1998. Neuroprotection by bromocriptine against 1-methyl-4-phenyl-1,2,3,6-tetrahydropyridine-induced neurotoxicity in mice. FASEB J. **12:** 905–912.

21. ZHANG, Y., V.L. DAWSON & T.M. DAWSON. 2000. Oxidative stress and genetics in the pathogenesis of Parkinson's disease. Neurobiol. Dis. **7:** 240–250.
22. SCHAPIRA, A.H. 2001. Causes of neuronal death in Parkinson's disease. Adv. Neurol. **86:** 155–162.
23. MCNAUGHT, K.S. & P. JENNER. 2000. Extracellular accumulation of nitric oxide, hydrogen peroxide, and glutamate in astrocytic cultures following glutathione depletion, complex I inhibition, and/or lipopolysaccharide-induced activation. Biochem. Pharmacol. **60:** 979–988.
24. KNOTT, C., G. STERN & G. P. WILKIN. 2000. Inflammatory regulators in Parkinson's disease: iNOS, lipocortin-1, and cyclooxygenases-1 and -2. Mol. Cell Neurosci. **16:** 724–739.
25. KUIPER, M.A., *et al.* 2000. L-Glutamate, L-arginine, and L-citrulline levels in cerebrospinal fluid of Parkinson's disease, multiple system atrophy, and Alzheimer's disease patients. J. Neural Transm. **107:** 183–189.
26. SCHULZ, J.B., *et al.* 1995. Inhibition of neuronal nitric oxide synthase by 7-nitroindazole protects against MPTP-induced neurotoxicity in mice. J. Neurochem. **64:** 936–939.
27. ROYLAND, J.E., *et al.* 1999. 7-Nitroindazole prevents 1-methyl-4-phenyl-1,2,3, 6-tetrahydropyridine-induced ATP loss in the mouse striatum. Brain Res. **839:** 41–48.
28. ESTEVEZ, A.G., *et al.* 1998. Role of endogenous nitric oxide and peroxynitrite formation in the survival and death of motor neurons in culture. Prog. Brain Res. **118:** 269–280.
29. CHIUEH, C.C. 1999. Neuroprotective properties of nitric oxide. Ann. N.Y. Acad. Sci. **890:** 301–311.
30. RAUHALA, P., *et al.* 1998. Apparent role of hydroxyl radicals in oxidative brain injury induced by sodiumnitroprusside. Free Rad. Biol. Med. **24:** 1065–1073.
31. RAUHALA, P., *et al.* 1996. *S*-nitrosothiols and nitric oxide, but not sodium nitroprusside, protect nigrostriatal dopamine neurons against iron induced oxidative stress *in vivo*. Synapse **23:** 58–60.
32. GENC, S., *et al.* 2001. Erythropoietin exerts neuroprotection in 1-methyl-4-phenyl-1,2,3,6-tetrahydropyridine-treated C57/BL mice via increasing nitric oxide production. Neurosci. Lett. **298:** 139–141.
33. TSAI, M.J. & E.H. LEE. 1998. Nitric oxide donors protect cultured rat astrocytes from 1-methyl-4-phenylpyridinium-induced toxicity. Free Rad. Biol. Med. **24:** 705–713.
34. DI MONTE, D.A., *et al.* 1997. Inhibition of monoamine oxidase contributes to the protective effect of 7-nitroindazole against MPTP neurotoxicity. J. Neurochem. **69:** 1771–1773.
35. MATTHEWS, R.T., L. YANG & M.F. BEAL. 1997. *S*-methylthiocitrulline, a neuronal nitric oxide synthase inhibitor, protects against malonate and MPTP neurotoxicity. Exp. Neurol. **143:** 282–286.
36. SMITH, T.S., *et al.* 1994. Reduction of MPP(+)-induced hydroxyl radical formation and nigrostriatal MPTP toxicity by inhibiting nitric oxide synthase. Neuroreport **20:** 2598–2600.
37. STEWART, V.C., *et al.* 2000. Astrocyte-derived nitric oxide causes both reversible and irreversible damage to the neuronal mitochondrial respiratory chain. J. Neurochem. **75:** 694–700.
38. DEHMER, T., *et al.* 2000. Deficiency of inducible nitric oxide synthase protects against MPTP toxicity *in vivo*. J. Neurochem. **74:** 2213–2216.
39. ITZHAK, Y., J.L. MARTIN & S.F. ALI. 1999. Methamphetamine- and 1-methyl-4-phenyl-1,2,3, 6-tetrahydropyridine-induced dopaminergic neurotoxicity in inducible nitric oxide synthase-deficient mice. Synapse **34:** 305–312.

Nitric Oxide Triggers Classic Ischemic Preconditioning

AMANDA LOCHNER,[a,b] ERNA MARAIS,[b]
EUGENE DU TOIT,[b] AND JOHAN MOOLMAN[b]

[a]*Diabetes Research Group, MRC of South Africa, Republic of South Africa*

[b]*Department of Medical Physiology, University of Stellenbosch, Tygerberg, Republic of South Africa*

ABSTRACT: The role of NO in the classic ischemic preconditioning phenomenon of the myocardium is not well defined, and was investigated by using the isolated perfused rat heart as a model. Hearts were preconditioned with 3 × 5 minute ischemia in the presence and absence of the NOS inhibitors L-NAME (50 μM) and L-NNA (50 μM), and the guanylyl cyclase inhibitor ODQ (20 μM). These inhibitors significantly attenuated the protective effect of preconditioning against 25-min global ischemia (as measured by functional recovery), specifically if administered during the triggering phase. Cyclic infusions (3 × 5 min) of the NO-donors SNAP (50 μM) and SNP (100 μM) elicited protection against both 25-min global or low-flow ischemia. Hearts preconditioned with NO donors displayed significantly superior functional reserve, if stimulated with adrenaline, compared to hearts preconditioned with ischemia. Although the NO donors SNAP and SNP both activated p38 MAPK during the preconditioning protocol, protection was accompanied by significantly decreased p38 MAPK activity during sustained ischemia, as was the case in ischemic preconditioning. We conclude that (1) NO is a trigger for classic preconditioning, (2) cGMP generation plays an important role in its protection, (3) attenuation of p38 MAPK during sustained ischemia accompanies NO preconditioning and may mediate cardiac protection, and (4) preconditioning with NO may be more advantageous than using ischemia.

KEYWORDS: nitric oxide; classic ischemic preconditioning

INTRODUCTION

The phenomenon of ischemic preconditioning, by which a short episode of ischemia and reperfusion protects the tissue of that organ against a subsequent prolonged episode of ischemia, offers both short and long term protection. The immediate protection by ischemic preconditioning (i.e., within minutes to a few hours following a short episode of ischemia/reperfusion) is called classic preconditioning, whereas the protection that can be observed 24 hours after the preconditioning incident is called late preconditioning or *second window of protection*. Preconditioning is the most potent form of myocardial protection against ischemia that has yet been described,[1] and because of its potential clinical use its mechanism has been the

Address for correspondence: A. Lochner, P.O. Box 19063, 7505 Tygerberg, Republic of South Africa.
alo@gerga.sun.ac.za

subject of intense research during the past 15 years. Numerous activators of the preconditioning pathway have been described, such as adenosine, angiotensin, noradrenaline and opioids.[2] It is, however, likely that they share a common final signal transduction pathway leading to, amongst other things, activation of PKC and p38 MAP kinase.[3,4] The roles of the mitochondrial K^+-ATP channel[5] and the downstream effectors of p38 MAP kinase, such as MAPKAPK2, are currently being investigated in this respect.[6]

Several observations suggest that nitric oxide (NO) may act as a trigger for preconditioning: (1) activators of preconditioning, such as adenosine, acetylcholine, and bradykinin, can generate NO via cNOS;[7] (2) cyclic increases in cGMP occur during a multicycle preconditioning protocol;[8] and (3) NO synthase activity is increased by five minutes of global ischemia.[9] Indeed, evidence strongly implicates a role for NO in the activation of late preconditioning.[10,11] However, its role in classic preconditioning is less well defined.

We present evidence to show that NO is involved as a trigger in classic preconditioning, that the generation of cGMP may play an important part in its protective effect, and that the attenuated activity of the stress kinase, p38 MAP kinase, is involved in NO-mediated classic preconditioning. Furthermore, our results suggest that eliciting preconditioning with NO may offer advantages to eliciting preconditioning with ischemia. This may have important clinical implications.

NO IS A TRIGGER BUT NOT MEDIATOR OF CLASSIC PRECONDITIONING

The role of NO in classic preconditioning was mostly investigated in the context of reperfusion dysrhythmias; conflicting results were published. Some studies reported that the antiarrhythmic effects of preconditioning were independent of NO,[12,13] and that NOS-inhibition by L-NAME had no effect on preconditioning-induced reduction in infarct size[14] or improvement in functional recovery.[15] Other studies showed that NO donors or NOS-inhibitors could, respectively, mimic[16] or abolish[17] the protective effects of preconditioning on dysrhythmias. The role of NO in classic preconditioning is thus ambiguous.

We previously observed that a preconditioning protocol of 3×5-min cycles of ischemia and reperfusion elicits cyclic increases in both cAMP and cGMP,[8] and that the percentage increase in cGMP levels is significantly higher than that of cAMP. We subsequently showed that the increases in cAMP are β-adrenergic mediated, and *partially* participate in eliciting the protection of preconditioning.[18] Since NO mediates activation of guanylyl cyclase, which results in the production of cGMP, it was logical to investigate if these cyclical increases in cGMP during the preconditioning protocol indicate involvement of the NO pathway. Experimental evidence supported such a notion: although no increase in NO could be demonstrated during an ischemic preconditioning protocol, NOS activation and cGMP increases were observed within five minutes of onset of ischemia,[8,9] and cardiac biosynthesis of NO was shown to be essential as a trigger for preconditioning.[19]

We, therefore, investigated if the cyclic cGMP increases elicited by multiple short episodes of ischemia/reperfusion (such as occurs during a multicycle preconditioning protocol) indicated involvement of NO as a trigger of classic preconditioning in the isolated working rat heart model of preconditioning.

Our concept is that events occurring during the preconditioning protocol serve to activate (i.e., trigger) the protective mechanism of preconditioning. Biochemical events that take place during the subsequent phase of sustained ischemia reflect the protection that is elicited by prior preconditioning. Accordingly, abolishment or attenuation of biochemical events, during either the preconditioning phase or during the sustained ischemia and causing inhibition of protection during reperfusion as reflected by mechanical recovery, may indicate a role as trigger or mediator respectively.

The isolated perfused rat heart was used as model, with functional recovery as an endpoint, assessed by measuring aorta and coronary flow, aorta pressure, and calculation of total work as described elsewhere.[18] Hearts were subjected to a preconditioning protocol of 3×5-min ischemia interspersed with five minutes reperfusion, and subjected to 25-min normothermic global ischemia. Functional recovery was assessed after 30-min reperfusion. To investigate the role of NO as a trigger or donor, the effect of inhibition of NOS activation during the entire experiment (both triggering and mediating phase), and by inhibition of NOS activity during either the triggering or mediator phase alone were evaluated (see FIGURE 1).

Ischemic preconditioning was significantly attenuated by the NOS inhibitors N^{ω}-nitro-L-arginine methyl ester (L-NAME) (50 μM) (see FIGURE 2) and *N*-nitro-L-arginine (L-NNA) (50 μM), as well as the guanylyl cyclase inhibitor 1H-[1,2,4] oxadiazolol-[4,3-a]quinoxaline-1one (ODQ) (20 μM) if administered during both the triggering and mediating phases (see FIGURE 3). Inhibition of NOS activity with L-NAME and L-NNA resulted in a cardiac output of 16.6 ± 3.6 and 14.85 ± 2.1 ml/min, respectively, whereas inhibition of guanylyl cyclase activity resulted in 21.6 ± 1.7 ml/min compared with 33.9 ± 2.2 ml/min in preconditioned hearts ($p < 0.05$ in all cases). These results proved that NO does indeed play a role in classic preconditioning in our model.

To investigate whether NO was involved as trigger or mediator, hearts were either preconditioned with ischemia in the presence of NOS inhibitors, which were subsequently washed out before sustained ischemia, or the drugs were added after the preconditioning protocol but before the onset of sustained ischemia. Preconditioning was significantly attenuated by administering L-NAME and ODQ during the triggering phase, resulting in a cardiac output of 16.8 ± 2.8 and 18.6 ± 2.5 ml/min, respectively ($p < 0.05$ compared to preconditioned hearts). NO inhibitors infused after the triggering phase had no significant effect on preconditioning (as indicated by cardiac output during reperfusion). These findings suggest that NO acts as trigger, but not as mediator of preconditioning.

CLASSIC PRECONDITIONING CAN BE MIMICKED BY NO DONORS

The validity of the results described in the previous section was investigated by determining if NO donors could mimic preconditioning. Isolated perfused rat hearts were treated with intermittent infusions (3×5 min) of *S*-nitroso-*N*-penicillamine (SNAP) (50 and 100 μM) and sodium nitroprusside (SNP) (100 μM) or L-Arginine (L-Arg) (10 mM) prior to 25-min sustained ischemia to mimic cyclic preconditioning

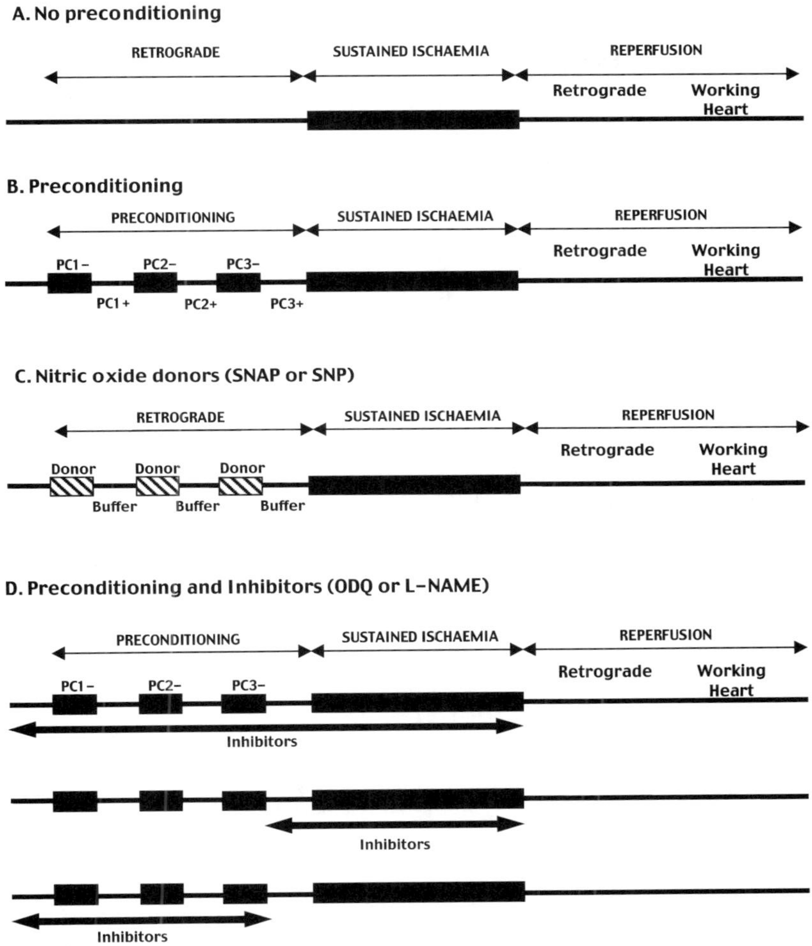

FIGURE 1. Experimental protocols: all hearts were stabilized by retrograde perfusion (15 min) and anterograde perfusion (i.e., as working hearts) for 15 min. All interventions shown followed on stabilization. Cardiac function was assessed by measuring coronary flow, aortic output, peak systolic pressure, and calculation of the total work at the end of stabilization and at the end of reperfusion following sustained global ischemia. **A.** Non-preconditioned hearts were perfused with Krebs-Henseleit buffer for a similar period to that of the preconditioning protocol, and than subjected to 25 min global ischemia and 30 min reperfusion. **B.** Hearts were preconditioned with 3×5-minute global ischemia interspersed with five minutes perfusion with Krebs-Henseleit buffer. Hearts were then subjected to 25-min global ischemia. **C.** Nitric oxide (NO) donors SNAP, SNP, and L-Arg were administered for 3×5-minutes, alternating with perfusion with normal buffer, before global ischemia of 25-min and 30-min reperfusion. **D.** The NO inhibitors ODQ, L-NAME, and L-NNA were administered in both preconditioned hearts (as indicated by the *arrows*) and non-preconditioned control hearts (not shown).

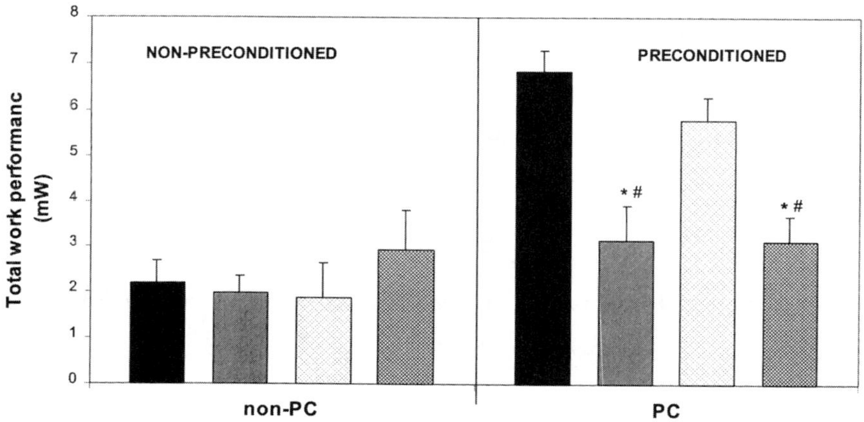

FIGURE 2. Effect of L-NAME (50 μM) treatment on functional recovery as determined by calculation of total work during reperfusion of non-preconditioned and preconditioned hearts: ■, untreated; ▩, 4×5-minutes L-NAME, administered until onset of sustained ischemia; ▢, L-NAME administered for five minutes before onset of sustained ischemia; ▩, L-NAME administered for 3×5-minutes, washed out for 10 minutes before onset of sustained ischemia; *$p < 0.05$ versus preconditioned (untreated); #$p < 0.05$ versus L-NAME administered for five minutes. (Figure reproduced with permission of the *American Journal of Physiology.*)

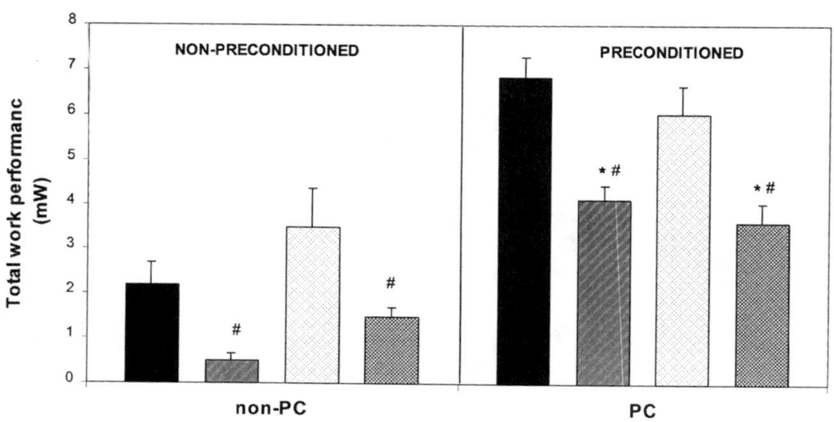

FIGURE 3. Effect of ODQ (20 μM) treatment on functional recovery as determined by calculation of total work during reperfusion of non-preconditioned and preconditioned hearts: ■, untreated; ▩, 4×5-minutes ODQ, administered until onset of sustained ischemia; ▢, ODQ administered for five minutes before onset of sustained ischemia; ▩, ODQ administered for 3×5-minutes, washed out for 10 minutes before onset of sustained ischemia; *$p < 0.05$ versus preconditioned (untreated); #$p < 0.05$ versus ODQ administered for five minutes in non-preconditioned hearts. (Figure reproduced with permission of the *American Journal of Physiology.*)

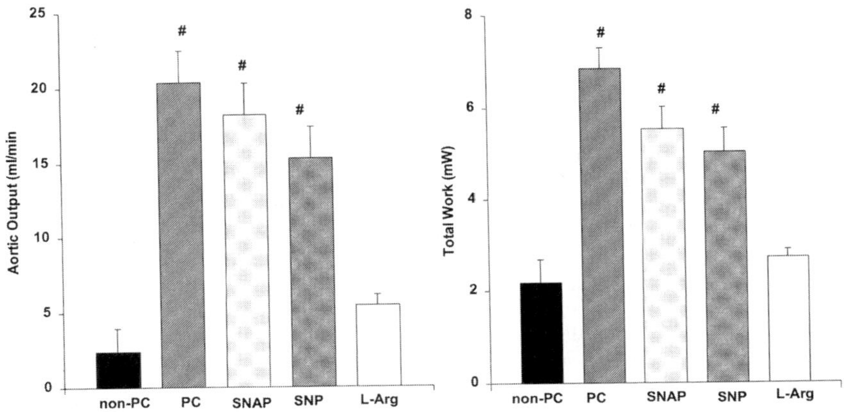

FIGURE 4. Effect of NO donors on functional recovery as determined by aortic output and calculation of total work during reperfusion after 25 min global ischemia. PC, preconditioned; non-PC, non-preconditioned. Doses were 100 μM SNP, 50 μM SNAP, and 10 mM L-Arg; #$p < 0.05$ versus non-PC. (Figure reproduced with permission of the *American Journal of Physiology.*)

(FIG. 1). SNAP and SNP caused cyclic increases in cGMP during administration, returning to control levels during reperfusion with normal buffer (results not shown). L-Arg, however, was without effect. SNAP (50 μM) and SNP (100 μM) elicited protection equal to that afforded by ischemic preconditioning (cardiac output 29.0 ± 2.6 and 30.5 ± 1.3 ml/min, respectively) but L-Arg (10 mM) failed to mimic the effect of ischemic preconditioning (cardiac output 14.8 ± 0.8 and 32.2 ± 1.9 ml/min, respectively; see FIGURE 4). These results confirm the role of NO in classic preconditioning, and suggest the potential to use NO administration to elicit preconditioning clinically.

THE MECHANISM OF NO-MEDIATED PROTECTION

Relationship Between Protection and the Generation of cGMP

There are numerous mechanisms by which the NO/cGMP system could elicit protection (for a recent review, see Rakhit and coworkers[7]). One mechanism may be NO-mediated attenuation of β-adrenergic receptor–mediated response,[20] which may diminish energy depletion during sustained ischemia, or decrease cytosolic calcium levels during ischemia or calcium overload during reperfusion. Another possible mechanism is the generation of the free radical peroxynitrite, which results in activation of PKC in the preconditioning scenario.[21]

The profile of cGMP generation in hearts preconditioned with ischemia strongly suggested that an increase in cGMP may be important to elicit protection. Significantly increased cGMP levels were observed after each five-minute ischemia episode—after the first, second, and third episode of ischemia cGMP levels increased to 12.18 ± 0.71, 20.22 ± 0.91, and 16.79 ± 1.12, respectively, compared to control levels of 8.42 ± 0.35 pmol/g wet weight. cGMP levels did not return to baseline control values upon reperfusion after each of these episodes of ischemia—in fact, it was still elevated to 12.54 ± 0.69 pmol/g wet weight ($p < 0.05$ vs. controls) when the hearts entered sustained ischemia. The significance of these cyclic changes in cGMP was demonstrated by the fact that the guanylyl cyclase inhibitor ODQ abolished protection, as indicated by functional recovery if given during the triggering phase (as shown above).[22] These findings lend further credence to the hypothesis that cGMP generation during the triggering phase of preconditioning is involved in the mechanism of NO mediated preconditioning.

Interestingly the magnitude of cGMP increase during NO preconditioning did not relate directly to the protection induced—the functional recovery of the hearts preconditioned with NO donors was similar, irrespective of the magnitude of cGMP increase during the triggering phase. Intermittent administration of the NO donors SNAP (10 μM and 50 μM) and SNP (100 μM) that also elicited protection as described above, caused a ±4–5-fold, ±6-fold, and ±2-fold increase in cGMP upon cyclic infusion, respectively. It must also be noted that on cessation of infusion of the NO donors, cGMP normalized rapidly and no elevation of cGMP was observed immediately preceding sustained ischemia in any of the donor groups. Protection against sustained ischemia by NO donors was, therefore, associated with intermittent increases in cGMP prior to sustained ischemia and protection was still elicited despite the fact that the cGMP levels had returned to control levels at the onset of sustained ischemia.

Although our initial studies could not demonstrate a mediator action of cGMP during sustained ischemia, tissue cGMP levels at the end of 25-min global ischemia were also significantly increased by both ischemic and NO-mediated preconditioning: for example, cGMP levels of ischemic preconditioned and SNAP treated hearts averaged 11.92 ± 0.62 and 11.86 ± 1.12, respectively, compared to 8.20 ± 0.58 pmol/g wet weight in non-preconditioned hearts ($p < 0.05$ vs. non-preconditioned hearts).

To investigate the importance of a NO-mediated increase in cGMP in the protection against sustained ischemia, we performed experiments in which cGMP could be manipulated during ischemia. For this purpose isolated perfused rat hearts were subjected to low-flow ischemia (coronary flow of 0.2 ml/min) for 25 min, followed by normoxic reperfusion for 25 min.[23] Drugs inhibiting or stimulating the NO/cGMP pathway were administered during the ischemic phase only.

At the end of reperfusion the left ventricular developed pressure of untreated hearts was 58 ± 5 mmHg; that is, about 55% of the preischemic control value. Infusion of the NO donor SNP (100 and 1,000 μM) improved left ventricular developed pressure to about 80% of preischemic control value. Ischemia increased cGMP levels 1.8-fold in untreated controls, but SNP (1,000 μM), which protected upon reperfusion, increased cGMP levels 6-fold (see FIGURE 5). However, the guanylyl cyclase inhibitor ODQ significantly reduced both tissue cGMP levels during ischemia and functional recovery during subsequent reperfusion (results not shown). These results indicate that the cardioprotective effects of NO donors may be due to their ability to

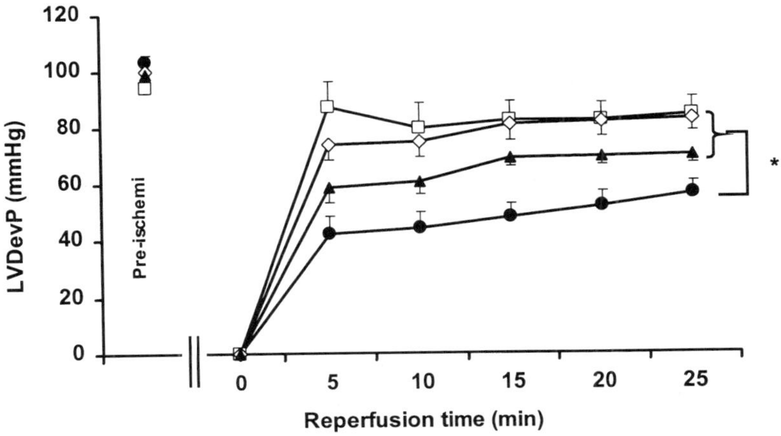

FIGURE 5. Effect of NO donors on recovery of left ventricular developed pressure during 25 min of reperfusion as determined with an intraventricular balloon. Hearts were perfused with vehicle (control), nitroprusside (SNP 1 μM, 100 μM, and 1,000 μM) or *S*-nitroso-*N*-acetyl-DL-penicillamine (SNAP 100 μM) during the ischemic phase: ●, control; □, SNP 0.1 mM; ◇, SNP 1.0 mM; ▲, SNAP 0.1 mM; $*p < 0.05$ versus control. (Figure reproduced with permission of the *British Journal of Pharmacology*.)

increase cGMP during ischemia in this model, and that guanylyl cyclase inhibition reverses these cardioprotective properties.

These experimental findings suggest that the protective effect of ischemic preconditioning may well be related to the elevation of cGMP at the beginning of sustained ischemia in ischemic preconditioned heart, as well as the increased levels seen at the end of 25 min ischemia in hearts preconditioned with either ischemia or NO donors.

Activation of p38 MAP Kinase

The role of both the stress kinase, p38 MAP kinase, and that of the mitochondrial K_{ATP} channel in the mechanism of ischemic preconditioning is currently being elucidated. Activation of the mitochondrial K_{ATP} channel by NO has indeed been demonstrated,[24] but to our knowledge the relationship between NO and p38 MAP kinase in preconditioning has not been investigated previously. There is also no data concerning the downstream effects of cGMP in the context of preconditioning.

To investigate the involvement of p38 MAP kinase activation in NO donor mediated preconditioning, we preconditioned isolated perfused rat hearts as described above (FIG. 1). Hearts were freeze-clamped during both the ischemic or NO-donor preconditioning protocol and at the end of 25 min ischemia. After Western blotting, filters were probed with a dual phospho-p38 MAPK (Thr180/Tyr182) antibody (New England Biolabs) and the activated p38 MAPK visualized used the ECL system. The NO donors SNAP (50 μM) and SNP (100 μM) activated p38 MAP kinase during the preconditioning protocol, and significantly attenuated the increase in activity during sustained ischemia as seen in non-preconditioned hearts (see FIGURE 6).

We recently investigated the role of p38 MAP kinase in ischemic preconditioning, and our results indicate that p38 MAP kinase plays a role as mediator of protection in ischemic preconditioning.[25] Short episodes of ischemia caused activation of p38 MAP kinase, and preconditioning-induced improvement in functional recovery was associated with a decrease in activation of p38 MAP kinase during the sustained period of ischemia in the isolated perfused rat heart. To establish whether p38 MAP kinase acts as trigger or mediator of preconditioning, isolated adult ventricular myocytes were exposed to the specific p38 MAP kinase inhibitor, SB 203580. Inhibition of p38 MAP kinase by SB 203580 protected isolated rat ventricular cardiomyocytes against sustained hypoxia to the same degree as hypoxic preconditioning, whereas administration of SB 203580 during the triggering phase did not abolish preconditioning.[25] These results indicate that p38 MAP kinase plays a role as mediator

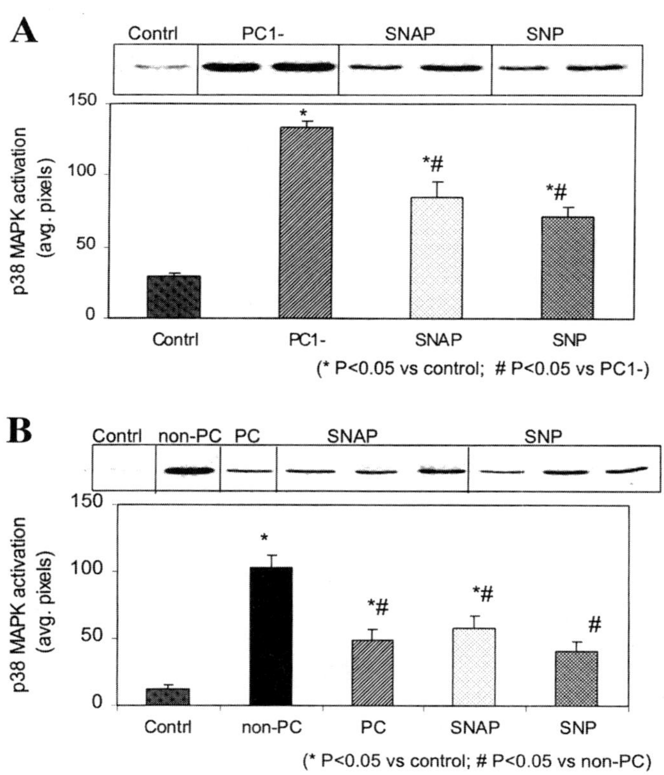

FIGURE 6. Effect of the NO donors SNAP (50 μM) and SNP (100 μM) on p38 MAP kinase activation as determined with antibodies against dual phosphorylated p38 MAP kinase of: **(A)** hearts freeze-clamped after five minutes infusion of SNAP and SNP or five minutes global ischemia (PC1-); **(B)** heart freeze-clamped at the end of 25 minutes global ischemia after no prior treatment (non-preconditioned, non-PC), pre-treatment with ischemic preconditioning, SNAP or SNP.

of protection in ischemic preconditioning, but not as trigger. These findings are supported by others: attenuated activation of p38 MAP kinase following preconditioning was also observed in a model of regional ischemia in the rabbit.[26] A role for p38 MAP kinase as mediator but not trigger of preconditioning has also been proposed based on the findings similar to ours in rat hearts.[27] These observations are also in keeping with findings indicating that the activation of p38 MAP kinase is associated with the detrimental effects of ischemia in neonatal cell cultures, as inhibition of p38 MAP kinase protected against ischemia.[28] The mechanism by which attenuated p38 MAP kinase activity could protect is suggested by Saurin *et al.*,[29] who showed that preconditioning protects by suppressing the activation of p38-alpha, an activator of apoptosis. Our results are opposite to those of Downey *et al.*, who hypothesized that PKC or oxygen radical–mediated activation of p38 MAP kinase during sustained ischemia results in the activation of a downstream effector of protection, such as MAPKAPK2 or HSP 27.[30] These discordant results may be due to differences in species (rat vs. rabbits) or in experimental models used in the studies. In summary, our results suggest that NO-mediated attenuation of activation of p38 MAP kinase during sustained ischemia plays a role in the mechanism of protection. Additional studies are required to elucidate the exact role of p38 MAP kinase and the final effector of protection in preconditioning.

POTENTIAL CLINICAL VALUE OF NO-INDUCED PRECONDITIONING: FUNCTIONAL RESERVE IS SUPERIOR TO USING ISCHEMIA TO PRECONDITION HEARTS

Despite the obvious potential benefit of ischemic preconditioning, the risk of subjecting patients to ischemia to elicit this protection remains. It is also not yet clear if the pathologic myocardium responds favorably to ischemic preconditioning—hearts of hypertensive rats[31] and aging rats[32] and pathologic human myocardium[33] did not precondition successfully with a preconditioning protocol that elicited protection in normal myocardial tissue. It is, therefore, important to find pharmacologic agents that can elicit preconditioning safely and to circumvent the potential dangers inherent in using ischemia.

In our studies as cited above, there was no difference in the protection induced by either ischemia or NO donors (see FIG. 4). We investigated whether there was any difference in the contractile reserve of hearts preconditioned with either ischemia or NO. This was assessed by infusion of adrenaline10^{-6}M at the end of reperfusion. The increase in functional capability of hearts preconditioned with the NO donors SNAP and SNP was superior to that of hearts preconditioned with ischemia—cardiac output increased by 122.1 ± 18.0% and 114.8 ± 13.3% in SNAP and SNP treated hearts respectively versus 40.7 ± 4.0% in hearts preconditioned with ischemia (see FIGURE 7). This indicates that the superior protection induced by NO was unmasked by adrenaline, and that in fact, it offered a major advantage above preconditioning with ischemia. These data suggest that NO-induced preconditioning may be a useful clinical manipulation, and makes investigation of its clinical usefulness imperative.

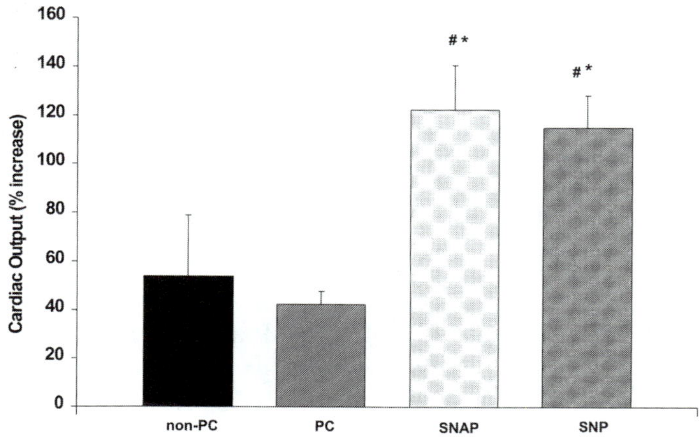

FIGURE 7. Effect of adrenaline on mechanical performance during reperfusion of non-preconditioned hearts, and hearts preconditioned with ischemia, SNAP- or SNP as described in the perfusion protocol in FIGURE 1. At the end of reperfusion, hearts were perfused with buffer containing adrenaline 10^{-6} M, and mechanical activity determined after an additional 10 minutes. The increase in cardiac output indicated is the percentage increase above the numerical value shown in FIGURE 4. #$p < 0.05$ versus non-PC; *$p < 0.05$ versus PC.

CONCLUSIONS

To determine if nitric oxide (NO) is involved in classic preconditioning, the effect of NO inhibitors donors on cGMP tissue levels and functional recovery during reperfusion of the isolated perfused working rat heart were evaluated. The nitric oxide synthase inhibitor L-NAME and the guanylyl cyclase inhibitor ODQ inhibited functional recovery of preconditioned hearts if administered during the triggering phase of the preconditioning protocol (i.e., during and before the preconditioning protocol) but not if given after the ischemic preconditioning protocol and before sustained ischemia only. This suggests a role for NO as a trigger, but not as mediator of preconditioning. Pharmacologic manipulation of cGMP by SNAP and SNP by intermittent infusion before sustained global ischemia resulted in concomitant increases in cGMP which were accompanied by functional recovery, thus mimicking preconditioning.

The role of cGMP in NO-mediated protection was investigated in a model of low-flow ischemia. cGMP levels were elevated during ischemia by infusion of the same NO donors. This resulted in a significant increase in cGMP levels during ischemia, and was accompanied by functional recovery upon reperfusion. These results indicate that the elevated cGMP levels observed in hearts preconditioned with either ischemia or NO donors could play a role in the improved recovery.

The role of p38 MAP kinase activation in the mechanism of NO-mediated preconditioning was also investigated. The NO donors SNAP and SNP activated p38 MAP kinase within five minutes of infusion, and hearts preconditioned in this way demonstrated significantly less p38 MAP kinase activation after 25 min global ischemia, as has also previously been observed by us in hearts preconditioned with ischemia or isoproterenol. However, attenuated p38 MAP kinase activation during sustained global ischemia, rather than activation of p38 MAP kinase during the triggering phase, may be involved in the protective effect of NO-mediated preconditioning.

The potential use of pharmacologic preconditioning with NO donors was further investigated by evaluating the functional reserve of hearts preconditioned with NO donors compared to preconditioning with ischemia. Adrenaline increased the function of NO preconditioned hearts to a greater extent than that of hearts preconditioned with ischemia, suggesting that it may be a safer alternative to elicit the beneficial effects of preconditioning.

REFERENCES

1. Kloner, R.A., R. Bolli, E. Marban, *et al.* 1998. Medical and cellular implications of stunning, hibernation and preconditioning: an NHLBI Workshop. Circulation **97:** 1848–1867.
2. Cohen, M.V., C.P. Baines & J.M. Downey. 2000. Ischemic preconditioning: from adenosine receptor to K_{ATP} channel. Annu. Rev. Physiol. **62:** 79–109.
3. Ytrehus, K., Y. Liu & J.M. Downey. 1994. Preconditioning protects ischemic rabbit heart by protein kinase C activation. Am. J. Physiol. **266:** 1145–1152.
4. Weinbrenner, C., G.S. Liu, M.V. Cohen & J.M. Downey. 1997. Phosphorylation of tyrosine 182 of p38 mitogen-activated protein kinase correlates with the protection of preconditioning in the rabbit heart. J. Mol. Cell Cardiol. **29:** 2383–2391.
5. O'Rourke, B. 2000. Myocardial K(ATP) channels in preconditioning. 2000. Circ. Res. **87:** 845–855.
6. Ping, P. & E. Murphy. 2000. Role of p38 mitogen-activated protein kinases in preconditioning: a detrimental factor or a protective kinase? Circ. Res. **86:** 921–922.
7. Rakhit, D., R.J. Edwards & M.S. Marber. 1999. Nitric oxide, nitrates and ischaemic preconditioning. Cardiovasc. Res. **43:** 621–647.
8. Lochner, A., S. Genade, E. Tromp, *et al.* 1995. Role of cyclic nucleotide phosphodiesterases in ischemic preconditioning. Mol. Cell. Biochem. **186:** 169–175.
9. Depré, C., L. Fiérain & L. Hue. 1997. Activation of nitric oxide synthase by ischemia in the perfused heart. Cardiovasc. Res. **33:** 82–87.
10. Takano, H., S. Manchikalapudi, X.L. Tang, *et al.* 1998. Nitric oxide synthase is the mediator of late preconditioning against myocardial infarction in conscious rabbits. Circulation **98:** 441–449.
11. Bolli, R., B. Dawn, X.L. Tang, *et al.* 1998. The nitric oxide hypothesis of late preconditioning. Bas. Res. Cardiol. **93:** 325–338.
12. Lu, H.R., P. Remeysen & F. de Clerck. 1995. Does the antiarrhythmic effect of ischemic preconditioning in rats involve the L-arginine nitric oxide pathway? J. Cardiovasc. Pharmacol. **25:** 524–530.
13. Sun, W. & C.L. Wainright. 1997. The role of nitric oxide in modulating ischemia-induced arrhythmias in rats. J. Cardiovasc. Pharmacol. **29:** 554–562.
14. Woolfson, R.G., V.C. Patel, G.H. Neild & D.M. Yellon. 1995. Inhibition of nitric oxide synthesis reduces infarct size by an adenosine-dependent mechanism. Circulation **91:** 1545–1551.
15. Weselcouch, E.O., A.J. Baird, P. Sleph & G.J. Grover. 1995. Inhibition of nitric oxide synthesis does not affect ischemic preconditioning in isolated perfused rat hearts. Am. J. Physiol. **268:** H242–H249.
16. Bilinska, M., M. Maczewski & A. Beresewicz. 1996. Donors of nitric oxide mimic effects of ischaemic preconditioning on reperfusion induced arrhythmias in isolated rat hearts. Mol. Cell. Biochem. **160–161:** 265–271.
17. Vegh, A., L. Szekeres & J. Parrat. 1992. Preconditioning of the ischaemic myocardium; involvement of the L-arginine nitric oxide pathway. Br. J. Pharmacol. **107:** 648–652.

18. Lochner, A., S. Genade, E. Marais, *et al.* 1999. Ischemic preconditioning and the β-adrenergic signal transduction pathway. Circulation **100:** 985–966.
19. Csonka, C., Z. Szilvássy, F. Fülöp, *et al.* 1999. Classic preconditioning decreases the harmful accumulation of nitric oxide during ischemia and reperfusion in rat hearts. Circulation **100:** 2260–2266.
20. Ebihara, Y. & M. Karmazyn. 1996. Inhibition of β, but not α-mediated adrenergic responses in isolated hearts and cardiomyocytes by nitric oxide and 8-bromo-cyclic cGMP. Cardiovasc. Res. **32:** 622–629.
21. Ping, P., H. Takano, J. Zhang, *et al.* 1999. Isoform-selective activation of protein kinase C by nitric oxide in the heart of conscious rabbits: a signaling mechanism for both nitric oxide-induced and ischemia-induced preconditioning. Circ. Res. **84:** 587–604.
22. Lochner, E. Marais, S. Genade & J.A. Moolman. 2000. Nitric oxide: a trigger for classic preconditioning? Am. J. Physiol. **279:** H2752–2765.
23. Du Toit, E.F., J. McCarthy, J. Miyashiro, *et al.* 1998. Effect of nitrovasodilators and inhibitors of nitric oxide synthase on ischaemic reperfusion function of rat isolated hearts. Br. J. Pharmacol. **123:** 1159–1167.
24. Sasaki, N., T. Sato, A. Ohler, *et al.* 2000. Activation of mitochondrial ATP-dependent potassium channels by nitric oxide. Circulation **101:** 439–445.
25. Marais, E., S. Genade, B. Huisamen, *et al.* 2001. Activation of p38 MAPK Induced by a multi-cycle ischaemic preconditioning protocol is associated with attenuated p38 MAPK activity during sustained ischemia and reperfusion. J. Mol. Cell. Cardiol. **33:** 769–778.
26. Gysembergh, A., B.Z. Simkhovich, R.A. Kloner & K. Przyklenk. 2001. p38 Mapk activity is not increased early during sustained coronary artery occlusion in preconditioned versus control rabbit heart. J. Mol. Cell Cardiol. **33:** 681–690.
27. Schneider, S., W. Chen, J. Hou, *et al.* 2001. Inhibition of p38 MAPK α/β reduces ischemic injury and does not block protective effects of preconditioning. Am. J. Physiol. **280:** H499–H508.
28. MacKay, K. & D. Mochly-Rosen. 1999. An inhibitor of p38 mitogen-activated protein kinase protects neonatal cardiac myocytes from ischemia. J. Biol. Chem. **274:** 6272–6279.
29. Saurin, A.T., J.L. Martin, R.J. Heads, *et al.* 2000. The role of differential activation of p38-mitogen-activated protein kinase in preconditioned ventricular myocytes. FASEB J. **14:** 2237–2246.
30. Nakano, A., C.P. Baines, S.O. Kim, *et al.* 2000. Ischemic preconditioning activates MAPKAPK2 in the isolated rabbit heart: evidence for involvement of p38 MAPK. Circ. Res. **86:** 144–151.
31. Moolman, J.A., S. Genade, E. Tromp, *et al.* 1997. Ischaemic preconditioning does not protect hypertrophic myocardium against ischaemia. SAMJ **87:** C151–C156.
32. Kersten, J.R., W.G. Toller, E.R. Gross, *et al.* 2000. Diabetes abolishes ischemic preconditioning: role of glucose, insulin and osmolality. Am. J. Phys. **278:** H1218–H1224.
33. Ghosh, S., N.B. Standen & M. Galiñanes. 2001. Failure to precondition pathological human myocardium. J. Am. Coll. Cardiol. **37:** 711–718.

Mechanisms of Hepatoprotection by Nitric Oxide

YINNA WANG, YORAM VODOVOTZ, PETER K.M. KIM, RUBEN ZAMORA, AND TIMOTHY R. BILLIAR

Department of Surgery, University of Pittsburgh, Pittsburgh, Pennsylvania, U.S.A.

ABSTRACT: Nitric oxide (NO) exerts numerous antiapoptotic effects on hepatocytes in settings of inflammation and tissue damage. These actions of NO are modulated by a variety of mechanisms under both physiologic and pathologic conditions. Nitric oxide inhibits cell death or apoptosis by modulation of heat shock proteins, *S*-nitrosylation of caspases at their catalytic site cysteine residue, triggering of the cGMP pathway, and prevention of mitochondrial dysfunction. Our preliminary studies also suggest that NO can modulate apoptosis-related genes in a manner consistent with an antiapoptotic effect. This review focuses on these molecular mechanisms of cytoprotection by NO.

INTRODUCTION

Nitric oxide (NO) plays important roles in liver homeostasis and disease, and its production is paradoxically implicated in both cytoprotection and cytotoxicity.[1,2] The cell types found in the liver are the hepatic parenchymal cells (hepatocytes) and the nonparenchymal cells, including endothelial cells, smooth muscle cells, Kupffer cells (resident hepatic macrophages), and Ito cells (fat-storage cells). We have focused our studies on hepatocytes, since they play a predominant role in the regulation of liver function.[3] Nitric oxide is synthesized in hepatocytes by inducible nitric oxide synthase (iNOS; NOS-2), which is rapidly upregulated following stimulation by endotoxin and cytokines.[4] Hepatocytes require a combination of cytokines to upregulate iNOS expression, although high levels of IL-1β alone can also induce iNOS in this cell type.[5]

Nitric oxide appears to have at least two generally beneficial effects on hepatocytes: cytoprotection against inflammation and tissue damage, and direct cytotoxic effects on invading microorganisms or tumor cells.[6] Our laboratory and others initially viewed iNOS as detrimental when expressed in non-immune cells. We described the suppression of numerous functions of hepatocytes, such as protein synthesis, by NO. These effects of NO mimicked known effects of inflammatory states, such as sepsis, on the function of hepatocytes.[1] More recently, however, we have begun to appreciate that in addition to these detrimental effects, NO also benefits hepatocytes.

Address for correspondence: Timothy R. Billiar, M.D., George V. Foster Professor of Surgery and Chair Department of Surgery, F-1281 Presbyterian University Hospital, Pittsburgh, PA 15213 U.S.A. Voice: 412-647-1749; fax: 412-647-3247.
billiartr@msx.upmc.edu

Ann. N.Y. Acad. Sci. **962: 415–422 (2002).**

BENEFICIAL EFFECTS OF NO IN HEPATOCYTES

In Vitro *Models*

We have developed several models of hepatocyte cell damage and apoptosis in response to endotoxin, proinflammatory factors, cytokines, such as tumor necrosis factor-α (TNF-α), and the transcriptional inhibitor actinomycin D (ActD) in order to investigate the molecular mechanisms underlying the effects of NO on hepatocytes. Both exogenous and endogenous NO protected hepatocytes against cell damage and apoptosis induced by treatment with TNF-α or transforming growth factor-β (TGF-β), ligation of Fas, or growth factor withdrawal. The chemical NO donor, *S*-nitroso-*N*-acetyl-D, L-penicillamine (SNAP), caused dose-dependent resistance to TNF-α–induced hepatotoxicity, whereas the specific iNOS inhibitor L-N^6-(1-iminoethyl)-lysine (L-NIL) caused increased hepatocyte apoptosis.[7–14] In physiologic concentration, NO also acts as an antioxidant to protect many cell types against oxidative killing.[15] Addition of low doses of NO from the chemical NO donor (Z)-1-[*N*-(3-ammoniopropyl)-*N*-(*n*-propyl)-amino]-diazen-1-ium-1,2-diolate (PAPA/NO) or SNAP to either isolated rat hepatocytes or cultured FTO2B rat hepatoma cells attenuated the cytotoxicity of the oxidant *t*-butylhydroperoxide (*t*BH).[16]

In Vivo *Models*

Several groups have attempted to determine the effects of suppressing NO production *in vivo* on the viability of hepatocytes. We had demonstrated that non-selective inhibition of NOS using N^G-monomethyl-L-arginine (L-NMMA) increased organ damage in rats.[17] This inhibitor was also deleterious in a rabbit model of lipopolysaccharide (LPS)-induced shock, whereas the antiinflammatory glucocorticoid dexamethasone beneficially suppressed the expression of iNOS.[18] In a direct comparative study, nonspecific inhibition of NOS (using L-NMMA or N^G-nitro-L-arginine-methyl ester [L-NAME]) in LPS-induced endotoxemia caused hepatic necrosis by a mechanism independent of neutrophils, whereas specific inhibitors of iNOS (L-NIL or aminoguanidine [AG]) resulted in hepatocyte apoptosis.[19] Gene transfer of human iNOS using an adenoviral vector (Ad-iNOS) to cultured rat hepatocytes resulted in overexpression of iNOS, sustained NO production, and protection of hepatocytes from spontaneous and TNFα/ActD-induced apoptosis.[7]

MECHANISMS BY WHICH NO PROTECTS HEPATOCYTES

We have defined several distinct pathways that are independent of these antioxidant effects, by which NO protects hepatocytes. These mechanisms involve the induction of heat shock proteins, suppression of caspase activity, activation of soluble guanylyl cyclase, and modulation of mitochondrial respiration. Additionally, we recently found that NO can modulate the expression of both proapoptotic and antiapoptotic genes, leading to a profile of gene expression that appears to favor suppression of apoptosis. We discuss each of these mechanisms.

Effects of NO on Caspases

Caspases comprise a family of cysteine proteases that act to propagate death signals by cleaving a number of cellular protein substrates, leading ultimately to apoptosis.[20] Caspases-8, -9, and -10 are initiator caspases that can transduce apoptotic signals by directly activating the downstream executioner caspases-3, -6, and -7.[21]

There is growing evidence that NO blocks apoptosis by inhibiting caspases in hepatocytes.[22,23] Our group has shown that NO, generated either by an NO donor or by iNOS transduction, suppresses apoptosis induced by TNFα/ActD, TNFα plus D-galactosamine, ligation of Fas, and growth factor withdrawal. This antiapoptotic effect on NO was shown, both *in vitro* and *in vivo*, to be a result of suppression of caspase activity and proteolytic activation.[7–9,11,24]

Subsequently, we further explored the mechanisms underlying the inhibition of caspases by NO. Protein *S*-nitrosylation reactions can either alter protein structure directly or result in oxidation of thiols, modulating protein function in a manner analogous to protein phosphorylation.[25] Our studies indicate that *S*-nitrosylation of a cysteine thiol at the catalytic site of the enzyme plays a critical role in caspase activity and activation. We have found that NO suppresses apoptosis, at least in part, by inhibiting caspases-3 and -8. This inhibitory effect can be reversed by dithiothreitol (DTT), a reducing agent known to break down *S*-nitrosothiol modification in proteins.[8,9,11,24] In addition, NO has the potential to act as an electron donor or an electron acceptor due to its varied biological properties. Nitrosonium (NO^+) or its equivalent can interact with iron (both protein-bound and non protein-bound) as an electron acceptor. *S*-nitrosothiols most likely occur through NO^+ species generated by the loss of one electron from NO.[25–28]

We have confirmed that caspase *S*-nitrosylation is related to intracellular iron content. High levels of non-heme iron in hepatocytes facilitate caspase *S*-nitrosylation and prevent apoptosis. This effect was not observed in RAW264.7 macrophage-like cells, which have low levels of non-heme iron. Iron loading of RAW264.7 cells delayed NO-induced cell death and inhibited caspase-3 activation, suggesting that non-heme iron content may determine whether NO exerts cytoprotective or cytotoxic effects.[29]

Effects of NO on Mitochondria

Nitric oxide affects mitochondrial function in numerous ways.[30] More recently, mitochondria have been shown to mediate cell death and apoptosis. Cytochrome *c*, a basic component in the mitochondrial respiratory chain, plays a secondary role in propagation of the caspase cascade, subsequently amplifying apoptosis signal transduction.[31] The mitochondrial permeability transition pore (MPTP) is a protein complex consisting of the mitochondrial voltage-dependent anion channel, the adenine nucleotide translocase, and cyclophilin D. The opening of the MPTP is associated with release of cytochrome *c*, AIF (apoptosis inducing factor), and caspases.[32,33] In isolated rat liver mitochondria, exposure to NO caused significantly lower cytochrome *c* release as well as MPTP opening, and low concentrations of NO suppressed the opening of MPTP by depolarizing the mitochondrial membrane and inhibiting Ca^{2+} fluxes.[34]

Nitric oxide protects hepatocytes from apoptosis by another mitochondrial-related mechanism. Members of the Bcl-2 family of proteins, including Bid and Bax, modulate mitochondrial-mediated apoptotic processes. Bid, a 26-kDa cytosolic protein found in more specific cells as an inactive precursor, is cleaved by caspase-8 to produce a 15-kDa carboxy-terminal fragment that binds to the mitochondria and induces the release of cytochrome *c* by an unknown mechanism. In turn, this release activates the downstream caspases.[35,36] Our laboratory has found that NO blocks TNFα/ActD-induced apoptosis and caspase activation by inhibition of cytochrome *c* release from mitochondria, reducing the decrease in mitochondrial transmembrane potential and maintaining Bcl-2 levels in hepatocytes.[9,37] More recently, we have confirmed that the increased caspase activity in TNFα/ActD-induced apoptosis is associated with Bid cleavage and accumulation of cytochrome *c* in the cytosol. Nitric oxide, produced either by chemical NO donors or generated by adenovirally-transferred iNOS, prevented Bid cleavage and cytochrome *c* release *in vitro* and *in vivo* via *S*-nitrosylation of caspase-8, which blocked propagation of the apoptotic signal cascade in rat hepatocytes.[8]

Effects of NO on Heat Shock Proteins

Stress-inducing agents often increase the expression of heat shock proteins that can protect the liver by functioning to restore or refold proteins.[38] At high doses of the NO, donor SNAP induced the expression of heat shock protein-70 (Hsp70) in cultured rat hepatocytes treated with TNFα/ActD, and an antisense oligonucleotide to hsp70 abolished the hepatoprotective effect of SNAP.[13] However, the exact mechanism by which heat shock proteins modulate resistance to cytotoxicity is still elusive and controversial. Some studies suggest that Hsp70 prevents apoptosis by inhibiting the stress-activated protein kinase signaling cascade.[39] Studies in hepatocytes of overexpressing Hsp70 suggest that this protein acts by modulating processes downstream of cytochrome c release and upstream of caspase-3 activation.[40] *In vitro,* hsp70 inhibited cytochrome *c*-mediated procaspase 9 processing, whereas hsp70 mutant protein did not prevent procaspase processing.[41] We have also shown that NO regulates the expression of Hsp32 and heme oxygenase 1 (HO-1) in hepatocytes, which may help to protect these cells from death.[42]

Antiapoptotic Effects Mediated via the NO/cGMP Signaling Pathway

The best characterized mechanism for NO-mediated signal transduction is through an increase in intracellular cyclic guanosine 3′,5′-monophosphate (cGMP) by binding and activation of soluble guanylyl cyclase (sGC).[43,44] Nitric oxide-mediated hepatoprotective effects appear to depend partially on cGMP production. In an *in vivo* model of fulminant hepatic failure of mice, in which ActD/LPS induced hepatocyte cell death or apoptosis was inhibited by administration of NO donors (SNAP, hydroxylamine, 2,2′-[hydroxynitrosohydrazino]bis-ethanamine) [NOC 18]) or two different cGMP analogs (dibutyryl-cGMP and 8-Br-cGMP). NOC18 also reduced the plasma alanine aminotransferase (ALT) levels and hepatic DNA fragmentation, enzymes whose release is indicative of liver damage. These protective effects of NO were blocked by the sGC inhibitor LY83583.[45] Another inhibitor of sGC—1,H-[1,2,4] oxadiazole [4,3-a]quinoxalon-1-one (ODQ)—was partially able to block the suppression of apoptosis and caspase-3 activity by NO in primary rat cultured hepatocytes treated with TNFα/ActD.[11]

Effects of NO on Expression of Apoptosis-Related Genes

We hypothesized that some of the antiapoptotic effects of NO may be mediated by the modulation of gene expression in hepatocytes. We identified genes regulated by NO in mouse hepatocytes using the technique of differential display polymerase chain reaction (DD-PCR).[46] Since iNOS may be induced during manipulation of hepatocytes,[47,48] we decided to use cells derived from iNOS null ($iNOS^{-/-}$) mice.[49] Since iNOS is usually induced in hepatocytes by treatment with cytokines that can modulate gene expression, we instead infected hepatocytes with Ad-iNOS. As a control, we used an adenovirus expressing the bacterial β-galactosidase gene (Ad-LacZ).

Using this analysis, we identified a gene fragment suppressed by NO that was 92% homologous to the proapoptotic Bcl-2 binding protein, BNIP3.[50] This protein localizes to mitochondria and other cytoplasmic membrane structures.[51] BNIP3 appears to have an overall resemblance to several BH3-containing Bcl-2 family proteins, such as BIK, BID, HRK, and BAD.[52] Indeed, BNIP3 sensitizes Rat-1 fibroblasts to apoptosis by granzyme B and topoisomerase inhibitors, and overcomes Bcl-2 suppression of apoptosis.[53] Northern analysis confirmed the NO-dependent suppression of BNIP3 in cultured cells. Similarly, SNAP reduced the expression of BNIP3 in both $iNOS^{-/-}$ and wildtype hepatocytes. This effect of NO was reversed by ODQ, suggesting the involvement of the sGC/cGMP pathway in the modulation of BNIP3 by NO. Ongoing studies in our laboratory are aimed at elucidating the effects of modulation of BNIP3 on the antiapoptotic role of NO in hepatocytes.

We also sought to address, on a large scale, the genetic changes caused by the expression of iNOS in hepatocytes.[50] Accordingly, we carried out a DNA microarray analysis using mRNA derived from $iNOS^{-/-}$ mouse hepatocytes infected either with Ad-iNOS or with Ad-LacZ. Out of the approximately 6,500 gene fragments contained on this array, only approximately 200 were significantly modulated, either up or down, by infection with Ad-iNOS as compared to infection with Ad-LacZ.

We observed a reduction in certain proapoptotic genes following infection with Ad-iNOS, including sox-4, which was found to be the predominant upregulated gene during prostaglandin-mediated apoptosis of hepatocellular carcinoma cells.[54] The calpain gene was also reduced; activation of calpain by NO has been reported to induce apoptosis in pancreatic islet cells.[55] The proapoptotic R-ras gene[56] was also reduced.

We have also initiated the next step in study of the effect of NO on hepatocytes: large-scale proteomic analysis in order to correlate and extend the results of the DNA array. Proteins whose expression is altered by NO can then be analyzed for functional changes and their role in apoptosis.

SUMMARY

As we have outlined, NO acts largely as cytoprotective molecule in hepatocytes. This effect may be beneficial in various physiologic and pathologic conditions, such as acute inflammation and liver damage. We have explored the molecular mechanisms by which NO exerts this hepatoprotective action, and have determined that NO suppresses caspase activity and activation, induces heat shock proteins, activates the cGMP signal pathway, and reduces mitochondrial dysfunction. Additionally, NO appears to modulate the expression of genes associated with apoptosis. Understanding the beneficial actions of NO in hepatocytes will hopefully allow us to devise potential diagnostic and therapeutic strategies for various pathologic conditions involving liver failure. To this end, further mechanistic studies on the hepatoprotective actions of NO must still be proposed.

REFERENCES

1. TAYLOR, B.S., L.H. ALARCON & T.R. BILLIAR. 1998. Inducible nitric oxide synthase in the liver: regulation and function. Biochemistry (Moscow) **63:** 766.
2. LI, J. & T.R. BILLIAR. 1999. Nitric oxide. IV. Determinants of nitric oxide protection and toxicity in liver. Am. J. Physiol. **276:** G1069.
3. MURIEL, P. 2000. Regulation of nitric oxide synthesis in the liver. J. Appl. Toxicol. **20:** 189.
4. ZAMORA, R., Y. VODOVOTZ & T.R. BILLIAR. 2000. Inducible nitric oxide synthase and inflammatory diseases. J. Molec. Med. **6:** 347.
5. KITADE, H., K. SAKITANI, K. INOUE, *et al.* 1996. Interleukin 1 beta markedly stimulates nitric oxide formation in the absence of other cytokines or lipopolysaccharide in primary cultured rat hepatocytes but not in Kupffer cells. Hepatology **23:** 797.
6. SUZUKI, H., M. MENEGAZZI, A. CARCERERI DE PRATI, *et al.* 1995. Nitric oxide in the liver: physiopathological roles. Adv. Neuroimmunol. **5:** 379.
7. TZENG, E., T.R. BILLIAR, D.L. WILLIAMS, *et al.* 1998. Adenovirus-mediated inducible nitric oxide synthase gene transfer inhibits hepatocyte apoptosis. Surgery **124:** 278.
8. KIM, Y.M., T.H. KIM, H.T. CHUNG, *et al.* 2000. Nitric oxide prevents tumor necrosis factor alpha-induced rat hepatocyte apoptosis by the interruption of mitochondrial apoptotic signaling through *S*-nitrosylation of caspase-8. Hepatology **32:** 770.
9. LI, J., C.A. BOMBECK, S. YANG, *et al.* 1999. Nitric oxide suppresses apoptosis via interrupting caspase activation and mitochondrial dysfunction in cultured hepatocytes. J. Biol. Chem. **274:** 17325.
10. CURRAN, R.D., F.K. FERRARI, P.H. KISPERT, *et al.* 1991. Nitric oxide and nitric oxide-generating compounds inhibit hepatocyte protein synthesis. FASEB J. **5:** 2085.
11. KIM, Y.M., R.V. TALANIAN & T.R. BILLIAR. 1997. Nitric oxide inhibits apoptosis by preventing increases in caspase-3-like activity via two distinct mechanisms. J. Biol. Chem. **272:** 31138.
12. CURRAN, R.D., T.R. BILLIAR, D.J. STUEHR, *et al.* 1990. Multiple cytokines are required to induce hepatocyte nitric oxide production and inhibit total protein synthesis. Ann. Surg. **212:** 462.
13. KIM, Y.M., M.E. DE VERA, S.C. WATKINS & T.R. BILLIAR. 1997. Nitric oxide protects cultured rat hepatocytes from tumor necrosis factor-alpha-induced apoptosis by inducing heat shock protein 70 expression. J. Biol. Chem. **272:** 1402.
14. SAAVEDRA, J.E., T.R. BILLIAR, D.L. WILLIAMS, *et al.* 1997. Targeting nitric oxide (NO) delivery *in vivo*. Design of a liver-selective NO donor prodrug that blocks tumor necrosis factor-alpha-induced apoptosis and toxicity in the liver. J. Med. Chem. **40:** 1947.
15. WINK, D.A., Y. VODOVOTZ, M.B. GRISHAM, *et al.* 1999. Antioxidant effects of nitric oxide. Meth. Enzymol. **301:** 413.
16. JOSHI, M.S., J.L. PONTHIER & J.R. LANCASTER, JR. 1999. Cellular antioxidant and pro-oxidant actions of nitric oxide. Free Rad. Biol. Med. **27:** 1357.
17. BILLIAR, T.R., R.D. CURRAN, B.G. HARBRECHT, *et al.* 1990. Modulation of nitrogen oxide synthesis *in vivo*: N^{G}-monomethyl-L-arginine inhibits endotoxin-induced nitrate/nitrate biosynthesis while promoting hepatic damage. J. Leukocyte Biol. **48:** 565.

18. WRIGHT, C.E., D.D. REES & S. MONCADA. 1992. Protective and pathological roles of nitric oxide in endotoxin shock. Cardiovasc. Res. **26:** 48.
19. OU, J., T.M. CARLOS, S.C. WATKINS, *et al.* 1997. Differential effects of nonselective nitric oxide synthase (NOS) and selective inducible NOS inhibition on hepatic necrosis, apoptosis, ICAM-1 expression, and neutrophil accumulation during endotoxemia. Nitric Oxide **1:** 404.
20. WOLF, B.B. & D.R. GREEN. 1999. Suicidal tendencies: apoptotic cell death by caspase family proteinases. J. Biol. Chem. **274:** 20049.
21. THORNBERRY, N.A. & Y. LAZEBNIK. 1998. Caspases: enemies within. Science **281:** 1312.
22. LI, J. & T.R. BILLIAR. 1999. The anti-apoptotic actions of nitric oxide in hepatocytes [see comments]. Cell Death Diff. **6:** 952.
23. LIU, L. & J.S. STAMLER. 1999. NO: an inhibitor of cell death. Cell Death Diff. **6:** 937.
24. LI, J., T.R. BILLIAR, R.V. TALANIAN & Y.M. KIM. 1997. Nitric oxide reversibly inhibits seven members of the caspase family via *S*-nitrosylation. Bioch. Biophys. Res. Commun. **240:** 419.
25. D.A WINK, M. FEELISCH, Y. VODOVOTZ, *et al.* 1999. The chemical biology of nitric oxide. *In* Reactive Oxygen Species in Biological Systems: An Interdisciplinary Approach. C.A. Colton & D.L. Gilbert, Eds.: 245-291. Kluwer Academic/Plenum Publishing, New York.
26. COOPER, C.E. 1999. Nitric oxide and iron proteins. Biochim. Biophys. Acta **1411:** 290.
27. LIPTON, S.A., Y.B. CHOI, Z.H. PAN, *et al.* 1993. A redox-based mechanism for the neuroprotective and neurodestructive effects of nitric oxide and related nitrosocompounds. Nature **364:** 626.
28. LIPTON, S.A. 1999. Neuronal protection and destruction by NO. Cell Death Diff. **6:** 943.
29. KIM, Y.M., H.T. CHUNG, R.L. SIMMONS & T.R. BILLIAR. 2000. Cellular non-heme iron content is a determinant of nitric oxide-mediated apoptosis, necrosis, and caspase inhibition. J. Biol. Chem. **275:** 10954.
30. BROWN, G.C. 1999. Nitric oxide and mitochondrial respiration. Biochim. Biophys. Acta **1411:** 351.
31. ZOU, H., W.J. HENZEL, X. LIU, *et al.* 1997. Apaf-1, a human protein homologous to C. elegans CED-4, participates in cytochrome c-dependent activation of caspase-3. Cell **90:** 405.
32. CROMPTON, M. 1999. The mitochondrial permeability transition pore and its role in cell death. Biochem. J. **341:** 233.
33. CROMPTON, M. 2000. Bax, Bid and the permeabilization of the mitochondrial outer membrane in apoptosis. Curr. Opin. Cell Biol. **12:** 414.
34. BROOKES, P.S., E.P. SALINAS, K. DARLEY-USMAR, *et al.* 2000. Concentration-dependent effects of nitric oxide on mitochondrial permeability transition and cytochrome c release. J. Biol. Chem. **275:** 20474.
35. LI, H., H. ZHU, C.J. XU & J. YUAN. 1998. Cleavage of BID by caspase 8 mediates the mitochondrial damage in the Fas pathway of apoptosis. Cell **94:** 491.
36. LUO, X., I. BUDIHARDJO, H. ZOU, *et al.* 1998. Bid, a Bcl2 interacting protein, mediates cytochrome c release from mitochondria in response to activation of cell surface death receptors. Cell **94:** 481.
37. KIM, Y.M., T.H. KIM, D.W. SEOL, *et al.* 1998. Nitric oxide suppression of apoptosis occurs in association with an inhibition of Bcl-2 cleavage and cytochrome c release. J. Biol. Chem. **273:** 31437.
38. JOLLY, C. & R.I. MORIMOTO. 2000. Role of the heat shock response and molecular chaperones in oncogenesis and cell death. J. Natl. Cancer Inst. **92:** 1564.
39. MOSSER, D.D., A.W. CARON, L. BOURGET, *et al.* 1997. Role of the human heat shock protein hsp70 in protection against stress-induced apoptosis. Molec. Cell. Biol. **17:** 5317.
40. LI, C.Y., J.S. LEE, Y.G. KO, *et al.* 2000. Heat shock protein 70 inhibits apoptosis downstream of cytochrome c release and upstream of caspase-3 activation. J. Biol. Chem. **275:** 25665.

41. MOSSER, D.D., A.W. CARON, L. BOURGET, *et al.* 2000. The chaperone function of hsp70 is required for protection against stress-induced apoptosis. Molec. Cell. Biol. **20:** 7146.
42. KIM, Y.M., H. BERGONIA & J.R. LANCASTER. 1995. Nitrogen oxide-induced autoprotection in isolated rat hepatocytes. FEBS Lett. **374:** 228.
43. DENNINGER, J.W. & M.A. MARLETTA. 1999. Guanylate cyclase and the ·NO/cGMP signaling pathway. Biochim. Biophys. Acta **1411:** 334.
44. LUCAS, K.A., G.M. PITARI, S. KAZEROUNIAN, *et al.* 2000. Guanylyl cyclases and signaling by cyclic GMP. Pharmacol. Rev. **52:** 375.
45. AKAHORI, M., S. YAMADA, N. TAKEYAMA & T. TANAKA. 1999. Nitric oxide ameliorates actinomycin D/endotoxin-induced apoptotic liver failure in mice. J. Surg. Res. **85:** 286.
46. LIANG, P. & A.B. PARDEE. 1995. Recent advances in differential display. Curr. Opin. Immunol. **7:** 274.
47. WANG, H., X. GAO, S. FUKUMOTO, *et al.* 1998. Post-isolation inducible nitric oxide synthase gene expression due to collagenase buffer perfusion and characterization of the gene regulation in primary cultured murine hepatocytes. J. Biochem. **124:** 892.
48. TIRMENSTEIN, M.A., F.A. NICHOLLS-GRZEMSKI, T.D. SCHMITTGEN, *et al.* 2000. Characterization of nitric oxide production following isolation of rat hepatocytes. Toxicol. Sci. **53:** 56.
49. MACMICKING, J.D., C. NATHAN, G. HOM, *et al.* 1995. Altered responses to bacterial infection and endotoxic shock in mice lacking inducible nitric oxide synthase. Cell **81:** 641.
50. ZAMORA, R., L. ALARCON, B. BETTEN, *et al.* 2001. Nitric oxide suppresses the expression of Bcl-2 binding protein BNIP3 in hepatocytes. J. Biol. Chem. **276:** 46887.
51. BOYD, J.M., S. MALSTROM, T. SUBRAMANIAN, *et al.* 1994. Adenovirus E1B 19 kDa and Bcl-2 proteins interact with a common set of cellular proteins. Cell **79:** 341.
52. YASUDA, M., P. THEODORAKIS, T. SUBRAMANIAN & G. CHINNADURAI. 1998. Adenovirus E1B-19K/BCL-2 interacting protein BNIP3 contains a BH3 domain and a mitochondrial targeting sequence. J. Biol. Chem. **273:** 12415.
53. CHEN, G., R. RAY, D. DUBIK, *et al.* 1997. The E1B 19K/Bcl-2-binding protein Nip3 is a dimeric mitochondrial protein that activates apoptosis. J. Exp. Med. **186:** 1975.
54. AHN, S.G., G.H. CHO, S.Y. JEONG, *et al.* 1999. Identification of cDNAs for Sox-4, an HMG-Box protein, and a novel human homolog of yeast splicing factor SSF-1 differentially regulated during apoptosis induced by prostaglandin A2/delta12-PGJ2 in Hep3B cells. Biochem. Biophys. Res. Commun. **260:** 216.
55. NAKATA, M., N. UTO, I. MARUYAMA & T. YADA. 1999. Nitric oxide induces apoptosis via Ca2+-dependent processes in the pancreatic beta-cell line MIN6. Cell Struct. Funct. **24:** 451.
56. WANG, H.G., J.A. MILLAN, A.D. COX, *et al.* 1995. R-Ras promotes apoptosis caused by growth factor deprivation via a Bcl-2 suppressible mechanism. J. Cell Biol. **129:** 1103.

Protective Effects of Glial Cell Line-Derived Neurotrophic Factor in Ischemic Brain Injury

Y. WANG,[a] C.F. CHANG,[a,b] M. MORALES,[a] Y.H. CHIANG,[b] AND J. HOFFER[a]

[a]*National Institute on Drug Abuse, National Institutes of Health, Baltimore, Maryland, U.S.A.*

[b]*Triservice General Hospital, National Defense Medical Center, Taiwan*

ABSTRACT: Glial cell line-derived neurotrophic factor (GDNF), a member of the transforming growth factor-β (TGF-β) superfamily, has been shown to have trophic activity on dopaminergic neurons. Recent studies indicate that GDNF can protect the cerebral hemispheres from damage induced by middle cerebral arterial ligation. We found that such neuroprotective effects are mediated through specific GDNF receptor alpha-1 (GFRα1). Animals with a deficiency in GFRα-1 have less GDNF-induced neuroprotection. Ischemia also enhances nitric oxide synthase (NOS) activity, which can be attenuated by GDNF. These.data suggest that GDNF can protect against ischemic injury through a GFRα-1/NOS mechanism. We also found that the receptor for GDNF, GFRα1, and its signaling moiety c-Ret were upregulated, starting immediately after ischemia. This upregulation suggests that activation of an endogenous neuroprotective mechanism occurs so that responsiveness of GDNF can be enhanced at very early stages during ischemia.

KEYWORDS: cerebral ischemia; MCA ligation; GDNF

INTRODUCTION

Ischemic brain injury is a major cause of mortality worldwide.[1–3] Cell injury and death after cerebral vascular occlusion has been postulated to result from a number of interacting pathophysiologic factors, including depolarization-induced calcium entry via *N*-methyl-D-aspartate (NMDA) receptors, intracellular free radical generation, damage to mitochondrial respiratory enzymes, and induction of programmed cell death, to name but a few proposed mechanisms.[4–7]

Glial cell line-derived neurotrophic factor (GDNF), a protein in transforming growth factor-β (TGF-β) superfamily, was initially considered to be a specific trophic factor and neuroprotective agent for dopaminergic neurons. Pretreatment with GDNF reduces 6-hydroxydopamine (6-OHDA),[8,9] 1-methyl-4-phenyl-1,2,3,6-tetrahydropyridine (MPTP),[10] or axotomy-induced neuronal damage in the nigrostriatal dopaminergic pathway.[11] It was later found that GDNF has effects on other systems, in addition to the dopaminergic pathway. GDNF diminishes damage produced by kainate-induced tonic-clonic convulsions[12] and axotomy injury in

Address for correspondence: Yun Wang, M.D., Ph.D., National Institute on Drug Abuse, I.R.P., 5500 Nathan Shock Drive, Baltimore, MD 21224, U.S.A. Voice: 410-550-1753; fax: 410-550-1645.
ywang@intra.nida.nih.gov

spinal cord motor neurons.[13–15] Since GDNF has an antiapoptotic effect in dopamine and non-dopaminergic neurons after injury,[16–19] it is possible that GDNF can attenuate ischemic brain injury through the same mechanism.

GDNF REDUCES ISCHEMIC INSULTS IN BRAIN

GDNF Protein

Pretreatment with GDNF reduces the cerebral infarction induced by middle cerebral artery (MCA) ligation in rats.[20] Animals receiving intracerebral administration of GDNF showed a significant reduction in the volume of infarction (see FIGURE 1, saline: $100.0 \pm 18.3\%$, $n = 7$ vs. GDNF: $35.6 \pm 27.5\%$, $n = 6$, $p < 0.05$, Mann-Whitney rank sum test). These data suggest that intracerebral administration of GDNF protects against ischemia/reperfusion–induced brain injury. Similarly, it was later reported that application of GDNF reduces cerebral infarction in animals given transient or permanent MCA occlusion.[21–23] The protective of GDNF can be achieved if GDNF is given before[20] ischemia and up to one hour after the onset of reperfusion.[24] There is no neuroprotection if GDNF is given three hours after transient MCA occlusion.[24]

GDNF-Containing Cells

It has been demonstrated that GDNF is a very diffusible protein. The content of GDNF in the brain decreases markedly after a single intraventricular injection.[25]

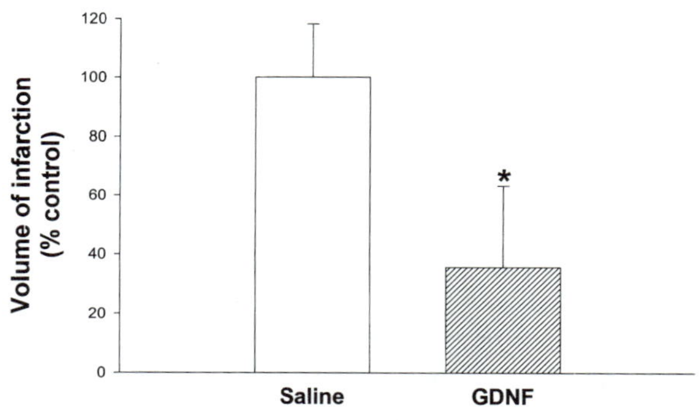

FIGURE 1. Intracerebral administration of GDNF attenuates cortical infarction induced by 90 minute-ischemia and 24 hour-reperfusion in rats. GDNF was injected intracerebrally at a dose of 10 µg/rat before MCA ligation. The ligation of the right MCA and bilateral common carotids (CCAs) was performed as described elsewhere.[64] Twenty-four hours after reperfusion, animals were sacrificed. The brain tissue was then removed and sliced into 2.0-mm sections. The brain slices were incubated in a 2% triphenyl-tetrazolium chloride (TTC) dissolved in saline and then transferred to 5% formaldehyde solution for fixation. The volumes of infarction were measured in each slice and summed. The volume of infarction from rats pretreated with GDNF was normalized by comparison to the mean level in animals receiving PBS injection. The volume of infarction was significantly reduced by pretreatment with GDNF as compared to control.

Consequently, we found that a high dose of GDNF was required to protect against ischemic insult. In a similar vein, such doses of GDNF are also needed to "rescue" neurotoxically injured DA neurons.[8,9] It is possible that these high doses of GDNF may induce side effects after intracranial administration.[26] Because GDNF is highly expressed in fetal kidney tissue,[27] it is possible that fetal kidney tissue could serve as a cellular reservoir for such molecules and protect against ischemic injury in the cerebral cortex.[28,29] Fetal kidneys obtained from rat embryos at gestational day 16, and adult kidney cortex, were dissected, cut into small pieces, and stereotactically transplanted into three cortical areas adjacent to the right MCA in rats. We found that the stroke animals that received adult kidney transplantation developed motor imbalance and marked cerebral infarction. However, animals that received fetal kidney grafts showed significant behavioral improvement and reduced infarction volume (see FIGURE 2). These data indicate that fetal kidney transplantation reduces ischemia/reperfusion–induced cortical infarction and behavioral deficits in adult rats, and that such tissue grafts could serve as an unique cellular reservoir for trophic factor application to the brain. Similar to these findings, it has been reported that transplantation of kidney cell lines, which contain a high level of GDNF, also reduces cerebral infarction volume in stroke rats.[30]

Pharmacologic Manipulation

Since GIDNF is known not to be permeable across the blood–brain barrier, its use for neuroprotection might be limited to central administration. It has been demonstrated that trophic factors can be upregulated by various pharmacology treatments. For example, 1,25 dihydroxyvitamin D_3 (D3), an active metabolite of vitamin D, is a potent inducer of trophic factors *in vitro.* D3 augments GDNF expression in C6 glioma cells[31] and GDNF release in human U-87 MG glioblastoma cells,[32] increases NGF[33] and TGF-β2[34] expression in neuroblastoma cells, and elevates NT3/NT4 mRNA levels in astrocytes.[35] D3 has synergic effects with TGF-β1–induced NGF release from astrocytes.[36] We found that systemic administration with D3 for eight days increased endogenous concentrations of immunoreactive GDNF protein in the cerebral cortex. The levels of GDNF in D3-treated rats (276.6 ± 51.6 pg/mg protein, $n = 6$) were significantly increased, as compared to those in saline-treated animals (143.6 ± 27.7 pg/mg protein, $n = 6$, $p < 0.05$, *t*-test). These data suggest that administration of D3 for eight days increases brain GDNF levels *in vivo.* Furthermore, rats receiving systemic D3 injections for eight days showed a significant reduction in infarction volume (see FIGURE 3, control: $100.0 \pm 21.3\%$, $n = 11$ vs. D3: $40.2 \pm 19.3\%$, $n = 14$; $p < 0.05$, *t*-test). These data suggest that D3 can upregulate various trophic factors, including GDNF, *in vivo.* Moreover, since D3, unlike GDNF, is able to cross the blood brain barrier. It is, thus, possible that systemic administration of D3 could protect against ischemic brain injury indirectly via an elevation of trophic factors.

UPREGULATION OF GDNF AND ITS RECEPTORS

Recent studies have indicated that expression of GDNF is associated with brain injury.[37,38] Systemic administration of kainate or pilocarpine, which elicits seizures, induces GDNF[39,40] and GDNF receptor-alpha (GFRα1) expression.[41] Similarly,

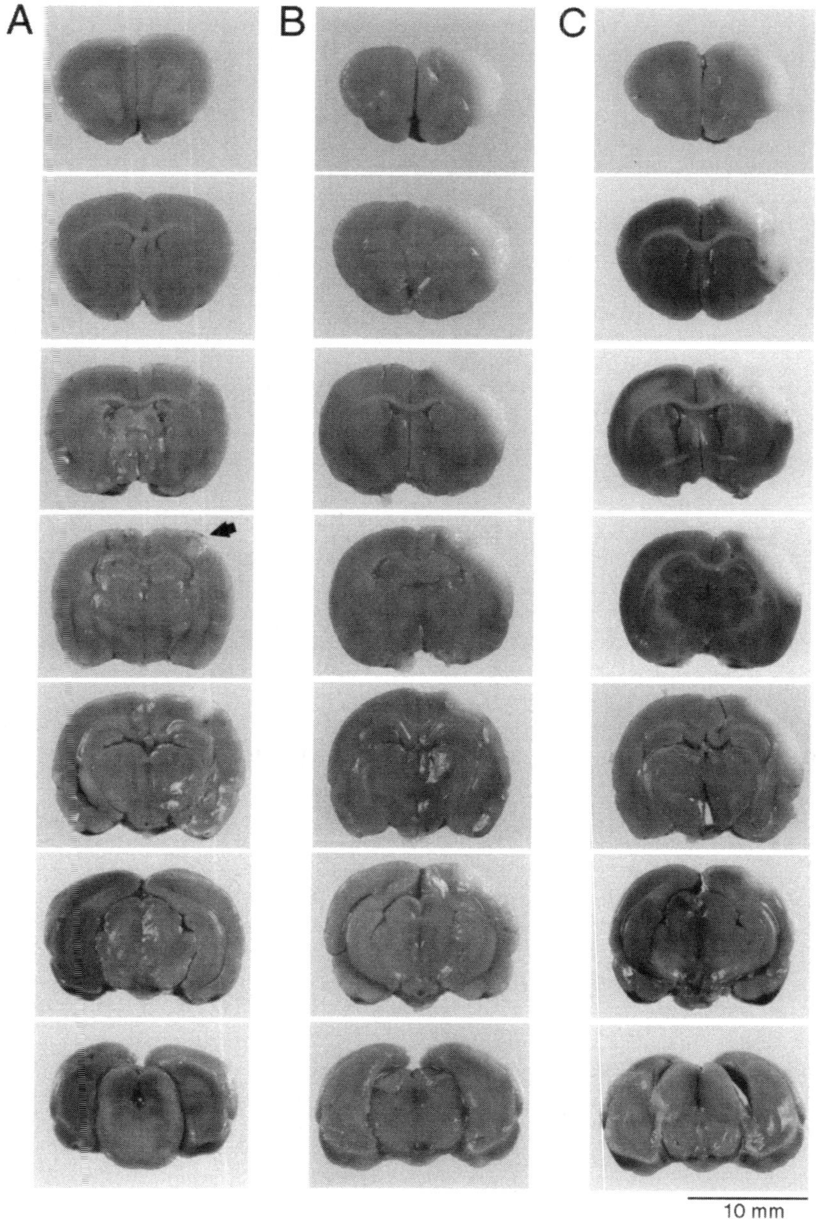
A
B
C
10 mm

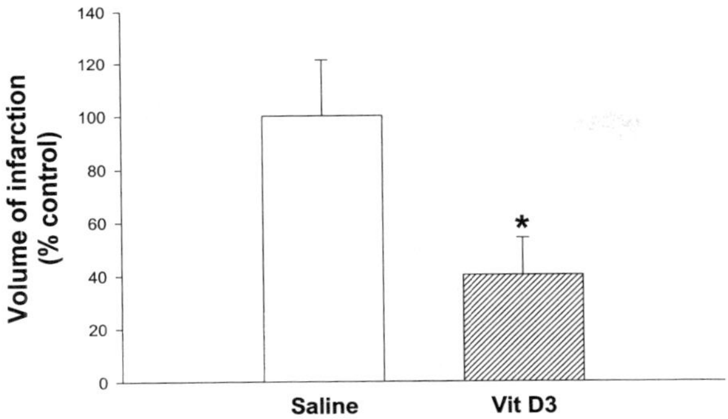

FIGURE 3. Pretreatment with 1,25 dihydroxyvitamin D3 for eight days attenuates cortical infarction induced by 90 minutes of MCA occlusion and reperfusion in rats. The volume of infarction and the area of the largest infarction slice from rats pretreated with D3 were normalized by comparison to the mean levels in animals receiving eight-day saline injection. Volume of infarction was significantly reduced by pretreatment with D3 for eight days as compared to control. The area of the largest infarction in one brain slice from a given rat was significantly attenuated by an eight-day D3 pretreatment. $^*p < 0.05$, *t*-test.

GDNF mRNA expression in the striatum is increased in animals treated with MPTP for seven days[42] or within six hours following mechanical injury to adult mouse striatum.[38] In a complementary fashion, GDNF and associated proteins are upregulated by ischemia. Transient focal ischemia and reperfusion increases GDNF mRNA expression in the ischemic cortex.[43] Global forebrain ischemia increases the expression of GFRα1 in dentate granule cell layer, CA3 pyramidal layer and GFRα2 mRNA in parietal cortex.[44] These data suggest that the GDNF family molecules may be endogenous neuroprotective agents that can be activated during ischemia or reperfusion.

Previous studies have shown that biological effects of GDNF are mediated by GFRα1 and c-Ret. Using *in situ* hybridization, we found that GFRα1 mRNA was upregulated in CA3, dentate gyrus (DG), cortex, and striatum after 90 min MCA ligation in rats (see FIGURE 4). The peak of upregulation in DG was six hours after

---

FIGURE 2. Grafting of fetal kidney tissue into the cerebral cortex markedly reduces cortical infarction induced by middle cerebral arterial ligation and reperfusion. **(A)**. The right middle cerebral artery was ligated for 90 minutes after bilateral common carotid arterial ligation in a rat pregrafted with fetal kidney tissue (*arrow*) 30 min before MCA ligation. The brain was removed 24 hours later and sliced coronally at 2-mm thickness. In this animal, TTC staining showed significantly less infarction in the cerebral cortex. **(B)**. In a similarly anesthetized rat, adult kidney cortical tissue was transplanted in right cerebral cortex 30 min before MCA ligation. An infarcted (*white*) zone in the right cerebral cortex was found in almost all slices after a 90-minute MCA ligation and 24-hour reperfusion. **(C)**. In a vehicle control rat, saline (30 μl) was injected into right cortex prior MCA ligation. Almost all slices showed cortical infarction 24 hour after the onset of reperfusion. *Calibration bar*, 10 mm.

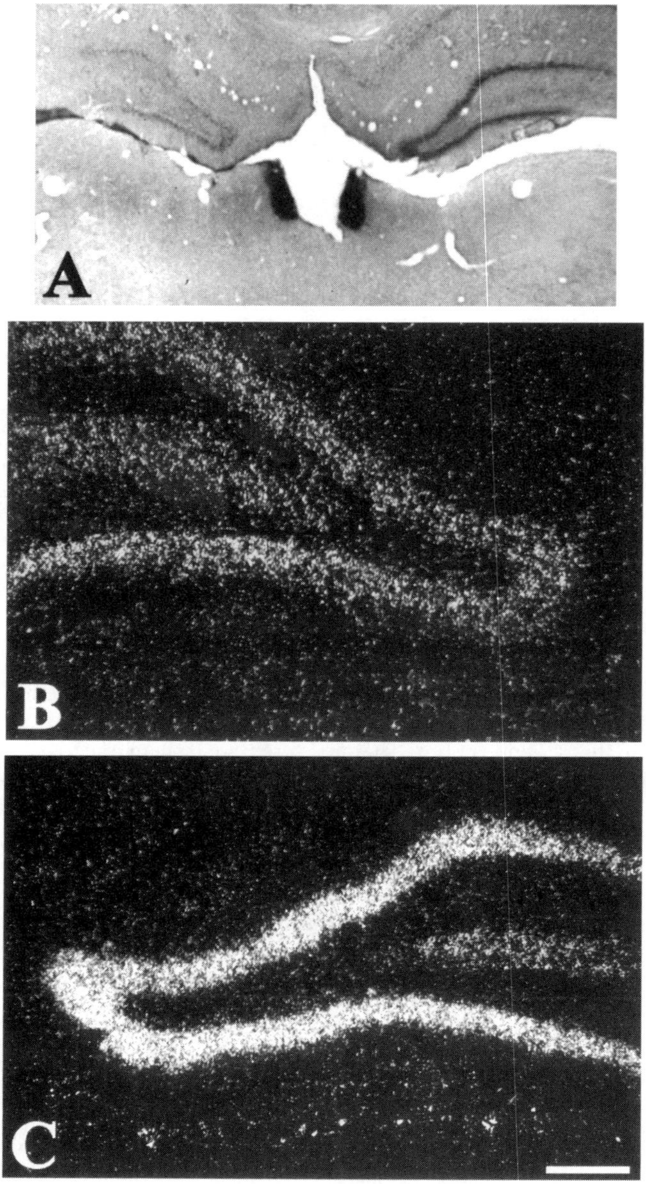

FIGURE 4. GFRα1 mRNA upregulation in dentate gyrus six hours after onset of reperfusion. **A.** Under bright field and low magnification, upregulation of GFRα1 mRNA was clearly seen on the ischemic side (right dentate gyrus) compared to the non-ischemic side (left dentate gyrus). **B** and **C.** Dark field microscopy. **B.** Low levels of GFRα1 mRNA were detected in the non-ischemic dentate gyrus. **C.** High levels of GFRα1 mRNA were detected in the ischemic dentate gyrus. *Calibration bar*, 900 μm in **A**; 225 μm in **B** and **C**.

reperfusion. GFRα1 mRNA levels in CA3 were gradually upregulated over the 24-hour reperfusion period. In cortex, GFRα1 mRNA was upregulated at all measurement times, however, the peak of upregulation was observed at 0 and 24 hours after reperfusion. In striatum, an initial upregulation of GFRα1 was found immediately after ischemia and upregulation of c-Ret mRNA was detected immediately after reperfusion. A gradual increase was found at 6, 12, and 24 hours after reperfusion. Thus, our results indicate that there are both regional and temporal differences in upregulation of GFRα1 and c-Ret after ischemia. Since GDNF is neuroprotective, upregulation of GFRα1 and c-Ret could enhance the responsiveness to GDNF and reduce neuronal damage. The differential upregulation in different brain areas suggests that there may be regional differences in GDNF-induced neuroprotection in stroke.

MECHANISMS OF NEUROPROTECTIVE RESPONSES

GDNF Receptor Alpha 1

In order to determine the role of GFRαl in mediating the protective effects of GDNF, we used targeting gene mutations to generate mice lacking functional GFRα1.[45] Confirming previous reports, we found that these animals have phenotypic disturbances similar to those seen in $GDNF^{-/-}$ as well as $Ret^{-/-}$ mice. We found that mice homozygous for the GFRα1 null mutation (–/–) die within 24 h of birth because of uremia secondary to renal agenesis. Heterozygous animals (+/–) have fewer GFRα1 receptors but live to adulthood. There is a significantly reduced neuroprotective effect of GDNF or fetal kidney tissue transplantation in the +/– animals,

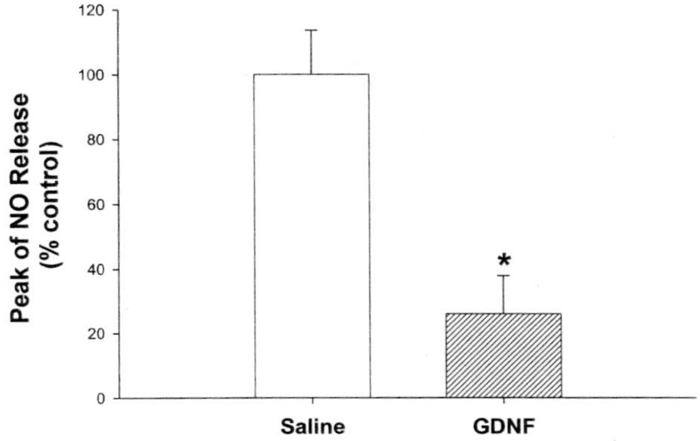

FIGURE 5. GDNF markedly reduces NO release from the cortex during cerebral ischemia. Ischemia was induced by MCA ligation for 40 minutes in urethane-anesthetized rats. The peak of extracellular NO concentration during ischemia was normalized by comparison with the mean value in all stroke animals (n = 10). The production of NO during ischemia was reduced in animals received GDNF pretreatment ($n = 7$, $p < 0.05$).

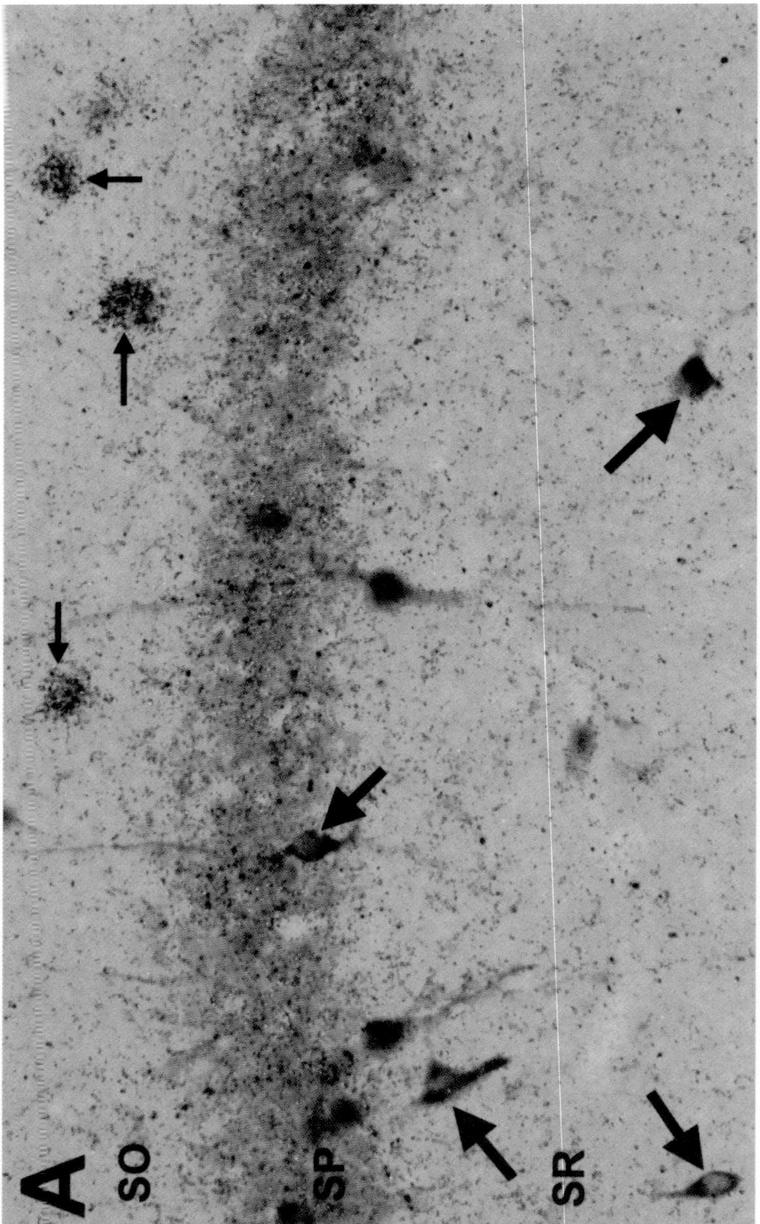
A
SO
SP
SR

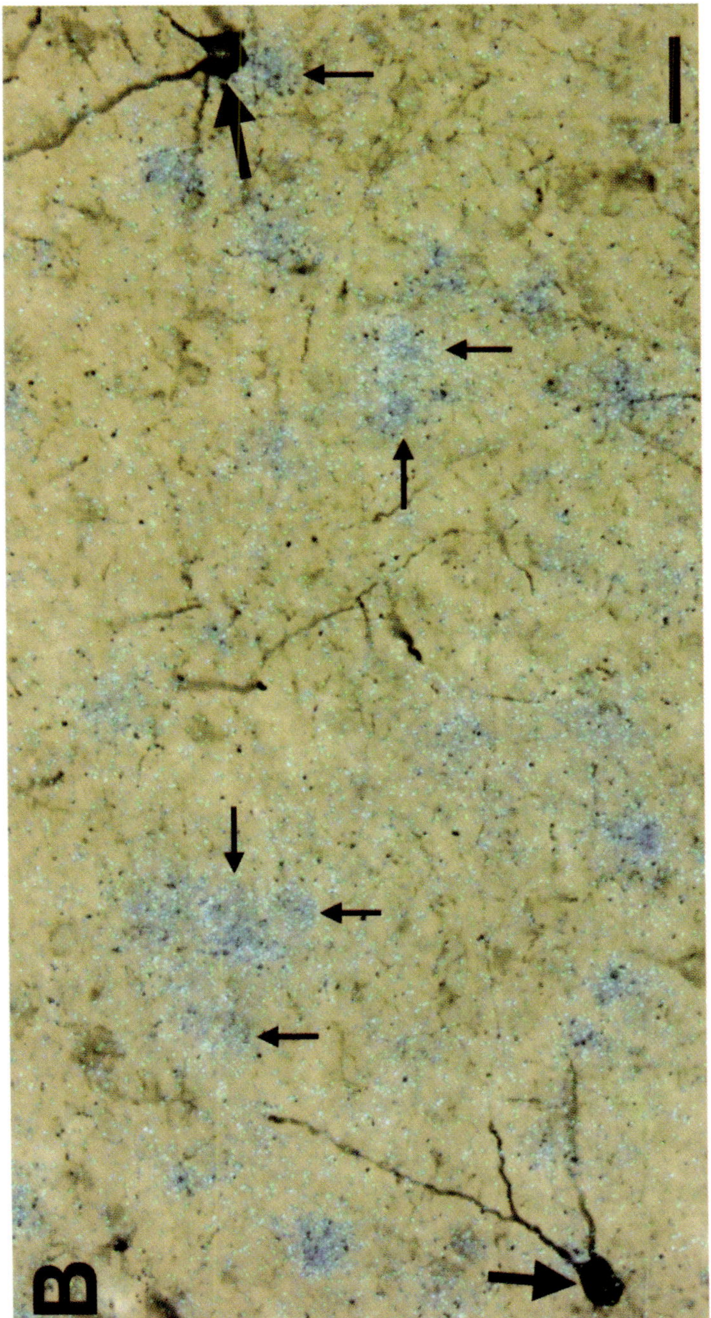

FIGURE 6. Simultaneous detection of GFRα1 transcripts and nNOS immunoreactivity in the hippocampus **(A)** and cortex **(B)**. GFRα1/nNOS double-labeled cells were not detected either in the hippocampus or cortex. In contrast, single label neurons were found in these brain areas. GFRα1 expressing neurons (*small arrows*) were found in the hippocampus and cortex. Similarly, nNOS immunoreactive neurons (*large arrows*) were observed in hippocampus and cortex. SP, stratum pyramidale; SO, s. oriens; SR, s. radiatum. *Scale bar*, 30 μm. *In situ* hybridization combined with immunolabeling was performed as described elsewhere.[65]

compared with wildtype (+/+), after cerebral ischemia. These data suggest that neuroprotective effects of GDNF or fetal kidney tissue transplantation are mediated, at least partially, through GFRαl.

NMDA

Recent studies demonstrated that GDNF reduces NMDA-induced calcium influx and excitotoxic neuronal death in cultured cortical neurons through the activation of MAP kinase pathways.[46] Pretreatment with GDNF also reduces NMDA-induced neuronal cell death in hippocampal brain slices.[47] Administration of subseizure doses of NMDA and kainic acid, similar to ischemic insults, enhanced the expression of GDNF.[48] Since ischemic injury involves the release of excitatory amino acids and activation of their receptors, it is possible that the protective effect of GDNF may involve the attenuation of NMDA responses *in vivo.*

Nitric Oxide

Nitric oxide is a second important messenger involved in ischemia/reperfusion injury. There are three isoforms of NO synthase: neuronal NOS (nNOS or NOS-1), inducible or immunologic NOS (iNOS, NOS-2), and endothelial NOS (eNOS or NOS-3). Both NOS-1 and NOS-3 are constitutive and require calcium for activity. NOS-2 is calcium-independent and is responsible for NO release during inflammation. Of these NOS species, NOS-1 has been found to participate in glutamate/NMDA-induced toxicity.[49] For example, cortical cultures from n-NOS deficient mice are resistant to NMDA toxicity.[50] Using porphrine-coated nitric oxide sensors, we found that local application of NMDA induces NO release in cerebral cortex *in vivo* or in neuron/glial cortical cultures *in vitro.* This response can be antagonized by pretreatment with the NO synthase inhibitor L-NAME.[51] Glial-enriched cultures were less responsive to NMDA, suggesting that NMIDA-induced NO release in cortical cultures is mainly derived from the activation of NOS in neurons.[52] We also found that ischemia induced by MCA occlusion, activates NO release from the ischemic cortex in rats. NO release during the ischemic phase was antagonized by pretreatment with the NO synthase inhibitor L-NAME, or by the NMDA antagonists ketamine or MK801. Local application of nitroprusside also induced NO production. However, this effect was not antagonized by ketamine. Our preliminary studies also indicate that pretreatment with nNOS inhibitor 7-nitroindazole monosodium (7-NINA) significantly reduces ischemia-induced NO production and brain infarction. These data suggest that NO release, induced by middle cerebral artery occlusion, involves activation of NMDA receptors and nNOS.[20]

GDNF has been found to suppress the activation of nitric oxide synthase during neuronal injury[53] and to reduce toxicity induced by nitric oxide itself.[54] Administration of adenoviral vectors encoding GDNF, but not adenovirus encoding beta-galactosidase, increased the survival of lesioned adult rat facial and spinal motoneurons. [53,55] The treatment with GDNF/viral vector after avulsion significantly reduced loss of lesioned facial motoneurons, normalized choline acetyltransferase immunoreactivity, and suppressed the activity of nitric oxide synthase in these neurons. These results indicate that adenovirus-mediated gene transfer of GDNF may prevent the degeneration of motoneurons with peripheral nerve injury and motor neuron diseases through the inhibition of NOS.[53,55]

Similar to these protection experiments, the increase in nitric oxide that accompanies MCA occlusion and subsequent reperfusion is antagonized by GDNF (see FIGURE 5). Since GFRαl is not colocalized with nNOS in non-ischemic brain (see FIGURE 6), it is also possible that GDNF may indirectly inhibit nNOS activity through interaction with NMDA. On the other hand, nNOS immunoreactivity and nNOS (+) cells can be enhanced as early as 15 min after the onset of reperfusion in stroke animals.[56] We also found that the upregulation of GFRα1[57] in the lesioned regions occurs early during ischemia (FIG. 4). It is possible that the activity of nNOS and GFRαl can be colocalized in the same neurons after ischemia and that GDNF may directly inhibit nNOS activity in the same neurons during this insult.

Apoptosis and Necrosis

There is much evidence indicating that antiapoptotic mechanisms are involved in GDNF-induced neuroprotection in dopaminergic neurons.[58,59] Using terminal deoxynucleotidyl transferase-mediated dUTP-biotin *in situ* nick labeling (TUNEL) staining, it has been demonstrated that administration GDNF protein before permanent middle cerebral artery occlusion or transient four-vessel occlusion reduced ischemia-induced DNA fragmentation[21,22] and brain edema in rats.[60] Similarly, pretreatment with replication-defective adenoviral vector or recombinant adeno-associated viral vector containing GDNF gene before middle cerebral artery occlusion reduced TUNEL (+) cells, immunoreactive caspase-3, and cytochrome c-positive neurons in the ipsilateral cerebral cortex or striatum.[61–63] These data suggest that antiapoptotic and antinecrotic mechanisms are involved in GDNF -induced neuroprotection in stroke animals.

SUMMARY

GDNF is an endogenous neuroprotective protein that can be activated during ischemic injury. Exogenous administration of GDNF reduces the toxic effects of excitatory amino acids, attenuates NO production and lowers apoptosis/cell death in stroke animals.

ACKNOWLEDGMENTS

This work was supported by the National Institute on Drug Abuse, NIH (U.S.A.)

REFERENCES

1. BROWN, R.D., J.P. WHISNANT, J.D. SICKS, *et al.* 1996. Stroke incidence, prevalence, and survival: secular trends in Rochester, Minnesota, through 1989. Stroke **27:** 373–380.
2. MENOTTI, A., D.R. JACOBS, H. BLACKBURN, *et al.* 1996. Twenty-five-year prediction of stroke deaths in the seven countries study: the role of blood pressure and its changes. Stroke **27:** 381–387.

3. Rastenyte, D., J. Tuomilehto, C. Sarti, *et al.* 1996. Increasing trends in mortality from cerebral infarction and intracerebral hemorrhage in Kaunas, Lithuania. Cerebrovasc. Dis. **6:** 216–221.
4. Dalkara, T., C. Ayata, M. Demirci, *et al.* 1996. Effects of cerebral ischemia on *N*-methyl-D-aspartate and dihydropyridine-sensitive calcium currents: an electrophysiological study in the rat hippocampus *in situ.* Stroke **27:** 127–133.
5. Hajimohammadreza, I., A.W. Probert, L.L. Coughenour, *et al.* 1995. A specific inhibitor of calcium/calmodulin-dependent protein kinase-11 provides neuroprotection against NMDA- and hypoxia/hypoglycemia-induced cell death. J. Neurosci. **15:** 4093–4101.
6. Strijbos, P.J.L.M., M.J. Leach & J. Garthwaite. 1996. Vicious cycle involving Na^+ channels, glutamate release, and NMDA receptors mediates delayed neurodegeneration through nitric oxide formation. J. Neurosci. **16:** 5004–5013.
7. Vornov, J.J. 1995. Toxic NMDA-receptor activation occurs during recovery in a tissue culture model of ischemia. J. Neurochem. **65:** 1681–1691.
8. Hoffer, B.J., A. Hoffman, K. Bowenkamp, *et al.* 1994. Glial cell line-derived neurotrophic factor reverses toxin-induced injury to midbrain dopaminergic neurons *in vivo.* Neurosci. Lett. **182:** 107–111.
9. Bowenkamp, K.E., A.F. Hoffman, G.A. Gerhardt, *et al.* 1995. Glial cell line-derived neurotrophic factor supports survival of injured midbrain dopaminergic neurons. J. Comp. Neurol. **355:** 479–489.
10. Tomac, A., E. Lindqvist, L.F. Lin, *et al.* 1995. Protection and repair of the nigrostriatal dopaminergic system by GDNF *in vivo.* Nature **373:** 335–339.
11. Beck, K.D., J. Valverde, T. Alexi, *et al.* 1995. Mesencephalic dopaminergic neurons protected by GDNF from axotomy-induced degeneration in the adult brain. Nature **373:** 339–341.
12. Martin, D., G. Miller, M. Rosendahl & D.A. Russell. 1995. Potent inhibitory effects of glial derived neurotrophic factor against kainic acid mediated seizures in the rat. Brain Res. **683:** 172–178.
13. Henderson, C.E., H.S. Phillips, R.A. Pollock, *et al.* 1994. GDNF: a potent survival factor for motoneurons present in peripheral nerve and muscle. Science **266:** 1062–1064.
14. Trok, K., B. Hoffer & L. Olson. 1996. Glial cell line-derived neurotrophic factor enhances survival and growth of prenatal and postnatal spinal cord transplants. Neuroscience **71:** 231–241.
15. Yan, Q., C. Matheson & O.T. Lopez. 1995. *In vivo* neurotrophic effects of GDNF on neonatal and adult facial motor neurons. Nature **373:** 341–344.
16. Aarnisalo, A.A., U. Pirvola, X.Q. Liang, *et al.* 2000. Apoptosis in auditory brainstem neurons after a severe noise trauma of the organ of Corti: intracochlear GDNF treatment reduces the number of apoptotic cells. ORL J. Otorhinolaryngol. Relat. Spec. **62:** 330–334.
17. McAlhany, R.E., Jr., J.R. West & R.C. Miranda. 2000. Glial-derived neurotrophic factor (GDNF) prevents ethanol-induced apoptosis and JUN kinase phosphorylation. Brain Res. Dev. Brain Res. **119:** 209–216.
18. Ferrari, G., M.C. Minozzi, A.M. Zanellato & B. Silvestrini. 1998. GM1, like IGF-I and GDNF, prevents neuronal apoptosis. Ann. N.Y. Acad. Sci **845:** 408.
19. Clarkson, E.D., W.M. Zawada & C.R. Freed. 1995. GDNF reduces apoptosis in dopaminergic neurons *in vitro.* Neuroreport **7:** 145–149.
20. Wang, Y., S.Z. Lin, A.L. Chiou, *et al.* 1997. GDNF protects against ischemia-induced injury in the cerebral cortex. J. Neurosci. **17:** 4341–4348.
21. Miyazaki, H., Y. Okuma, Y. Fujii, *et al.* 1999. Glial cell line-derived neurotrophic factor protects against delayed neuronal death after transient forebrain ischemia in rats. Neuroscience **89:** 643–647.
22. Kitagawa, H., T. Hayashi, Y. Mitsumoto, *et al.* 1998. Reduction of ischemic brain injury by topical application of glial cell line-derived neurotrophic factor after permanent middle cerebral artery occlusion in rats. Stroke **29:** 1417–1422.

23. Kitagawa, H., K. Abe, T. Hayashi, *et al.* 1998. Ameliorative effect of glial cell line-derived neurotrophic factor on brain edema formation after permanent middle cerebral artery occlusion in rats. Neurol. Res. **20:** 333–336.
24. Zhang, W.R., T. Hayashi, M. Iwai, *et al.* 2001. Time dependent amelioration against ischemic brain damage by glial cell line-derived neurotrophic factor after transient middle cerebral artery occlusion in rat. Brain Res. **903:** 253–256.
25. Lapchak, P.A., S. Jiao, F. Collins & P.J. Miller. 1997. Glial cell line-derived neurotrophic factor: distribution and pharmacology in the rat following a bolus intraventricular injection. Brain Res. **747:** 92–102.
26. Giehl, K.M., C.M. Schacht, Q. Yan & P. Mestres. 1998. Infusion of GDNF into the cerebral spinal fluid through two different routes: effects on body weight and corticospinal neuron survival. Neuroreport **9:** 2809–2813.
27. Tomac, A.C., A. Grinberg, S.P. Huang, *et al.* 2000. GFR1 availability regulates GDNF signaling: evidence from mice carrying one or two mutated alleles. Neuroscience **95:** 1011–1123.
28. Borlongan, C.V., F.C. Zhou, T. Hayashi, *et al.* 2001. Involvement of GDNF in neuronal protection against 6-OHDA-induced Parkinsonism following intracerebral transplantation of fetal kidney tissues in adult rats. Neurobiol. Dis. **8:** 636–646.
29. Chiang, Y.H., S.Z. Lin, C.V. Borlongan, *et al.* 1999. Transplantation of fetal kidney tissue reduces cerebral infarction induced by middle cerebral artery ligation. J. Cerebral Blood Flow Metab. **19:** 1329–1335.
30. Johnston, R.E., O. Dillon-Carter, W.J. Freed & C.V. Borlongan. 2001. Trophic factor secreting kidney cell lines: *in vitro* characterization and functional effects following transplantation in ischemic rats. Brain Res. **900:** 268–276.
31. Naveilhan, P., I. Neveu, D. Wion & P. Brachet. 1996. 1,25-Dihydroxyvitamin D3, an inducer of glial cell line-derived neurotrophic factor. Neuroreport **7:** 2171–2175.
32. Verity, A.N., T.L. Wyatt, W. Lee, *et al.* 1999. Differential regulation of glial cell line-derived neurotrophic factor (GDNF) expression in human neuroblastoma and glioblastoma cell lines. J. Neurosci. Res. **55:** 187–197.
33. Veenstra, T.D., J.M. Londowski, A.J. Windebank, *et al.* 1997. Effects of 1,25-dihydroxyvitamin D3 on growth of mouse neuroblastoma cells. Brain Res. Dev. Brain Res. **99:** 53–60.
34. Veenstra, T.D., A.J. Windebank & R. Kumar. 1997. 1,25-dihydroxyvitamin D3 regulates the expression of N-myc, c-myc, protein kinase C, and transforming growth factor-beta2 in neuroblastoma cells. Biochem. Biophys. Res. Commun. **235:** 15–18.
35. Neveu, I., P. Naveilhan, C. Baudet, *et al.* 1994. 1,25-dihydroxyvitamin D3 regulates NT-3, NT-4 but not BDNF mRNA in astrocytes. Neuroreport **6:** 124–126.
36. Hahn, M., H. Lorez & G. Fischer. 1997. Effect of calcitriol in combination with corticosterone, interleukin-1 beta, and transforming growth factorbeta1 on nerve growth factor secretion in an astroglial cell line. J. Neurochem. **69:** 102–109.
37. Hoffer, B. & L. Olson. 1997. Treatment strategies for neurodegenerative diseases based on trophic factors and cell transplantation techniques. J. Neural Transm. Suppl. **49:** 1–10.
38. Liberatore, G.T., J.Y.F. Wong, M.J. Porritt, *et al.* 1997. Expression of glial cell line-derived neurotrophic factor (GDNF) mRNA following mechanical injury to mouse striatum. Neuroreport **8:** 3097–3101.
39. Humpel, C., B. Hoffer, I. Stromberg, *et al.* 1994. Neurons of the hippocampal formation express glial cell line- derived neurotrophic factor messenger RNA in response to kainate-induced excitation. Neuroscience **59:** 791–795.
40. Schmidt-Kastner, R., A. Tomac, B. Hoffer, *et al.* 1994. Glial cell-line derived neurotrophic factor (GDNF) mRNA upregulation in striatum and cortical areas after pilocarpine-induced status epilepticus in rats. Brain Res. Mol. Brain Res. **26:** 325–330.
41. Reeben, M., A. Laurikainen, J.O. Hiltunen, *et al.* 1998. The messenger RNAs for both glial cell line-derived neurotrophic factor receptors, c-ret and GDNFR alpha, are induced in the rat brain in response to kainate-induced excitation. Neuroscience **83:** 151–159.

42. TANG, Y.P., Y.L. MA, C.C. CHAO, *et al.* 1998. Enhanced glial cell line-derived neurotrophic factor mRNA expression upon (–)-deprenyl and melatonin treatments. J. Neurosci. Res. **53:** 593–604.
43. ABE, K. & T. HAYASHI. 1997. Expression of the glial cell line-derived neurotrophic factor gene in rat brain after transient MCA occlusion. Brain Res. **776:** 230–234.
44. KOKAIA, Z., M.S. AIRAKSINEN, A. NANOBASHVILI, *et al.* 1999. GDNF family ligands and receptors are differentially regulated after brain insults in the rat. Eur. J. Neurosci. **11:** 1202–1216.
45. TOMAC, A.C., A. GRINBERG, S.P. HUANG, *et al.* 1999. GFR alpha 1 availability regulates GDNF signaling: evidence from mice carrying one or two mutated alleles. Neuroscience **95:** 1011–1023.
46. NICOLE, O., C. ALI, F. DOCAGNE, *et al.* 2001. Neuroprotection mediated by glial cell line-derived neurotrophic factor: involvement of a reduction of NMDA-induced calcium influx by the mitogen-activated protein kinase pathway. J. Neurosci. **21:** 3024–3033.
47. BONDE, C., B.W. KRISTENSEN, M. BLAABJERG, *et al.* 2000. GDNF and neublastin protect against NMDA-induced excitotoxicity in hippocampal slice cultures. Neuroreport **11:** 4069–4073.
48. HO, A., A.C. GORE, C.S. WEICKERT & M. BLUM. 1995. Glutamate regulation of GDNF gene expression in the striatum and primary striatal astrocytes. Neuroreport **6:** 1454–1458.
49. DAWSON, V.L. & T.M. DAWSON. 1996. Nitric oxide neurotoxicity. J. Chem. Neuroanat. **10:** 179–190.
50. DAWSON, V.L., V.M. KIZUSHI, P.L. HUANG, *et al.* 1996. Resistance to neurotoxicity in cortical cultures from neuronal nitric oxide synthase-deficient mice. J. Neurosci. **16:** 2479–2487.
51. LIN, S.Z., A.L. CHIOU & Y. WANG. 1996. Ketamine antagonizes NO release from cerebral cortex after middle cerebral artery ligation. Stroke **27:** 747–752.
52. LIU, D.M., J.N. WU, A.L. CHIOU, *et al.* 1997. NMIDA induces NO release from primary cell cultures of human fetal cerebral cortex. Neurosci. Lett. **223:** 145–148.
53. SAKAMOTO, T., K. WATABE, T. OHASHI, *et al.* 2000. Adenoviral vector-mediated GDNF gene transfer prevents death of adult facial motoneurons. Neuroreport **11:** 1857–1860.
54. LINGOR, P., K. UNSICKER & K. KRIEGISTEIN. 2000. GDNF and NT-4 protect midbrain dopaminergic neurons from toxic damage by iron and nitric oxide. Exp. Neurol. **163:** 55–62.
55. WATABE, K., T. SAKAMOTO, T. OHASHI, *et al.* 2001. Adenoviral gene transfer of glial cell line-derived neurotrophic factor to injured adult motoneurons. Hum. Cell **14:** 7–15.
56. HOLTZ, M.L., S.D. CRADDOCK & L.C. PETTIGREW. 2001. Rapid expression of neuronal and inducible nitric oxide synthases during post-ischemic reperfusion in rat brain. Brain Res. **898:** 49–60.
57. SARABI, A., C.F. CHANG, Y. WANG, *et al.* 2001. Time course study of GFR alpha-1 expression in an animal model of stroke. Exp. Neurol. **170:** 283–289.
58. CLARKSON, E.D., W.M. ZAWADA & C.R. FREED. 1997. GDNF improves survival and reduces apoptosis in human embryonic, dopaminergic neurons *in vitro.* Cell Tissue Res. **289:** 207–210.
59. SAWADA, H., M. IBI, T. KIHARA, *et al.* 2000. Neuroprotective mechanism of glial cell line-derived neurotrophic factor in mesencephalic neurons. J. Neurochem. **74:** 1175–1184.
60. ABE, K., T. HAYASHI & Y. ITOYAMA. 1997. Amelioration of brain edema by topical application of glial cell line-derived neurotrophic factor in reperfused rat brain. Neurosci. Lett. **231:** 37–40.
61. KITAGAWA, H., C. SASAKI, K. SAKAI, *et al.* 1999. Adenovirus-mediated gene transfer of glial cell line-derived neurotrophic factor prevents ischemic brain injury after transient middle cerebral artery occlusion in rats. J. Cereb. Blood Flow Metab. **19:** 1336–1344.

62. TSAI, T.H., S.L. CHEN, Y.H. CHIANG, *et al.* 2000. Recombinant adeno-associated virus vector expressing glial cell line-derived neurotrophic factor reduces ischemia-induced damage. Exp. Neurol. **166:** 266–275.
63. HERMANN, D.M., E. KILIC, S. KUGLER, *et al.* 2001. Adenovirus-mediated GDNF and CNTF pretreatment protects against striatal injury following transient middle cerebral artery occlusion in mice. Neurobiol. Dis. **8:** 655–666.
64. CHEN, S.T., C.Y. HSU, E.L. HOGAN, *et al.* 1986. A model of focal ischemic stroke in the rat: reproducible extensive cortical infarction. Stroke **17:** 738–743.
65. MORALES, M. & F.E. BLOOM. 1997. The 5-HT3 receptor is present in different subpopulations of GABAergic neurons in the rat telencephalon. J. Neurosci. **17:** 3157–3167.

Index of Contributors

